Oxford University Press,
Great Clarendon Street, Oxford OX2 6DP

Oxford New York
Athens Auckland Bangkok Bogota Bombay
Buenos Aires Calcutta Cape Town Dar es Salaam Delhi
Florence Hong Kong Istanbul Karachi
Kuala Lumpur Madras Madrid Melbourne
Mexico City Nairobi Paris Singapore
Taipei Tokyo Toronto Warsaw

and associated companies in
Berlin Ibadan

Oxford is a trade mark of Oxford University Press

First published 1998
Reprinted 2000
0 19 914708 6 (Student's edition)
0 19 914709 4 (Bookshop edition)

Acknowledgement
Oxford University Press is grateful to London Examinations, a division of Edexcel Foundation, for the permission to reproduce questions from specimen examination papers.

Ideal for revision, *Oxford Revision Guide: GCSE Mathematics* includes

- all the topics required for **intermediate** and **higher** levels of all syllabuses
- hundreds of worked examples
- specimen exam questions and solutions with full cross-referencing to where concepts are explained.

Typesetting, design and illustration by Andrew Edmondson
Cartoons by Hardlines, Charlbury, Oxford
Printed in Great Britain

CONTENTS

NOTE Higher level material is indicated by **H**

Syllabuses

Foundation, higher and intermediate – three levels of examination

The scale of grades at GCSE is A*, A, B, C, D, E, F and G. It would be impossible to grade pupils from A* to G based on the same examination, as pupils expecting an A would find all the questions too easy, or pupils expecting an F would find them too hard. To overcome this problem, there are three **tiers** or levels of exam called **Foundation, Intermediate** and **Higher**. Pupils entered for the Foundation tier can achieve grades G to D; those entered for the Intermediate tier can achieve grades E to B; those entered for the Higher tier can achieve grades C to A*.

Examining groups and syllabuses – which one are you studying?

There are several examining groups or boards that set syllabuses and examinations, which are listed below with their abbreviations. Make sure you know which group and syllabus you are studying for. You can obtain a copy of the syllabus, sample examination papers and other useful publications from your examining group; these will give you a good idea of what to expect when your examination comes. Full details of how to contact the examining groups are given on page 9.

EDEXCEL: London Examinations

IGCSE: University of Cambridge Local Examinations Syndicate

MEG: Midland Examining Group

NEAB: Northern Examinations and Assessment Board

NICCEA: Northern Ireland Council for the Curriculum, Examinations and Assessment

SEG: Southern Examining Group

WJEC: Welsh Joint Education Committee

The IGCSE is an international GCSE. One of the two available syllabuses is entirely assessed by two end of year examinations, there being no coursework, terminal task or aural test.

Assessment – how they award your grade

Your GCSE grade can be based on several kinds of assessment: written **examinations**, **coursework**, **terminal tasks** and **aural tests**.

There is one set of examinations for the Foundation tier, another set for the Intermediate tier and another for the Higher tier; so you will sit one of these. You normally have to take two examination papers but see Modular Syllabuses below. Most of these examinations can be retaken in November.

Investigative work is assessed either by coursework completed during the course or by terminal tasks taken under examination conditions. Coursework and terminal tasks are explained in more detail on pages 6 and 7.

Some syllabuses require you take an aural test in which you answer questions read out by your teacher. See page 7 for more details.

In general, the examinations account for 80% of the final grade and the coursework/terminal tasks for 20% (if there is an aural test, it accounts for 5% or 10%). The details of the assessment structure for each examining group and syllabus are given in the table on the opposite page. Find your own syllabus on the table and make sure you know what you are expecting. Percentages indicate the contribution of each component to the final grade.

Modular Syllabuses

The SEG and MEG both offer modular syllabuses in which several shorter written examinations can be taken during the course.

The MEG modular syllabus is mainly designed for students taking their mathematics course over 2 years. In March of their first year, students take a module test; this is a practice test and gives the student and teacher feedback. In November and April of their second year, students take two further module tests; the better mark obtained counts towards their final grade. If a student is on a 1 year course, he or she will take the November and April module tests, but only the mark obtained from the last test will count towards his or her final grade. A longer final examination is taken in May or June. Coursework is completed during the course. An advantage of this syllabus is that most module tests can be taken at two levels (stages); students can decide which stage to take, depending on their past performance and the grade they are aiming for.

MEG also offers the MEI modular syllabus, designed for post-16 students. Students take three written papers. The first paper may be taken in January or June of the first or second year of the course. The other two papers are taken in June of the final year. One syllabus has a mixture of coursework and terminal tasks. The other allows intermediate level students to take terminal tasks only. An advantage of this flexible syllabus is that successful completion of the first written paper at the foundation or intermediate level provides credit towards a student's GNVQ.

The SEG modular syllabus is also designed for post-16 students, including adults returning to education after a break. Students take three module examinations, an aural test and submit coursework. The three module examinations can be taken all at once or spread over 2 years, offering great flexibility.

Examinations can be taken in the Winter, Spring and Summer, as shown.

	Modules
Winter	1, 2, 3
Spring	1, 2
Summer	1, 2, 3

Assessment structure

Examination board and syllabus	Written papers	Aural test	Coursework/terminal tasks
EDEXCEL			
Syllabus A (1385)	2 papers (40% each)	No	Coursework (20%). Minimum of 2 extended tasks selected from a bank of tasks supplied by EDEXCEL and marked by the teacher
Syllabus B (1386)	2 papers (40% each)	No	Terminal tasks (20%). 1 short and 1 long task in a single examination
IGCSE			
Syllabus 0580	1 paper with short questions (35%) and 1 paper with longer questions (65%)	No	No coursework or terminal task
Syllabus 0581	1 paper with short questions (30%) and 1 paper with longer questions (50%)	No	Coursework (20%). 1 or 2 tasks set and marked by the teacher
MEG			
Syllabus A (1662)	2 papers (40% each)	No	Coursework (20%). 2 tasks set and marked by MEG. Written up in 3 hours of classwork
Syllabus B (1664)	2 papers (37.5% each)	Yes (5%)	Coursework (20%). A range of tasks set and marked by the teacher
Syllabus C (1666)	Two year course: 3 module tests (30%) and 1 terminal paper (50%) One year course: 2 module tests (30%) and 1 terminal paper (50%)	No	Coursework (20%). As above
Syllabus D (1668)	1 paper with short questions (40%) and 1 paper with longer questions (40%)	No	1 piece of coursework (10%) set and marked by the teacher and 3 terminal tasks (10%) in a single examination
Syllabus E (1668)	1 paper with short questions (40%) and 1 paper with longer questions (40%)	No	As above for foundation and higher levels. Intermediate level students do 3 terminal tasks in a single examination
NEAB			
Syllabus A (1131)	2 papers (40% each)	No	Coursework (20%). A minimum of 2 tasks set and marked by the teacher
Syllabus B (1132)	2 papers (40% each)	No	Terminal tasks (20%). 2 tasks in a single examination
NICCEA			
Syllabus A	2 papers (35% each)	Yes (10%)	Coursework (20%). 2 extended tasks set and marked by the teacher: one practical, the other a mathematical investigation
Syllabus B	2 papers (35% each)	Yes (10%)	Terminal task (20%). 1 task set and marked by the teacher
SEG			
Syllabus 2500T	2 papers (37.5% each)	Yes (5%)	Coursework (20%). Two tasks set and marked by the teacher
Syllabus 2500X	2 papers (37.5% each)	Yes (5%)	Coursework (20%). Choice of 2 tasks from a set of 4 supplied and marked by SEG
Modular (2520T)	2 module tests (18.75% each) and 1 terminal module test (37.5%)	Yes (5%)	Coursework (20%). Choice of 1 task from 2 tasks supplied and marked by SEG, plus 1 task set and marked by the teacher
Modular (2520X)	2 module tests (18.75% each) and 1 terminal module test (37.5%)	Yes (5%)	Coursework (20%). Choice of 2 tasks from a set of 4 supplied and marked by SEG
WJEC			
Syllabus A	2 papers (40% each)	No	Coursework (20%). A range of tasks set and marked by teacher
Syllabus B	2 papers (40% each)	No	Terminal tasks (20%). 2 tasks in a single examination

Coursework

Practical work, including investigations, will be marked by your teacher throughout the course and will count 20% towards your final grade. There are three separate skill areas in which you will be awarded a mark:

- Making and monitoring decisions to solve problems.
- Communicating mathematically.
- Developing skills of mathematical reasoning.

These three areas are explained in more detail below. You will be awarded a mark out of 8 in each area. You will have opportunities throughout the course to practise and improve your skills. Your teacher will tell you which tasks are to be assessed. Your best mark in each of the three areas will normally be used, though the marks must come from at least two different pieces of coursework.

Some syllabuses require just two extended tasks. Extended tasks are longer, involve a wide range of skills and are open ended. At least one of these must be of a practical nature, e.g. performing an experiment, a statistical investigation, designing a children's playground. You may also be given a mathematical investigation in which you investigate a purely mathematical problem or puzzle. This may also be an extended task, allowing you to pursue further lines of enquiry.

You can order a coursework guide from your examination board; the addresses are given on page 9. This contains good and bad examples of students' coursework with comments and assessment by the examiners.

Your teacher may use tasks provided by the examination board or devise his or her own tasks. It is important that the tasks you are given allow you to achieve maximum marks; for example, an intermediate task may not be suitable for a higher level student.

You will probably spend 2 to 5 weeks of classroom and homework time completing your tasks. A large part of your coursework will be done in the classroom, especially writing your report. Your teacher will be available to help you and monitor your progress; but if you need a lot of help your marks will be reduced accordingly. Your teacher has to authenticate your work and will take steps to make sure that it is your own.

If you are resitting your GCSE, you will probably be able to carry forward your best coursework marks from last year.

Making and monitoring decisions to solve problems

First read the problem carefully and think about what you are trying to find out. Then write a short plan, which could include some of the following:

- Title and description of the task.
- Materials and equipment required, e.g. stopwatch, cardboard, scissors.
- Sources of information, e.g. magazines, textbooks
- Your intended approach, e.g. 'first, I will draw 3 intersecting lines; then I will count the number of intersections, inner regions and outer regions; then I will repeat this for 4 lines, 5 lines etc.; I will record my results in a table and look for patterns and relationships; etc.'.
- The mathematical topics you expect to use, e.g. calculating areas, percentages, drawing graphs.
- Your predicted results, e.g. 'I expect to find that men prefer listening to jazz'.

Record all your results carefully and clearly, either in a list or in a table with precise headings. Remember to include units. Look over your working regularly, checking for inaccuracies and errors. Repeat an experiment to obtain a second set of readings.

Communicating mathematically

You should write up your investigation clearly and neatly in the form of a report. Here is a skeleton report:

- Title of task. Include a description of the project.
- Your plan.
- Results. Use a table if necessary.
- Comments. Make comments on any patterns you noticed, ideas you had, predictions of what you thought would happen next.
- Illustrations. Use appropriate diagrams to illustrate your results, e.g. bar charts, graphs, designs.
- Conclusions. Make your conclusions about the results. This means describing what your results actually show, not what you expected them to show! Generalise your conclusion. Use algebra where possible.
- Extension. Suggest ways in which the task may be extended and/or improved. Carry out the extension if time allows.
- Final summary. Summarise the entire task and findings in a single paragraph.

Your report should be self-contained and clear enough for the average person to understand; give it to a friend or member of your family to read.

Make sure you describe any problems you encountered and how you overcame them.

Your teacher may ask you questions about your task and will record your ability to discuss it mathematically.

Developing skills of mathematical reasoning

You are expected to apply mathematical reasoning during your task in addition to the more mechanical aspects of collecting information, making tables, drawing diagrams, performing calculations etc.

At the planning stage and whilst you are carrying out your task, write down and then test any predictions, hypotheses or ideas that you have. Look for possible patterns, rules and relationships and try to express them in words and algebra. If your prediction turns out to be wrong, try and find another explanation for the results and test it further. Use logical reasoning to explain your results.

Here are a few more tips to bear in mind:

- Choose a task that interests you and stick to it.
- Make sure your task contains enough mathematics.
- Rely on your teacher for guidance; he or she may be the one who is assessing your work.
- Explain everything you have done, including your diagrams, measurements, tables, calculations, graphs etc.
- Be original.
- Double check your results for accuracy.
- Don't leave your coursework to the last minute. It counts up to 20% of your marks! Put the coursework deadlines in your diary.
- Don't copy other people's work. Don't copy from books without acknowledging it in your report (even then, don't copy much). This is called plagiarism.

Terminal tasks

Terminal tasks are mini-coursework tasks taken under examination conditions at the end of your course and are an alternative to ordinary coursework. They require many of the skills needed for coursework. Your teacher will have practice tasks that can be worked on in the classroom and at home and which can be used for mock tests. You can order more terminal tasks from your examination board; the addresses are given on page 9.

Aural test

A few syllabuses require you to take an aural test. This will take place about a month before your final examinations during a normal lesson. Your teacher will read out each question twice. You then write your working and answer on the answer paper provided. You will only be allowed to use a pen: no calculator, ruler etc.

You should have lots of practice throughout your course so that you can develop the necessary skills and build confidence. Your teacher will have a supply of past papers. You can order more test papers from your examination board so that someone can test you at home; the addresses are given on page 9.

Here are a few tips on taking aural tests:

- Arrive for the test with time to spare so that you can relax.
- Listen carefully. Let your teacher know if you have a hearing difficulty well in advance of the test date.
- Write down any values or important words during the first reading of a question. Fill in any gaps at the second reading.
- Work out all but the simplest of calculations on paper rather than in your head. Use your pen!
- Check that your answers make sense.
- As soon as a new question begins, stop working on the previous question. You need to hear each question both times and cannot afford to miss any information.
- Look ahead to the next question if you finish a question early.
- Turn the page in good time for the next question.

Tips on how to study

Think about how you are going to study. Work out a routine and stick to it. Here are a few tips:

- Get organised from the beginning. Buy all the equipment you need at the beginning of your course: textbooks, scientific calculator, ringbinder folder and paper, pens, pencils, eraser, highlighter pens, ruler, protractor, compasses, graph paper etc. Keep these items in your school/college bag. Read the instructions that come with your calculator. Organise your folder.
- Take notes from the board. When you get home read through them again and rewrite them neatly. Make a shorter summary or start an alphabetical 'results' notebook if you have a bad memory; writing helps your memory. Learn formulas by repeating them for a few days and giving yourself short memory tests; this saves you time looking up formulas in the examination.
- Work through your homework while the lesson is fresh in your mind, the same day if possible. Read through the relevant topics in this book for more information and worked examples.
- Take regular breaks when working at home. Don't spend a long time stuck on a question; make a note of it with your highlighter pen and ask your teacher for help.
- Get into the habit of presenting your work clearly and logically. When writing your answer to a question, use your notes as a guide or find a similar worked example in this book.
- Keep a diary to record test dates, deadlines for homework and coursework, term times etc.

Success in examinations

If you have worked hard during your course and put in a lot of effort, there are some steps you can take to ensure that your examination result reflects this.

Be prepared

If the thought of an examination fills you with dread, then you are probably not fully prepared. Of course everybody will be a bit nervous about an important examination, but there is no reason why you should fall apart and fail to do yourself justice if you know what to expect before you go in. The best preparation is working through past examination papers and looking at your syllabus to avoid any nasty surprises. A positive attitude will help you - try to look on the examination as an opportunity to show the examiner all the things you have learnt during your course.

Make sure you have everything ready the night before: pens, pencils, ruler, protractor, compasses, calculator etc. Ask your teacher if tracing paper will be provided; if not, take your own.

Plan your time

Work out how much time you should spend on each question, based on how many marks it has. There are about 100 marks available in most final examinations. If the examination lasts 2 hours, you can spend just over one minute per mark. Also, the space available for your answer gives you a rough idea how much working is necessary; however, if there is a large space, you don't have to fill it.

If you get stuck on one part of a question, don't abandon the whole question; you may be able to do some of the remaining parts, even if this means making up an answer to the part you could not do.

Don't spend ages racking your brains to think of the answer to one part of a question - go on and start the next question instead. You will probably get more marks for attempting *all* the questions than for finishing a few questions completely. If you have time left at the end, you can go back and think about the more difficult parts when you will be feeling more relaxed. You can also use this time to check your working.

Read carefully

Read each question more than once to make sure it really does say what you think it says. Follow the instructions to the letter, as marks can only be awarded for precise answers to the question asked, not for any other information about the subject.

Present your work clearly

Remember that the examiner will have a large pile of papers to mark in a short space of time. Struggling to read your writing or follow an illogical argument will not help, so write as clearly as you possibly can in the time available and think through what you are going to write before you start. The examiner is *trying* to award you marks - make it easy to find them.

Show all of your working. Get into the habit of writing out your working for every question you answer. Then, if your answer is wrong, you could still get most of the available marks.

Use a few words to help the examiner understand what you are doing. For example, when calculating the total area of a shape, instead of just writing the calculation $5\times7 + 3\times4 = 47\ \text{cm}^2$ you could write:
Total area = $5\times7 + 3\times4 = 47\ \text{cm}^2$

Accuracy

Look at the question to see if a particular degree of accuracy is required. If not, you are expected to give your answers to a reasonable degree of accuracy. You will lose marks if you are consistently inaccurate.

Work to a greater degree of accuracy than required for the final answer. For example, if a question asks for an answer correct to 3 significant figures, write down your working calculations correct to 4 significant figures, then round your final answer to 3 significant figures.

Use a calculator for all but the simplest of calculations.

When measuring lengths, you are expected to work to the nearest 1 mm. When measuring angles, you are expected to work to the nearest 1° or 2°. Always use your protractor to draw right angles. Using a sharp pencil for drawing.

Plot the points of graphs as accurately as you can; use crosses (not dots) to mark them. If graph paper makes you dizzy, use a ruler to help you line up the points and also to read off values.

Stay calm

If you find a question you have no idea about, do not panic! Breathe slowly and deeply and have another look. It's likely that if you stay calm and think clearly, it will start to make more sense, or at least you may be able to answer part of it. If not, then don't agonise about it but concentrate first on the questions you *can* answer.

How to revise

You will probably have found by talking to other pupils that everybody has a different way of revising. It's important to make sure that your method is an effective one.

Working through past examination papers

The best way to revise mathematics is to work through past examination papers. Work through as many as time allows so that you feel comfortable when presented with your final examination papers.

Throughout the last year of your course you should become familiar with the examination papers you are going to take. Ask your teacher for some past examination papers. Order more papers from your examination board; the addresses are given on this page. You should have at least 5 of each examination paper plus the specimen papers for the year of your examination.

It's a good idea to practise more examination questions in any areas of the syllabus that you are unsure of. Ask your teacher for practice questions or find some in your textbook.

When working through past examination papers:

- Look at the marks available and work out roughly how much time you should aim to spend. Practise more questions you are taking too long over.
- Get used to writing your working in the space provided.
- If you get stuck, find a similar worked example in this book. If you still cannot solve the problem, make a note of it and ask your teacher. Don't waste your time agonising over a question: it makes you unnecessarily anxious and leaves you with a false impression of how well you are doing.
- Use some of your later papers as mock examinations. Work through some or all of a paper under timed conditions, without taking a break.
- Try working with a friend. You may be able to help each other when you get stuck.

Make a revision timetable

Plan your revision period; don't leave revising until the last minute. A blank revision timetable is provided overleaf. You should have a bunch of past examination papers and examination questions to work through. Put these in order and arrange them on your timetable. Allocate say 4 hours to work through a 2 hour paper; you can always modify this estimate once you get started.

You could also put those topics you are unsure of on the timetable; use your syllabus to help you. Make sure you cover all of these by the end of your revision period.

Once you have made the timetable, though, make sure you really do start revising - don't kid yourself that planning your revision is all you need to do!

Find a suitable place to revise

You need somewhere quiet where you will not be interrupted. Some people can concentrate with music playing, but if you can't, then turn it off. If you find it difficult to work at home, then try going to a library or some other quiet place.

Take regular breaks

Your concentration will start to wander and you will be less effective if you work solidly for hours without a break. Divide up your study periods into chunks of about 40 minutes, and then go and make a drink, or have a brief walk, or do something completely different for about 10 minutes. Then you can go back for the next session feeling refreshed, and sometimes you will find that you understand things more clearly after a break.

Of course, don't do this during a mock examination.

Don't panic at the last minute

If you have stuck to your revision timetable, you should feel as relaxed and confident as you can about the examination. There is then no point in panicking and trying to learn it all again the night before. Just have a quick look through some of the past examination papers you worked through, and then relax. If you are nervous and feel convinced you have forgotten everything, then take some exercise or do something else to clear your thoughts and stop worrying. You will probably find that when you come back and have another look, you have remembered far more than you thought.

Exam Organisation Addresses

EDEXCEL (formerly **ULEAC)**
Stewart House, 32 Russell Square, LONDON, WC1B 5DN.
Tel: 0171 331 4000 Fax: 0171 331 4044

Midland Examining Group (MEG) and International General Certificate of Education (IGCSE)
Head Office, Syndicate Buildings, 1 Hills Road, CAMBRIDGE, CB1 2EU.
Tel: 01223 553311 Fax: 01223 460278

Northern Examinations and Assessment Board (NEAB)
12 Harter Street, Manchester, M1 6HL.
Tel: 0161 953 1170 Fax: 0161 953 1177

Northern Ireland Council for the Curriculum Examinations and Assessment (NICCEA)
Beechill House, 42 Beechill Road, BELFAST, BT8 4RS.
Tel: 01232 704666 Fax: 01232 799913

Southern Examining Group (SEG)
Associated Examining Board, Stag Hill House, GUILDFORD, Surrey, GU2 5XJ.
Tel: 01483 506506 Fax: 01483 300152

Welsh Joint Committee (WJEC)
Welsh Joint Education Committee, 245 Western Avenue, CARDIFF, CF5 2YX
Tel: 01222 265000 Fax: 01222 575994

Revision timetable

Date							
Day	**Monday**	**Tuesday**	**Wednesday**	**Thursday**	**Friday**	**Saturday**	**Sunday**
Topic/Exam paper							
Time/Comments							
Date							
Day	**Monday**	**Tuesday**	**Wednesday**	**Thursday**	**Friday**	**Saturday**	**Sunday**
Topic/Exam paper							
Time/Comments							
Date							
Day	**Monday**	**Tuesday**	**Wednesday**	**Thursday**	**Friday**	**Saturday**	**Sunday**
Topic/Exam paper							
Time/Comments							
Date							
Day	**Monday**	**Tuesday**	**Wednesday**	**Thursday**	**Friday**	**Saturday**	**Sunday**
Topic/Exam paper							
Time/Comments							
Date							
Day	**Monday**	**Tuesday**	**Wednesday**	**Thursday**	**Friday**	**Saturday**	**Sunday**
Topic/Exam paper							
Time/Comments							

Whole Numbers

1. Basic Arithmetic

Make sure you can answer these questions *without* using a calculator.

1. *There are 857 boys and 946 girls in a school. How many children are there altogether?* $\begin{array}{r} 857 \\ +\ 946 \\ \hline \end{array}$
2. *The battle of Waterloo took place in 1815. The battle of Hastings took place in 1066. How many years apart are these events?* $\begin{array}{r} 1815 \\ -\ 1066 \\ \hline \end{array}$
3. *Boxes contain 17 tiles each. How many tiles are there in 12 boxes?* $\begin{array}{r} 17 \\ \times\ 12 \\ \hline \end{array}$
4. *Lollipops cost 13p each. How many can be bought for 234p?* $13\overline{)234}$

2. Sum, Difference, Product, Quotient

The **sum** of 6 and 2 is written 6+2 or 2+6. Both calculations give the same answer, 8.

The **difference** of 2 and 6 is written 6−2 or 2−6. The **positive difference** is 6−2 = 4. The **negative difference** is 2−6 = -4.

The **product** of 2 and 6 is written 2×6 or 6×2. Both calculations give the same answer, 12.

The **quotient** of 2 and 6 is written 6÷2 or 2÷6. These calculations give different answers.

NOTE Another way of writing 6÷2 is $\frac{6}{2}$.

3. Brackets

A bracket contains a *single* number. For example, the bracket (6+2) contains the single number 8.

When a bracket is multiplied by a number, the × sign is usually omitted. For example:

3×(4+1) is usually written 3(4+1)

(2+3)×(5−2) is usually written (2+3)(5−2)

4. BoDMAS

Always use the following order of working when performing a calculation.

① Work out **B**rackets
② **D**ivide and **M**ultiply
③ **A**dd and **S**ubtract

Remember this using the word BoDMAS

Examples

3(4+1) = 3×(4+1) = 3×(5) = 3×5 = 15
brackets first.......... drop brackets......then multiply

4+3×2 = 4+6 = 10
multiply first........then add

3+4(3−1)+6÷3−5 = 3+4×2+6÷3−5 = 3+8+2−5 = 8
brackets first.................................multiply and divide.............add and subtract

4×3×2 = 12×2 = 24
4×3×2 = 4×6 = 24
Multiply any two numbers first; you will always get the same answer

12÷2÷3 = 6÷3 = 2
6÷3×4 = 2×4 = 8
Work from left to right with combinations of × and ÷ signs

5. Natural Numbers

Whole positive numbers are sometimes called **natural** numbers:

1, 2, 3, 4, 5, 6, ...

6. Multiplying and Dividing by Zero

Multiplying numbers by 0 always gives 0.

For example, 5×0 = 0, 0×5 = 0, 3×5×0 = 0

Dividing 0 by any number (other than 0) always gives 0.

For example, 0÷5 = 0 or $\frac{0}{5} = 0$

Dividing by 0 is impossible.

For example, 5÷0 or $\frac{5}{0}$ is impossible

7. Powers

4^3 is shorthand for 4×4×4 and is called the **3rd power** of 4. It is commonly called the **cube** of 4. We also say that 4 has been **raised to the 3rd power**. So, $4^3 = 4\times4\times4 = 64$.

NOTE 4^3 does ***not*** mean the same as 4×3. Here is the difference:
$4^3 = 4\times4\times4$ whereas $4\times3 = 4+4+4$

Similarly, 4^2 is shorthand for 4×4 and is called the **2nd power** of 4, or more commonly the **square** of 4 (or 4 squared). So, $4^2 = 4\times4 = 16$.

4^1 is another way of writing 4

Question *Calculate the value of* 2^4+3^2.

$2^4+3^2 = 2\times2\times2\times2 + 3\times3 = 16+9 = 25$

Base ⟶ 4^3 ⟵ Index (plural: indices)

8. Power of a Bracket

$(\ )^2$ means $(\ )\times(\ )$ $(\ )^3$ means $(\ )\times(\ )\times(\ )$

So, $(5+3)^2 = (5+3)\times(5+3) = 8\times8 = 64$

And $(6-1)^3 = (6-1)\times(6-1)\times(6-1) = 5\times5\times5 = 125$

Also $(5\times3)^2 = (5\times3)\times(5\times3) = 15\times15 = 225$.

You get the same answer by squaring each number in the bracket and multiplying the results together:

$(5\times3)^2 = 5^2\times3^2 = 5\times5\times3\times3 = 25\times9 = 225$

So we have the result: $(5\times3)^2 = 5^2\times3^2$

Similarly: $(6\div3)^2 = 6^2\div3^2 = \left(\frac{6}{3}\right)^2$

NOTE $(5+3)^2$ is ***not*** equal to 5^2+3^2. Here's why:
$(5+3)^2 = 8^2 = 64$ whereas $5^2+3^2 = 25+9 = 34$

Whole Numbers (Contd)

9. Roots

The opposite process of calculating the power of a number is finding its **root**.

The 3rd power (cube) of 2 is written 2^3 and is equal to 8. The **3rd root (cube root)** of 8 is written $\sqrt[3]{8}$ and equals 2.

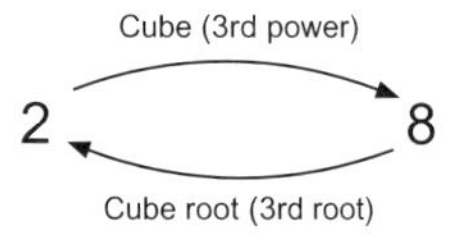

The 2nd power (square) of 5 is written 5^2 and equals 25. The **2nd root (square root) of 25** is written $\sqrt[2]{25}$, or simply $\sqrt{25}$, and is equal to 5.

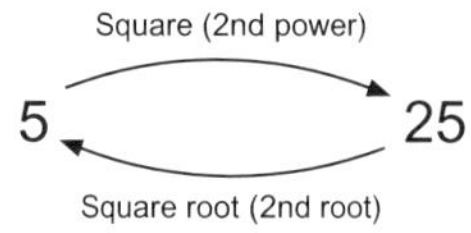

Question *Calculate $\sqrt[4]{16}$ without using a calculator.*

$\sqrt[4]{16}$ is the 4th root of 16.

The answer must be 2, because:

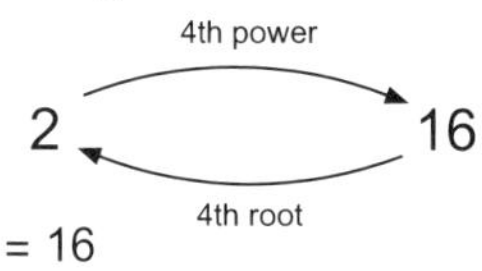

4th power of 2 = $2^4 = 2\times2\times2\times2 = 16$

So, $\sqrt[4]{16} = 2$ because $2\times2\times2\times2 = 16$

NOTE $\sqrt[4]{16}$ does ***not*** mean $16\div4$ (see above).

$\sqrt{16+9}$ is ***not*** equal to $\sqrt{16} + \sqrt{9} = 4+3 = 7$.

$\sqrt{16+9} = \sqrt{25} = 5$ is correct.

10. Powers and Roots

Powers and roots cancel each other out.

For example:

$$6 \xrightarrow{\text{Square}} 6^2 = 36 \xrightarrow{\text{Square root}} \sqrt{36} = 6$$

Another way of writing this is $\sqrt{6^2} = 6$

Similarly, $(\sqrt{6})^2 = 6$ or $\sqrt{6}\times\sqrt{6} = 6$

11. Multiples

1×3 = 3	These numbers are called **multiples** of 3.	1×4 = 4	These numbers are called **multiples** of 4.
2×3 = 6		2×4 = 8	
3×3 = 9		3×4 = 12	
4×3 = 12		4×4 = 16	
5×3 = 15		5×4 = 20	
6×3 = 18		6×4 = 24	
7×3 = 21		7×4 = 28	
8×3 = 24		8×4 = 32	
9×3 = 27		9×4 = 36	

Some of the multiples of 3 and 4 are the same. They are called **common multiples**.

3, 6, 9, 12, 15, 18, 21, 24, 27, ... Multiples of 3
4, 8, 12, 16, 20, 24, 28, 32, 36, ... Multiples of 4

12 and 24 are common multiples of 3 and 4.

12 is the **lowest common multiple (LCM)** of 3 and 4. It is the smallest number that both 3 and 4 divide into exactly.

12. Factors

The **factors** of 12 are those numbers that divide exactly into 12: 1, 2, 3, 4, 6, 12
(Don't forget to include 1 and 12.)

The factors of 18 are: 1, 2, 3, 6, 9, 18

Some of the factors of 12 and 18 are the same; they are called **common factors**.

1, 2, 3, 4, 6, 12 Factors of 12
1, 2, 3, 6, 9, 18 Factors of 18

The common factors of 12 and 18 are 1, 2, 3, and 6.

The **highest common factor (HCF)** of 12 and 18 is 6.

13. Prime Numbers

A **prime number** is a number that can be divided exactly only by itself and 1. For example, 11 is prime. Here are the first few prime numbers:

2, 3, 5, 7, 11, 13, 17, 19, 23, 29

1 is not considered to be a prime number.

14. Prime Factors

The factors of 12 are: 1, 2, 3, 4, 6, 12

Of these, 2 and 3 are prime numbers and so are called **prime factors**.

Question *What are the prime factors of 2^5?*

$2^5 = 2\times2\times2\times2\times2 = 32$

The factors of 32 are: 1, 2, 4, 8, 16, 32
So, 2 is the only prime factor of 2^5.

15. Product of Prime Factors

Any number can be written as a product of its prime factors, i.e. can be broken down into its prime factors.

Question *Express 36 as the product of its prime factors*

① Write down the first few prime numbers:
2, 3, 5, 7, 11, 13, ...

② Divide 36 by the first prime number, 2, as many times as possible (see opposite).

③ Then divide by the next prime number, 3, as many times as possible, and so on until you get to 1.

2) 36
2) 18
3) 9
3) 3
1 stop

④ Write down the product of all the prime numbers you divided by:
$36 = 2\times2\times3\times3$

⑤ Write any repeated prime numbers as powers (i.e. using index form):
$36 = 2^2\times3^2$

This is called **expressing a number as a product of primes using index form**.

Number Patterns (Sequences)

1. Number Patterns (Sequences)

A **number pattern** or **sequence** is an ordered list of numbers connected by a rule.

The numbers in the pattern below start with 3 and are connected by the rule 'add 2', i.e. adding 2 to one number gives the next number.

Add 2   Add 2   Add 2   Add 2

3 → 5 → 7 → 9 → 11

Some rules involve several steps. For example, starting with 1 and using the rule 'double then add 3' gives the number pattern:

Double, add 3   Double, add 3   Double, add 3

1 → 5 → 13 → 29

Some rules involve several previous numbers in the pattern. For example, starting with the numbers 2, 3 and using the rule 'add the two previous numbers' gives the number pattern:

2   3   5   8   13 (Add, Add, Add, Add)

Some number patterns involve negative numbers. For example, starting with 6 and using the rule 'subtract 2' gives the number pattern:

6, 4, 2, 0, −2, −4, ...

Other sequences involve fractions. For example, starting with $\frac{1}{2}$ and using the rule "increase the numerator by 1, increase the denominator by 1" gives the number pattern below:

$\frac{1}{2}$ $\frac{2}{3}$ $\frac{3}{4}$ $\frac{4}{5}$ $\frac{5}{6}$ ....

Number patterns often arise in coursework investigations. For example, the numbers of matches in these triangles form a number pattern:

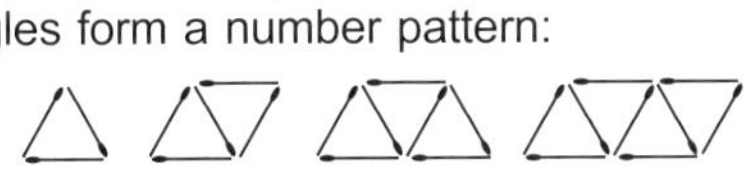

Number of matchsticks   3   5   7   9   .....

2. Number Patterns You Should Know

Even numbers	2, 4, 6, 8, 10, ...
Odd numbers	1, 3, 5, 7, 9, 11, ...
Prime numbers	2, 3, 5, 7, 11, 13, 15, 17, 19, 23, ...
Natural numbers	1, 2, 3, 4, 5, 6, 7, 8, 9, ...
Square numbers	$1^2, 2^2, 3^2, 4^2, 5^2, 6^2$, ... ↓ 1, 4, 9, 16, 25, 36, ...
Cube numbers	$1^3, 2^3, 3^3, 4^3, 5^3$, ... ↓ 1, 8, 27, 64, 125, ...
Powers of 2	$2^0, 2^1, 2^2, 2^3, 2^4, 2^5, 2^6$, ... ↓ 1, 2, 4, 8, 16, 32, 64, ...

3. Finding the Rule for a Number Pattern

Follow these steps to find the rule for your number pattern.

① **Is the pattern one you know?**

9, 16, 25, 36, ... Square numbers, starting with 3^2

② **Is the pattern a modification of one you know?**

2, 5, 10, 17, 26, ... Square numbers + 1

③ **Try adding consecutive numbers:**

2   4   6   10   16 .... (Add, Add, Add, Add)

RULE Start with 2 and 4; add two consecutive numbers to get the next

④ **Look at the differences between consecutive numbers:**

7   10   13   16   19 .... (differences 3, 3, 3, 3)

Same difference

RULE Start with 7; add 3

1   2   4   7   11 .... (differences 1, 2, 3, 4)

Increasing differences

RULE Start with 1; add 1, then add 2, then add 3, etc.

⑤ **Try multiplying consecutive numbers:**

1   2   2   4   8 .... (Multiply, Multiply, Multiply, Multiply)

RULE Start with 1 and 2; multiply two consecutive numbers to get the next

⑥ **Try dividing consecutive numbers:**

2   6   18   54 ....

3 (6÷2)   3 (18÷6)   3 (54÷18)

RULE Start with 2; multiply by 3

48   24   12   6 ....

2 (48÷24)   2 (24÷12)   2 (12÷6)

RULE Start with 48; divide by 2

If none of these steps worked, try the methods in Box 4

4. Terms of a Sequence

Each number in a sequence (number pattern) is called a **term** and occupies a certain **position** in the sequence.

Position	1st,	2nd,	3rd,	4th,	5th, ...
Term	3,	6,	9,	12,	15, ...

You can often find the value of a term by knowing its position. In the above sequence, each term can be found by multiplying its position by 3:

Position	1	2	3	4	5 ....
	×3	×3	×3	×3	×3
Term	3	6	9	12	15 ....

So we can now also find the 20th and 100th terms:

20th term = 20×3 = 60   100th term = 100×3 = 300

more

Number Patterns (Contd)

4. Terms of a Sequence (Contd)

Question *Find the 7th, 20th and 100th terms of the sequence: 3, 7, 11, 15, 19, 23, ...*

First, write down the positions of the terms:

Position 1, 2, 3, 4, 5, 6, ...
Term 3, 7, 11, 15, 19, 23, ...

Then find the rule connecting each term and its position number:

Position	1	2	3	4	5	....
Term	3	7	11	15	19	....
Difference		4	4	4	4	

A difference of 4 means that the rule involves multiplying the position number by 4:

Position 1, 2, 3, 4, 5, 6, ...
4× Position 4, 8, 12, 16, 20, 24, ...
Term 3, 7, 11, 15, 19, 23, ...

You can see that each term is calculated by multiplying each position by 4 and then subtracting 1.

So, 7th term $= 4\times7-1 = 28-1 = 27$
20th term $= 4\times20-1 = 80-1 = 79$
100th term $= 4\times100-1 = 400-1 = 399$

Question *As part of his coursework, Philip drew the following sequence of patterns:*

Pattern Number 1 2 3

(a) Complete the table below.

Pattern number	*1*	*2*	*3*	*4*	*5*
Number of dots	*5*	*8*	*11*	*14*	

(b) What is the number of the pattern with 137 dots?

(a) The number of dots can be found using the rule 'multiply the pattern number by 3 and add 2'.
Number of dots in pattern 5 is: $3\times5+2 = 17$

(b) *Reverse the rule* to find the number of the pattern with 137 dots, i.e. 'subtract 2, then divide by 3':
$137-2 = 135$ then $135\div3 = 45$
So, the 45th pattern has 137 dots.

Question *Find the next term in the sequence below.*

Position 1st, 2nd, 3rd, 4th, 5th, ...
Term 2, 5, 10, 17, 26, ...

One of the terms of this sequence is 226. What is the position of this term?

Try squaring the position numbers.

Position number	1,	2,	3,	4,	5, ...
Square of position number	1,	4,	9,	16,	25, ...
Term	2,	5,	10,	17,	26, ...

You can see that each term is found using the rule 'square the position number and add 1'.

So, the next term = 6th term $= 6^2+1 = 36+1 = 37$

To find the position number of the term 226, *reverse the rule,* i.e. 'subtract 1 and then square root'.

Subtract 1 $226-1 = 225$
Square root $\sqrt{225} = 15$

So, the term 226 has position number 15.

5. Describing Sequences Using Algebra

Sequences can be briefly described using algebra.

n represents the position of any term
u_1 represents the 1st term
u_2 represents the 2nd term, etc.
u_n represents the nth term (the **general term**)

For the sequence 3, 6, 9, 12, 15, ... we have:

Position	1,	2,	3,	4,	5,	6, ...,	n, ...
Term	3,	6,	9,	12,	15,	18, ...,	nth term, ...
	u_1,	u_2,	u_3,	u_4,	u_5,	u_6, ...,	u_n, ...

Each term in this sequence is found using the rule 'multiply the position number by 3'. So,

1st term $= u_1 = 3\times1 = 3$
2nd term $= u_2 = 3\times2 = 6$
3rd term $= u_3 = 3\times3 = 9$, etc.
nth term $= u_n = 3\times n = 3n$

Question *Write down the first three terms of the sequence where $u_n = n^2+1$.*

1st term $= u_1 = 1^2+1 = 1+1 = 2$
2nd term $= u_2 = 2^2+1 = 4+1 = 5$
3rd term $= u_3 = 3^2+1 = 9+1 = 10$

Question *Write down the nth term, u_n, for the sequence 1, 3, 5, 7, 9, ...*

First, write down the positions of the terms:

Position 1, 2, 3, 4, 5, 6, ..., n, ...
Term 1, 3, 5, 7, 9, 11, ..., u_n, ...

Find the rule connecting the terms and their positions.

Position	1	2	3	4	5	6	u_n ....
Term	1	3	5	7	9	11	n ....
Difference		2	2	2	2	2	

Same difference

A difference of 2 means that the rule involves multiplying the position number by 2. Each term is found using the rule 'multiply the position number by 2, then subtract 1':

Term = $2\times$(Position number)-1

So, $u_n = n$th term $= 2\times n-1 = 2n-1$

Question *For the sequence 2, 6, 12, 20, 30, ...*
(a) Find the next term.
(b) Write down an expression for u_n.
(c) Find the value of n when $u_n = 4970$.

(a) Write down the positions of the terms:

Position 1 2 3 4 5 6 7 ....
Term 2 6 12 20 30 ? ....

Each term is found by multiplying its position number by the next position number:

	Multiply (1, 2)	Multiply (2, 3)	Multiply (3, 4)		Multiply (n, $n+1$)	
Position	1	2	3	4	n	$n+1$ ....
Term	2	6	12	20	u_n	u_{n+1} ....

So, the next term is $6\times7 = 42$

(b) The position of u_n is n and the next position is $n+1$.
So, $u_n = n\times(n+1) = n(n+1)$

(c) We must find n such that $n(n+1) = 4970$. Since $n+1$ is close to n, then $n(n+1)$ is close to $n\times n$ or n^2. So, $n^2 \approx 4970$, giving $n \approx \sqrt{4970} = 70.5$.
Try $n = 70$: $n(n+1) = 70\times71 = 4970$. Correct

Negative Numbers

1. Positive and Negative Numbers

Positive and negative numbers can be represented as points on a straight line called the **number line**.

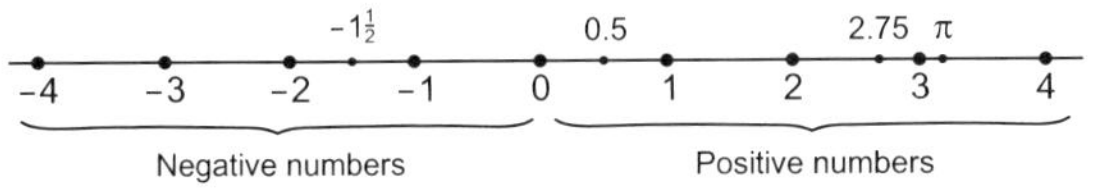

There are several ways of writing positive and negative numbers. For example,

2 can be written +2, (+2) or $^+2$

–2 can be written (–2) or $^-2$ and is called *negative 2*

NOTE $^+2$ and $^-2$ are called **directed numbers**

2. Integers

Whole numbers are sometimes called **integers**. They include the positive and negative numbers and zero.

.... –3, –2, –1, 0, 1, 2, 3, ...

3. Plus and Minus

Minus is the name of the sign –

Plus is the name of the sign +

So, 3+4–2 expressed in words is 'three plus four minus 2'

4. The Sign –

In calculations, the sign – can mean subtract or take away, or it can mean negative. For example,

5–2 means '5 subtract 2' or '5 take away 2'

–2+5 means 'negative 2 add 5' (the negative number –2 added to 5)

5. Distance Between Numbers

The distance between two numbers on the number line is always positive. For example, the distance between the numbers –2 and 3 is 5, as shown on the number line below.

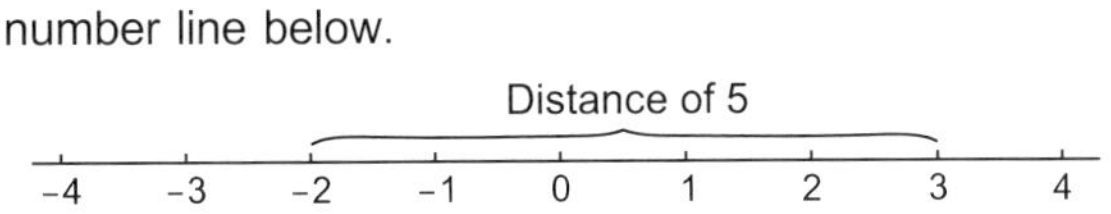

Question *The temperature in a thawing freezer increased from –15 °C to –3 °C. What was the rise in temperature?*

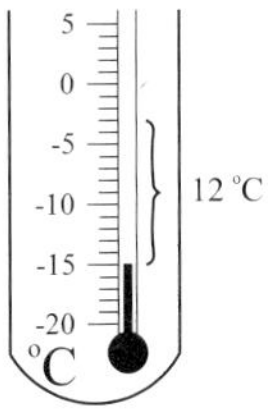

The diagram shows that the temperature rose by 12 °C (the distance between –15 and –3 is 12).

6. Adding and Subtracting

Use the number line to add/subtract two numbers.

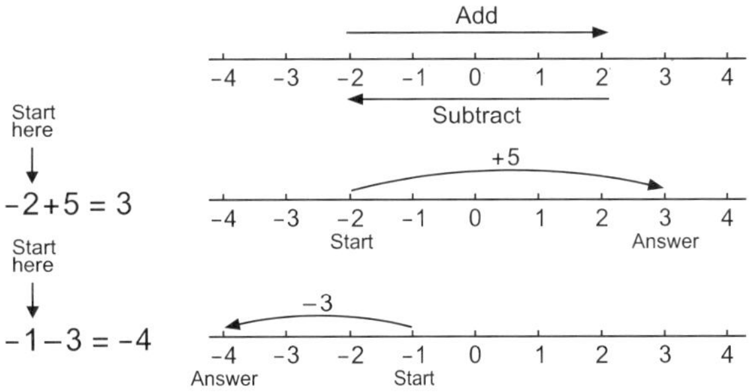

Sometimes you will see two + or – signs next to each other, e.g. 2+–3. Use the rules on the right to replace them with a single sign.

++ gives +
–– gives +
+– gives –
–+ gives –

Examples

2+–3 = 2–3 = –1

–(–5)–2 = ––5–2 = +5–2 = 3

7. Rearranging Numbers

–2+5 and +5–2 both give the same answer of 3.

So, –2+5 can be rearranged to give +5–2. Notice how **the signs stay with the numbers**.

Similarly, –2+5–1+4 = +5+4–2–1 = 9–3 = 6

8. Adding/Subtracting Several Numbers

When adding/subtracting more than two numbers, first combine the + numbers into a single number and then the – numbers into a single number:

–3+2+4–1–6+3+5

= +2+4+3+5–3–1–6   Rearrange numbers first

= +14–10   Combine numbers

= 4

9. Multiplying and Dividing

Use these rules for multiplying numbers.

+×+ = +   –×– = +
+×– = –   –×+ = –

Examples

–2×–5 = +10 = 10 $(-3)^2 = (-3)\times(-3) = +9 = 9$

–2+–3×4 = –2+–12 = –2–12 = –14 or alternatively

–2+–3×4 = –2–3×4 = –2–12 = –14

Use the same rules for dividing numbers. For example:

–6÷–2 = +3 = 3 or equivalently $\frac{-6}{-2} = +3 = 3$

$\frac{-8}{2} = \frac{-8}{+2} = -4$ and $\frac{8}{-2} = \frac{+8}{-2} = -4$ and $-\frac{8}{2} = -4$

The last example shows that $\frac{-8}{2} = \frac{8}{-2} = -\frac{8}{2} = -4$

Fractions

1. Making Fractions

Question *A whole pie is divided into 4 equal parts. Brian eats 3 parts. What fraction of the pie does he eat?*

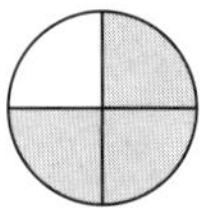

Brian eats 3 parts **out of** 4 parts $= \frac{\text{3 parts}}{\text{4 parts}}$ (3 parts **out of** 4 parts) $= \frac{3}{4}$

So, he eats $\frac{3}{4}$ (three quarters) of the pie.

2. The Divide Sign ÷

The divide sign looks like a fraction, with dots instead of numbers. The divide sign can be used as an alternative way of writing a fraction. For example:

$\frac{3}{4}$ can be written as $3 \div 4$

3. Mixed Numbers and Improper (Top Heavy) Fractions

4 quarters of a pie + 4 quarters of a pie + 3 quarters of a pie = 11 quarters $= \frac{11}{4}$ ← **Improper fraction** (top heavy fraction)

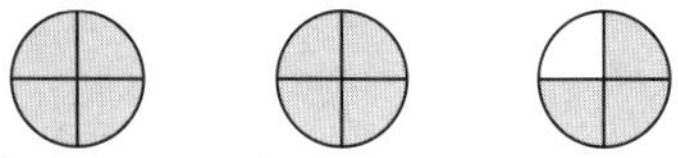

1 whole pie + 1 whole pie + $\frac{3}{4}$ of a pie $= 2 + \frac{3}{4} = 2\frac{3}{4}$ ← **Mixed number**

So, $2\frac{3}{4} = \frac{11}{4}$

π is a special number approximately equal to $3\frac{1}{7}$.

π is often written as an improper fraction, i.e. $\frac{22}{7}$.

Question *Convert $2\frac{3}{4}$ to an improper fraction.*

$2\frac{3}{4}$ — Multiply, Add

$4 \times 2 + 3 = 8 + 3 = 11$

So, $2\frac{3}{4} = \frac{11}{4}$

Denominators must be the same

Question *Convert $\frac{11}{4}$ to a mixed number.*

Divide 4 into 11

$4 \overline{)11}$ = ② rem ③

So, $\frac{11}{4} = 2\frac{3}{4}$

Circled numbers make the mixed number

4. Whole Numbers Written as Fractions

When working with fractions it is helpful to write a whole number as a fraction. For example:

$\frac{4}{1} = 4 \div 1 = 4$.

So we can write the whole number 4 as the fraction $\frac{4}{1}$

5. Rational Numbers

Fractions are **rational numbers**, which also include all positive and negative whole numbers, zero and all finite decimals. For example:

$-2\frac{3}{4}$, $-1\frac{1}{3}$, $-\frac{3}{4}$, 0, $\frac{1}{2}$, $\frac{2}{3}$, 2.34, $3\frac{1}{2}$, $\frac{19}{5}$

NOTE $2.34 = \frac{234}{100}$

6. Equivalent Fractions

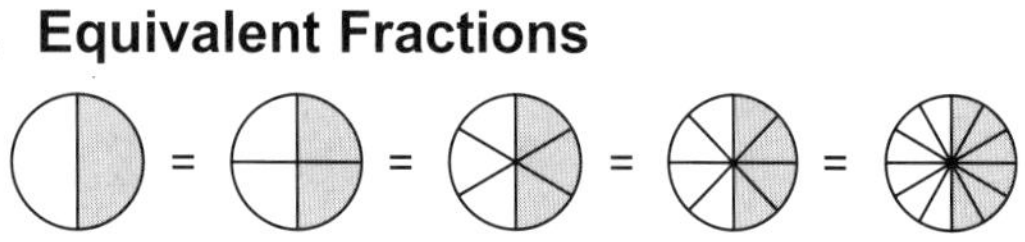

$\frac{1}{2} = \frac{2}{4} = \frac{3}{6} = \frac{4}{8} = \frac{6}{12}$

Equivalent fractions

Multiplying both the top and bottom of a fraction by the *same* number gives an **equivalent fraction**.

$\frac{2}{4} \xrightarrow{\times 3} \frac{6}{12}$

Question *Fill in the missing numbers.*

$\frac{3}{5} = \frac{}{10} = \frac{}{20} = \frac{300}{}$

$\frac{3}{5} \xrightarrow{\times 2} \frac{6}{10} \xrightarrow{\times 2} \frac{12}{20} = \frac{300}{500}$ (and $\frac{3}{5} \xrightarrow{\times 100} \frac{300}{500}$)

7. Cancelling

Dividing both the top and bottom of a fraction by the *same* number also gives an equivalent fraction. This is called **cancelling**.

$\frac{6}{12} \xrightarrow{\div 3} \frac{2}{4}$

Question *Reduce the fraction $\frac{30}{48}$ to its lowest terms.*

$\frac{30}{48} \xrightarrow{\div 2} \frac{15}{24} \xrightarrow{\div 3} \frac{5}{8}$

Since 5 and 8 cannot be further divided, the fraction has been **reduced to its lowest terms**.

A quicker way of writing this is to cross out the cancelled numbers and write in the new numbers.

$\frac{\cancel{30}^{\cancel{15}\,5}}{\cancel{48}_{\cancel{24}\,8}} = \frac{5}{8}$

8. Adding and Subtracting Fractions

$\frac{1}{5} \quad + \quad \frac{2}{5} \quad = \quad \frac{3}{5}$

When fractions have the *same* denominator, simply add/subtract their numerators.

The above addition can also be written:

$$\frac{1}{5}+\frac{2}{5}=\frac{1+2}{5}=\frac{3}{5}$$

When fractions have *different* denominators, change them to equivalent fractions with the *same* denominator.

Question *Express $\frac{4}{5}-\frac{2}{3}$ as a single fraction.*

First, change the denominators 5 and 3 into the same number.

The smallest number that both 5 and 3 divide into exactly is 15 (their lowest common multiple; see Box 11, p12).

Next, change $\frac{4}{5}$ and $\frac{2}{3}$ into equivalent fractions with 15 as denominator:

$$\frac{4}{5}\xrightarrow{\times 3}\frac{12}{15} \quad \text{and} \quad \frac{2}{3}\xrightarrow{\times 5}\frac{10}{15}$$

(numerator and denominator of $\frac{4}{5}$ are both multiplied by 3; numerator and denominator of $\frac{2}{3}$ are both multiplied by 5)

So, $\frac{4}{5}-\frac{2}{3}=\frac{12}{15}-\frac{10}{15}=\frac{12-10}{15}=\frac{2}{15}$

Question *Jane spends $\frac{1}{4}$ of her pocket money on a magazine, $\frac{1}{3}$ on sweets and $\frac{1}{6}$ on make-up.*
(a) What fraction of her pocket money does she spend?
(b) What fraction of her pocket money is left?
(c) If 80p is left, how much pocket money does she get?

(a) Total spent $=\frac{1}{4}+\frac{1}{3}+\frac{1}{6}$

$=\frac{3}{12}+\frac{4}{12}+\frac{2}{12}$ LCM of 4, 3 & 6 is 12

$=\frac{\cancel{9}^{3}}{\cancel{12}_{4}}$ Cancel to lowest terms

$=\frac{3}{4}$

(b) Jane's whole pocket money is represented by 1.
So, remaining pocket money $=1-\frac{3}{4}=\frac{1}{4}$

(c) If $\frac{1}{4}$ of Jane's pocket money is 80p then her whole pocket money $=4\times80\text{p}$
$=320\text{p}=£3.20$

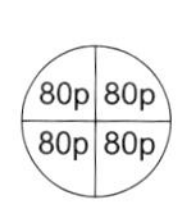

11. Sequences of Fractions

$\frac{1}{2}$	$\frac{1}{2}$	Notice that each numerator is 1 less than the denominator.
$\frac{1}{2}+\frac{1}{4}=\frac{2}{4}+\frac{1}{4}=\frac{3}{4}$	$\frac{3}{4}$	
$\frac{1}{2}+\frac{1}{4}+\frac{1}{8}=\frac{4}{8}+\frac{2}{8}+\frac{1}{8}=\frac{7}{8}$	$\frac{7}{8}$	
$\frac{1}{2}+\frac{1}{4}+\frac{1}{8}+\frac{1}{16}=\frac{8}{16}+\frac{4}{16}+\frac{2}{16}+\frac{1}{16}=\frac{15}{16}$	$\frac{15}{16}$	What is the next fraction?

9. Multiplying Fractions

Cancel numerators and denominators if possible. Then multiply numerators together and denominators together.

Examples

$\frac{4}{3}\times\frac{5}{6}=\frac{\cancel{4}^{2}}{3}\times\frac{5}{\cancel{6}_{3}}=\frac{2}{3}\times\frac{5}{3}=\frac{2\times5}{3\times3}=\frac{10}{9}=1\frac{1}{9}$

$2\frac{3}{4}\times1\frac{1}{5}=\frac{11}{4}\times\frac{6}{5}$ Change to improper fractions

$=\frac{11}{\cancel{4}_{2}}\times\frac{\cancel{6}^{3}}{5}$ Cancel numerators and denominators

$=\frac{11\times3}{2\times5}$ Multiply numerators and denominators

$=\frac{33}{10}=3\frac{3}{10}$ Change to a mixed number

Question *Calculate $\frac{3}{5}$ of 170 m.*

$\frac{3}{5}$ of $170=\frac{3}{5}\times170$ **of** means $\times$

$=3\div5\times170=102$ m

Question *Ahmed runs $\frac{2}{3}$ of a kilometre every day. How far does he run in 5 days?*

$5\times\frac{2}{3}=\frac{5}{1}\times\frac{2}{3}=\frac{10}{3}=3\frac{1}{3}$ km NOTE Change 5 into $\frac{5}{1}$

10. Dividing Fractions

Change $\div$ to $\times$ and turn the second fraction upside down.

Example

$\frac{7}{2}\div\frac{3}{4}=\frac{7}{2}\times\frac{4}{3}$ Now multiply as usual

$=\frac{7}{\cancel{2}_{1}}\times\frac{\cancel{4}^{2}}{3}=\frac{7\times2}{1\times3}=\frac{14}{3}=4\frac{2}{3}$

Question *A dog eats $\frac{2}{3}$ of a tin of food each day. How long will 8 tins last?*

Method 1 Each tin contains 3 thirds.
So, 8 tins contain $8\times3=24$ thirds.
Each day the dog eats 2 thirds.
So, 8 tins will last $24\div2=12$ days.

Method 2 Calculate how many $\frac{2}{3}$ there are in 8

$8\div\frac{2}{3}=\frac{8}{1}\div\frac{2}{3}=\frac{8}{1}\times\frac{3}{2}=\frac{8\times3}{1\times2}=\frac{24}{2}=12$

12. BoDMAS and Fractions

Question *Simplify $2\frac{1}{4}+1\frac{1}{4}\times1\frac{2}{5}$.*

$2\frac{1}{4}+1\frac{1}{4}\times1\frac{2}{5}=\frac{9}{4}+\frac{5}{4}\times\frac{7}{5}$ Change to improper fractions

$=\frac{9}{4}+\frac{\cancel{5}^{1}}{4}\times\frac{7}{\cancel{5}_{1}}$ Cancel before multiplying

$=\frac{9}{4}+\frac{1\times7}{4\times1}$ BoDMAS: multiply first

$=\frac{9}{4}+\frac{7}{4}$

$=\frac{16}{4}$

$=4$

Decimals

1. Decimal System

The decimal system is based on columns. Each column is worth ten times the column to its right.

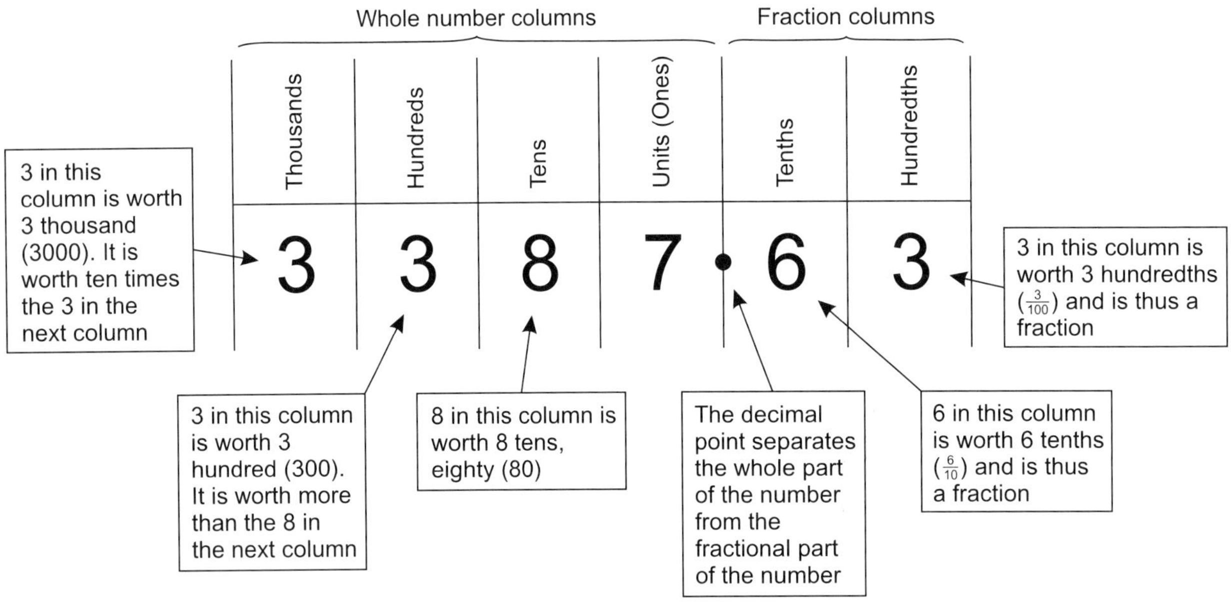

2. Digits

The numbers 0, 1, 2, 3, 4, 5, 6, 7, 8, 9 are called **digits**.

3. Reading and Writing Decimals

Question *Write in words the number 692 023 907.*

Long numbers are written in groups of three digits, starting from the decimal point. This helps us to read the number

6 9 2 (Millions)	0 2 3 (Thousands)	9 0 7
Six hundred and ninety two (million)	Twenty three (thousand)	Nine hundred and seven

So, 692 023 907 in words is six hundred and ninety two million, twenty three thousand, nine hundred and seven.

4. Decimal Places

The number of digits after the decimal point is called the number of **decimal places (dp)**. Here are some examples:

15.634	0.002	4.00
3 decimal places	3 decimal places	2 decimal places

5. Arranging Decimals in Order of Size

Question *Arrange these decimals in order of size, from smallest to largest:*

0.0101 0.0099 0.0109 0.1 0.011

Write down the numbers in a column, keeping the decimal points in line.

0.0101
0.0099
0.0109
0.1
0.011

Fill in any spaces in the columns with zeros (see the bold zeros).

0.0101
0.0099
0.0109
0.1**000**
0.011**0**

Imagine the decimal points are missing and ignore any initial zeros (see the bold numbers). Now the numbers can easily be compared.

0.0**101**
Smallest 0.00**99**
0.0**109**
Largest 0.**1000**
0.0**110**

We can now write the numbers from smallest to largest (this is called **ranking** the numbers).

0.0099 0.0101 0.0109 0.0110 0.1000

6. Adding and Subtracting Decimals

Keep the decimal points in line with each other, then add/subtract as for whole numbers.

Examples

$$\begin{array}{r} 2.3 \\ +\,4.5 \\ \hline 6.8 \\ \hline \end{array} \qquad \begin{array}{r} 8.3 \\ -\,3.2 \\ \hline 5.1 \\ \hline \end{array} \qquad \begin{array}{r} 13.04 \\ +\ \ 0.592 \\ \hline 13.632 \\ \hline \end{array} \qquad \begin{array}{r} 0.532 \\ -\,0.096 \\ \hline 0.436 \\ \hline \end{array}$$

Question *Ms Parker spent £2.38 on stamps, £3.94 on birthday cards and 65p on a pen. How much did she spend altogether?*

First, change 65p to £0.65 then add up the numbers.

So, Ms Parker spent £6.97 altogether.

$$\begin{array}{r} 2.38 \\ 3.94 \\ +\,0.65 \\ \hline 6.97 \\ \hline \end{array}$$

7. Multiplying and Dividing by 10

Multiplying by 10 moves the decimal point one place to the right.

$$10 \times 3.6894 = 36.894$$

Similarly

$$100 \times 3.6894 = 368.94$$

and

$$1000 \times 3.6894 = 3689.4$$

Dividing by 10 moves the decimal point one place to the left.

$$4567.8 \div 10 = 456.78$$

Similarly

$$4567.8 \div 100 = 45.678$$

and

$$4567.8 \div 1000 = 4.5678$$

8. Multiplying Decimals

You should know how to multiply simple numbers involving decimals *without* using a calculator.

Question *Calculate the value of 0.2×0.03 without using a calculator.*

① Count the total number of decimal places of the numbers being multiplied:

0.2   0.03

Total of 3 decimal places

8. Multiplying Decimals (Contd)

② Write the numbers without their decimal points:

02×003 or simply 2×3

③ Multiply these numbers together:

2×3 = 6 which can be written 6.

④ Move the decimal point of your answer to the left and insert 0s to make a number with the decimal places found in step 1:

.006.

So, 0.2×0.03 = 0.006

Question *A plastic moulding weighs 0.03 g. What is the weight of 50 mouldings?*

Calculate 50×0.03 using the above method. There is a total of 2 decimal places in 50 and 0.03

50×003 = 50×3 = 150 = 150.

So, 50×0.03 = 1.50 g or simply 1.5 g

Question *Salmon costs £5.63 per kg. What is the cost of 0.6 kg correct to the nearest p?*

Calculate 0.6×5.63 using the above method. There are a total of 3 decimal places in 0.6 and 5.63:

06×563 = 6×563 = 3378 = 3378.

So, 0.6×5.63 = 3.378

The cost of the salmon is £3.378 ≈ £3.38

9. Dividing Decimals

You should know how to divide simple numbers involving decimals *without* using a calculator.

Question *A roll of tape, 3.36 m long, is cut into 7 equal pieces. What is the length of each piece?*

Length of each piece = 3.36 m ÷ 7

Divide as with whole numbers. Write the decimal point of the answer above the decimal point of the number being divided,

$$\begin{array}{r} 0.48 \\ 7\,\overline{)\,3.36} \\ 2\,8 \\ \hline 56 \end{array}$$

The length of each piece of tape is 0.48 m,

Question *The thickness of a business card is 0.4 mm. A pile of cards is 50 mm high. How many cards are in the pile?*

Number of cards in pile = 50÷0.4

When dividing by a decimal use the following method:

① Move the decimal points of *both* numbers the same number of decimal places to the right until the number doing the dividing (the **divisor**) becomes a whole number.

0.4 becomes 4. or simply 4

50 = 50.0 and so becomes 500. or simply 500

② Divide as usual using these numbers.

$$\begin{array}{r} 125 \\ 4\,\overline{)\,500} \end{array}$$

So, 50÷0.4 = 125

There are 125 business cards in the pile.

more

Decimals (Contd)

10. Changing Decimals to Fractions

Question *Express 0.125 as a fraction in its lowest terms.*

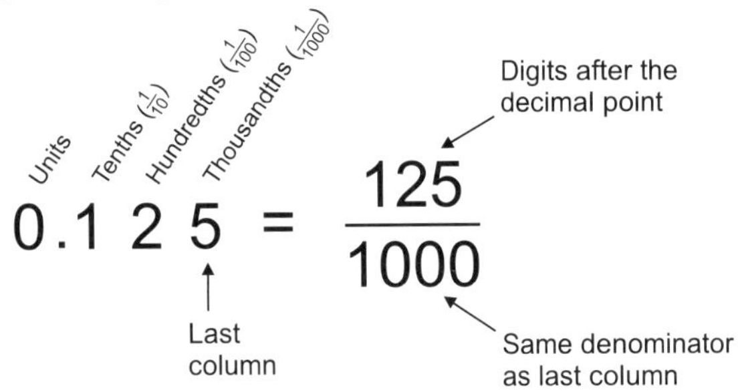

Question *Express 7.08 as a mixed fraction.*

Separate the whole part, 7, from the fraction part, 0.08, i.e. 7.08 = 7+0.08.

Now, $0.08 = \frac{8}{100} = \frac{2}{25}$

So, $7.08 = 7+0.08 = 7+\frac{2}{25} = 7\frac{2}{25}$

11. Changing Fractions to Decimals

Another way of writing the fraction $\frac{3}{5}$ is $3\div5$. So we can use a calculator to convert $\frac{3}{5}$ to a decimal:

$\frac{3}{5} = 3\div5 = 0.6$ 3 ÷ 5 = 0.6

Question *Express $2\frac{5}{8}$ as a decimal.*

Now, $2\frac{5}{8}$ can be written $2+\frac{5}{8}$

And $\frac{5}{8} = 5\div8 = 0.625$

So, $2\frac{5}{8} = 2+\frac{5}{8} = 2+0.625 = 2.625$

Question *Write down the numbers below in increasing order of size*

$\frac{3}{4}$ 0.7 $\frac{13}{20}$

Convert the fractions to decimals:

$\frac{3}{4} = 3\div4 = 0.75$ and $\frac{13}{20} = 13\div20 = 0.65$

Now compare the decimals (see Box 5, p18):

Largest 0.75
0.70
Smallest 0.65

Write down the original numbers in order of size:

$\frac{13}{20}$ 0.7 $\frac{3}{4}$

12. Recurring Decimals

Often a calculation results in a never-ending decimal. For example:

$\frac{4}{7} = 4\div7 = 0.571\ 428\ ....$

Sometimes, the digits after the decimal point form a pattern. For example:

$\frac{123}{999} = 123\div999 = 0.123\ 123\ ....$ (Try this yourself)

This is called a **recurring decimal**, and written briefly as $0.\dot{1}2\dot{3}$ (the recurring pattern is between the dots).

more

12. Recurring Decimals (Contd)

Here are two more examples:

$\frac{1}{3} = 1\div3 = 0.333\ 333\ ... = 0.\dot{3}$ Only one dot needed

$\frac{467}{180} = 467\div180 = 2.594\ 444\ ... = 2.59\dot{4}$

Question *Express the recurring decimal 0.345 345 345 ... as a fraction.*

$0.345\ 345\ 345\ ... = 0.\dot{3}4\dot{5} = \frac{345}{999}$ ← Write down a 9 for each digit in the recurring pattern.

Question *Express the recurring decimal 0.071 717 1... as a fraction. Hence, express 0.271 717 1... as a fraction.*

Write down a 9 for each digit of the recurring pattern, followed by any 0s after the decimal point.

$0.071\ 717\ 171\ ... = 0.0\dot{7}\dot{1} = \frac{71}{990}$

Now, $0.271\ 717\ ... = 0.2+0.071\ 717\ ...$

$= \frac{2}{10}+\frac{71}{990} = \frac{198}{990}+\frac{71}{990} = \frac{269}{990}$

13. Rational and Irrational Numbers **H**

A **rational number** is a number that can be written as a common fraction (Box 5, p16). Rationals include all whole numbers (integers). Here are some examples:

2 (because it can be written $\frac{2}{1}$ or $\frac{6}{3}$ for example)

-2 $(=-\frac{2}{1})$ 0 $(=\frac{0}{1})$ $\frac{4}{7}$ $-\frac{2}{3}$ $3\frac{1}{7}$ $-1\frac{2}{5}$

2.75 $(=2\frac{3}{4})$ $0.\dot{1}2\dot{3}$ $(=\frac{123}{999})$

An **irrational number** *cannot* be written as a fraction. Here are some examples:

$\sqrt{5}$ $\sqrt{\frac{1}{5}}$ $\sqrt{\frac{9}{5}}$ $\sqrt[3]{7}$	Roots that cannot be calculated exactly
$5+\sqrt{2}$ $4\times\sqrt{5}$	Numbers involving these roots
π 2π $\pi+2$	Numbers involving π
sin 20° tan 30°	Many trigonometric values cannot be calculated exactly

Some numbers look like irrational numbers but are not. For example, $\sqrt{9}$ is the rational number 3.

Question *Find an irrational number between 3 and 4.*

Square 3 to get $3^2 = 9$
Add 1 to get $9+1 = 10$
Square root the answer $= \sqrt{10}$.
Check: $\sqrt{10} = 3.16....$

Question *n is a positive integer such that $\sqrt{n} \approx 15.4$, correct to 1 decimal place. Find the value of n and explain why $\sqrt{n}$ is irrational.*

$\sqrt{n} \approx 15.4$. So, $n \approx 15.4^2 = 237.16$ (see Box 5, p39)

So, $n = 237$ (n is given as a positive whole number).

Check: $\sqrt{237} = 15.394.... \approx 15.4$ (1 dp)

Since $\sqrt{237}$ does not work out exactly, it is irrational.

Accuracy and Approximation

1. Rounding Off Numbers

Rounding off a number means replacing it by a simpler but less accurate number. For example, instead of saying that there were 283 fans at a concert, it is simpler to say there were about 300.

We always round off to a certain **degree of accuracy**. In the above example, 283 was rounded off *correct to the nearest 100 people*. We chose 300 instead of 200 because 283 is closer to 300 than 200. So, 283 was **rounded up** to 300.

Similarly, if 283 were rounded off to the nearest 10, it would be **rounded down** to 280, because 283 is closer to 280 than 290.

Using this rule, 285 would be rounded *up* to 290.

Less than 5, round *down*.
5 or more, round *up*.

So, 12.65 g $\approx$ 12.7 g (correct to the nearest $\frac{1}{10}$ g)
and 12.64 g $\approx$ 12.6 g (correct to the nearest $\frac{1}{10}$ g)

Question *A small cake weighs 33 g. What is this weight correct to the nearest 5 g?*

33 g is closer to 35 g than 30 g.
So, 33 g is approximately 35 g, to the nearest 5 g.

We write 33 g $\approx$ 35 g

$\simeq$ and $\approx$ both mean **approximately equal to**

Question *The diameter of Jupiter is 142 800 km. Write this distance correct to the nearest 1000 km.*

142 800 km is closer to 143 000 km than 142 000 km.
So, 142 800 km $\approx$ 143 000 km, to the nearest 1000 km.

2. Significant Figures

In Box 1, we replaced the number 283 with the simpler number 300. The reason that 300 is simpler than 283 is because it only has one **significant figure (sf)**, 3: the other figures (digits) are 0

$$\underbrace{285}_{\text{3 significant figures}} \approx \underbrace{3}_{\text{1 significant figure}}00$$

We say that 283 is approximately equal to 300, **correct to 1 significant figure**, and write:
285 $\approx$ 300 (1 sf)

Here are more examples of significant figures:

58 000	5.7	0.0009	508 000
2 sf	2 sf	1 sf	3 sf (0 is a significant figure because it is in between non-zero digits)

Significant figures can be used to approximate both large and small numbers.

Question *The distance from the Earth to the Moon is 384 400 km. Express this distance correct to 1 significant figure.*

Replace the unwanted significant figures by 0:

$\underbrace{3844}_{\text{4 sf}}00 \approx \underbrace{4}_{\text{1 sf}}00\,000$ km (correct to 1 sf)

3. Decimal Places

Decimal places (dp) can be used to approximate *small* numbers.

For example, the number below has 6 decimal places (6 digits after the decimal point).

$$5.\underbrace{483\,268}_{\text{6 decimal places}}$$

This number can be simplified (approximated) by using fewer decimal places:

5.483 268 (6 dp)
$\approx$ 5.48 (2 dp)

Similarly 0.057
$\approx$ 0.06 (2 dp) — Round the number *up* because 7 is greater than 5 (see Box 1)

and 4.239
$\approx$ 4.2 (1 dp) — Round the number *down* because 3 is less than 5

and 0.0499
$\approx$ 0.0 (1 dp) — This shows that an answer of 0.0 does not mean the same as 0

4. Accuracy

You are expected to express your answers to a reasonable degree of accuracy. For example, if you are calculating the area of a circular lawn and your calculator gives you the answer 1325.694 31 m^2, a reasonable degree of accuracy would be correct to the nearest 10 m^2, giving an answer of 1330 m^2.

Question *What would be a reasonable degree of accuracy when describing (a) the weight of a person; (b) the thickness of a pencil; (c) the number of people attending a first-division football match.*

(a) Correct to the nearest kg.
(b) Correct to the nearest mm.
(c) Correct to the nearest 1000 people.

Always check the question to see if a certain degree of accuracy is required, e.g. to the nearest 10 m^2.

5. Estimation

The result of a calculation can be estimated by replacing the numbers involved by simpler, approximate numbers. For example:

$59 \times 3.2 \approx 60 \times 3 = 180$ 59 replaced by 60, 3.2 by 3

There are often several ways of estimating a calculation. For example:

$$\frac{40.21+19.75}{0.3\times 12} \approx \frac{40+20}{0.25\times 12} = \frac{60}{\frac{1}{4}\text{ of }12} = \frac{60}{3} = 20$$

$$\frac{40.21+19.75}{0.3\times 12} \approx \frac{40+20}{0.\dot{3}\times 12} = \frac{60}{\frac{1}{3}\text{ of }12} = \frac{60}{4} = 15$$

The second estimate is more accurate because 0.3 is closer to $\frac{1}{3}$ than to $\frac{1}{4}$.

Make sure your estimate is accurate enough:
$0.6\times 21 \approx 1\times 20 = 20$ Poor
$0.6\times 21 \approx 0.5\times 20 = 10$ Good

Accuracy and Approximation (Contd)

6. Error

Kulvinder measured the height of her friend Paula to be 140 cm. Paula's true height was 136 cm. So, Kulvinder made an **error** of 140 – 136 = 4 cm.

All of the lengths below are approximately 140 cm, correct to the nearest 10 cm (see Box 1).

135 cm 136 cm 137 cm 138 cm 139 cm 140 cm
141 cm 142 cm 143 cm 144 cm 144.9 cm 144.99 cm

So, if a person was measured to be 140 cm tall correct to the nearest 10 cm, the *true* height could be anywhere between 135 cm and 145 cm (the **lower** and **upper bounds**).

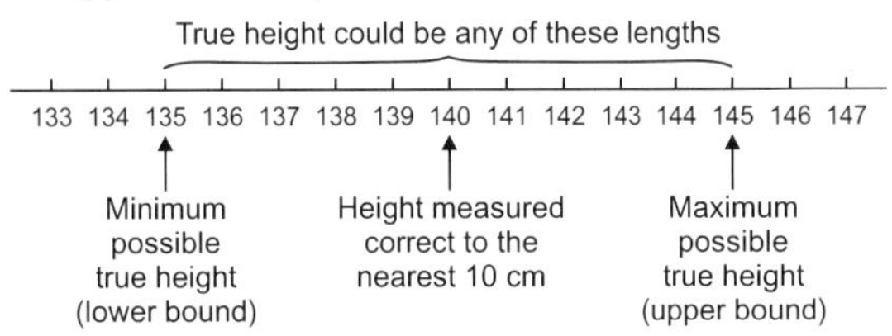

This can be written briefly as:

135 cm ≤ True height < 145 cm

NOTE Strictly speaking, the true height cannot be 145 cm (hence the < sign), because a height of 145 cm would be rounded *up* to 150 cm.

The greatest error would occur if the person's true height were 135 cm. So, the greatest possible error is 140 – 135 = 5 cm, half the degree of accuracy, 10 cm.

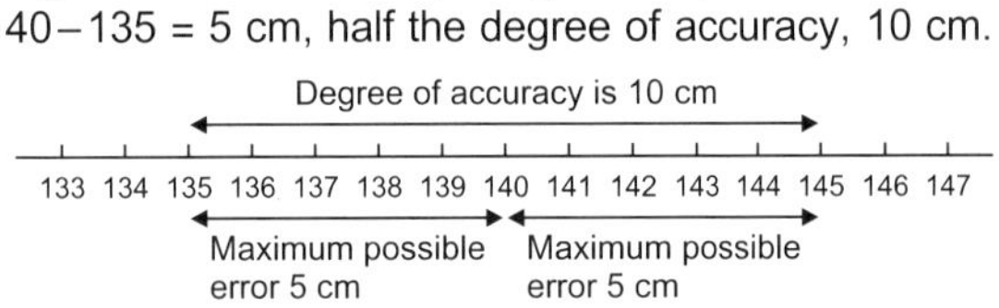

The maximum possible error of a measurement is half its degree of accuracy.

Question *The weight of a packet of tea is accurate to the nearest 5 g. What is the minimum possible weight of a 185 g packet?*

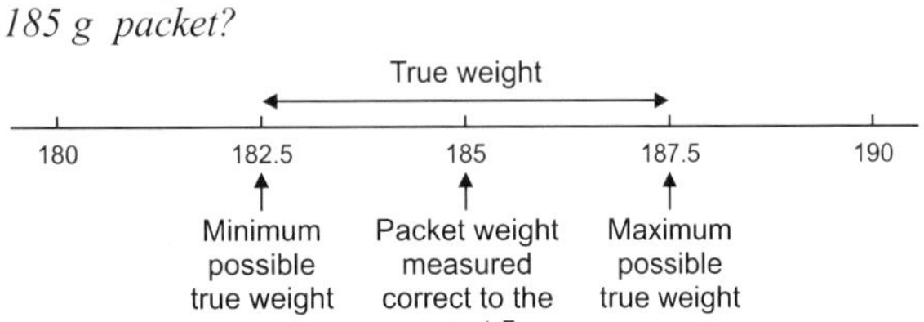

The diagram shows that the minimum possible true weight of the packet of tea is 182.5 g

Question *A machine is set to produce metal bolts of length 18 mm. The greatest possible error made by the machine is 0.2 mm. What are the upper and lower bounds of the length of bolt produced by the machine?*

The machine could produce a bolt whose length was up to 0.2 mm larger or smaller than the set value of 18 mm. So, the maximum possible length of a bolt is 18+0.2 = 18.2 mm and the minimum possible length is 18 – 0.2 = 17.8 mm

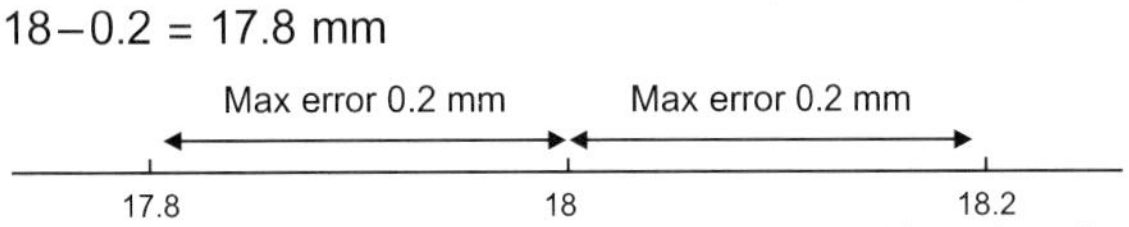

7. Upper and Lower Bounds

H

When we use approximate numbers in a calculation, we obtain an approximate result. The true result lies somewhere between the lower and upper bounds of the calculated value. We can find these lower and upper bounds when we know the degrees of accuracy of the numbers used in the calculation.

Calculations Involving Addition and Subtraction

Question *The length and width of the rectangle are correct to the nearest tenth of a centimetre. Calculate the upper and lower bounds for the perimeter of the rectangle.*

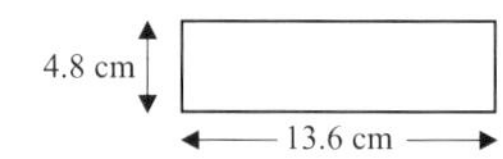

First, calculate the lower and upper bounds of the length and width (see Box 6)

13.55 cm ≤ Length < 13.65 cm
4.75 cm ≤ Width < 4.85 cm

Using the upper bounds will give the greatest possible perimeter, i.e. the upper bound of the perimeter:

Upper bound = 13.65+13.65+4.85+4.85 = 37 cm

Using the lower bounds will give the smallest possible perimeter, i.e. the lower bound of the perimeter:

Lower bound = 13.55+13.55+4.75+4.75 = 36.6 cm

Write the lower and upper bounds of the perimeter as follows:

36.6 cm ≤ Perimeter < 37 cm

Question *A heap of sand has a volume of 1.75 m^3, correct to 2 decimal places. John removes some sand and estimates its volume to be 0.4 m^3, correct to 1 decimal place. What is the upper bound of the volume of sand remaining?*

Firstly, calculate the lower and upper bounds of the heap of sand and the amount removed:

1.745 m^3 ≤ Heap < 1.755 m^3
0.35 m^3 ≤ Removed sand < 0.45 m^3

Greatest amount of sand remaining (upper bound) is

$\begin{pmatrix}\text{Upper bound}\\ \text{of heap}\end{pmatrix} - \begin{pmatrix}\text{Lower bound of}\\ \text{removed sand}\end{pmatrix}$

= 1.755 – 0.35
= 1.405 m^3

Calculations Involving Multiplication

Question *The diameter of a circle is 4.73 m, measured correct to 2 dp Calculate the upper and lower bounds of (a) the diameter, (b) the radius.*

(a)

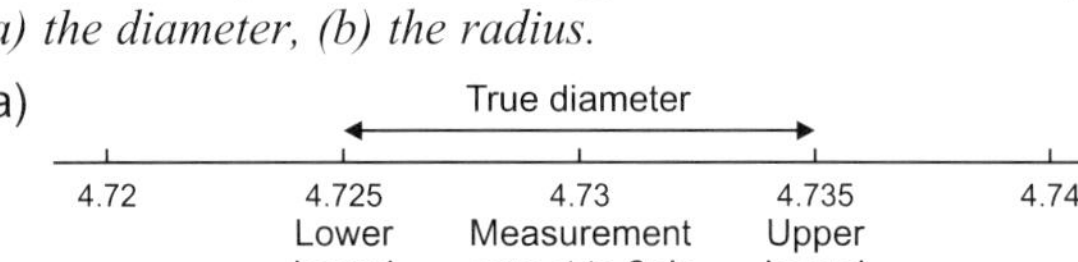

The diagram shows that the true value of the diameter lies somewhere between 4.725 m and 4.735 m. So we can write:

4.725 m ≤ Diameter < 4.735 m

(b) The radius is half the diameter and so its lower and upper bounds can be found by halving the bounds of the diameter:

4.725÷2 = 2.3625 and 4.735÷2 = 2.3675

2.3625 m ≤ Radius < 2.3675 m

Using a Calculator

1. The Kind of Calculator You Need

For GCSE mathematics you need a **scientific calculator**. Buy an inexpensive, basic model that can display only one number at a time (more sophisticated models often come with complicated, thick instruction manuals). Make sure it has trigonometry keys [sin] [cos] [tan] and statistics keys [Σx] [Σx²] [$\bar{x}$] [n] [σ]. The calculator steps given in this book are for a basic calculator.

2. The Keys of a Basic Calculator

Here are the keys found on a basic calculator. Some of the symbols may look different on your calculator or may be printed above or below another key. Check your instruction manual.

Numerical	[0] [1] [2] [3] [4] [5] [6] [7] [8] [9] [·] [π] [x]	Used to enter numbers into the calculator.
Arithmetical	[+] [−] [×] [÷] [(] [)]	Used to combine two or more numbers into a single number.
Functions	[+/−] [√] [x^2] [x^y] [$x^{1/y}$] [$\frac{1}{x}$] [INV] [sin] [cos] [tan] [EXP] [Σx] [Σx²] [$\bar{x}$] [n] [σ]	Used to modify the displayed number in some way.
Memory	[Min] [M+] [MR]	Used to store, change and retrieve the memory.
Others	[=] [C] [AC]	Used to finish and start a calculation, correct a mistake, etc.

3. Before You Start a Calculation

Calculator Modes

Calculators can work in several different **modes**, depending on the type of calculation. For example, **statistics mode** for statistical calculations; **degrees mode** for calculating with angles measured in degrees. Make sure your calculator is in the *correct working mode* before you begin a calculation; the mode is usually displayed on the screen, e.g. DEG for degrees mode (read your instruction manual).

Memory

The memory of your calculator starts with 0 when you switch it on. More sophisticated calculators can store numbers after they are switched off.

If you are performing several calculations involving the memory, *clear* the memory each time or replace it with a number you want to store (see Box 12, p24)

4. Entering Numbers into a Calculator

To enter the number 52.63 into your calculator, press

[5] [2] [·] [6] [3] which will be displayed as (52.63)

To enter the number 0.58 into your calculator, press

[0] [·] [5] [8] which will be displayed as (0.58)

or simply press [·] [5] [8]

To enter the negative number −40, use [+/−] to change 40 into −40. Press

[4] [0] [+/−] which will be displayed as (−40.)

5. If You Make a Mistake ....

[C] undoes the last number or function entered.

[AC] clears the entire calculation (but ***not*** the memory), ready to start a new calculation.

6. Addition and Subtraction

To calculate 16+31, press the following keys:

[1] [6] enters the number 16	(16.)
[+] ready for the next number	(16.)
[3] [1] enters the number 31	(31.)
[=] displays the answer	(47.)

To calculate −15+12, press the following keys:

[1] [5] enters the number 15	(15.)
[+/−] changes 15 to −15	(−15.)
[+] [1] [2] [=] displays the answer	(−3.)

To calculate 5−7+4−11+3, press the following keys:

[5] [−] [7] [+] [4] [−] [1] [1] [+] [3] [=] gives (−6.)

7. Multiplication

To calculate 16.3×0.04, press the following keys:

[1] [6] [·] [3] [×] [0] [·] [0] [4] [=] (0.652)

To calculate −7×3×2, press the following keys:

[7] [+/−] [×] [3] [×] [2] [=] (−42.)

8. Division

To calculate 16.3÷0.04, press the following keys:

[1] [6] [·] [3] [÷] [0] [·] [0] [4] [=] (407.5)

To convert $\frac{7}{8}$ to a decimal, press the following keys:

[7] [÷] [8] [=] (0.875)

See also Box 13, p24

Using a Calculator (Contd)

9. Accuracy

Your calculator works to a high degree of accuracy. The displayed answer is often more accurate than you need. For example, if £52.39 is divided equally between 7 people, each person would receive £52.39÷7. Using a calculator,

[5] [2] [·] [3] [9] [÷] [7] [=] gives the answer (7.4842857)

The displayed answer is too accurate, since money is usually expressed correct to 2 decimal places, i.e. correct to the nearest p.
So we would write £52.39÷7 = £7.48
or more strictly £52.39÷7 ≈ £7.48

10. BoDMAS and Brackets

Your calculator obeys the rules of BoDMAS (see Box 4, p11). This means it will correctly calculate any allowable combination of numbers and arithmetical symbols +, −, ×, ÷, (,) when the corresponding keys are pressed.

To calculate 2+3×4, press the following keys:

[2] [+] [3] [×] [4] [=] giving the answer (14.)

To calculate 15÷(2+3), press the following keys:

[1] [5] [÷] [(] [2] [+] [3] [)] [=] (3.)

To calculate 16+4(−5−12÷2)+6×4, press:

[1] [6] [+] [4] [×] [(] [5] [+/−] [−] [1] [2] [÷] [2] [)] [+] [6] [×] [4] [=] (-4.)

4(means 4×(← Changes 5 to −5

Combinations of × and ÷ signs are carried out by the calculator in the order they are entered

To calculate 24÷2×3÷9, press the following keys:

[2] [4] [÷] [2] [×] [3] [÷] [9] [=] (4.)

11. Function Keys

Most function keys modify the number on the calculator display.

To calculate 15^2, press the following keys:

[1] [5] enters the number 15 (15.)

[x^2] squares the displayed number (225.)

To calculate $\sqrt{16}$, press the following keys:

[1] [6] enters the number 16 (16.)

[√] square roots the displayed number (4.)

To calculate $\frac{1}{8}$, press the following keys:

[8] [$\frac{1}{x}$] (0.125)

To calculate sin 30°, press the following keys:

[3] [0] [sin] (0.5)

12. Using the Calculator Memory

You can temporarily store a number in the memory of your calculator. This saves having to write the number down and is also more accurate.

Question *Bricks cost 87p each and are sold in packs of 856.*
(a) How much does a pack cost?
(b) How many packs can a builder buy with £5000?

(a) Cost of one pack = 856×£0.87 = £744.72

[8] [5] [6] [×] [0] [·] [8] [7] [=] (744.72)

[Min] 744.72 is now stored in the memory

(b) Number of packs that can be bought for £5000
= 5000÷744.72

[5] [0] [0] [0] [÷] (5000.)

[MR] recalls 744.72 from the memory (744.72)

[=] gives the answer (6.7139328)

So, the builder can buy 6 complete packs.

To clear the memory, press [0] [Min]. This puts 0 into the memory.

Use the [M+] key to add the displayed number to the contents of the memory.

Question *Pens cost 60p each, pencils 45p, erasers 80p. Calculate the cost of 12 pens, 4 pencils and 2 erasers.*

Keys		Display
[0] [Min]	clear the memory	0.
[1] [2] [×] [6] [0]	cost of pens	720.
[M+]	.... added to the memory	720.
[4] [×] [4] [5] [M+]	cost of pens added	180.
[2] [×] [8] [0] [M+]	cost of erasers added	160.
[MR]	retrieves the total cost	1060.

Total cost is £10.60.

13. Calculations Involving Fractions

For calculations involving simple fractions, replace the dividing line by the divide sign ÷

To calculate $\frac{28}{5} - \frac{2}{0.63}$, write as 28÷5−2÷0.63 and press:

[2] [8] [÷] [5] [−] [2] [÷] [0] [·] [6] [3] [=] (2.4253968)

So, $\frac{28}{5} - \frac{2}{0.63}$ = 2.43 (correct to 2 dp)

For more complicated fractions, bracket the numerator and denominator.

To calculate $\frac{17.4-2.8}{3.6+4.9}$, write as $\frac{(17.4-2.8)}{(3.6+4.9)}$, then as (17.4−2.8)÷(3.6+4.9) and press:

[(] [1] [7] [·] [4] [−] [2] [·] [8] [)] [÷] [(] [3] [·] [6] [+] [4] [·] [9] [)] [=]

So, $\frac{17.4-2.8}{3.6+4.9}$ = 1.72 (2 dp) (1.7176471)

Alternatively, calculate the denominator first and store it in the memory for later:

[3] [·] [6] [+] [4] [·] [9] [=] [Min] [1] [7] [·] [4] [−] [2] [·] [8] [=] [÷] [MR] [=]

Denominator — Numerator

Percentages

1. Percentages, Fractions and Decimals

75% is a short way of writing the fraction $\frac{75}{100}$. So,

$$75\% = \frac{75}{100} = 75 \div 100 = 0.75$$

Percentage ↑ Fraction ↑ Decimal ↑

To convert a percentage to a fraction, divide by 100.

Question *Express 60% as a fraction in its lowest terms.*

$$60\% = \frac{60}{100} = \frac{6^3}{10_5} = \frac{3}{5}$$

Question *Place the numbers 88%, $\frac{7}{8}$ and 0.808 in increasing order of size.*

Convert the non-decimals to decimals.

88% = 88÷100 = 0.88 $\frac{7}{8}$ = 7÷8 = 0.875

Place the decimals in increasing order of size (see Box 5, p18).

0.808   0.875   0.88

Largest 0.88
0.875
Smallest 0.808

Write down the *original* numbers in increasing order of size.

0.808 $\frac{7}{8}$ 88%

To convert a decimal or fraction to a percentage, multiply by 100.

Question *Convert 0.45 to a percentage.*

0.45 = 0.45×100% = 45%

Question *Convert 0.05 to a percentage.*

0.05 = 0.05×100% = 5%

Question *$\frac{5}{14}$ of Joan's teeth have fillings. What percentage of her teeth are filled?*

$\frac{5}{14} = \frac{5}{14} \times 100\% = 5 \div 14 \times 100\% = 35.7\%$ (1 dp)

NOTE You get the same answer by calculating 5×100÷14 or 100÷14×5.

2. Common Percentages and Fractions

$100\% = \frac{100}{100} = 1$	**100%** represents a **whole**
$25\% = \frac{25^1}{100_4} = \frac{1}{4}$	**25%** represents a **quarter**
$50\% = \frac{50^1}{100_2} = \frac{1}{2}$	**50%** represents a **half**
$75\% = \frac{75^3}{100_4} = \frac{3}{4}$	**75%** represents **three quarters**
$10\% = \frac{10^1}{100_{10}} = \frac{1}{10}$	**10%** represents **one tenth**
$20\% = \frac{20^2}{100_{10}} = \frac{2}{10}$	**20%** represents **two tenths**
etc.	
$1\% = \frac{1}{100}$	**1%** represents **one hundredth**
$5\% = \frac{5^1}{100_{20}} = \frac{1}{20}$	**5%** represents **one twentieth**
$\frac{1}{3} = \frac{1}{3} \times 100\% = \frac{100}{3}\% = 33\frac{1}{3}\%$	**$33\frac{1}{3}$%** is **one third**
$\frac{2}{3} = \frac{2}{3} \times 100\% = \frac{200}{3}\% = 66\frac{2}{3}\%$	**$66\frac{2}{3}$%** is **two thirds**

3. Expressing One Quantity as a Percentage of Another Quantity

Question *A bottle contains 70 cl of wine. Bob drinks 15 cl. What percentage of the wine does Bob drink?*

Bob drinks 15 *cl out of* 70 *cl*.

= 15 *cl* out of 70 *cl* = $\frac{15}{70}$ As a fraction

= $\frac{15}{70} \times 100\%$ Converting to a percentage

= 15÷70×100

= 21.4% (1 dp)

Question *Ahmed buys a car for £6500 (the cost price) and sells it 2 years later to a dealer for £4000 (the selling price). Express the selling price as a percentage of the cost price.*

Selling price *as a percentage of* cost price

= £4000 as a percentage of £6500

= £4000 as a percentage of £6500 = $\frac{4000}{6500}$

= $\frac{4000}{6500} \times 100\%$

= 4000÷6500×100

= 61.5% (1 dp)

Question *A dealer buys a car for £4000 (the cost price) and sells it for £4600 (the selling price). Express the selling price as a percentage of the cost price.*

£4600 as a percentage of £4000

= $\frac{£4600}{£4000} = \frac{4600}{4000} \times 100\% = 115\%$

4. Finding a Percentage of a Quantity

Question *Calculate 20% of 150 kg.*

20% of 150 kg

= $\frac{20}{100} \times 150$ kg   **of** means ×

= 20÷100×150

= 30 kg

Question *A salesman of double glazed windows earns $6\frac{1}{4}$% commission. How much does he earn on a sale of £1700?*

Commission = $6\frac{1}{4}$% of £1700

= 6.25% of £1700

= $\frac{6.25}{100} \times £1700$

= 6.25÷100×£1700

= £106.25

Question *The share price of a gold mine was 128p. Share prices rose by 400% when a new gold vein was discovered. By how much did the share price rise?*

Rise in price = 400% of 128p

= $\frac{400}{100} \times 128$p

= 4×128p

= 512p

(Each share would then be worth 128+512 = 640p)

Percentages (Contd)

5. Finding the Percentage Change of a Quantity

$$\text{Percentage change} = \frac{\text{Change in quantity}}{\text{Original quantity}} \times 100\%$$

Question *During a fever, the temperature of a child increased from 98.4°C to 104°C. Find the percentage increase in temperature.*

NOTE *Percentage increase* can be written briefly as *% increase*

Original temperature = 98.4 °C

Change in temperature (increase) = 104 − 98.4
= 5.6 °C

$$\% \text{ increase} = \frac{\text{Increase in temperature}}{\text{Original temperature}} \times 100\%$$
$$= \frac{5.6}{98.4} \times 100\%$$
$$= 5.6 \div 98.4 \times 100\%$$
$$= 5.69\% \text{ (2 dp)}$$

The child's temperature increased by 5.69%.

Question *Alex and Grace bought their house in 1982 for £67 000 and sold it in 1994 for £98 000. Calculate:*
(a) the profit on the sale
(b) the percentage profit.

(a) Profit = Selling price − Cost price
= £98 000 − £67 000
= £31 000

$$\text{(b) \% profit} = \frac{\text{Profit (change in price)}}{\text{Cost price (original price)}} \times 100\%$$
$$= \frac{31\,000}{67\,000} \times 100\%$$
$$= 31\,000 \div 67\,000 \times 100\%$$
$$= 46.3\% \text{ (1 dp)}$$

7. Reversing an Increase or Decrease

Question *In a store sale all prices are discounted by 15%. If a customer pays £2300 for a three-piece suite, what would the price of the suite have been before the discount?*

NOTE A **discount** is a reduction in price.

Let 100% represent the original price of £2300, *before* the discount. Then the original price has been reduced by 15% to 85% of itself, as shown in the diagram

Original price	−	Discount	=	£2300
100%	−	15%	=	85%

We need to find the original price, i.e. 100%.

From the diagram, 85% of original price = £2300

So, 1% of original price = £2300 ÷ 85 = £27.06 (2 dp)

Then, original price = 100% = 100 × 1%
= 100 × £27.06
= £2705.88

6. Increasing or Decreasing a Quantity by a Given Percentage

Question *The shop price of a television set is £350. Value added tax (VAT) at $17\frac{1}{2}$% is added to this price. How much does the TV cost?*

VAT = $17\frac{1}{2}$% of £350

= 17.5% of £350 = $\frac{17.5}{100}$ × £350 = £61.25

So, cost of television set = Shop price + VAT
= £350 + £61.25
= £411.25

Question *£3000 is invested in a bank account and earns 7% interest per annum. The interest is added to the account at the end of each year. How much money will be in the account*
(a) at the end of the first year
(b) at the end of the second year.

(a) Interest earned at end of first year
= 7% of £3000
= $\frac{7}{100}$ × £3000 = 7 ÷ 100 × £3000 = £210
Account after 1 yr = £3000 + £210 = £3210

(b) Interest earned at end of second year
= 7% of £3210
= $\frac{7}{100}$ × £3210 = 7 ÷ 100 × £3210 = £224.70
Account after 2 yr = £3210 + £224.70 = £3434.70

The above calculations can be shortened using a **multiplier**.

(a) Let 100% represent the initial amount of £3000, i.e. 100% of £3000 is £3000.
This increases by 7% to 107% at the end of the year, as shown in the diagram:

£3000	+	Interest	=	Amount after 1 year
100%	+	7%	=	107%

So, increased amount
= 107% of £3000
= $\frac{107}{100}$ × £3000 = 1.07 × £3000 = £3210
The number 1.07 is called a **multiplier**. It increases whatever it multiplies by 7%.

(b) During the second year, £3210 increases by 7%.
So, increased amount = 1.07 × £3210 = £3434.70

Question *A pond contains 50 litres of water. Each day 3% of its water evaporates. What will its volume be in 3 days?*

Let 100% represent the original 50 litres.
In 1 day, 100% reduces by 3% to 97%.

50 ℓ	−	Evaporation	=	Volume after 1 day
100%	−	3%	=	97%

Reduced amount = 97% of 50 = 0.97 × 50 = 48.5 ℓ
Here the multiplier is 0.97.
So, after 2 days, volume = 0.97 × 48.5 = 47.045 ℓ
and after 3 days, volume = 0.97 × 47.045 ≈ 45.6 ℓ

An even quicker method is:

Volume after 3 days = $0.97^3 \times 50 \approx 45.6$ ℓ

0 . 9 7 x^y 3 × 5 0 = 45.63365

Ratio

1. Ratio

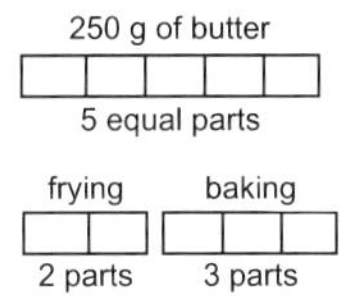

The wrapper of a 250 g pack of butter is marked into 5 equal parts. A cook uses 2 parts for frying and 3 parts for baking. We say that the butter has been divided in the **ratio** 2 parts to 3 parts, written 2 : 3.
We can also say that the ratio of frying butter to baking butter is 2 : 3 and write

Frying butter : baking butter = 2 : 3

We can find the amount of butter used for frying by first *finding* 1 *part*.

We know 5 parts = 250 g
and so 1 part = 250 g ÷ 5 = 50 g
So, frying butter = 2 parts = 2 × 50 g = 100 g
and baking butter = 3 parts = 3 × 50 g = 150 g

Question *Yogurt flip is made from a mixture of yogurt, orange juice and sparkling water in the ratio 7:3:2.*
(a) How much orange juice does a 30 cℓ glass contain?
(b) A jug of flip contains 20 cℓ of sparkling water. How much yogurt does it contain?

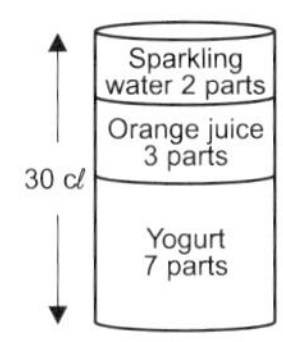

(a) 7:3:2 means 7 parts : 3 parts : 2 parts, giving a total of 7 + 3 + 2 = 12 parts (the whole glass).
So, 12 parts = 30 cℓ
Thus, 1 part = 30 cℓ ÷ 12 = 2.5 cℓ
Orange juice = 3 parts
= 3 × 2.5 cℓ = 7.5 cℓ

(b) Sparkling water = 2 parts = 20 cℓ
So, 1 part = 20 cℓ ÷ 2 = 10 cℓ
Yogurt = 7 parts = 7 × 10 cℓ = 70 cℓ

2. Equivalent Ratios

Multiplying or dividing the numbers of a ratio by the *same* number gives an **equivalent** ratio.

Examples

2 : 5 = 6 : 15 (×3) 12 : 20 = 6 : 10 = 3 : 5 (÷2, ÷2) $\frac{1}{2} : \frac{1}{4}$ = 2 : 1 (×4)

Question *A builder mixes together $1\frac{1}{4}$ buckets of sand, $\frac{1}{2}$ of a bucket of cement and $\frac{1}{4}$ of a bucket of water. Express the ratio of sand to cement to water, using whole numbers.*

Sand : Cement : Water = $1\frac{1}{4} : \frac{1}{2} : \frac{1}{4} = 2\frac{1}{2} : 1 : \frac{1}{2} = 5 : 2 : 1$ (×2, ×2)

Question *Lemonade is sold in small bottles costing 90p and large bottles costing £1.35. Express the ratio of these prices in its lowest terms.*

First, convert £1.35 to 135p.

When comparing quantities, make sure they have the *same* units of measurement.

Then, Small : Large = 90 : 135 = $\cancel{90}^{\cancel{18}\,2} : \cancel{135}^{\cancel{45}\,3}$ = 2 : 3

3. Increasing/Decreasing a Quantity in a Given Ratio

Question *Mandy was given this recipe for rock cakes by her grandmother. These quantities make 6 rock cakes.*

Rock Cakes
240 g flour
150 g sugar
150 g margarine
210 g dried fruit
2 large eggs

(a) How much sugar does Mandy need to make 8 rock cakes?
(b) Mandy has only 100 g of margarine but has plenty of all the other ingredients. How many rock cakes can she make?

(a) To increase the recipe from 6 to 8 cakes, we must increase each ingredient in the ratio 6 : 8, i.e. 3 : 4

Before increase	:	After increase
3 parts	:	4 parts
150 g sugar	:	Sugar needed?

3 parts = 150 g and so 1 part = 150 g ÷ 3 = 50 g
4 parts = 4 × 1 part = 4 × 50 g = 200 g
Mandy needs 200 g of sugar.

(b) Decreasing the margarine from 150 g to 100 g will decrease the number of rock cakes in the ratio 150 : 100, i.e. 3:2

Before decrease	:	After decrease
3 parts	:	2 parts
6 cakes	:	How many cakes?

3 parts = 6 cakes and so 1 part = 6 ÷ 3 = 2 cakes
2 parts = 2 × 1 part = 2 × 2 cakes = 4 cakes
Mandy can make 4 cakes.

4. Scale Factors

Question *A photograph measuring 18 cm by 12 cm was enlarged to 45 cm by 30 cm. Write down the ratio of the enlargement in the form 1:n.*

	Photo	:	Enlargement	
	18 cm	:	45 cm	
=	$\frac{18}{18}$	:	$\frac{\cancel{45}\,5}{\cancel{18}\,2}$	Divide both sides by 18 to make the left-hand side equal to 1
=	1	:	$\frac{5}{2}$	
=	1	:	$2\frac{1}{2}$ ←	This number is called the **scale factor** of the enlargement

The height and width of the enlargement are $2\frac{1}{2}$ times the height and width of the original photograph.

5. Ratios as Fractions

A ratio can be written as a fraction, e.g. 5 : 7 = $\frac{5}{7}$

Question *The ratio of the volume of a cup to a mug is 2 : 3. If the volume of a mug is 45 cℓ, calculate the volume of a cup.*

Cup : Mug = 2 : 3, which can be written as a fraction:

$\frac{\text{Cup}}{\text{Mug}} = \frac{2}{3}$ and so Cup = $\frac{2}{3}$ × Mug = $\frac{2}{3}$ × 45 = 30 cℓ

Units of Measurement

1. Metric Units

These are the basic units of measurement used in the UK today.

Length	Weight (Mass)	Capacity (Volume)	Money
kilometre (km)	kilogram (kg)	litre (ℓ)	pound (£)
metre (m)	gram (g)	centilitre (cℓ)	penny (p)
centimetre (cm)	milligram (mg)	millilitre (mℓ)	
millimetre (mm)			

2. Imperial Units

These are the most common units of measurement used before the introduction of the metric system.

Length	Weight (Mass)	Capacity	Money
inch (in *or* ")	ounce (oz)	fluid ounce (fl oz)	pound (£)
foot (ft *or* ')	pound (lb)	pint (pt)	shilling (s)
yard (yd)	stone (st)	quart (qt)	penny (d)
mile (m)	hundredweight (cwt)	gallon (gal)	
	ton (t)		

3. Prefixes

A **prefix** is a word or letter placed before a basic unit to make a smaller or larger new unit.

kilo (k) means **thousand (1000)**

So, kilometre (km) means thousand metres (1000 m)
kilogram (kg) means thousand grams (1000 g)
kilowatt (kW) means thousand watts (1000 W)

centi (c) means **hundredth** ($\frac{1}{100}$)

So, centimetre (cm) means hundredth of a metre ($\frac{1}{100}$ m)
centilitre (cℓ) means hundredth of a litre ($\frac{1}{100}$ ℓ)

milli (m) means **thousandth** ($\frac{1}{1000}$)

So, milliimetre (mm) means thousandth of a metre ($\frac{1}{1000}$ m)
millilitre (mℓ) means thousandth of a litre ($\frac{1}{1000}$ ℓ)
milligram (mg) means thousandth of a gram ($\frac{1}{1000}$ g)

4. Estimating

You may be asked to estimate the measurement of an everyday object. Here are a few examples:

Height of a door	2 m
Length of a car	3 m
Height of a work surface	1 m
Capacity of a wine glass	10 cℓ
Weight of an apple	150 g

Take a good look at a ruler to estimate the following

Width of this page	20 cm
Diameter of a 2p coin	2 cm
Thickness of a credit card	1 mm

5. Equivalent Metric Units

Length	Weight	Capacity
1 km = 1000 m	1 kg = 1000 g	1 ℓ = 100 cℓ
1 m = 100 cm	1 g = 1000 mg	1 ℓ = 1000 mℓ
1 m = 1000 mm	1 tonne (t) = 1000 kg	1 cℓ = 10 mℓ
1 cm = 10 mm		

6. Converting Metric Units

To convert a big unit to a small unit, *multiply*

km —×1000→ m —×100→ cm —×10→ mm
km ←÷1000— m ←÷100— cm ←÷10— mm

To convert a small unit to a big unit, *divide*

Question *Convert 0.13 kg to g.*

kg —×1000→ g   Big unit to small unit, *multiply*.

So, 0.13 kg = 0.13×1000 g = 130 g

Question *Convert 15 mm to cm.*

cm ←÷10— mm   Small unit to big unit, *divide*.

So, 15 mm = 15÷10 cm = 1.5 cm

Convert measurements to the *same* unit before starting a calculation. Look at the question to see which unit is required.

Question *Patrick made some ginger ale by mixing together 1.25 ℓ of lemonade, 30 cℓ of ginger syrup and 250 mℓ of rum. Calculate the volume of ginger ale he made, giving your answer in cℓ.*

Change 1.25 ℓ to cℓ: ℓ —×100→ cℓ

1.25 ℓ = 1.25×1 ℓ = 1.25×100 cℓ = 125 cℓ

Change 250 mℓ to cℓ: cℓ ←÷10— mℓ

250 mℓ = 250÷10 cℓ = 25 cℓ

So, total volume of ginger ale = 30+125+25 = 180 cℓ.

7. Converting Metric and Imperial Units

Length	Weight	Capacity
1 m ≈ 3 ft = 1 yd	1 kg ≈ 2 lb	1 ℓ ≈ $1\frac{3}{4}$ pt
1 in ≈ 25 mm	500 g ≈ 1 lb	1 gallon ≈ 4.5 ℓ
1 ft ≈ 30 cm	250 g ≈ $\frac{1}{2}$ lb	
5 miles ≈ 8 km		

Question *Suna is 5 ft 2 in tall. Given that 1 in ≈ 2.5 cm and 1 ft = 12 in, calculate Suna's height in metres.*

5 ft 2 in = 5 ft + 2 in = 5×12 in + 2 in = 62 in
62 in = 62×1 in ≈ 62×2.5 cm = 155 cm
155 cm = 155÷100 m = 1.55 m
Suna is approximately 1.55 m tall

Question *Jeremy weighs 134 lb. Estimate his weight in kilograms.*

Rough estimate	Better estimate
1 kg ≈ 2 lb	1 kg ≈ 2.2 lb
134 lb ≈ 134÷2 kg	134 lb ≈ 134÷2.2 kg
= 67 kg	≈ 61 kg

8. Currency Conversion

The exchange rates in this table are used in the questions that follow.

Currency Conversion Table
£1 = DM 2.68 (German marks)
£1 = FFr 9.03 (French francs)
£1 = L 2685 (Italian liras)
£1 = BFr 55.37 (Belgian francs)
£1 = \$ 1.59 (US dollars)

Question *Convert £25 to German marks (DM).*

The exchange rate for German marks is £1 = DM 2.68.

So, £25 = 25×£1 = 25×DM 2.68 = DM 67

Question *Convert FFr 165 to pounds sterling.*

The exchange rate for French francs is £1 = FFr 9.03.

So, FFr 165 = £165÷9.03 = £18.27

Question *Convert L 500 000 to US dollars (\$)*

First, convert liras to £ and then £ to \$.

The exchange rate for Italian liras is £1 = L 2685.

So, L 500 000 = £500 000÷2685 = £186.22

The exchange rate for US dollars is £1 = \$1.59.

So, £186.22 = 186.22×\$1.59 = \$296.09

Question *Mary converted £300 to German marks (DM) and £400 to Belgian francs (BFr). She spent DM 535 in Germany and BFr 13 400 in Belgium and converted the rest back to pounds sterling. How many pounds did she receive?*

Mary started with £300+£400 = £700

The quickest method is to convert the spent DM and BFr to £ and subtract both from £700

DM 535 = £535÷2.68 = £199.63
BFr 13 400 = £13 400÷55.37 = £242.01

Mary received £700−£199.63−£242.01 = £258.36

Question *Gunter changed £200 into DM 548. What was the exchange rate from £ to DM?*

£200 = DM 548 and so £1 = DM 548÷200 = DM 2.74
The exchange rate is £1 = DM 2.74

9. Compound Units

Examples of compound units are:

Speed 5 km/h means 5 km travelled in 1 hour

Cost 85p/kg means 1 kg costs 85p

Density 5 g/cm^3 means 1 cm^3 weighs 5 g

Fuel efficiency 15 km/ℓ means 15 km travelled on 1 ℓ of fuel

10. Best Buy

Question *Baked beans are sold in small and large cans, as shown in the diagram. Which can is better value for money?*

Small can 15p buys 125 g
So, 1p buys 125 g÷15 = 8.33g (2 dp)
1p buys 8.33 g

Large can 38p buys 275 g
So, 1p buys 275 g÷38 = 7.24g (2 dp)
1p buys 7.24 g

The small can is better value since you get more weight per penny.

Question *Suggest a size and corresponding price for a medium box of washing powder. You must give reasons for your answers and show your working.*

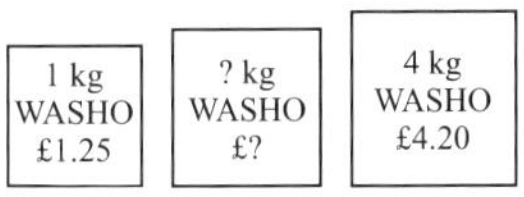

The weight of a medium box could be halfway between 1 kg and 4 kg, i.e. $\frac{1+4}{2} = \frac{5}{2} = 2.5$ kg

To find a reasonable price of a 2.5 kg pack, we must first find a reasonable cost per kg.

Small size 1 kg costs £1.25
So, cost per kg is £1.25/kg

Large size 4 kg costs £4.20
So, cost of 1 kg = £4.20÷4 = £1.05
So, cost per kg is £1.05/kg

A reasonable price per kg would be halfway between £1.25p/kg and £1.05p/kg, i.e. $\frac{1.25+1.05}{2} = \frac{2.30}{2} =$ £1.15p/kg

Price of medium box = 2.5×£1.15 = £2.875 ≈ £2.88

Powers, Roots, Reciprocals and Surds

1. Calculating Powers

To calculate 1.7^2, press:

1 · 7 x^2 → 2.89

To calculate 5^7, press:

5 x^y 7 = → 78125

x^y Power key

2. Calculating Roots

To calculate $\sqrt{2.9}$, press:

2 · 9 $\sqrt{\ }$ → 1.7029386

To calculate $\sqrt[5]{32}$, press:

3 2 $x^{1/y}$ 5 = → 2.

$x^{1/y}$ Root key $\sqrt[y]{x}$

3. Two Special Powers

5^0 means 1

5^1 means 5

4. Negative Indices

5^{-1} means $\frac{1}{5}$ 5^{-2} means $\frac{1}{5^2}$ 5^{-3} means $\frac{1}{5^3}$

Question *Write 6^{-2} as a fraction.*

$6^{-2} = \frac{1}{6^2} = \frac{1}{36}$

Question *Write 10^{-3} as a decimal.*

$10^{-3} = \frac{1}{10^3} = \frac{1}{1000} = 0.001$

Question *Calculate the value of 1.5^{-3} correct to 2 dp.*

1 · 5 x^y 3 +/− = → 0.2962963

1.5 Power −3

So, $1.5^{-3} = 0.30$ correct to 2 dp

See also Box 5

5. Fractional Indices

H

$9^{\frac{1}{2}}$ means $\sqrt{9}$ $8^{\frac{1}{3}}$ means $\sqrt[3]{8}$ $16^{\frac{1}{4}}$ means $\sqrt[4]{16}$

Question *Find the value of $8^{\frac{1}{3}}$ without using a calculator.*

$8^{\frac{1}{3}} = \sqrt[3]{8} = 2$ because $2\times2\times2 = 8$ (see Box 9, p12)

Question *Write $81^{-\frac{1}{2}}$ as a fraction.*

$81^{-\frac{1}{2}} = \frac{1}{81^{\frac{1}{2}}}$ (see Box 4)

$= \frac{1}{\sqrt{81}} = \frac{1}{9}$

$8^{\frac{2}{3}}$ means $(\sqrt[3]{8})^2$ or equivalently $\sqrt[3]{8^2}$ $8^{\frac{2}{3}}$ ← 2 power, ← 3 root

Question *Find the value of $8^{\frac{2}{3}}$ without using a calculator.*

$8^{\frac{2}{3}} = (\sqrt[3]{8})^2$ Finding the root first is easier

$= (2)^2 = 2^2 = 4$

6. More Difficult Calculations

Question *Calculate the value of $\sqrt{1-0.8^2}$.*

$\sqrt{1-0.8^2}$ means $\sqrt{(1-0.8^2)}$. So, calculate the bracket, then square root the answer.

1 − 0 · 8 x^2 = $\sqrt{\ }$ → 0.6

So, $\sqrt{1-0.8^2} = 0.6$

Question *Calculate the value of $(2\pi)^{\frac{3}{4}}$, correct to 3 decimal places.*

2 × π = x^y 3 = $x^{1/y}$ 4 = → 3.9685778

3rd power 4th root

So, $(2\pi)^{\frac{3}{4}} = 3.969$ correct to 3 dp.

7. Reciprocals

The **reciprocal** of a number is $\frac{1}{\text{number}}$.

Examples

The reciprocal of 2 is $\frac{1}{2}$.

The reciprocal of 0.4 is $\frac{1}{0.4}$.

Question *Calculate the value of $\frac{1}{0.4} + \frac{1}{0.5}$.*

0 · 4 $\frac{1}{x}$ + 0 · 5 $\frac{1}{x}$ = → 4.5

Reciprocal key

So, $\frac{1}{0.4} + \frac{1}{0.5} = 4.5$

8. Summary

$8^0 = 1$

$8^1 = 8$

$8^3 = 8\times8\times8$

$8^{-1} = \frac{1}{8}$

$8^{-3} = \frac{1}{8^3}$

$8^{\frac{1}{2}} = \sqrt[2]{8} = \sqrt{8}$

$8^{\frac{1}{3}} = \sqrt[3]{8}$

$8^{\frac{2}{3}} = (\sqrt[3]{8})^2 = \sqrt[3]{8^2}$

$8^{-\frac{1}{3}} = \frac{1}{8^{\frac{1}{3}}} = \frac{1}{\sqrt[3]{8}}$

$8^{-\frac{2}{3}} = \frac{1}{8^{\frac{2}{3}}} = \frac{1}{(\sqrt[3]{8})^2}$

9. Surds

H

A **surd** is the root of a whole number which produces an irrational number (it also refers to an expression that involves one). Here are a few examples:

$\sqrt{2}$ $\quad 3+\sqrt{2}$ $\quad 4\sqrt{2}$ $\quad 5-3\sqrt{2}$ $\quad \frac{4}{\sqrt{2}}$ $\quad \frac{1}{4+3\sqrt{2}}$

NOTE Surds are *irrational* numbers (see Box 13, p20).
So, $\sqrt{9}$ is ***not*** a surd, because $\sqrt{9} = 3$

Surds involving the *same* root can be simplified by adding, subtracting, multiplying and dividing.

Adding/Subtracting Surds

$3\sqrt{5}$ means $3\times\sqrt{5}$

$3\sqrt{5}+4\sqrt{5} = 7\sqrt{5}$ This is like $3a+4a = 7a$

$4+5+6\sqrt{7} = 9+6\sqrt{7}$ This cannot be further simplified

$(3+6\sqrt{3})+(5-2\sqrt{3}) = 3+5+6\sqrt{3}-2\sqrt{3}$ Rearrange
$= 8+4\sqrt{3}$ Combine

Multiplying Surds

$\sqrt{3}\sqrt{3} = 3$ or equivalently $(\sqrt{3})^2 = 3$ (see Box 10, p12)

$5(2-4\sqrt{7}) = 5\times2-5\times4\sqrt{7} = 10-20\sqrt{7}$

$(5-2\sqrt{3})(4-\sqrt{3}) = 20-5\sqrt{3}-8\sqrt{3}+2\sqrt{3}\sqrt{3}$
$= 20-13\sqrt{3}+2\times3$ Since $\sqrt{3}\sqrt{3} = 3$
$= 26-13\sqrt{3}$

Dividing Surds

Only simple examples will be set at this level

$\frac{6}{\sqrt{3}} = \frac{6\times\sqrt{3}}{\sqrt{3}\times\sqrt{3}}$ Multiply top and bottom by $\sqrt{3}$ to eliminate the root on the bottom

$= \frac{\cancel{6}^{2}\times\sqrt{3}}{\cancel{3}_{1}}$

$= 2\sqrt{3}$

Question *Simplify* $\sqrt{48}+2\sqrt{3}$.

$\sqrt{48} = \sqrt{16\times3} = \sqrt{16}\times\sqrt{3} = 4\sqrt{3}$ (see Box 11, p34)

So, $\sqrt{48}+2\sqrt{3} = 4\sqrt{3}+2\sqrt{3} = 6\sqrt{3}$

Standard Form

1. What is Standard Form Used For?

Standard form is a convenient way of writing *very small* and *very large* numbers. It is sometimes called **standard index form** because it is based on powers of 10.

2. Multiplying Decimals by Powers of 10

Multiplying a number by a *positive* power of 10 gives a *bigger* number.

2.8×10^5

$= 280\,000$ Move the decimal point 5 places right

Multiplying a number by a *negative* power of 10 gives a *smaller* number.

2.8×10^{-5}

$= 0.000\,028$ Move the decimal point 5 places left

Question *The mass of a meteorite is* 3.61×10^7 *kg. Write this as an ordinary number.*

3.61×10^7

$= 36\,100\,000$ kg

3. Standard Form

These numbers are written in **standard form**.

2.8×10^5
1.76×10^{-5}

(Number between 1 and 10; Power of 10)

The power of 10 is called the **order of magnitude** and indicates the overall size of the number.

The number between 1 and 10 gives more detail about the number.

NOTE The number 29.6×10^3 is ***not*** in standard form because 29.6 does not lie between 1 and 10.

4. Writing Numbers in Standard Form

Question *Write the number 150 000 in standard form*

150 000. Put in the decimal point if it is missing

1.50 000. Move the decimal point 5 places to make a number between 1 and 10

So, $150\,000 = 1.5\times10^5$ ← Number of decimal places moved
(Big number; Positive power)

Question *Write the number 0.000 004 in standard form.*

0.000 004. Move the decimal point 6 places to make a number between 1 and 10

So, $0.000\,004 = 4\times10^{-6}$ ← Number of decimal places moved
(Small number; Negative power)

Standard Form (Contd)

5. Comparing Numbers in Standard Form

The powers of 10 below are arranged in increasing order of size.

10^{-3}	10^{-2}	10^{-1}	10^{0}	10^{1}	10^{2}	10^{3}
$\frac{1}{1000}$	$\frac{1}{100}$	$\frac{1}{10}$	1	10	100	1000

To compare numbers in standard form, compare their powers of 10.
For example, 5.8×10^{3} is bigger than 7.2×10^{2} because 10^{3} is bigger than 10^{2}.
Similarly, 2.84×10^{-1} is bigger than 9.75×10^{-3} because 10^{-1} is bigger than 10^{-3}.

Question *Place the following numbers in increasing order of magnitude (size):*

5.4×10^{6} $\quad 9.4\times10^{4}$ $\quad 3.6\times10^{6}$ $\quad 2.4\times10^{-9}$

The correct order is shown below

2.4×10^{-9} $\quad 9.4\times10^{4}$ $\quad 3.6\times10^{6}$ $\quad 5.4\times10^{6}$

Although 5.4×10^{6} and 3.6×10^{6} have the same power of 10, 5.4×10^{6} is bigger than 3.6×10^{6} because 5.4 is bigger than 3.6.

6. Standard Form and Calculators

To enter the number 5.8×10^{5} into the calculator:

[5] [·] [8] [EXP] [5] — display: 5.8 05

Standard form key — Power of 10

NOTE The [EXP] key is pressed *instead of* the [×] [1] [0] keys.

To enter the number 3.4×10^{-12} into the calculator:

[3] [·] [4] [EXP] [1] [2] [+/-] — display: 3.4 -12

NOTE The display looks like the power 3.4^{-12} but actually means 3.4×10^{-12}.

7. Standard Form and Significant Figures

Question *The mass of the Moon is 7.38×10^{19} tonnes. The mass of the Earth is 81 times the mass of the Moon. Calculate the mass of the Earth, expressing your answer in standard form correct to 2 significant figures.*

Mass of Earth = $81\times7.38\times10^{19}$

[8] [1] [×] [7] [·] [3] [8] [EXP] [1] [9] [=] — display: 5.978 21

So, mass of Earth = 6.0×10^{21} tonnes, correct to 2 significant figures.

8. Questions Involving Standard Form

Question *The Sun is approximately 149.6 million km from the Earth.*
(a) Write this number in standard form.
(b) If light travels 1.80×10^{7} km in 1 minute, how many minutes does light from the Sun take to reach the Earth? Give your answer to the nearest minute.

(a) 149.6 million = $149.6\times1\ 000\ 000$

[1] [4] [9] [·] [6] [×] [1] [0] [0] [0] [0] [0] [0] [=] — display: 149600000

So, 149.6 million km = 149 600 000 km
= 1.496×10^{8} km

(b) Number of minutes = $1.496\times10^{8}\div1.80\times10^{7}$

[1] [·] [4] [9] [6] [EXP] [8] [÷] [1] [·] [8] [0] [EXP] [7] [=] — display: 8.3111111

So, it takes approximately 8 minutes.

Question *The number 10^{100} is called a googol. Write 50 googols in standard form.*

50 googols = 50×10^{100}

Now, this number is too big for your calculator. Try pressing [5] [0] [EXP] [1] [0] [0] [=]. A basic calculator will not have space for three-digit indices like 100. So we must do without a calculator:

$50\times10^{100} = 5\times10\times10^{100} = 5\times10^{1}\times10^{100}$
$= 5\times10^{1+100} = 5\times10^{101}$

Basic Algebra

1. Algebra and Arithmetic

Algebra uses the same rules as arithmetic.

Always remember that letters represent numbers.

An **algebraic expression** is any allowable combination of letters (*a, b, c, x, y, C, N*), numbers (1, 2, $\frac{1}{2}$, 5.7) and arithmetical symbols (+, ×, −, ÷, (,), √). Here are a few simple examples:

$a+b$	Sum of the numbers *a* and *b*
$N-2$	Number 2 subtracted from the number *N*
$5y$	Product of 5 and *y* NOTE 5*y* is an abbreviation of $5\times y$.
ab	Product of the numbers *a* and *b* (*ab* is short for $a\times b$ and is the same as *ba*)
$\frac{m}{n}$	Number *m* divided by the number *n* NOTE write $\frac{m}{n}$ rather than $m \div n$.
$\sqrt{n}$	Square root of the number *n*

1*x* is usually written simply as *x*.

2. Making Algebraic Expressions

Question *Apples cost 15p each. Write down an expression for the cost of n apples.*

First, write down the expression using words.

Cost = Number of apples × Price of an apple

= $n\times 15$ pence — Replace words with symbols

= $n15$ pence — Omit the × sign

= $15n$ pence — Write numbers before letters

Question *Mary starts with a number, x, doubles it, adds 6 and divides the result by 2. Write down an expression for the number she ends up with.*

Mary starts with *x*.

She doubles *x* to give 2*x*.

Then she adds 6, giving $2x+6$ or $6+2x$

She divides the result by 2, giving $\frac{2x+6}{2}$

NOTE If you really must use the ÷ sign, this answer could also be written $(2x+6)\div 2$. But ***not*** $2x+6\div 2$, where only the 6 is being divided by 2

3. Powers

Powers of letters have the same meaning as powers of numbers (see Box 7, p.11 and Boxes 3, 4, 5, p.30).

Since $2^3 = 2\times 2\times 2$ then $a^3 = a\times a\times a$

Similarly $2^1 = 2$ and so $a^1 = a$

$2^0 = 1$ and so $a^0 = 1$

$2^{-3} = \frac{1}{2^3}$ and so $a^{-3} = \frac{1}{a^3}$

See also Box 9, p34.

4. Substituting Numbers for Letters

Replacing letters by numbers enables us to calculate the **value** of an algebraic expression.

Question *Find the value of the expression 2(b+w) given that b = 4 and w = 6.*

Replacing *b* by 4 and *w* by 6 gives

$2(b+w) = 2\times(b+w) = 2\times(4+6) = 2\times(10) = 2\times 10 = 20$

Question *Find the value of u+at given that u = 4, a = −3 and t = 2.5.*

Put negative numbers in brackets to avoid mistakes.

$u+at = u+a\times t = 4+(-3)\times 2.5 = 4+-7.5 = 4-7.5 = -3.5$

Multiply first

Question *Find the value of* $2\pi\sqrt{\frac{L}{g}}$ *when L = 15, g = 9.81.*

$2\pi\sqrt{\frac{L}{g}} = 2\times\pi\times\sqrt{\frac{15}{9.81}} = 7.77$ (correct to 2 dp)

1 5 ÷ 9 . 8 1 = √ × 2 × π = 7.769462685

Start with the most difficult part of the calculation

5. Multiplying Numbers and Letters

Simplify by multiplying numbers together.

Examples

$2\times 5a = 10a$

$-3(2xy) = -3\times 2xy = -6xy$

Group together numbers before multiplying.

Examples

$2a3b = 2\times a\times 3\times b = 2\times 3\times a\times b = 6\times a\times b = 6ab$

$(4pq)(-3t) = 4\times p\times q\times -3\times t = 4\times -3\times p\times q\times t = -12pqt$

Multiplying letters and numbers by 0 gives 0.

Examples

$0\times 3\times 2 = 0$ $\quad 0ab = 0\times a\times b = 0$ $\quad 0(x-y) = 0$

6. Multiplying Powers

$a^2a^3 = a^2\times a^3 = a\times a \times a\times a\times a = a^5$

So, $a^2a^3 = a^{2+3}$

RULE When **multiplying** powers of the *same* letter or number, **add** their indices.

Examples

$m^2m^3 = m^2\times m^3 = m^{2+3} = m^5$

$xx^2x^3 = x^1x^2x^3 = x^{1+2+3} = x^6$ — *x* can be written x^1

$v^5v^{-2} = v^{5+-2} = v^{5-2} = v^3$ — Remember: +− gives −

$p^{12}p^n = p^{12+n}$

$a^2b^5a^3b^{-4} = a^2a^3b^5b^{-4}$ — Group the same letters

$= a^{2+3}\times b^{5+-4}$

$= a^5b^{5-4} = a^5b^1 = a^5b$

$10^3\times 10^5 = 10^{3+5} = 10^8$ $\quad 10^3\times 10^{-5} = 10^{3+-5} = 10^{-2}$

Question *Without using a calculator, find the value of* $(3\times 10^7)\times(2\times 10^{-4})$ *in standard form.*

$(3\times 10^7)\times(2\times 10^{-4}) = 3\times 2\times 10^7\times 10^{-4} = 6\times 10^{7+-4} = 6\times 10^3$

Basic Algebra (Contd)

7. Dividing Powers

Just as $\frac{\not{5}^1}{\not{5}_1} = \frac{1}{1} = 1$ so $\frac{\not{a}^1}{\not{a}_1} = \frac{1}{1} = 1$

That is, letters can also be cancelled.

So, $\frac{a^5}{a^3} = \frac{\not{a}\times\not{a}\times\not{a}\times a\times a}{\not{a}\times\not{a}\times\not{a}} = \frac{1\times1\times1\times a\times a}{1\times1\times1} = \frac{a^2}{1} = a^2$

So, $\frac{a^5}{a^3} = a^{5-3} = a^2$ or $a^5 \div a^3 = a^{5-3} = a^2$

A common way of writing cancelling is: $\frac{a^{\not{5}\,2}}{\not{a^3}\,1} = \frac{a^2}{1} = a^2$

RULE When **dividing** powers of the *same* letter or number, **subtract** their indices.

Examples

$\frac{y^5}{y^7} = y^5 \div y^7 = y^{5-7} = y^{-2}$ Which can be written $\frac{1}{y^2}$

Alternatively:

$\frac{\not{y^5}\,1}{y^{\not{7}\,2}} = \frac{1}{y^2}$ Which can be written y^{-2}

$10^{15} \div 10^{11} = 10^{15-11} = 10^4$

$a^6 \div a^{-2} = a^{6--2} = a^{6+2} = a^8$

Special case:

$\frac{1}{x^{-3}} = \frac{x^0}{x^{-3}} = x^{0--3} = x^{0+3} = x^3$ So, $\frac{1}{x^{-3}} = x^3$

Question *Evaluate* $\frac{1}{5^{-2}}$ *without using a calculator.*

$\frac{1}{5^{-2}} = 5^2 = 25$

8. Power of a Power

$(\text{Number})^3 = (\text{Number})\times(\text{Number})\times(\text{Number})$

So, $(a^2)^3 = (a^2)\times(a^2)\times(a^2)$
$= (a\times a)\times(a\times a)\times(a\times a)$
$= a\times a \times a\times a \times a\times a$
$= a^6$

So, $(a^2)^3 = a^{2\times3} = a^6$

RULE When a power is **raised** to a power, **multiply** the indices.

Examples

$(p^4)^{-3} = p^{4\times(-3)} = p^{-12}$

$(10^2)^3 = 10^{2\times3} = 10^6$

$(m^{12})^{\frac{3}{4}} = m^{12\times\frac{3}{4}} = m^{\frac{12}{1}\times\frac{3}{4}} = m^{\frac{36}{4}} = m^9$

9. Powers and Roots

H

Since $9^{\frac{1}{2}} = \sqrt{9}$ then $a^{\frac{1}{2}} = \sqrt{a}$ (see Box 5, p30)

Similarly, $8^{\frac{1}{3}} = \sqrt[3]{8}$ and so $a^{\frac{1}{3}} = \sqrt[3]{a}$

$8^{-\frac{1}{3}} = \frac{1}{8^{\frac{1}{3}}} = \frac{1}{\sqrt[3]{8}}$ and so $a^{-\frac{1}{3}} = \frac{1}{a^{\frac{1}{3}}} = \frac{1}{\sqrt[3]{a}}$

$8^{\frac{2}{3}} = \sqrt[3]{8^2} = (\sqrt[3]{8})^2$ and so $a^{\frac{2}{3}} = \sqrt[3]{a^2} = (\sqrt[3]{a})^2$

Replace roots by fractional powers.

Examples

$\sqrt[5]{n^2} = n^{\frac{2}{5}}$ $\sqrt{n^4} = n^{\frac{4}{2}} = n^2$ $\sqrt[4]{x^2} = x^{\frac{2}{4}} = x^{\frac{1}{2}}$

$r\sqrt{r} = r^1 r^{\frac{1}{2}} = r^{1+\frac{1}{2}} = r^{1\frac{1}{2}} = r^{\frac{3}{2}}$

$\frac{x^2}{\sqrt[3]{x}} = \frac{x^2}{x^{\frac{1}{3}}} = x^2 \div x^{\frac{1}{3}} = x^{2-\frac{1}{3}} = x^{\frac{5}{3}}$

$x^{\frac{1}{2}}x^{\frac{1}{3}} = x^{\frac{1}{2}+\frac{1}{3}} = x^{\frac{3}{6}+\frac{2}{6}} = x^{\frac{5}{6}}$

Special cases

$\sqrt{n^2} = n$ and $(\sqrt{n})^2 = n$
$\sqrt[3]{n^3} = n$ and $(\sqrt[3]{n})^3 = n$

Powers and roots cancel each other

10. Powers and Brackets

$(ab)^3 = (ab)\times(ab)\times(ab) = a\times b\times a\times b\times a\times b$
$= a\times a\times a\times b\times b\times b$
$= a^3b^3$

NOTE ab^3 is ***not*** $(ab)^3$ because $ab^3 = a\times b\times b\times b$.

Similarly, $\left(\frac{a}{b}\right)^3 = \frac{a^3}{b^3}$

The power of a product (or quotient) is the product (or quotient) of powers.

$(2xy)^3 = 2^3x^3y^3 = 8x^3y^3$

$(p^2q)^4 = (p^2)^4q^4 = p^{2\times4}q^4 = p^8q^4$

$(8x^6)^{\frac{1}{3}} = 8^{\frac{1}{3}}(x^6)^{\frac{1}{3}} = \sqrt[3]{8}\, x^{6\times\frac{1}{3}} = 2x^2$

11. Roots of Products and Quotients

$\sqrt{9\times16} = \sqrt{9}\times\sqrt{16} = 3\times4 = 12$

Similarly, $\sqrt{ab} = \sqrt{a}\sqrt{b}$

The root of a product (or quotient) is the product (or quotient) of roots.

Examples

$\sqrt[3]{8p^6} = \sqrt[3]{8}\,\sqrt[3]{p^6} = 2(p^6)^{\frac{1}{3}} = 2p^{6\times\frac{1}{3}} = 2p^2$

$\sqrt{9x^2y^4} = \sqrt{9}\sqrt{x^2}\sqrt{y^4} = 3(x^2)^{\frac{1}{2}}(y^4)^{\frac{1}{2}} = 3x^1y^2 = 3xy^2$

$\sqrt[4]{\frac{m^8}{16}} = \frac{\sqrt[4]{m^8}}{\sqrt[4]{16}} = \frac{(m^8)^{\frac{1}{4}}}{2} = \frac{m^{8\times\frac{1}{4}}}{2} = \frac{m^2}{2}$

12. Index Equations

If two powers are equal, their indices are also equal.

Question *Find n when* $(x^n)^2 = \frac{1}{x^4}$.

$(x^n)^2 = x^{2n}$ and $\frac{1}{x^4} = x^{-4}$ and so $x^{2n} = x^{-4}$

For the powers x^{2n} and x^{-4} to be equal, their indices $2n$ and -4 must also be equal, i.e. $2n = -4$, so $n = -2$.

13. Terms of an Algebraic Expression

Terms are algebraic quantities that are separated from each other by + or − signs.

$2x^2-3x+4$ — Terms

$\frac{2a}{3}+a-4b$ — Terms

Like terms are multiples of the same algebraic quantity.

$a \quad 2a \quad 3a \quad \tfrac{1}{2}a$ — Like terms

$2xy \quad 3xy \quad \frac{3xy}{2}$ — Like terms

$3p \quad 2q \quad 5x \quad 7$ — Unlike terms

$1 \quad x \quad x^2 \quad x^3$ — Unlike terms

Like terms can be combined into a single term.

Examples

$2a+3a = 5a$

$4p-7p = -3p$ Because $4-7 = -3$

$5x+2+3x+4 = 5x+3x+2+4$ Group like terms

$= 8x+6$ Combine like terms

NOTE $8x+6$ cannot be simplified any further.

$2p-3q-7p+q = 2p-7p-3q+1q = -5p-2q$

$5x^2+7x-2x^2+4 = 5x^2-2x^2+7x+4 = 3x^2+7x+4$

14. Simplifying Algebraic Expressions

To **simplify** an algebraic expression means to combine terms and powers where possible. This usually results in a shorter expression.

Question *Simplify* $(3x)^2+4x-2x^2+x$ *as far as possible.*

$(3x)^2+4x-2x^2+x$

$= 3^2x^2+4x-2x^2+x$ Power of a bracket

$= 9x^2-2x^2+4x+1x$ Group like terms

$= 7x^2+5x$ Combine like terms

Question *Simplify* $(ab)^2+a^2b+4a^2b^2-3a(ba)+ab^2$

$(ab)^2+a^2b+4a^2b^2-3a(ba)+ab^2$

$= a^2b^2+a^2b+4a^2b^2-3aba+ab^2$ See Box 10, p34

$= 1a^2b^2+4a^2b^2-3aab+a^2b+ab^2$ Group like terms

$= 5a^2b^2-3a^2b+1a^2b+ab^2$ Combine terms

$= 5a^2b^2-2a^2b+ab^2$ Combine terms

Question *Simplify* $\frac{3}{6a}$.

$\frac{3}{6a} = \frac{\cancel{3}^1}{{}_2\cancel{6}a} = \frac{1}{2a}$ Cancel numbers

Question *Simplify* $\frac{4a^3b^2}{6ab^4}$.

$\frac{4a^3b^2}{6ab^4} = \frac{{}^2\cancel{4}a^3b^2}{{}_3\cancel{6}a^1b^4} = \frac{2a^{\cancel{3}2}\cancel{b^2}}{3\cancel{a}b^{\cancel{4}2}} = \frac{2a^2}{3b^2}$ Cancel numbers and powers

Question *Simplify* $(6p^2q)(2pq)^2$.

$(6p^2q)(2pq)^2$

$= 6p^2q^1\times2^2p^2q^2$ Power of a bracket

$= 6\times2^2\times p^2\times p^2\times q^1\times q^2$ Rearrange

$= 6\times4\times p^{2+2}q^{1+2}$ Combine powers

$= 24p^4q^3$

15. More Difficult Questions

H

Question *The algebraic expression* $2\pi\sqrt{\frac{L}{g}}$ *can be written in the form* kL^x. *Find the values of k and x given that* $g = 9.81$.

$2\pi\sqrt{\frac{L}{g}} = 2\times\pi\times\sqrt{\frac{L}{9.81}}$ Replace g by 9.81

$= 2\times\pi\times\frac{\sqrt{L}}{\sqrt{9.81}}$ Root of a quotient (see Box 11, p34)

$= 2\times\pi\times\frac{L^{\frac{1}{2}}}{3.13}$ Change root to a power

$= 2\times\pi\div3.13\,L^{\frac{1}{2}}$ Combine numbers

$= 2.01\,L^{0.5}$

So, $k = 2.01$ and $x = 0.5$.

Question *Given* $v = 3.2\times10^{-1}$ *calculate the value of the algebraic expression* $\frac{2\pi}{\sqrt{1-v^2}}$ *correct to 2dp.*

$\frac{2\pi}{\sqrt{1-v^2}} = \frac{2\times\pi}{\sqrt{1-(3.2\times10^{-1})^2}} = 6.63$ (6.6319073)

Calculator: 1 − 3 . 2 EXP 1 +/− x^2 = √ Min 2 × π = ÷ MR =

Calculate the denominator and store it in the memory — Calculate the numerator and divide by the recalled denominator

Question *The formula for the population, P, of mice with a growth rate of r% per year, in time t months, is given by the formula*

$$P = P_0\left(1+\frac{r}{100}\right)^t$$

where P_0 *is the population this month. Calculate the population of mice in 6 months given that the population this month is 24 and the growth rate is 12% per month.*

The question tells us that $P_0 = 24$, $t = 6$ and $r = 12$.

Substitute these values into the formula:

$P = P_0\left(1+\frac{r}{100}\right)^t = 24\left(1+\frac{12}{100}\right)^6 = 47.37 \approx 47$

Calculator: 2 4 × (1 + 1 2 ÷ 1 0 0) x^y 6 = (47.371745)

The population will have 47 mice in 6 months time.

Question *Find the value of y for which* $2\times4^y = 64$.

Notice that the numbers 2, 4 and 64 are all powers of 2, i.e. $2 = 2^1$, $4 = 2^2$ and $64 = 2^6$. So we can rewrite the equation using powers of 2:

$2\times4^y = 64$

$\therefore 2^1\times(2^2)^y = 2^6$ Rewrite using powers of 2

$\therefore 2^1\times2^{2y} = 2^6$ Power of a power: multiply indices

$\therefore 2^{1+2y} = 2^6$ Multiplying powers: add indices

$\therefore 1+2y = 6$ Indices are equal

$\therefore 2y = 5$

$\therefore y = \frac{5}{2} = 2.5$

Equations

1. Equations

$$1+2+3 = 6$$

This is an **equation** and tells us that the sum 1+2+3 is equal to the number 6.

An equation can be thought of as a balanced pair of weighing scales, with numbers as weights.

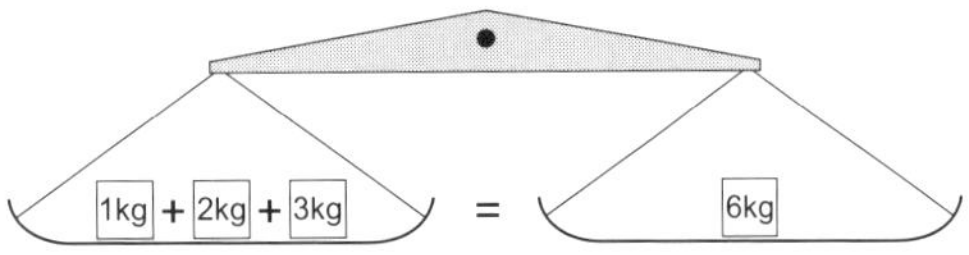

If we change one side of an equation, we must change the other side in exactly the same way in order for the equation to remain true.

For example, if we add 4 to the left-hand side (LHS) we must also add 4 to the right-hand side (RHS):

$$1+2+3+4 = 6+4$$

The scales are still balanced (the equation is still true).

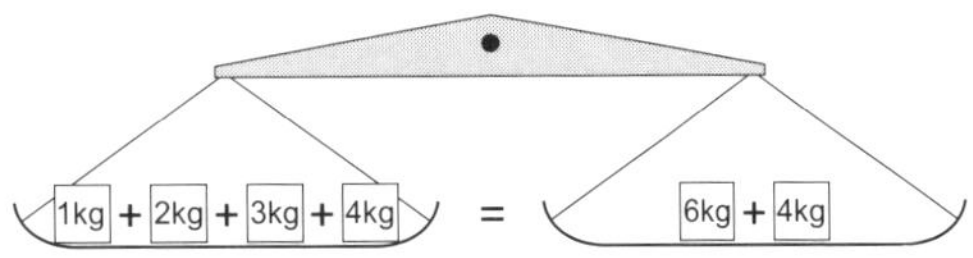

2. Solving Equations

$$x+2 = 5$$

This equation involves an **unknown** number x. **Solving** the equation means finding the number x which makes it true. The scales below show that x must be 3. So we can write $x = 3$.

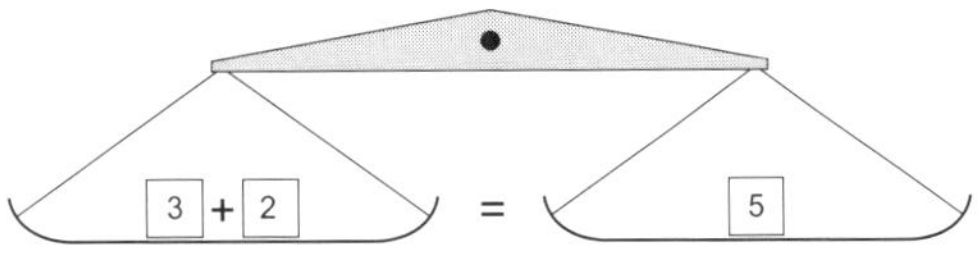

We say that $x = 3$ is the **solution** of the equation. We also say that $x = 3$ **satisfies** the equation.

3. Equations Requiring Addition or Subtraction

Question *Solve the equation $x+2 = 5$.*

The aim is to find the number x which makes the equation true. We do this by changing both sides of the equation until we get x = *number*.

$x+2 = 5$	Write down the equation
$\therefore x+2-2 = 5-2$	Subtract 2 from *both* sides,
$\therefore x+\cancel{2}-\cancel{2} = 5-2$	to leave x on its own
$\therefore x = 5-2$	Simplify RHS, to give the solution
$\therefore x = 3$	

The sign $\therefore$ means **therefore** and shows that the current line follows from the previous line

more

3. Equations Requiring Addition or Subtraction (Contd)

In practice, we simply move the +2 to the RHS and change it to −2

$x+2 = 5$	Rewrite the equation each time
$\therefore x = 5-2$	you change it. Write each equation
$\therefore x = 3$	on a new line if there is space

Question *Solve the equation $5 = y-2$.*

$5 = y-2$	Write down the equation
$\therefore 5+2 = y$	Move −2 to LHS and change it
$\therefore 7 = y$	to +2. Simplify LHS
$\therefore y = 7$	Turn the answer around. It does not matter which way around you write an equation

Question *Solve the equation $2+T = 7$.*

$2+T = 7$	2 is another way of writing +2 and
$\therefore T = 7-2$	so it changes to −2 when it moves
$\therefore T = 5$	to RHS

4. Equations Requiring Division

Question *Solve the equation $5x = 35$.*

The 5 multiplies the number x. So we must divide both sides by 5 to cancel it out, leaving x on its own.

$5x = 35$	Write down the equation
$\therefore \frac{5x}{5} = \frac{35}{5}$	Divide both sides by 5
$\therefore \frac{\cancel{5}x}{\cancel{5}} = \frac{35}{5}$	Cancel
$\therefore x = \frac{35}{5}$	Simplify RHS
$\therefore x = 7$	

In practice, we simply move the 5× to the RHS and change it to ÷5.

$5x = 35$	Move the 5× to RHS ....
$\therefore x = \frac{35}{5}$	.... and change it to ÷5
$\therefore x = 7$	

Question *Solve the equation $-4p = 7$.*

The negative number −4 is *multiplying* the number p (there is no subtraction on the LHS).

$-4p = 7$	Move the −4× to RHS ....
$\therefore p = \frac{7}{-4}$	.... and change it to ÷(−4)
$\therefore p = -1.75$	

Question *Solve the equation $\frac{1}{4}m = 2$.*

$\frac{1}{4}m = 2$	Move the $\frac{1}{4}$× to RHS ....
$\therefore m = 2\div\frac{1}{4}$	.... and change it to $\div\frac{1}{4}$
$\therefore m = 2\times4$	
$\therefore m = 8$	

5. Equations Requiring Multiplication

Question *Solve the equation* $\frac{x}{2} = 7$

The 2 divides the number x. So we must multiply both sides by 2 to cancel it out, leaving x on its own.

$\frac{x}{2} = 7$	Write down the equation
$\therefore 2\times\frac{x}{2} = 2\times7$	Multiply both sides by 2
$\therefore \cancel{2}\times\frac{x}{\cancel{2}} = 2\times7$	Cancel
$\therefore x = 2\times7$	Simplify
$\therefore x = 14$	

In practice, we simply move the $\div2$ to the RHS and change it to $\times2$.

$\frac{x}{2} = 7$	Move $\div2$ to the other side ....
$\therefore x = 7\times2$	.... changing to $\times2$
$\therefore x = 14$	

6. Equations Requiring Multiplication and Division

Question *Solve the equation* $\frac{3t}{4} = 2$.

The 3 multiplies the number t and the 4 divides it. Moving the $3\times$ and the $\div4$ to the RHS changes them to $\div3$ and $\times4$, leaving t on its own.

$\frac{3t}{4} = 2$	Move $3\times$ and $\div4$ to RHS ....
$\therefore t = \frac{2\times4}{3}$	.... changing to $\div3$ and $\times4$
$\therefore t = \frac{8}{3}$	Simplify
$\therefore t = 2\frac{2}{3}$	

Cross Multiplying

Cross multiplying is a quick way of solving equations where one fraction equals another

Question *Solve the equation* $\frac{2p}{3} = \frac{5}{6}$.

$\frac{2p}{3} \times \frac{5}{6}$	Cross multiply
$\therefore 2p\times6 = 3\times5$	
$\therefore 12p = 15$	Simplify. Move $12\times$ to RHS ....
$\therefore p = \frac{\cancel{15}^{5}}{\cancel{12}_{4}}$	.... changing to $\div12$. Cancel
$\therefore p = \frac{5}{4}$	Simplify
$\therefore p = 1\frac{1}{4}$	

7. Equations Requiring Addition, Subtraction, Multiplication, Division

RULE Add and subtract numbers from both sides before multiplying and dividing.

Question *Solve the equation* $3y+2 = 11$.

$3y+2 = 11$	Move $+2$ to RHS and
$\therefore 3y = 11-2$	change it to -2.
$\therefore 3y = 9$	Simplify. Move $3\times$ to RHS ...
$\therefore y = \frac{9}{3}$	... and change it to $\div3$
$\therefore y = 3$	Simplify

Question *Solve the equation* $\frac{3x}{2} - 5 = 1$.

$\frac{3x}{2} - 5 = 1$	Move -5 to RHS ....
$\therefore \frac{3x}{2} = 1+5$	.... and change it to $+5$. Simplify
$\therefore \frac{3x}{2} = 6$	Move $3\times$ and $\div2$ to RHS ....
$\therefore x = \frac{2\times6}{3}$	.... and change to $\div3$ and $\times2$
$\therefore x = \frac{12}{3}$	Simplify
$\therefore x = 4$	

8. Equations with the Unknown in the Denominator

Question *Solve the equation* $\frac{3}{t} = 8$.

The 3 is divided by t. So we must move the $\div t$ to the RHS and change it to $\times t$.

$\frac{3}{t} = 8$	Move $\div t$ to RHS ....
$\therefore 3 = 8\times t$	.... and change it to $\times t$
$\therefore \frac{3}{8} = t$	Now move $8\times$ to LHS and change it to $\div8$
$\therefore t = \frac{3}{8}$	Turn the equation around

9. Equations Involving $-x$

Question *Solve the equation* $-x = 5$.

Here we need to eliminate the minus sign. The trick is to multiply both sides by -1.

$-x = 5$	
$\therefore -1\times-x = -1\times5$	Multiply both sides by -1
$\therefore +x = -5$	Remember: $-\times- = +$
$\therefore x = -5$	$+x$ can be written as x

A quicker method is to put a minus sign in front of the LHS and RHS.

$-x = 5$	
$\therefore --x = -5$	Put a minus sign on both sides
$\therefore +x = -5$	Remember: $-- = +$
$\therefore x = -5$	$+x$ can be written as x

Equations (Contd)

10. Terms of an Equation

The **terms** of an equation are separated from each other by +, − and = signs.

$3x+2 = 5$ (Terms: $3x$, 2, 5)

$\frac{5t}{6} - 4 = \frac{3}{5}$ (Terms: $\frac{5t}{6}$, 4, $\frac{3}{5}$)

11. Rearranging Terms of an Equation

Question *Solve the equation $3x+2x = 10$.*

$3x+2x = 10$ — Combine the x-terms

$\therefore 5x = 10$ — Move the $5\times$ to RHS ....

$\therefore x = \frac{10}{5}$ — .... changing it to $\div 5$

$\therefore x = 2$

Question *Solve the equation $9x+6 = 30-3x$.*

Move all the x-terms to one side and all the number terms to the other side.

$9x+6 = 30-3x$ — Move $-3x$ to LHS and ....

$\therefore 9x+6+3x = 30$ — .... change it to $+3x$. Move $+6$ to RHS and change it to -6

$\therefore 9x+3x = 30-6$

$\therefore 12x = 24$ — Combine terms

$\therefore x = \frac{24}{12}$

$\therefore x = 2$

12. Using Equations to Solve Problems

Question *Pencils cost 15 pence each and erasers cost 20 pence each. Paul buys n pencils and $2n$ erasers, costing £2.20 altogether. Write down an equation in n and solve it to find the number of pencils and erasers Paul bought.*

Write down the equation using words first. Then replace the words by symbols.

Cost of pencils + Cost of erasers = £2.20

Now, cost of pencils = Number × Price
= $n\times15$ pence = $15n$ pence

and cost of erasers = Number × Price
= $2n\times20$ pence = $40n$ pence

The equation can now be written using algebra:

$15n$ pence $+40n$ pence = £2.20

Convert £ to pence, to make all the units the same:

$15n$ pence + $40n$ pence = 220 pence

Now we can leave out the units (pence):

$15n+40n = 220$

$\therefore 55n = 220$

$\therefore n = \frac{220}{55}$

$\therefore n = 4$

So, Paul bought 4 pencils and $2\times n = 2\times4 = 8$ erasers

Formulas

1. Formulas

A **formula** is an equation involving several letters, each letter representing a different quantity. For example, the formula connecting speed (s), distance (d) and time (t) is:

$$s = \frac{d}{t}$$

The letter, s, on the LHS of a formula is called the **subject** of the formula. We say that s is **expressed in terms of *d* and *t***.

2. Making Formulas

You may be asked to find the formula connecting the quantities involved in a real situation (see also Box 12 above).

Question *Snak Bars cost 26 pence each. Write down the cost, C pence, of n Snak Bars.*

Write down the formula in words first:

Cost = Number of bars × Price of a Snak Bar

Now replace the words by letters:

C pence = $n \times 26$ pence

$\therefore C = n\times26$ — Drop the units, pence

$\therefore C = 26n$

3. Replacing Letters by Numbers (Substitution)

Question *The formula below can be used to convert temperatures measured in degrees Celsius (°C) into degrees Fahrenheit (°F)*

$$F = \frac{9}{5}C + 32$$

(a) Convert 25 °C to Fahrenheit.
(b) Convert −4 °F to Celsius.

(a) Replace C by 25 in the formula (this is called **substitution**).

$F = \frac{9}{5}C+32$ — Replace C by 25

$F = \frac{9}{5}\times25+32$

$= 45+32$

$= 77$

So, 25 °C is equivalent to 77 °F.

(b) Replace F by −4 in the formula:

$F = \frac{9}{5}C+32$

$-4 = \frac{9}{5}C+32$ — Solve equation to find C

$\therefore -4-32 = \frac{9}{5}C$

more

3. Replacing Letters by Numbers (Contd)

$\therefore -36 = \frac{9}{5}C$

$\therefore \frac{-\overset{4}{\cancel{36}}\times 5}{\cancel{9}_1} = C$

$\therefore -20 = C$

$\therefore C = -20$

So, -4 °F is equivalent to -20 °C.

4. Rearranging Formulas

Formulas can be rearranged (**transposed**) using the same methods as for equations.

Question *Rearrange the formula $F = 25G+250$ to make G the subject.*

We must rearrange the formula to get G on its own on the LHS:

$F = 25G+250$	Turn the formula around
$\therefore 25G+250 = F$	Move 250 to RHS
$\therefore 25G = F-250$	Move 25× to RHS ....
$\therefore G = \frac{F-250}{25}$	.... changing to ÷25

G is now the subject, on its own on the LHS.

NOTE Don't use the ÷ sign when rearranging equations and formulas. Use a dividing line, as in the above example.

Question *Transpose the formula $v = u+at$ to make t the subject.*

$v = u+at$	
$\therefore u+at = v$	Turn the formula around
$\therefore at = v-u$	Move u to RHS, changing to $-u$
$\therefore t = \frac{v-u}{a}$	Move a× to RHS, changing to $\div a$

Question *Given that $T = \frac{\lambda x}{L}$, express x in terms of λ, T and L.*

$T = \frac{\lambda x}{L}$	
$\therefore \frac{\lambda x}{L} = T$	Turn formula around to get x on LHS
$\therefore \lambda x = TL$	Move $\div L$ to RHS, changing to $\times L$
$\therefore x = \frac{TL}{\lambda}$	Move $\lambda\times$ to RHS, changing to $\div\lambda$

Question *Make A the subject of the formula $\frac{1}{A} + B = C$.*

$\frac{1}{A} + B = C$	
$\therefore \frac{1}{A} = C-B$	Move $+B$ to RHS, changing to $-B$
$\therefore \frac{1}{A} = \frac{C-B}{1}$	Make RHS a fraction
$\therefore \frac{A}{1} = \frac{1}{C-B}$	Turn both sides upside down
$\therefore A = \frac{1}{C-B}$	

5. Equations and Formulas Involving Powers and Roots

When rearranging formulas that involve powers or roots, remember that powers and roots cancel each other out, e.g. squaring cancels out a square root.

Question *Find the value of n if $\sqrt{4n} = 5$.*

$\sqrt{4n} = 5$	
$\therefore (\sqrt{4n})^2 = 5^2$	Square both sides to eliminate
$\therefore 4n = 25$	square root
$\therefore n = \frac{25}{4}$	
$\therefore n = 6.25$	

Question *Make M the subject of the formula $F = \sqrt{3M}$.*

$F = \sqrt{3M}$	
$\therefore F^2 = (\sqrt{3M})^2$	Square both sides to cancel
$\therefore F^2 = 3M$	out the square root
$\therefore 3M = F^2$	Turn the equation around
$\therefore M = \frac{F^2}{3}$	

Question *The time, T, for a pendulum of length, L, to swing back and forth once is given by the formula $T = 2\pi\sqrt{\frac{L}{g}}$. Express L in terms of T and g.*

$T = 2\pi\sqrt{\frac{L}{g}}$	
$\therefore T^2 = 2^2\pi^2\left(\sqrt{\frac{L}{g}}\right)^2$	Square *all* of RHS
$\therefore T^2 = 4\pi^2\frac{L}{g}$	Move g and $4\pi^2$ to LHS
$\therefore \frac{T^2 g}{4\pi^2} = L$	
$\therefore L = \frac{T^2 g}{4\pi^2}$	Turn the equation around

Question *The volume, V, of a sphere is given by the formula $V = \frac{4}{3}\pi r^3$.*

(a) Rearrange the formula to express r in terms of V.

(b) Find the radius of a sphere whose volume is 54 cm³, correct to 2 decimal places.

(a)

$V = \frac{4}{3}\pi r^3$	
$\therefore \frac{4}{3}\pi r^3 = V$	Turn the equation around
$\therefore r^3 = \frac{3V}{4\pi}$	Move 4π and 3 to RHS
$\therefore \sqrt[3]{r^3} = \sqrt[3]{\frac{3V}{4\pi}}$	Take 3rd root of both sides
$\therefore r = \sqrt[3]{\frac{3V}{4\pi}}$	

(b) Put $V = 54$ in the formula:

$\therefore r = \sqrt[3]{\frac{3\times 54}{4\pi}} = 2.34$ cm

(3 × 5 4) ÷ (4 × π) = $x^{\frac{1}{y}}$ 3 =

2.3447779

Expanding Brackets and Factorisation

1. Expanding Brackets

To **expand** brackets means to remove them.

We have already done this in arithmetic using the BoDMAS rule

$5(3+4) = 5\times(3+4) = 5\times(7) = 5\times7 = 35$

Calculate bracket first — Drop bracket

You can get the same answer by multiplying each number in the bracket by 5.

$5(3+4) = 5\times(3+4) = 5\times3 + 5\times4 = 15+20 = 35$

Similarly

$6(5-3) = 6\times(5-3) = 6\times5 - 6\times3 = 30-18 = 12$

This method can also be used to expand brackets in algebra.

$5(a+3) = 5\times(a+3) = 5\times a + 5\times3 = 5a+15$

Simplify the result after expanding.

Examples

$y(y-4) = y\times y - y\times4 = y^2-4y$

$2m(3m+4) = 2m\times3m + 2m\times4 = 6m^2+8m$

$\frac{1}{2}(4p+6) = \frac{1}{2}\times4p+\frac{1}{2}\times6 = 2p+3$

Question *Expand and simplify* $2(x-3)+4(2x+1)$.

$2(x-3)+4(2x+1)$	
$= 2x-6+8x+4$	Expand both brackets
$= 2x+8x+4-6$	Group like terms
$= 10x-2$	Combine terms

When multiplying a bracket by a negative number, use these rules

$- - = +$ $\quad - \times - = +$

$- + = -$ $\quad - \times + = -$

Examples

$-2(a+3) = -2\times a + -2\times3$	Expand bracket
$= -2a+-6$	Use the rule $-\times+ = -$
$= -2a-6$	Use the rule $+- = -$
$-y(1-y) = -y\times1 - -y\times y$	Expand bracket
$= -1y--y^2$	Use the rule $-\times+ = -$
$= -y+y^2$	Use the rule $-- = +$
$-(a-b) = -a--b$	Expand bracket
$= -a+b$	

NOTE The above example shows that a minus sign before a bracket changes the signs of the terms in the bracket, e.g. $-(-2-x) = +2+x$

Question *Expand and simplify* $8y-2(1-3y)$

$8y-2(1-3y)$	
$= 8y-2\times1--2\times3y$	Expand bracket
$= 8y-2+6y$	
$= 14y-2$	

Question *Expand and simplify* $4t(t+1)-t(4t-1)$.

$4t(t+1)-t(4t-1)$	
$= 4tt+4t1-t4t--t1$	Expand both brackets
$= 4t^2+4t-4tt+t$	Use the rule $-- = +$
$= \cancel{4t^2}+4t-\cancel{4t^2}+t$	Cancel
$= 4t+t$	
$= 5t$	

2. Equations and Formulas Involving Brackets

Question *Solve the equation* $3(2x-1) = 2(x-4)$.

$3(2x-1) = 2(x-4)$	
$\therefore 6x-3 = 2x-8$	Expand brackets
$\therefore 6x-2x = -8+3$	Rearrange terms
$\therefore 4x = -5$	Simplify
$\therefore x = \frac{-5}{4}$	Remember that $- \div + = -$
$\therefore x = -\frac{5}{4}$ (or -1.25)	

Question *Paper cups cost n pence each. Paper plates cost 2 pence more.*

(a) Write down an algebraic expression for the price of a plate in terms of n.

(b) George bought 4 cups and 2 plates for 34 pence. Write down an equation in n.

(c) Solve your equation to find the value of n.

(d) State the price of a cup and the price of a plate.

(a) Cup price = n pence

Plate price = Cup price + 2 pence
= n pence + 2 pence
= $(n+2)$ pence

Use brackets to contain the number of pence

(b) 4 cups plus 2 plates cost 34 pence

$\therefore 4\times n$ pence + $2\times(n+2)$ pence = 34 pence	
$\therefore 4\times n + 2\times(n+2) = 34$	Leave out the units
$\therefore 4n + 2(n+2) = 34$	

(c)

$\therefore 4n+2n+4 = 34$	Expand brackets
$\therefore 6n+4 = 34$	Simplify
$\therefore 6n = 34-4$	Rearrange terms
$\therefore 6n = 30$	Simplify
$\therefore n = 5$	

(d) Price of cup = n pence = 5 pence

Price of plate = $(n+2)$ pence = $(5+2)$ pence = 7 pence

Question *Rearrange the formula* $P = 2(L+B)$ *to make L the subject.*

Method 1

$P = 2(L+B)$	
$\therefore 2(L+B) = P$	Turn the formula around
$\therefore 2L+2B = P$	Expand bracket
$\therefore 2L = P-2B$	Rearrange terms
$\therefore L = \frac{P-2B}{2}$	Move $2\times$ to RHS, changing to $\div2$

Method 2

$P = 2(L+B)$	
$\therefore 2(L+B) = P$	Turn formula around
$\therefore 2\times(L+B) = P$	Move $2\times$ to RHS ....
$\therefore (L+B) = \frac{P}{2}$	.... changing to $\div2$
$\therefore L+B = \frac{P}{2}$	Drop the bracket
$\therefore L = \frac{P}{2} - B$	

NOTE These two methods give the same formula, even though they look different; one is a rearrangement of the other.

3. Multiplying Two Brackets

We do this in arithmetic using the BoDMAS rule:

$(5+2)(3+4) = (5+2)\times(3+4) = (7)\times(7) = 7\times7 = 49$

Calculate brackets first — Drop brackets

We can get the same answer by using each number in the first bracket to multiply the second bracket:

$$\begin{aligned}(5+2)(3+4) &= 5\times(3+4) + 2\times(3+4)\\ &= 5\times3+5\times4 + 2\times3+2\times4\\ &= 15+20+6+8\\ &= 49\end{aligned}$$

Another way is to multiply each number in the first bracket by each number in the second bracket:

$(5+2)(3+4) = 5\times3+5\times4+2\times3+2\times4 = 15+20+6+8 = 49$

These methods can also be used to expand brackets in algebra.

Examples

$(a+2)(b+c)$	a and 2 multiply bracket $(b+c)$
$= a(b+c)+2(b+c)$	
$= ab+ac+2b+2c$	Expand brackets
$(x+3)(x+4)$	x and 3 multiply bracket $(x+4)$
$= x(x+4)+3(x+4)$	
$= xx+x4+3x+3\times4$	Expand brackets
$= x^2+4x+3x+12$	Simplify
$= x^2+7x+12$	Simplify
$(y-3)(y-2)$	y and -3 multiply $(y-2)$
$= y(y-2)-3(y-2)$	
$= yy-y2-3y--3\times2$	Expand brackets
$= y^2-2y-3y+6$	Simplify
$= y^2-5y+6$	Simplify

4. Squaring Brackets

$(x+3)^2$ means $(x+3)(x+3)$ and so

$$\begin{aligned}(x+3)^2 &= (x+3)(x+3)\\ &= x(x+3)+3(x+3)\\ &= xx+x3+3x+3\times3\\ &= x^2+3x+3x+9\\ &= x^2+6x+9\end{aligned}$$

Examples

$$\begin{aligned}(y-4)^2 &= (y-4)(y-4)\\ &= y(y-4)-4(y-4)\\ &= yy-y4-4y--4\times4\\ &= y^2-4y-4y+16\\ &= y^2-8y+16\end{aligned}$$

$$\begin{aligned}(3a+2)^2 &= (3a+2)(3a+2)\\ &= 3a(3a+2)+2(3a+2)\\ &= 3a3a+3a2+2\times3a+2\times2\\ &= 3\times3aa+3\times2a+6a+4\\ &= 9a^2+6a+6a+4\\ &= 9a^2+12a+4\end{aligned}$$

Question *Simplify* $(x+3)^2-x^2$.

$$\begin{aligned}(x+3)^2-x^2 &= (x+3)(x+3)-x^2\\ &= x(x+3)+3(x+3)-x^2\\ &= x^2+3x+3x+9-x^2\\ &= x^2-x^2+3x+3x+9\\ &= 3x+3x+9\\ &= 6x+9\end{aligned}$$

5. Factors of an Algebraic Expression

Every number can be expressed as a product of its prime factors (see also Box 15, p12).
For example: $12 = 2\times2\times3$ (↑ ↑ ↑ Prime factors)

We can find the highest common factor of two numbers using this method. For example, to find the highest common factor of 12 and 18:

$12 = 2\times2\times3$
$18 = 2\times3\times3$

Common prime factors are 2 and 3.
So, HCF = $2\times3 = 6$

Some algebraic expressions can also be **broken down** into their constituent parts.
For example: $12a = 2\times2\times3\times a$
$9tw = 3\times3\times t\times w$
$10p^2q = 2\times5\times p\times p\times q$

This helps us to find the highest common factor of two algebraic expressions, such as $4ab$ and $6ac$.

$4ab = 2\times2\times a\times b$
$6ac = 2\times3\times a\times c$

The common factors are 2 and a.
So, the HCF = $2\times a = 2a$

Question *Find the highest common factor of* $6y^2$ *and* $8y$.

$6y^2 = 2\times3\times y\times y$
$8y = 2\times2\times2\times y$

The common factors are 2 and y.
So, the HCF = $2\times y = 2y$

Question *Find the highest common factor of 12 and* $4x$.

$12 = 2\times2\times3$
$4x = 2\times2\times x$

The common factors are 2 and 2.
So, the HCF = $2\times2 = 4$

6. Factorisation

Expanding removes brackets. **Factorising** does the opposite by *inserting* brackets. For example:

expanding $3(a+b)$ gives $3a+3b$, whereas factorising $3a+3b$ gives $3(a+b)$

Question *Factorise* $4a+6$.

① Find the HCF of the terms $4a$ and 6:

$4a = 2\times2\times a$
$6 = 2\times3$ HCF = 2

② Write down the HCF multiplying an empty bracket:

$4a+6 = 2(\ \ +\ \ )$

③ Remove the HCF from the original terms and put what is left in the bracket:

$4a = 2\times2\times a$ leaves $2a$
$6 = 2\times3$ leaves 3

So, $4a+6 = 2(2a+3)$

NOTE $2(2a+3)$ is the answer to the question.

④ Check your answer by expanding the bracket. You should get back to where you started from:

$2(2a+3) = 2\times2a+2\times3 = 4a+6$ ✓

more

Expanding Brackets and Factorisation (Contd)

6. Factorisation (Contd)

Question *Factorise* $2a-6a^2$.

$2a = 2 \times a$
$-6a^2 = -2 \times 3 \times a \times a$ HCF $= 2 \times a = 2a$

Notice that the HCF is the same as one of the terms.

So, $2a-6a^2 = 2a\,(1-3a)$

↑ HCF ↑ Must be 1 here so that $2a \times 1 = 2a$

Check: $2a(1-3a) = 2a \times 1 - 2a \times 3a = 2a-6a^2$ ✓

Question *Factorise* $-12-8b$.

$-12 = -2 \times 2 \times 3$
$-8b = -2 \times 2 \times 2 \times b$ HCF $= -2 \times 2 = -4$

So, $-12-8b = -4(3+2b)$

Check: $-4(3+2b) = -4 \times 3 + -4 \times 2b = -12-8b$ ✓

7. Factorising Quadratic Expressions

A **quadratic** expression contains x^2 as its highest power. Here are the different types:

$5x^2$	Contains just an x^2-term
$10x^2-15x$	Contains an x^2-term and an x-term
$4x^2-9$	Contains an x^2-term and a number term
x^2+5x+4	Contains an x^2-term, x-term and number term

The last three types can sometimes be factorised.

Expressions with an x^2-term and x-term

To factorise $10x^2-15x$, use the same method as shown in Box 6.

$10x^2 = 2 \times 5 \times x \times x$
$-15x = -3 \times 5 \times x$ HCF $= 5 \times x = 5x$

So, $10x^2-15x = 5x(2x-3)$

Check: $5x(2x-3) = 5x \times 2x - 5x \times 3 = 10x^2-15x$ ✓

Expressions with an x^2-term and a number term

To factorise $4x^2-9$:

Notice that 4, x^2 and 9 are squares, i.e. $4 = 2^2$, x^2 and then $9 = 3^2$.

① Write each term as the square of a quantity:
$4x^2 = 2x \times 2x = (2x)^2$
$9 = 3^2$

② Write down the product of two brackets that contain the sum and difference of these quantities:
$4x^2-9 = (2x)^2-3^2 = (2x+3)(2x-3)$
(Sum) (Difference)

③ Check by expanding the brackets:
$(2x+3)(2x-3) = 2x(2x-3)+3(2x-3)$
$= 2x \times 2x - 2x \times 3 + 3 \times 2x - 3 \times 3$
$= 4x^2-6x+6x-9$
$= 4x^2-9$ ✓

NOTE $(2x)^2-3^2$ is called the **difference of two squares**.
For any two squares a^2 and b^2, we have the general result:

$$a^2-b^2 = (a+b)(a-b)$$

more

7. Factorising Quadratic Expressions (Contd)

Question *Factorise* $9x^2-16y^2$.

$9x^2-16y^2 = 3^2x^2-4^2y^2 = (3x)^2-(4y)^2 = (3x+4y)(3x-4y)$

Compare with result: $a^2 - b^2 = (a+b)(a-b)$

Question *Factorise* a^4-25.

$a^4-25 = (a^2)^2-5^2 = (a^2+5)(a^2-5)$

Question *Factorise* $18-2t^2$.

Notice that 18 and $-2t^2$ have a common factor, 2.
Take out this common factor first:

$18-2t^2 = 2(9-t^2) = 2(3^2-t^2) = 2(3-t)(3+t)$

Expressions with an x^2-term, x-term and a number term

To factorise x^2+5x+4.

① Write down two empty brackets:
( )( )

② Fill in the first term of each bracket so that their product is the x-term (there may be several alternatives):
$(x\quad)(x\quad)$ Because $x \times x = x^2$

③ Fill in the last term of each bracket so that their product is the number term (there may be several alternatives):
$(x+2)(x+2)$ Because $2 \times 2 = 4$
or $(x+4)(x+1)$ Because $4 \times 1 = 4$

④ Expand the brackets to find which gives the original quadratic expression:
$(x+2)(x+2) = x(x+2)+2(x+2)$
$= xx+x2+2x+2 \times 2$
$= x^2+2x+2x+4$
$= x^2+4x+4$ ✗

$(x+4)(x+1) = x(x+1)+4(x+1)$
$= xx+x1+4x+4 \times 1$
$= x^2+1x+4x+4$
$= x^2+5x+4$ ✓

⑤ Write down your answer:
$x^2+5x+4 = (x+4)(x+1)$

Question *Factorise* $2x^2+5x-3$.

② $(2x\quad)(x\quad)$ Because $2x \times x = 2x^2$

③ $(2x+3)(x-1)$ Because $+3 \times -1 = -3$
$(2x-3)(x+1)$ Because $-3 \times +1 = -3$
$(2x+1)(x-3)$ Because $+1 \times -3 = -3$
$(2x-1)(x+3)$ Because $-1 \times +3 = -3$

④ $(2x+3)(x-1) = 2x(x-1)+3(x-1)$
$= 2x^2-2x+3x-3$
$= 2x^2+x-3$ ✗

$(2x-3)(x+1) = 2x(x+1)-3(x+1)$
$= 2x^2+2x-3x-3$
$= 2x^2-x-3$ ✗

$(2x+1)(x-3) = 2x(x-3)+1(x-3)$
$= 2x^2-6x+1x-3$
$= 2x^2-5x-3$ ✗

$(2x-1)(x+3) = 2x(x+3)-1(x+3)$
$= 2x^2+6x-1x-3$
$= 2x^2+5x-3$ ✓

⑤ So, $2x^2+5x-3 = (2x-1)(x+3)$

Quadratic Equations

1. Quadratic Equations

A **quadratic** equation has x^2 as its highest power. There are three different types:

$x^2 = 9$	Contains an x^2-term and a number term
$10x^2-15x = 0$	Contains an x^2-term and a x-term
$x^2+5x+4 = 0$	Contains an x^2-term, x-term and a number term

Quadratic equations may have no solution, one solution or two solutions.

There are five methods of solving quadratic equations that you need to know: direct, factorisation, formula, graphical, iteration (the last two methods are dealt with on p45 and p63).

Sometimes you will be told which to use. Otherwise you have to choose the most appropriate method.

2. Direct Method

This method can be used if the quadratic equation can be rearranged to look like: x^2 = *number*.

Question *Solve the equation* $x^2 = 9$.

$x^2 = 9$

$\therefore\ \sqrt{x^2} = \sqrt{9}$ — Square root both sides to change x^2 into x

$\therefore\ x = 3$

This method gives you one solution, *but there is another.*

3 is a solution because $3^2 = 9$

-3 is also a solution because $(-3)^2 = (-3)\times(-3) = 9$

So, the two solutions are: $x = 3$ and $x = -3$.

These two solutions are written briefly as: $x = \pm 3$. — $\pm$ means $+$ *or* $-$

The correct way of writing the solution is:

$x^2 = 9$

$\therefore\ \sqrt{x^2} = \pm\sqrt{9}$ — Write the $\pm$ sign when taking the square root of both sides.

$\therefore\ x = \pm 3$

3. Factorisation Method

If you have difficulty with this method, use the formula method described in Box 4.

Question *Solve the equation* $10x^2 = 15x$.

① Rearrange the equation, if necessary, to get 0 on the RHS:

$10x^2 = 15x$

$\therefore\ 10x^2-15x = 0$ — Move $15x$ to LHS

② Factorise the LHS:

$5x(2x-3) = 0$ — See Box 7, p42

③ When the product of several numbers is 0, one of them must be 0:

$5 \times x \times (2x-3) = 0$

Number × Number × Number

We know that 5 is not 0. So, either x is 0 or the number in the bracket $(2x-3)$ is 0.

$\therefore\ x = 0$ or $2x-3 = 0$

$\therefore\ x = 0$ or $2x = 0+3$

more

3. Factorisation Method (Contd)

$\therefore\ x = 0$ or $2x = 3$

$\therefore\ x = 0$ or $x = \frac{3}{2}$

Question *Solve the equation* $x^2 = 4x-3$.

① $x^2 = 4x-3$

$\therefore\ x^2-4x+3 = 0$ — Rearrange to get 0 on RHS

② $\therefore\ (x-1)(x-3) = 0$ — Factorise LHS (see Box 7, p42)

Number × Number

③ $\therefore\ x-1 = 0$ or $x-3 = 0$

$\therefore\ x = 0+1$ or $x = 0+3$

$\therefore\ x = 1$ or $x = 3$

Question *Solve the equation* $x^2-9 = 0$.

$x^2-9 = 0$

$\therefore\ x^2-3^2 = 0$ — Notice the LHS is the difference of two squares

$\therefore\ (x+3)(x-3) = 0$ — Factorise the LHS (see Box 7, p42)

$\therefore\ x+3 = 0$ or $x-3 = 0$

$\therefore\ x = 0-3$ or $x = 0+3$

$\therefore\ x = -3$ or $x = 3$

$\therefore\ x = \pm 3$

4. The Formula Method

H

This method works for *all* quadratic equations, provided they are arranged in the following form:

$$ax^2 + bx + c = 0$$

(ax^2: x^2-term; bx: x-term; c: Number term)

where a, b and c are numbers. For example, in the equation $3x^2+2x-6 = 0$, a is 3, b is 2 and c is -6.

Always use this method when the question asks for an approximate answer, e.g. correct to 2 dp.

Question *Solve the equation* $2x^2-6x = 1$ *correct to 2 decimal places.*

① Rearrange the equation to get 0 on the RHS:

$2x^2-6x = 1$

$\therefore\ 2x^2-6x-1 = 0$

② Write down the values of a, b and c:

$2x^2-6x-1 = 0$:

$a = 2$, $b = -6$ and $c = -1$ — Don't forget the $-$ signs

③ Substitute the values of a, b and c into the formula:

$$x = \frac{-b\pm\sqrt{b^2-4ac}}{2a}$$

This formula will be given on the question paper

Use brackets to avoid errors with minus signs:

$$x = \frac{-(-6)\pm\sqrt{(-6)^2-4(2)(-1)}}{2(2)}$$

Dividing line must go right across the fraction

④ Calculate the two values of x:

$x = \frac{6\pm\sqrt{36--8}}{4}$ — Because $4(2)(-1) = 8(-1) = -8$

$= \frac{6\pm\sqrt{44}}{4}$ — Remember: $-- = +$

$= \frac{6\pm 6.633}{4}$ — Store the answer to $\sqrt{44}$ in your calculator

$= \frac{6+6.633}{4}$ or $\frac{6-6.633}{4}$

$= 3.16$ or -0.16 — Correct to 2 dp

Quadratic Equations (Contd)

5. Solving Problems Using Quadratic Equations

H

Question *A golfer hits a golf ball a horizontal distance d metres. The ball rises to a height of h metres given by the formula* $h = d - 0.004d^2$.

(a) How high will the ball have risen when it has travelled a distance of 90 metres?

(b) How far will the ball have travelled when it first reaches a height of 50 metres?

(a) Put $d = 90$ into the formula to find the height, h.

$h = 90 - 0.004 \times 90^2 = 57.6$ metres

(b) Put $h = 50$ into the formula to find the distance, d.

$50 = d - 0.004d^2$

Now solve this quadratic equation to find d:

$0.004d^2 - d + 50 = 0$ Rearrange into the form

$\therefore\ 0.004d^2 - 1d + 50 = 0$ $ax^2 + bx + c = 0$

Here, $a = 0.004$, $b = -1$ and $c = 50$.

$$\therefore\ d = \frac{-b \pm \sqrt{b^2 - 4ac}}{2a}$$

$$= \frac{-(-1) \pm \sqrt{(-1)^2 - 4(0.004)(50)}}{2(0.004)}$$

$$= \frac{1 \pm \sqrt{1 - 0.8}}{0.008}$$

$$= \frac{1 \pm \sqrt{0.2}}{0.008}$$

$$= \frac{1 + 0.447}{0.008} \text{ or } \frac{1 - 0.447}{0.008}$$

$$\approx 181 \text{ m or } 69 \text{ m}$$

The ball reaches a height of 50 m when it has travelled 69 m and 181 m. So it first reaches this height when it has travelled 69 m.

Convergence and Iteration

1. Convergence of a Sequence

The terms of the sequence 1.9, 1.99, 1.999, 1.9999, ... are getting closer and closer to the number 2. We say that the sequence **converges** to the **limit** 2.

If a sequence is convergent, the differences between successive terms gets smaller. For example:

1.9 1.99 1.999 1.9999

0.09 0.009 0.0009

Differences getting smaller

A sequence that is not convergent is said to be **divergent**. Here are two examples:

2 4 6 8 10 ...

2 2 2 2

Differences are *not* getting smaller

1 −1 1 −1 1 ...

Numbers *oscillate* between −1 and 1

Question *Is the sequence below convergent? If so, what is its limit?*

$\frac{1}{2}$ $\frac{3}{4}$ $\frac{5}{6}$ $\frac{7}{8}$ $\frac{9}{10}$ $\frac{11}{12}$ ....

Convert the terms to decimals and calculate the differences between successive terms:

$\frac{1}{2}$ $\frac{3}{4}$ $\frac{5}{6}$ $\frac{7}{8}$ $\frac{9}{10}$ $\frac{11}{12}$ ...

0.5 0.75 0.83 0.88 0.90 0.91 ...

0.15 0.08 0.05 0.02 0.01

Differences are getting smaller

The differences are getting smaller, so the sequence is convergent.

To be sure of the limit, calculate a few more terms of the sequence:

$\frac{31}{32} \approx 0.97$ $\frac{47}{48} \approx 0.98$ $\frac{99}{100} = 0.99$ $\frac{999}{1000} = 0.999$

You can see that the terms are getting closer and closer to 1. So, the limit of the sequence is 1.

2. Iteration

H

Iteration is a way of finding any term of a sequence using the previous term(s). Each term in the sequence below can be found by adding 2 to the previous term.

5 →(Add 2) 7 →(Add 2) 9 →(Add 2) 11 →(Add 2) 13

This process can be described in words:

term = previous term + 2, with the first term 5

or algebraically using an **iterative formula**:

$u_{n+1} = u_n + 2$, starting with $u_1 = 5$

Using this formula we can generate the sequence:

$u_1 = 5$

$u_2 = u_1 + 2 = 5 + 2 = 7$

$u_3 = u_2 + 2 = 7 + 2 = 9$ etc.

Question *Investigate the sequence generated by the iterative formula* $u_{n+1} = \frac{u_n + 2}{5}$, *given that* $u_1 = 3$.

$u_1 = 3$ $\quad u_2 = \frac{u_1 + 2}{5} = \frac{3 + 2}{5} = \frac{5}{5} = 1$

$u_3 = \frac{u_2 + 2}{5} = \frac{1 + 2}{5} = \frac{3}{5} = 0.6$

$u_4 = \frac{u_3 + 2}{5} = \frac{0.6 + 2}{5} = \frac{2.6}{5} = 0.52$ etc.

The sequence converges because the differences become smaller:

u_1	u_2	u_3	u_4	u_5	....
3	1	0.6	0.52	0.504	....

2 0.4 0.08 0.016

Differences getting smaller

To find the limit of a convergent sequence, replace x_{n+1} and u_n by x and then solve the equation.

So, $u_{n+1} = \frac{u_n + 2}{5}$ becomes $x = \frac{x + 2}{5}$

$\therefore\ 5x = x + 2 \ \Rightarrow 5x - x = 2 \ \Rightarrow 4x = 2 \ \Rightarrow x = \frac{2}{4} = \frac{1}{2}$

So the limit of the sequence is 0.5.

Numerical Methods for Solving Equations

1. Trial and Improvement Method

With this method, you estimate the solution of an equation and then gradually improve your estimate.

Suppose that you have to find the solution of the equation $x^2+2x = 10$ correct to 1 decimal place. Your first guess might be $x = 2$. You will normally be given the first estimate in the question. Check how close this guess is by replacing x with 2 in the LHS of the equation. You are aiming for 10 on the RHS:

$x^2+2x = 2^2+2\times2 = 4+4 = 8$

Since the result is too low (less than 10), try increasing your estimate slightly, say to $x = 2.5$:

$x^2+2x = 2.5^2+2\times2.5 = 6.25+5 = 11.25$

Since the result is too high (greater than 10), the correct solution must lie somewhere between your two estimates of 2 and 2.5. Try $x = 2.3$:

$x^2+2x = 2.3^2+2\times2.3 = 5.29+4.6 = 9.89$

You are now getting much closer to 10. Since 9.89 is less than 10, the correct solution must lie somewhere between the last two estimates of 2.5 and 2.3. Try $x = 2.4$. Continue with more calculations until you obtain two estimates that are the same correct to 1 decimal place, one giving the RHS greater than 10, the other giving it as less than 10. The results are shown in the table below

Trial	Trial value (Guess)	x^2+2x (LHS)	Comment
1	2	8	Too low
2	2.5	11.25	Too high
3	2.3	9.89	Too low
4	2.4	10.56	Too high
5	{2.34	10.1556	Too high
6	{2.31	9.9561	Too low

The last two estimates are 2.34 and 2.31. These are both equal to 2.3, correct to 1 decimal place, and so the solution to the equation is $x = 2.3$ (1 dp).

Question *Given that $2^x = 20$, use the method of trial and improvement to find the value of x, correct to 1 dp.*

Try $x = 4$ to start with: $2^4 = 16$ Too low

So, try a larger value, say $x = 5$: $2^5 = 32$ Too high

Try a value in between 4 and 5, say $x = 4.5$. To calculate the value of $2^{4.5}$, use the power key x^y of your calculator:

2 x^y 4 . 5 = 22.627417

So, $2^{4.5} \approx 22.6$ Too high

The last two estimates in the table are $x = 4.34$ (too high) and $x = 4.32$ (too low). These are both equal to 4.3, correct to 1 decimal place, and so the solution to the equation is $x \approx 4.3$ (1 dp).

Trial	Trial value (Guess)	2^x (LHS)	Comment
1	4	16	Too low
2	5	32	Too high
3	4.5	22.6274	Too high
4	4.4	21.1121	Too high
5	4.3	19.6983	Too low
6	{4.34	20.2521	Too high
7	{4.32	19.9733	Too low

2. Using Iteration to Solve an Equation H

With this method, the equation is rearranged to produce an iterative formula (see Box 2, p44) that can be used to find an approximate solution.

Suppose we want to solve the equation $x^2+2x = 10$ correct to 2 decimal places. This is the same equation as in Box 1.

First, rearrange the equation to get x on its own on the LHS. There are several ways of doing this; some work better than others:

(a) $x^2+2x = 10$

$\therefore\ 2x = 10-x^2$

$\therefore\ x = \dfrac{10-x^2}{2}$

(b) $x^2+2x = 10$

$\therefore\ x(x+2) = 10$

$\therefore\ x = \dfrac{10}{x+2}$

(c) $x^2+2x = 10$

$\therefore\ x^2 = 10-2x$

$\therefore\ x = \sqrt{10-2x}$ Ignore the $\pm$ sign

Equations involving a fraction with x in the denominator often work well, so we will choose equation (b).

To change an equation into an iterative formula, replace x by x_{n+1} and x_n. For equation (b):

$$x = \frac{10}{x+2} \quad \text{becomes} \quad x_{n+1} = \frac{10}{x_n+2}$$

(You will probably be given the iterative formula in the question, as well as the first value, x_1).

If the values of x_n converge, they converge to the solution of the equation.

Question *Solve the equation $x^2+2x = 10$, correct to 2 dp, using the iterative formula $x_{n+1} = \dfrac{10}{x_n+2}$ starting with $x_1 = 2.3$.*

$x_1 = 2.3$

$$x_2 = \frac{10}{x_1+2} = \frac{10}{2.3+2} = \frac{10}{4.3} \approx 2.32558$$

NOTE Store this accurate value in the memory of your calculator, ready for the next step.

$$x_3 = \frac{10}{x_2+2} = \frac{10}{2.32558+2} = \frac{10}{4.32558} = 2.31183$$

Continue calculating values of x_n until the 3rd decimal place stops changing (one more decimal place than required in the question).

The values of x_n are summarised in the table below.

x_1	x_2	x_3	x_4	x_5	x_6	x_7	x_8
2.3	2.32558	2.31183	2.31920	2.31524	2.31737	2.31623	2.31684

The table shows that the values of x_n are converging. The last two values both begin 2.316.... So, we can safely say that the solution of the equation is $x = 2.32$, correct to 2 decimal places.

Flow Charts

1. Flow Charts

A **flow chart** is a series of instructions and questions arranged in the form of a diagram.

Instructions and questions are contained in shaped boxes, as shown below:

Arrows show you which box to go to next. Sometimes this depends on the answer to a question.

Question *Use this flow chart to calculate*
(a) the output, when the input is 7
(b) the two possible inputs when the output is 6.

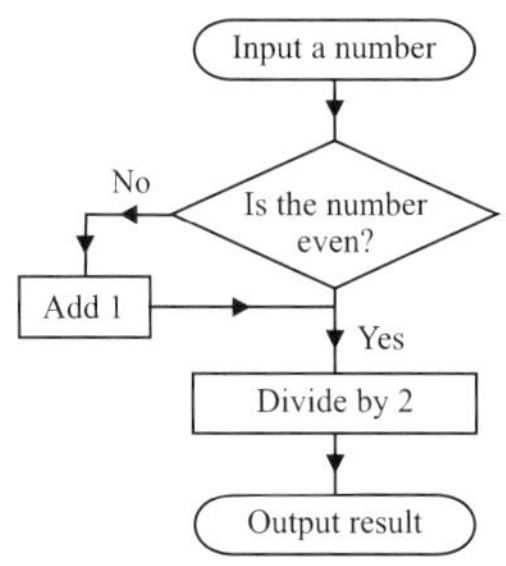

(a)

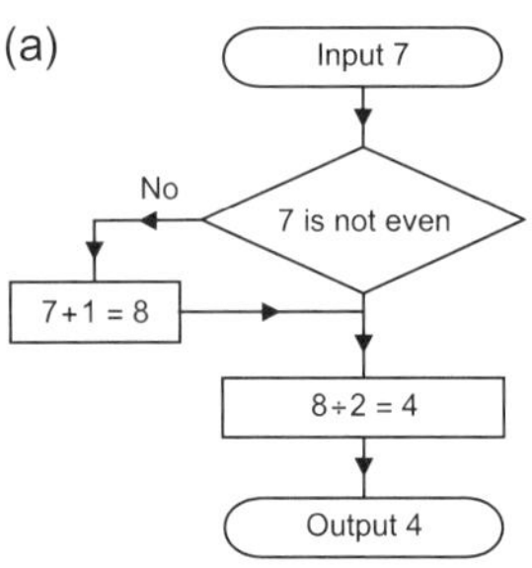

(b) Work from the bottom up, reversing the instructions.

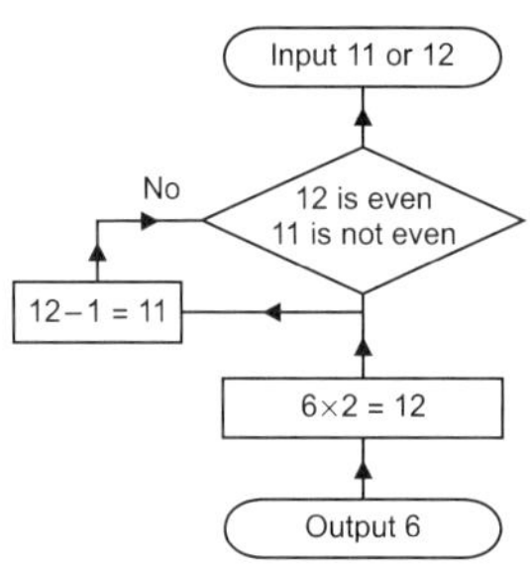

2. Flow Charts and Algebra

Most flow charts use algebra to describe the instructions and questions.

Question *Use the flow chart below to calculate the output when x = 2 and y = 3*

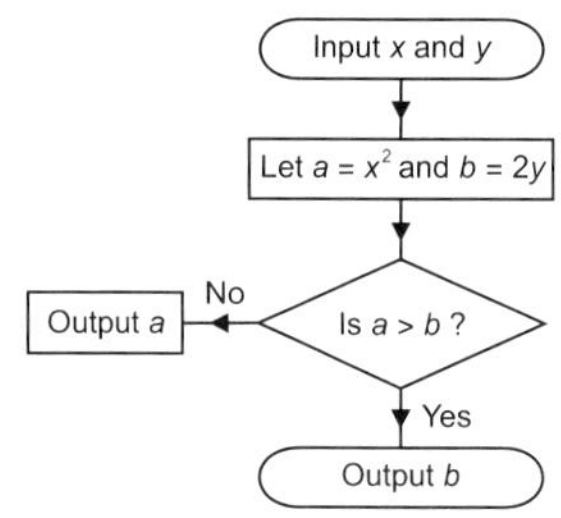

Solution

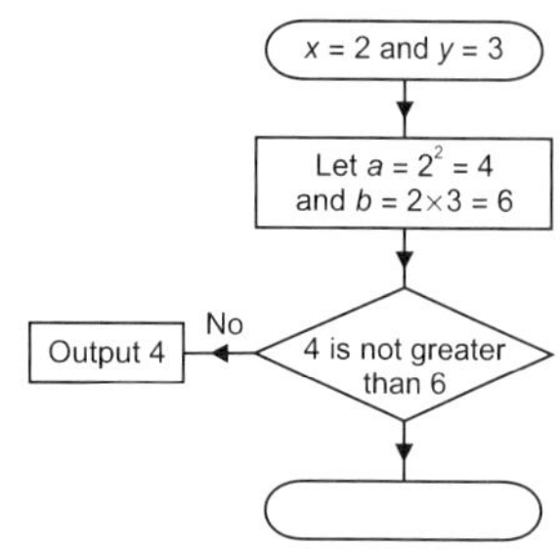

Question *What does the flow diagram below do?*

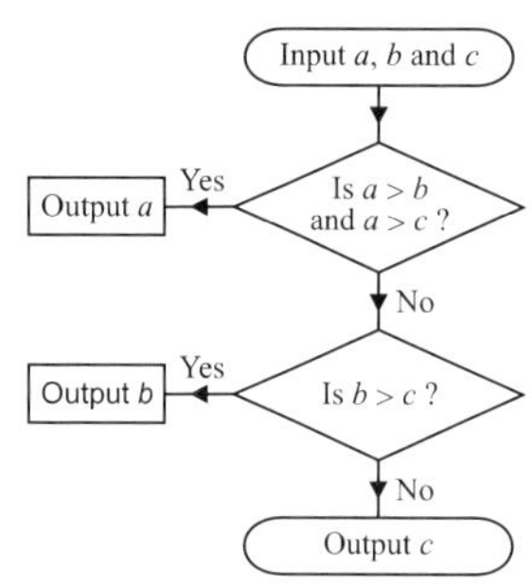

Input a few numbers to get some idea of what the flow diagram does.

Try $a = 2$, $b = 5$, $c = 3$.
a is not greater than both b and c, but b is greater than c. So, the output is $b = 5$. You can see that the flow diagram has output the biggest number.

Try a few more numbers to be sure.

So, the flow diagram finds the greatest of any three numbers.

3. Flow Charts with Loops

A **loop** repeats a set of instructions until a certain condition is met, e.g. keep on adding 2 until you reach 100.

Question *The flow chart below can be used to find the cube root of 24. Use the diagram to complete the table of values.*

x	y
2.5	

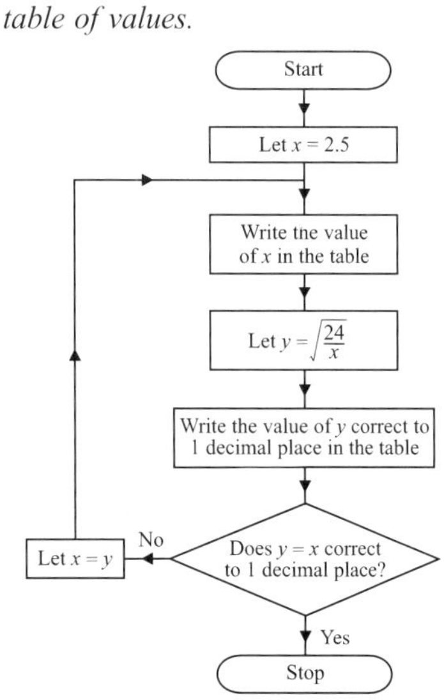

Solution

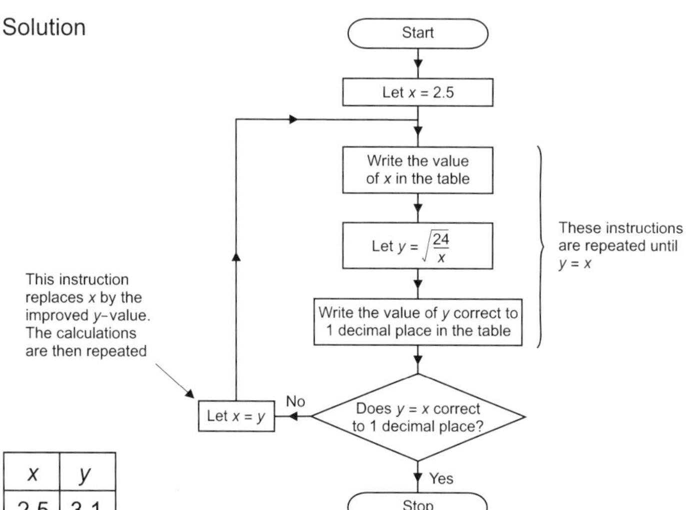

x	y
2.5	3.1
3.1	2.8
2.8	2.9
2.9	2.9

← $x = y$ here and so the process stops

So, $\sqrt[3]{24} \approx 2.9$ correct to 1 decimal place.

Simultaneous Equations

1. Simultaneous Equations

Simultaneous equations involve *two* unknown numbers, x and y say. You need two equations to be able to find the values of x and y. For example:

$$3x+2y = 11$$
$$2x+y = 6$$

Solving these equations means finding the values of x and y which make the LHS equal to the RHS in *both* equations.

The solution of the above two equations is $x = 1$ and $y = 4$. We can check this by substituting these values in the LHS of each equation:

$$3\times1+2\times4 = 3+8 = 11 \checkmark$$
$$2\times1+4 = 2+4 = 6 \checkmark$$

2. Solving by Subtracting Equations

The method of solving simultaneous equations involves eliminating the x-terms or y-terms by adding or subtracting the equations.

Question *Solve the simultaneous equations:*

$$2x+3y = 11$$
$$2x+y = 5$$

① Label the equations (A) and (B):

$$2x+3y = 11 \text{ .... (A)}$$
$$2x+y = 5 \text{ .... (B)}$$

② Subtract one equation from the other when either the x-terms or the y-terms are exactly the same.

Here, the x-terms are both $2x$:

$$2x+3y = 11 \text{ .... (A)}$$
$$2x+y = 5 \text{ .... (B)}$$
$$\text{(A)}-\text{(B) gives:} \quad 2y = 6$$

③ Solve the new equation:

$2y = 6$

$\therefore\ y = \frac{6}{2}$

$\therefore\ y = 3$

④ Substitute this solution into either of the original equations (A) or (B).

Putting $y = 3$ into equation (A) gives:

$2x+3\times3 = 11$

$\therefore\ 2x+9 = 11$

$\therefore\ 2x = 11-9$

$\therefore\ 2x = 2$

$\therefore\ x = 1$

⑤ Write down the solution to the simultaneous equations:

$x = 1$ and $y = 3$

⑥ Check the solution by substituting the values of x and y into the original equations:

Equation (A): $2\times1+3\times3 = 2+9 = 11 \checkmark$

Equation (B): $2\times1+3 = 2+3 = 5 \checkmark$

3. Solving by Adding Equations

Question *Solve the simultaneous equations*

$$4x-2y = 7$$
$$3x+2y = 7$$

① Label the equations:

$$4x-2y = 7 \text{ .... (A)}$$
$$3x+2y = 7 \text{ .... (B)}$$

② Add the equations when either the x-terms or the y-terms are the same except for their signs.

Here, the y-terms are $-2y$ and $+2y$

$$4x-2y = 7 \text{ .... (A)}$$
$$3x+2y = 7 \text{ .... (B)}$$
$$\text{(A)+(B):} \quad 7x = 14$$

NOTE $-2y$ and $+2y$ cancel each other when added together: $(-2y)+(+2y) = -2y+2y = 0$

③ Solve the new equation:

$7x = 14$

$\therefore\ x = 2$

④ Substitute this solution into equation (A) or (B).

Putting $x = 2$ into equation (B) gives:

$3\times2+2y = 7$

$\therefore\ 6+2y = 7$

$\therefore\ 2y = 7-6$

$\therefore\ 2y = 1$

$\therefore\ y = \frac{1}{2}$

⑤ Write down the solution to the simultaneous equations:

$x = 2$ and $y = \frac{1}{2}$

⑥ Check the solution by substituting the values of x and y into the original equations:

Equation (A): $4\times2-2\times\frac{1}{2} = 8-1 = 7 \checkmark$

Equation (B): $3\times2+2\times\frac{1}{2} = 6+1 = 7 \checkmark$

4. More Difficult Subtraction

Question *Solve the simultaneous equations*

$$3x+4y = 34$$
$$3x-2y = 10$$

The x-terms are the same and so we must subtract the equations:

$$3x+4y = 34 \text{ .... (A)}$$
$$3x-2y = 10 \text{ .... (B)}$$
$$\text{(A)}-\text{(B):} \quad 6y = 24$$

NOTE Take care when subtracting negative terms: $(+4y)-(-2y) = +4y+2y = 6y$

$\therefore\ y = \frac{24}{6}$

$\therefore\ y = 4$

Putting $y = 4$ into equation (A) gives:

$3x+4\times4 = 34$

$\therefore\ 3x+16 = 34$

$\therefore\ 3x = 18$

$\therefore\ x = 6$ and $y = 4$

Simultaneous Equations

5. Equations Requiring Multiplication

When neither the x-terms nor the y-terms of the equations are the same, we must make either pair of terms the same by multiplying the equations by numbers. Then we can use the previous methods.

Question *Solve the simultaneous equations*

$2x-y = 3$ .... (A)
$3x+2y = 22$ .... (B)

The x-terms are different and so are the y-terms. We can make the y-terms the same by multiplying equation (A) by 2:

$2x-y = 3$ .... (A)
2×(A): $4x-2y = 6$ .... (C) Label new equation
$3x+2y = 22$ .... (B)

Equations (B) and (C) now have the same y-terms, apart from their signs. Adding the equations eliminates y (see Box 3, p47):

$4x-2y = 6$ .... (C)
$3x+2y = 22$ .... (B)
(C)+(B): $7x = 28$

$\therefore x = 4$

Putting $x = 4$ into equation (B) gives:

$3\times4+2y = 22$
$\therefore 12+2y = 22$
$\therefore 2y = 22-12$
$\therefore 2y = 10$
$\therefore y = 5$

Solution to equation is $x = 4$, $y = 5$.

Check: Equation (A): $2\times4-5 = 8-5 = 3$ ✓
Equation (B): $3\times4+2\times5 = 12+10 = 22$ ✓

Question *Solve the simultaneous equations*

$2p-3q = 5$ .... (A)
$5p-2q = 40$ .... (B)

The p-terms are different and so are the q-terms. We can make both the p-terms $10p$ by multiplying equation (A) by 5 and equation (B) by 2:

$2p - 3q = 5$ .... (A)
5×(A): $10p - 15q = 25$ .... (C) Label new equation
$5p - 2q = 40$ .... (B)
2×(B): $10p - 4q = 80$ .... (D) Label new equation

Rewrite equations (C) and (D) and subtract (D) from (C) to eliminate the p-terms:

$10p - 15q = 25$ .... (C)
$10p - 4q = 80$ .... (D)
(C)−(D): $-11q = -55$

NOTE: $(-15q)-(-4q) = -15q+4q = -11q$

$\therefore q = \frac{-55}{-11} = 5$

Putting $q = 5$ into equation (A) gives:

$2p-3\times5 = 5$
$\therefore 2p-15 = 5$
$\therefore 2p = 20$
$\therefore p = 10$

Solution to equation is $p = 10$, $q = 5$.

Check: Equation (A): $2\times10-3\times5 = 20-15 = 5$ ✓
Equation (B): $5\times10-2\times5 = 50-10 = 40$ ✓

6. Solving Problems Using Simultaneous Equations

Question *Toffees cost x pence each and chocolates cost y pence each. Lee bought 7 toffees and 5 chocolates, paying £2.00 altogether. He spent 80 pence more on the toffees than the chocolates.*
(a) Write down two equations in x and y.
(b) Solve your equations to find the values of x and y.

Write down the equations in words first; then replace the words with symbols.

Cost of 7 toffees and 5 chocolates is £2.00

$7\times x$ pence + $5\times y$ pence = 200 pence
$7x + 5y = 200$

Label this first equation (A):

$7x+5y = 200$ .... (A)

Now make the second equation:

7 toffees cost 80 pence more than 5 chocolates

$7\times x$ pence = 80 pence + $5\times y$ pence
$7x = 80 + 5y$

Label this equation (B):

$7x = 80+5y$ .... (B)

Write down the equations above each other:

$7x+5y = 200$ .... (A)
$7x = 80+5y$ .... (B)

Rearrange the equations, if necessary, to get the x and y terms on the LHS and the number term on the RHS.

Equation (B) needs rearranging

$7x = 80+5y$
$7x-5y = 80$ .... (C)

Write down the equations again and solve them

$7x+5y = 200$ .... (A)
$7x-5y = 80$ .... (C)
(A)+(C): $14x = 280$

$\therefore x = \frac{280}{14} = 20$

Putting $x = 20$ into equation (A) gives:

$7\times20+5y = 200$
$\therefore 140+5y = 200$
$\therefore 5y = 200-140$
$\therefore 5y = 60$
$\therefore y = 12$

Solution to equation is $x = 20$, $y = 12$. So, toffees cost 20 pence each and chocolates cost 12 pence each.

Check: Eqn (A): $7\times20+5\times12 = 140+60 = 200$ ✓
Eqn (B): $7\times20 = 80+5\times12$ ✓

Algebraic Fractions

1. Multiplying and Dividing Fractions

Question *Simplify* $\frac{4a^2c}{3b} \times \frac{21b^3}{2ac^2}$.

$\frac{4a^2c}{3b} \times \frac{21b^3}{2ac^2}$ Cancel numbers

$= \frac{2a^2c}{b} \times \frac{7b^3}{ac^2}$ Cancel letters

$= \frac{2a}{1} \times \frac{7b^2}{c}$

$= \frac{2a \times 7b^2}{1 \times c}$ Multiply numerators and denominators

$= \frac{14ab^2}{c}$

Question *Simplify* $\frac{1}{6x} \div \frac{1}{3x^2}$.

$\frac{1}{6x} \div \frac{1}{3x^2}$

$= \frac{1}{6x} \times \frac{3x^2}{1}$ Change $\div$ to $\times$ and invert $\frac{1}{3x^2}$

$= \frac{1}{_2 6x} \times \frac{3x^{2\,1}}{1}$ Cancel numbers and letters

$= \frac{1 \times x}{2 \times 1}$ Multiply

$= \frac{x}{2}$

2. Adding and Subtracting Fractions

Fractions that have numbers as denominators can be added and subtracted the same way as in arithmetic (see Box 8, p17).

Question *Express* $\frac{7a}{15} - \frac{3a}{10}$ *as a single fraction in its lowest terms.*

The LCM of 15 and 10 is 30. So, we must convert each fraction to an equivalent fraction whose denominator is 30.

$\frac{7a}{15} \xrightarrow{\times 2} = \frac{14a}{30}$ $\qquad$ $\frac{3a}{10} \xrightarrow{\times 3} = \frac{9a}{30}$

Now we can add the fractions:

$\frac{7a}{15} - \frac{3a}{10} = \frac{14a}{30} - \frac{9a}{30} = \frac{14a - 9a}{30} = \frac{5a}{30} = \frac{a}{6}$

When the denominators contain letters, their LCM contains the highest power of each letter.

Question *Express* $\frac{7}{15a^2} - \frac{3}{10ab}$ *as a single fraction.*

The LCM of $15a^2$ and $10ab$ is $30a^2b$ because 30 is the LCM of 15 and 10, a^2 is the highest power of *a*, and *b* is the highest power of *b*. Now convert each fraction to an equivalent fraction whose denominator is $30a^2b$:

$\frac{7}{15a^2} \xrightarrow{\times 2b} = \frac{14b}{30a^2b}$ $\qquad$ $\frac{3}{10ab} \xrightarrow{\times 3a} = \frac{9a}{30a^2b}$

more

2. Adding•Subtracting Fractions (Contd)

Now we can subtract the fractions:

$\frac{7}{15a^2} - \frac{3}{10ab} = \frac{14b}{30a^2b} - \frac{9a}{30a^2b} = \frac{14b - 9a}{30a^2b}$

NOTE You cannot further simplify the answer by cancelling. For example, although 14 and 30 can be divided by 2, 9 cannot.

Question **H** *Express* $\frac{7}{(x+2)} - \frac{3}{(x-3)}$ *as a single fraction, simplifying your answer.*

When a denominator contains a bracket, the LCM also contains this bracket.

The LCM of $(x+2)$ and $(x-3)$ is $(x+2)(x-3)$. So, we must convert each fraction to an equivalent fraction whose denominator is $(x+2)(x-3)$:

$\frac{7}{(x+2)} \xrightarrow{\times(x-3)} = \frac{7(x-3)}{(x+2)(x-3)}$ $\qquad$ $\frac{3}{(x-3)} \xrightarrow{\times(x+2)} = \frac{3(x+2)}{(x+2)(x-3)}$

Now we can subtract the fractions

$\frac{7}{(x+2)} - \frac{3}{(x-3)} = \frac{7(x-3)}{(x+2)(x-3)} - \frac{3(x+2)}{(x+2)(x-3)}$

$= \frac{7(x-3) - 3(x+2)}{(x+2)(x-3)}$

$= \frac{7x - 21 - 3x - 6}{(x+2)(x-3)}$

$= \frac{4x - 27}{(x+2)(x-3)}$

3. Equations Involving Fractions **H**

Question *Solve the equation* $\frac{2x}{5} + \frac{x}{4} = \frac{3}{2}$.

Multiply each term by the LCM of the denominators (the lowest common denominator).

The LCM of 5, 4 and 2 is 20

$\frac{2x}{5} + \frac{x}{4} = \frac{3}{2}$ Write the equation

$\therefore\ 20^4 \times \frac{2x}{5_1} + 20^5 \times \frac{x}{4_1} = 20^{10} \times \frac{3}{2_1}$ Multiply each term by 20 and then cancel

$\therefore\ 4 \times 2x + 5 \times x = 10 \times 3$ Simplify

$\therefore\ 8x + 5x = 30$

$\therefore\ 13x = 30$

$\therefore\ x = \frac{30}{13}$

Question *Solve the equation* $\frac{t-2}{3} + \frac{t+1}{2} = \frac{3}{4}$.

The LCM of 3, 2 and 4 is 12.

$\frac{(t-2)}{3} + \frac{(t+1)}{2} = \frac{3}{4}$ Bracket numerators

$\therefore\ 12^4 \times \frac{(t-2)}{3_1} + 12^6 \times \frac{(t+1)}{2_1} = 12^3 \times \frac{3}{4_1}$ Multiply by 12 and then cancel

$\therefore\ 4 \times (t-2) + 6 \times (t+1) = 3 \times 3$

$\therefore\ 4t - 8 + 6t + 6 = 9$

$\therefore\ 10t - 2 = 9$

$\therefore\ 10t = 11$

$\therefore\ t = \frac{11}{10}$ (or 1.1)

Inequalities

1. Inequalities

An **inequality** states that one number is greater (or smaller) than another.

The inequality '5 is less than 6' can be written algebraically as $5 < 6$. The symbol $<$ means 'less than' when read from left to right.

'5 is less than 6' is the same as saying '6 is greater than 5', which can be written algebraically as $6 > 5$. The symbol $>$ means 'greater than' when read from left to right.

So, inequalities can be turned around

$5 < 6$ means the same as $6 > 5$

2. Rules of Inequalities

Inequalities can be manipulated in a similar way to equations.

A number can be added or subtracted from both sides of an inequality.

For example, we know $5 < 6$ is true. Adding 2 to both sides gives $5+2 < 6+2$ or $7 < 8$ which is also true.

A **positive** number can multiply or divide both sides of an inequality.

For example, we know $10 < 12$ is true. Dividing both sides by 2 gives $\frac{10}{2} < \frac{12}{2}$ or $5 < 6$, which is also true. Similarly, multiplying both sides by 3 gives $3\times10 < 3\times12$ or $30 < 36$, which is also true.

Multiplying or dividing both sides of an inequality by a **negative** number reverses the inequality sign.

For example, we know $5 < 6$ is true. Multiplying both sides by -2 gives $-2\times5 < -2\times6$ or $-10 < -12$, which is ***not*** true. However, reversing the inequality sign gives $-10 > -12$, which is true.

Similarly, we know $10 < 12$ is true. Dividing both sides by -2 and reversing the inequality sign gives $\frac{10}{-2} > \frac{12}{-2}$ or $-5 > -6$, which is also true.

3. Algebraic Inequalities

The inequality $x < 3$ tells us that x is a number less than 3, as shown on the number line below:

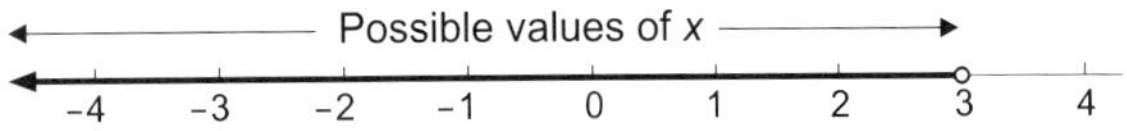

The empty circle indicates that x cannot be 3; x is **strictly less than** 3.

The inequality $x \leqslant 3$ tells us that x is a number less than or equal to 3, as shown on the number line below:

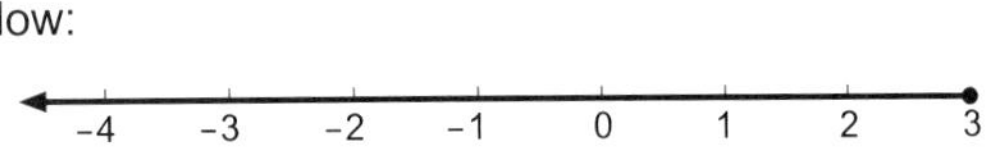

The solid circle indicates that x can be 3.

more

Smaller number $<$ Greater number

Greater number $>$ Smaller number

Smaller or equal number $\leqslant$ Greater or equal number

Greater or equal number $\geqslant$ Smaller or equal number

3. Algebraic Inequalities (Contd)

Sometimes x is on the RHS. You can either reverse the inequality or read the inequality from right to left.

The inequality $-2 < x$ states that -2 is less than the number x; or, equivalently, x is a number greater than -2, which could be written $x > -2$.

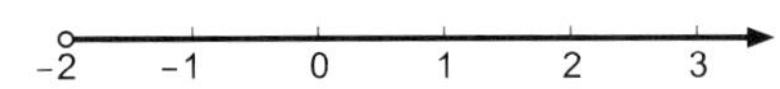

The inequality $-2 < x \leqslant 3$ states that x is a number greater than -2 (read the inequality from right to left) but less than or equal to 3, as shown on the number line below.

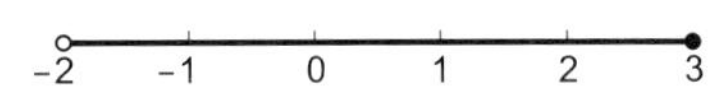

Question *Write down the positive integers, n, satisfying the inequality $n < 5$.*

We can see from the number line that the positive integers (whole numbers) less than 5 are: 1, 2, 3 and 4.

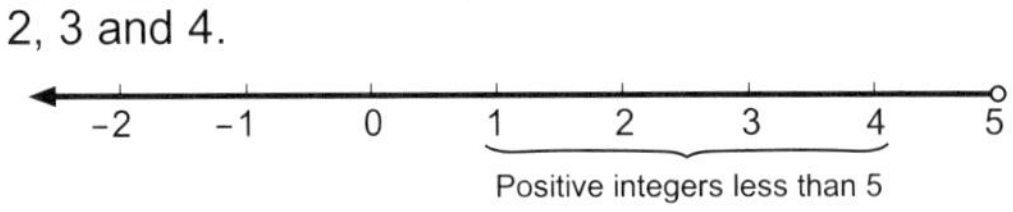

4. Solving Inequalities

To solve an inequality means to find the values of x which make it true. This involves rearranging the inequality to get x on its own on one side, using the rules in Box 2.

Question *Solve the inequality $x-2 < 5$.*

$x-2 < 5$	Write down the inequality
$\therefore\ x-2+2 < 5+2$	Add 2 to both sides to cancel out -2
$\therefore\ x < 5+2$	
$\therefore\ x < 7$	

As with equations, moving the -2 to the RHS and changing it to $+2$ gives the same result:

$x-2 < 5$	Move -2 to RHS and change to $+2$
$\therefore\ x < 5+2$	
$\therefore\ x < 7$	

Question *Solve the inequality $3x \geqslant 12$.*

$3x \geqslant 12$	Write down the inequality
$\therefore\ \frac{3x}{3} \geqslant \frac{12}{3}$	Divide both sides by 3 to leave x on its own on LHS
$\therefore\ x \geqslant 4$	

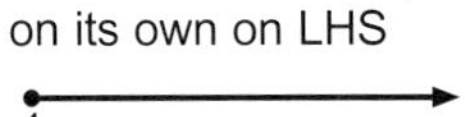

As with equations, moving the $3\times$ to the RHS and changing it to $\div3$ gives the same result:

$3x \geqslant 12$	Write down the inequality
$\therefore\ x \geqslant \frac{12}{3}$	Move $3\times$ to RHS and change to $\div3$
$\therefore\ x \geqslant 4$	

more

4. Solving Inequalities (Contd)

Question *Solve the inequality $-2x \geqslant 6$.*

$-2x \geqslant 6$ — Write down the inequality

$\therefore \frac{-2x}{-2} \leqslant \frac{6}{-2}$ — Divide both sides by −2 and reverse the $\geqslant$ sign

$\therefore x \leqslant \frac{6}{-2}$

$\therefore x \leqslant -3$

Question *Solve the inequality $4n+3 < 18$.*

$4n+3 < 18$ — Move +3 to RHS, change to −3

$\therefore 4n < 18-3$

$\therefore 4n < 15$ — Move 4× to RHS, change to ÷4

$\therefore n < \frac{15}{4}$

$\therefore n < 3\frac{3}{4}$

5. Inequalities Involving x^2

Question *Solve the inequality $x^2 \leqslant 9$.*

The largest value of x satisfying this inequality is $x = 3$ since $3^2 = 9$. This value is calculated by $\sqrt{9}$.

The smallest value of x satisying this inequality is $x = -3$ since $(-3)^2 = (-3)\times(-3) = 9$.

So, x can be any number between −3 and 3:

$-3 \leqslant x \leqslant 3$

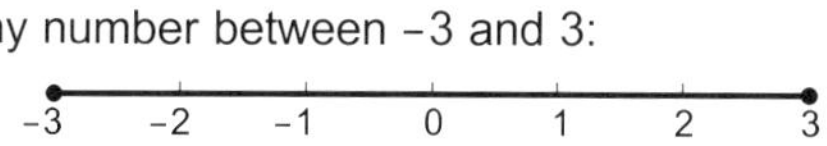

6. Using Inequalities to Solve Problems

Question *Pencils cost 27 pence each and erasers cost 30 pence each. Jim has £2.00 to spend on pencils and erasers. He buys n pencils and 2 erasers.*
(a) Write down an inequality in n.
(b) What is the largest number of pencils Jim bought?

(a) Cost of n pencils and 2 erasers is less than or equal to 200 pence.

We can write this algebraically as an inequality:

Cost of n pencils + Cost of 2 erasers $\leqslant$ 200 pence

$n\times27$ pence + 2×30 pence $\leqslant$ 200 pence

$\therefore 27n+2\times30 \leqslant 200$

(b) Solve the inequality as usual:

$27n+60 \leqslant 200$

$\therefore 27n \leqslant 200-60$

$\therefore 27n \leqslant 140$

$\therefore n \leqslant \frac{140}{27}$

$\therefore n \leqslant 5.19$ (2 dp)

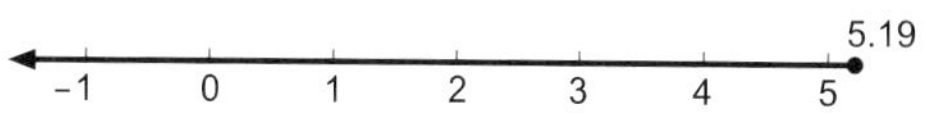

Now n is the number of pencils Jim bought and so it must be a whole number. From the number line we can see that n is 5. So, Jim bought 5 pencils.

Direct and Inverse Proportion

1. Direct Proportion

Suppose apples cost 10 pence each. Then:

Cost of 2 apples = 10×2 pence = 20 pence
Cost of 3 apples = 10×3 pence = 30 pence etc.

The more apples bought, the greater the cost.

Generally:

Cost of apples = 10 pence × Number of apples

If N apples cost C pence, we have the formula:

C pence = 10 pence × N

or simply $C = 10N$

As the number of apples, N, increases, their cost, C pence, also increases. We say that the cost, C pence, is **directly proportional** to the number of apples, N. To indicate that C is directly proportional to N, we write:

$$C \propto N$$

Two quantities, C and N, are said to be directly proportional if they are related by the formula:

C = Fixed number × N

The fixed number is called the **constant of proportionality**. In the above example, the constant of proportionality is 10.

1. Direct Proportion (Contd)

Question *The voltage, V, in an electrical circuit is directly proportional to the current flowing, A. Given that V = 12 when A = 5.*
(a) Write down an equation expressing V in terms of A.
(b) Find the value of V when A = 7.

(a) V is directly proportional to A and so:

V = Fixed number × A

We are told that $V = 12$ when $A = 5$. Substitute these values into the equation to find the fixed number.

12 = Fixed number × 5

$\therefore \frac{12}{5}$ = Fixed number

$\therefore$ Fixed number = $\frac{12}{5} = 2.4$

We can now rewrite the equation connecting V and A:

$V = 2.4\times A$ or simply $V = 2.4A$

(b) Putting $A = 7$ into the formula gives:

$V = 2.4\times7 = 16.8$

more

Direct and Inverse Proportion (Contd)

1. Direct Proportion (Contd)

Question *When a stone is thrown upwards with an initial speed s metres per second, it reaches a maximum height, h metres. Given that h varies directly as the square of s and that h = 5 when s = 10:*
(a) Work out a formula connecting h and s.
(b) Calculate the value of s when h = 20.

(a) h **varies directly** as the square of s is another way of saying that h is directly proportional to s^2 and so we can write:

h = Fixed number $\times s^2$

When $s = 10$, $h = 5$ and so:

5 = Fixed number $\times 10^2$

$\therefore$ Fixed number $= 5 \div 10^2 = 0.05$

So, the formula connecting h and s is: $h = 0.05s^2$.

(b) Putting $h = 20$ in the formula gives:

$20 = 0.05s^2$

$\therefore s^2 = 20 \div 0.05$

$\therefore s^2 = 400$

$\therefore s = \sqrt{400} = 20$ m/s

2. Inverse Proportion

Suppose it takes 8 men 6 hours to dig a hole. Common sense tells us that 4 men would take twice as long as 8 men, i.e. 12 hours. And 2 men would take twice as long as 4 men, i.e. 24 hours.

If we multiply the number of men by the number of hours they take to dig the hole, the result is always the same:

8 men taking 6 hours: $8 \times 6 = 48$
4 men taking 12 hours: $4 \times 12 = 48$
2 men taking 24 hours: $2 \times 24 = 48$

The fewer the men, the longer they take to dig the hole. Generally:

Number of men $\times$ Number of hours = 48

If M men take T hours to dig the hole, we have the formula:

$$M \times T = 48$$

This formula can be rearranged as:

$$M = \frac{48}{T}$$

As the number of men, M, decreases, the number of hours, T, they take to dig the hole increases. We say that the number of men, M, is **inversely proportional** to the number of hours, T, they take to dig the hole. To indicate that M is inversely proportional to T, we write:

$$M \propto \frac{1}{T}$$

Two quantities, M and T, are said to be inversely proportional if they are related by the formula:

$M \times T$ = Fixed number   or $M = \frac{\text{Fixed number}}{T}$

Here also, the fixed number is called the **constant of proportionality**, and is 48 in the above example.

more

2. Inverse Proportion (Contd)

Question *The pressure, P, of a gas is inversely proportional to its volume, V. Given that P = 50 when V = 200, express P in terms of V and find the value of P when V = 250.*

P is inversely proportional to V and so we can write:

$P \times V$ = Fixed number

We are told that $P = 50$ when $V = 200$. Substitute these values into the equation to find the fixed number:

50×200 = Fixed number

$\therefore$ Fixed number $= 50 \times 200 = 10\,000$

We can now rewrite the equation connecting P and V:

$P \times V = 10\,000$

We can use this formula to find the value of P when $V = 250$:

$P \times 250 = 10\,000$

$\therefore P = \frac{10\,000}{250}$

$\therefore P = 40$

3. Inverse Square Law **H**

Two quantities, A and B, satisfy the **inverse square law** if A is inversely proportional to B^2, written briefly as:

$$A \propto \frac{1}{B^2}$$

This means that A and B are related by the formula:

$$A = \frac{k}{B^2} \quad \text{or} \quad A \times B^2 = k$$

where k is the constant of proportionality.

Question *The force, F, between two objects is inversely proportional to the square of the distance, d, between them. Given that F = 0.75 when d = 3.8, find the value of d when F = 12.*

The force F is inversely proportional to d^2 and so:

$F \times d^2 = k$

We are given that $F = 0.75$ when $d = 3.8$. Put these values in the equation to find k:

$\therefore 0.75 \times 3.8^2 = k$

$\therefore k = 0.75 \times 3.8^2 = 10.83$

We can now rewrite the equation connecting F and d:

$F \times d^2 = 10.83$

So, when $F = 12$:

$12 \times d^2 = 10.83$

$\therefore d^2 = 10.83 \div 12$

$\therefore d^2 = 0.9025$

$\therefore d = \sqrt{0.9025}$

$\therefore d = 0.95$ (2 dp)

Plotting and Reading Graphs

1. Coordinates and Graphs

A **graph** is a diagram that shows the relationship between two quantities. Each quantity is represented by a straight line called an **axis**.

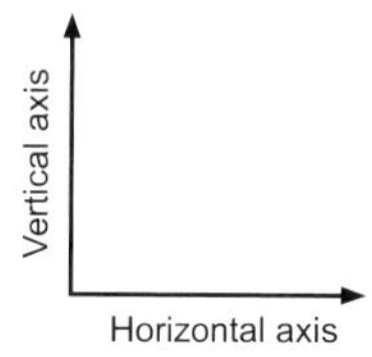

Suppose we are interested in the relationship between Ann's age and weight. We could let the horizontal axis represent her age and the vertical axis represent her weight.

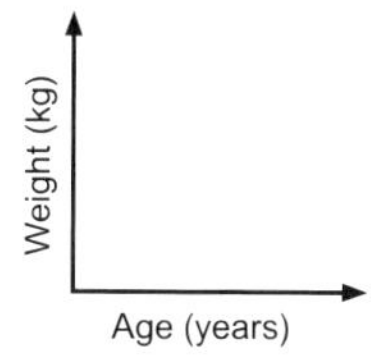

When Ann was 5 years old she weighed 20 kg. The point on the graph represents Ann's age and weight.

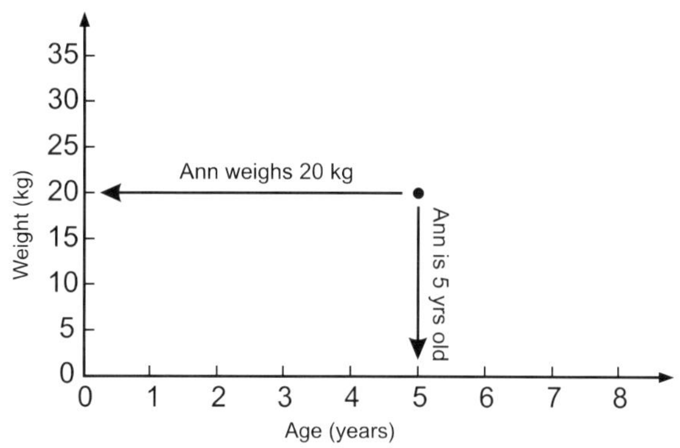

Ann's age and weight can be summarised as (5 yr, 20 kg) or more simply as (5, 20). These numbers are called the **coordinates** of the point on the graph.

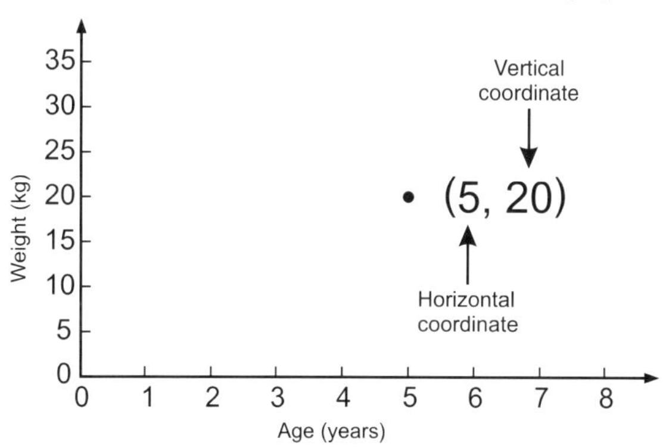

The table below shows Ann's weight at different ages. This is called a **table of values**.

Age (years)	0 (birth)	1	2	3	4	5	6	7	8
Weight (kg)	4	10	13	15	17	20	24	28	32

Each (age, weight) pair of values corresponds to a point on the graph. These points have been plotted on the graph below.

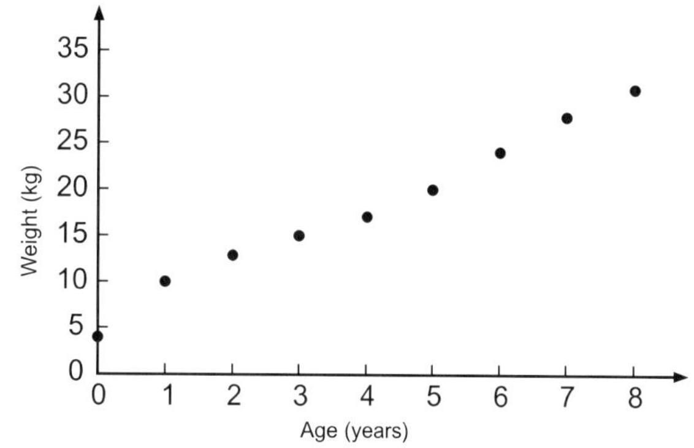

Joining up the dots with a smooth line improves the appearance of the graph and can be very useful (see Boxes 2 and 5).

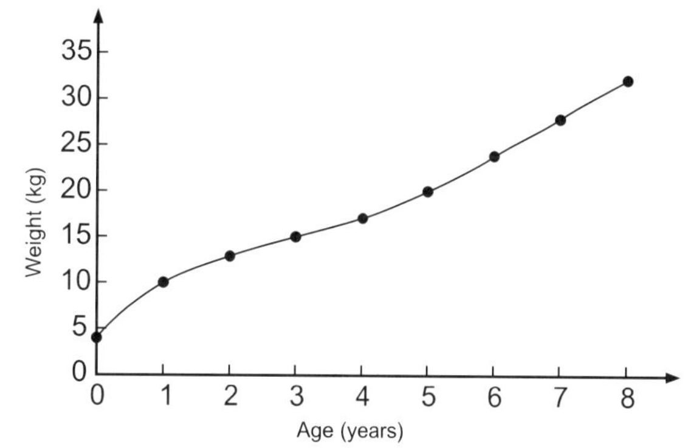

2. Reading Graphs

The graph opposite is called a **currency converter**. It can be used to convert German marks (DM) into pounds sterling (£), and vice-versa.

The dashed lines show that:

£20 is equivalent to 60 DM

120 DM is equivalent to £40

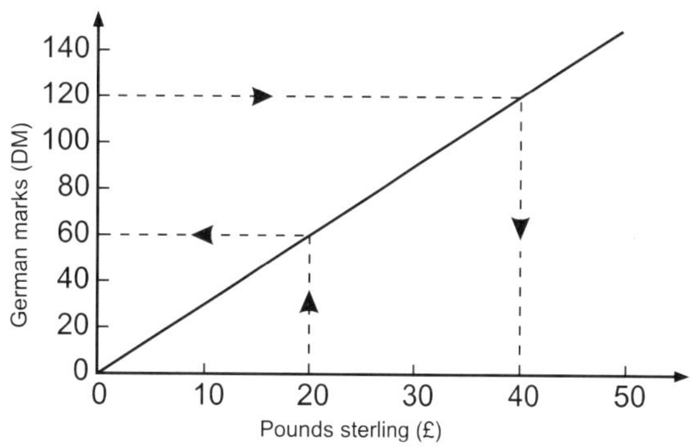

Plotting and Reading Graphs (Contd)

3. The Origin

The axes usually cross at a point called the **origin**, whose coordinates are (0, 0).

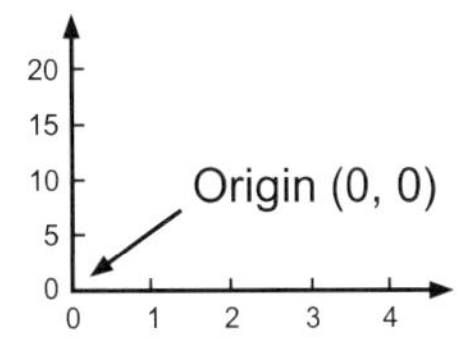

The axes do not need to cross at the origin. For example, the pulse rate of a person at rest is usually somewhere between 50 and 90.

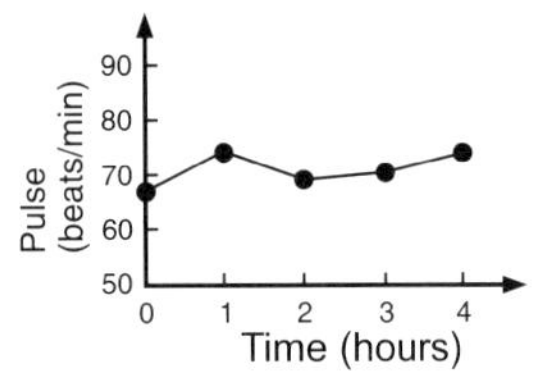

4. Graph Paper

Graph paper is usually divided into 1 cm squares.

Each 1 cm square is further divided into either 5 or 10 smaller squares. There are 5 or 10 small squares along each edge of a 1 cm square.

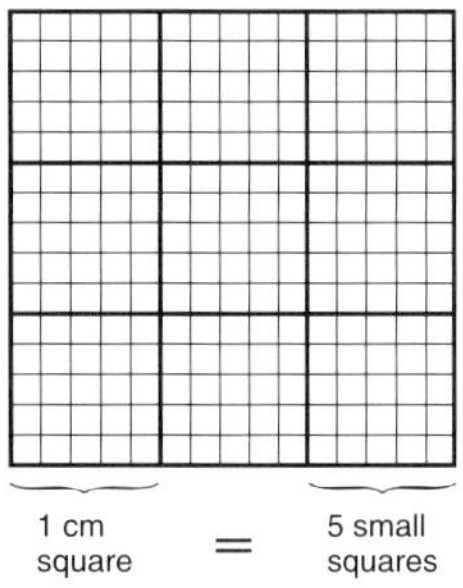

5. Scales

A **scale** states what 1 cm represents on an axis.

The scale for the horizontal axis is:

1 cm represents 1 gallon of petrol

The scale for the vertical axis is:

1 cm represents 100 km

So, 5 small squares represents 100 km

So, 1 small square represents 100 km ÷ 5 = 20 km

From the dashed line we can see that the Vista travels 160 km on 2 gallons of petrol.

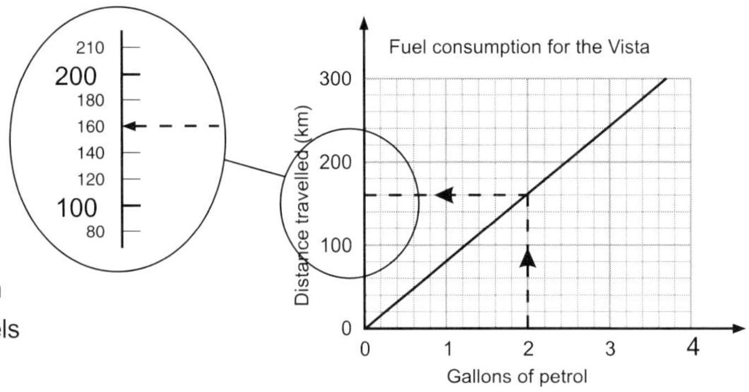

6. *x* and *y* Coordinates

You will come across abstract graphs whose axes do not represent real-life quantities

The horizontal axis is called the ***x*–axis**.

The vertical axis is called the ***y*–axis**.

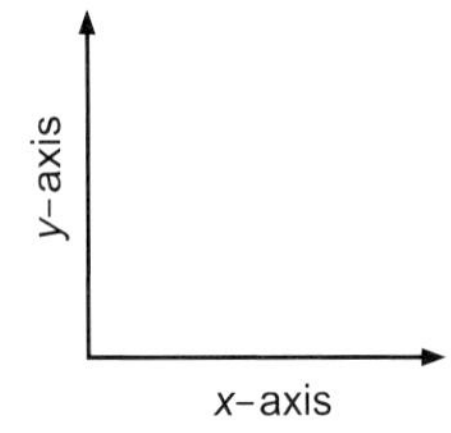

Every point on the graph has an *x* and *y* coordinate.

Remember the order for writing coordinates:

(*x*–coordinate, *y*–coordinate)

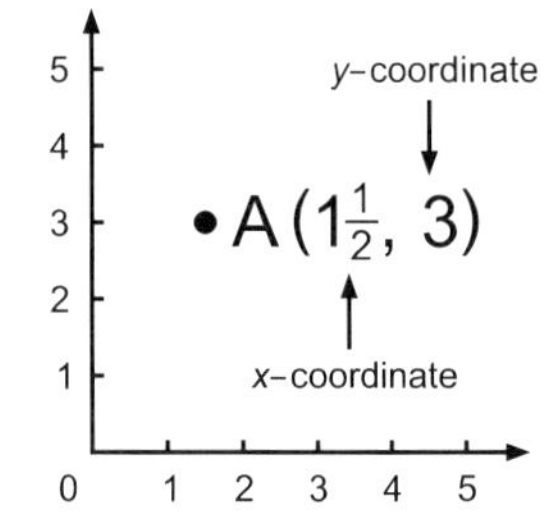

7. Negative Coordinates

The *x* and *y* axes can be extended to include negative numbers.

Point A has coordinates (–2, 4).

Point B has coordinates (1, –4).

Point C has coordinates (–3, –4).

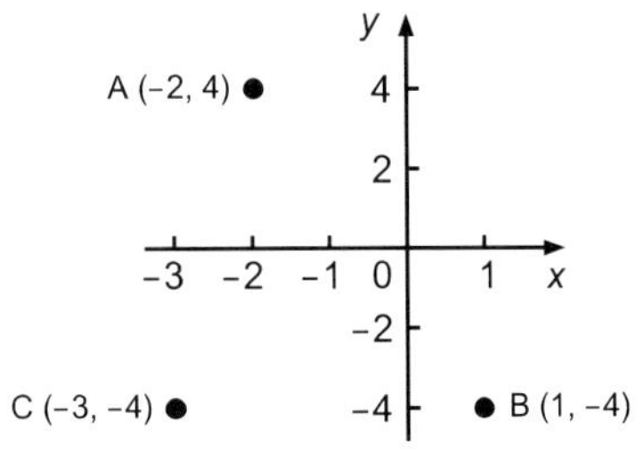

Question *Plot the points A (–3, 2), B (1, 2) and C (1, –2). If ABCD is a rectangle, what are the coordinates of point D?*

There is only one possible rectangle that has A, B and C as corners. The fourth corner is D and has coordinates (–3, –2).

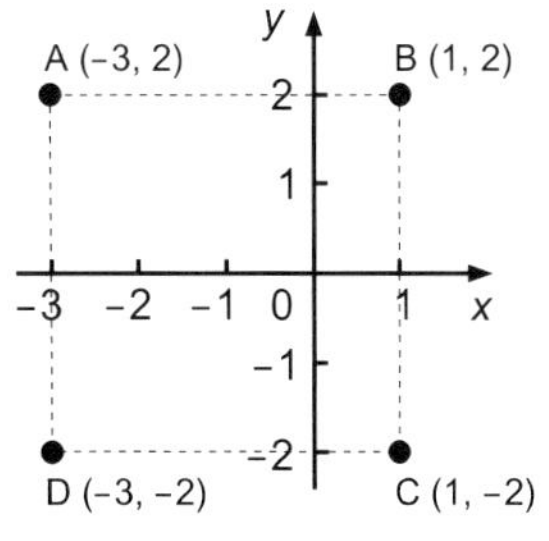

Graphs of Equations

1. Graphs of Equations

Equations can be used to calculate the coordinates of a graph.

Question *Complete the table of coordinates below using the equation* $y = 3x^2$.

x	-2	-1	0	1	2	3
y	12			3		27

Draw the graph of the equation.

The x-coordinates in the table range from $x = -2$ to $x = 3$. Some of the y-coordinates are given; the others must be calculated using the equation $y = 3x^2$

When $x = -1$ $\quad y = 3\times(-1)^2 = 3\times1 = 3$
When $x = 0$ $\quad y = 3\times0^2 = 3\times0 = 0$
When $x = 2$ $\quad y = 3\times2^2 = 3\times4 = 12$

NOTE Use a calculator if you are not very good at mental arithmetic.

The completed table of coordinates is shown below.

x	-2	-1	0	1	2	3
y	12	3	0	3	12	27

The question will include graph paper with labelled axes. Carefully plot the points and join them with a *smooth* curve.

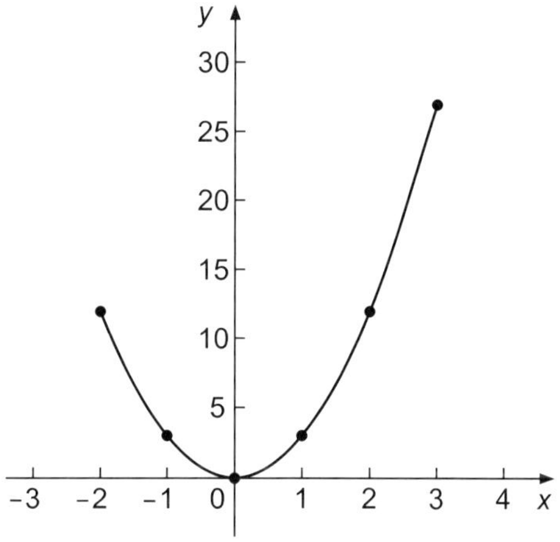

The y-coordinate of each point on the curve can be obtained using the equation of the curve $y = 3x^2$. For example, when $x = 2.5$, $y = 3\times(2.5)^2 = 18.75$. See the dashed line on the graph below.

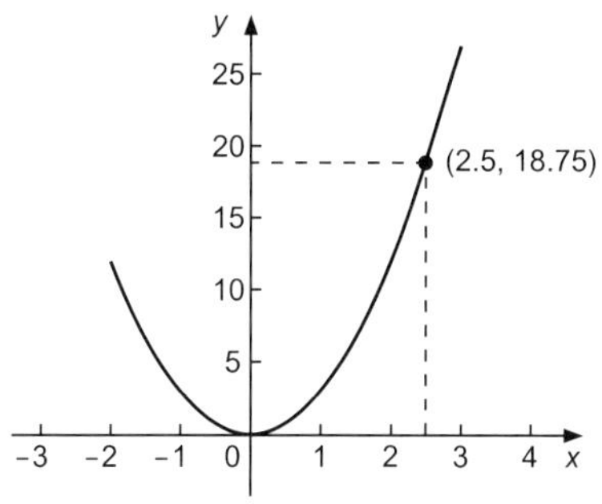

We say that every point on the line **satisfies** the equation $y = 3x^2$. In this case, the point (2.5, 18.75) satisfies the equation $y = 3x^2$.

2. Straight–Line Graphs

You are sometimes expected to make your own table of coordinates, especially for a straight-line graph.

Question *Use the axes provided on the graph paper below to draw the graph of the straight line* $y = 2x+2$.

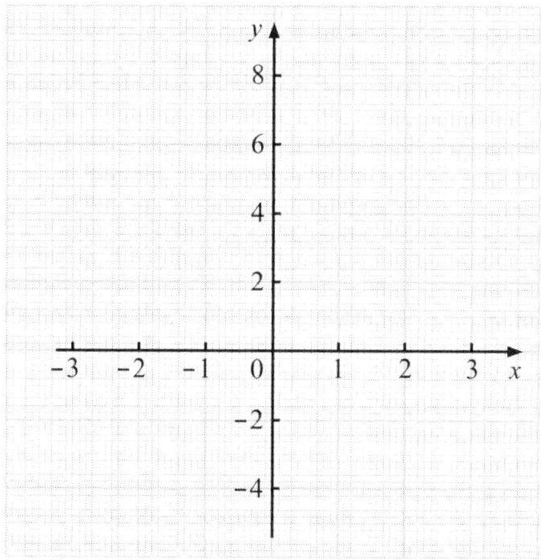

Notice that the x-axis values range from $x = -3$ to $x = 3$. Use the equation to calculate the y-coordinates for these two extreme values of x and for a third value somewhere in between, say $x = 0$.

When $x = -3$ $\quad y = 2\times(-3)+2 = -6+2 = -4$
When $x = 0$ $\quad y = 2\times0+2 = 0+2 = 2$
When $x = 3$ $\quad y = 2\times3+2 = 6+2 = 8$

NOTE Only two points are needed to draw a straight line, but calculating a third point acts as a check, just in case you made a mistake with one of the other points.

You can make a table of these coordinates, but it is not necessary.

x	-3	0	3
y	-4	2	8

Plot the points and join them using a straight line. If they cannot be joined using a straight line, go back and check your calculations.

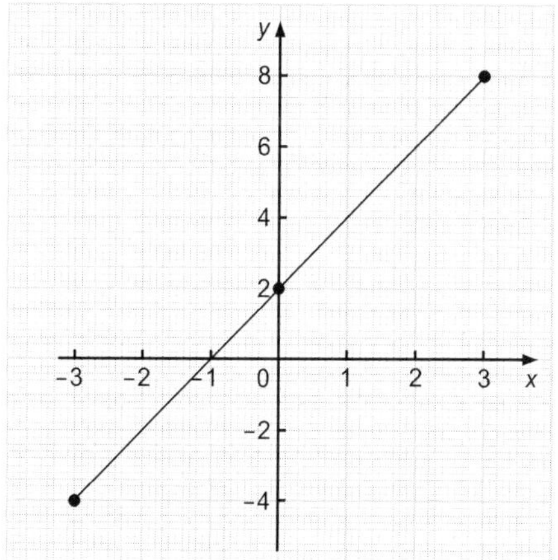

Graphs of Equations (Contd)

3. Using Graphs to Solve Problems

Question *The stopping distance, S m, for a car travelling at a speed of V km/h is given by the formula:*

$$S = 2V + \frac{V^2}{17}$$

Calculate the missing values of S in the table below (to the nearest metre) and plot the graph of S against V. Use your graph to find the speed when the stopping distance is 75 m.

V km/h	5	10	15	20	25	30	35
S m		26		64		113	142

When $V = 5$, $S = 2\times5+\frac{5^2}{17} = 10+\frac{25}{17} = 11.47 \approx 11$

When $V = 15$, $S = 2\times15+\frac{15^2}{17} = 30+\frac{225}{17} = 43.24 \approx 43$

When $V = 25$, $S = 2\times25+\frac{25^2}{17} = 50+\frac{625}{17} = 86.76 \approx 87$

Here is the completed table:

V km/h	5	10	15	20	25	30	35
S m	11	26	43	64	87	113	142

Plot the points and join them with a smooth curve.

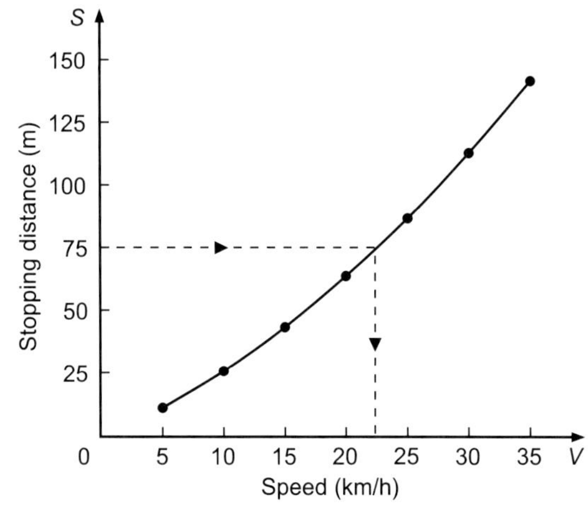

From the graph we can see that $V \approx 22.5$ km/h when $S = 75$ m.

4. Graphs You Should Know

You may be asked to recognise a graph and its equation. Here are a few common examples. Try to remember the general shape of each graph. If you forget, calculate a few values like those in the tables and make a quick sketch.

$y = x^2$

x	-3	-2	-1	0	1	2	3
y	9	4	1	0	1	4	9

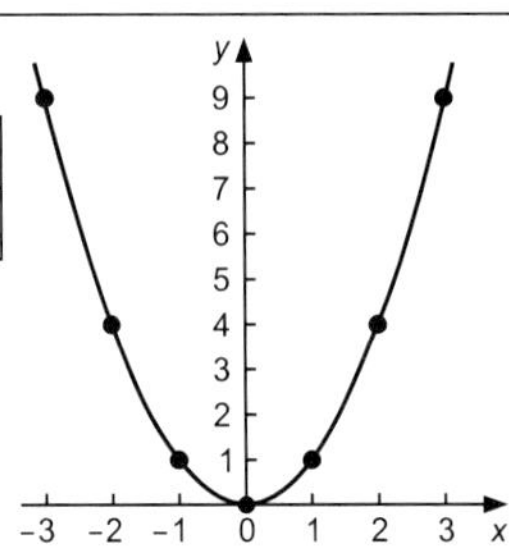

$y = x^3$

x	-2	-1	0	1	2
y	-8	-1	0	1	8

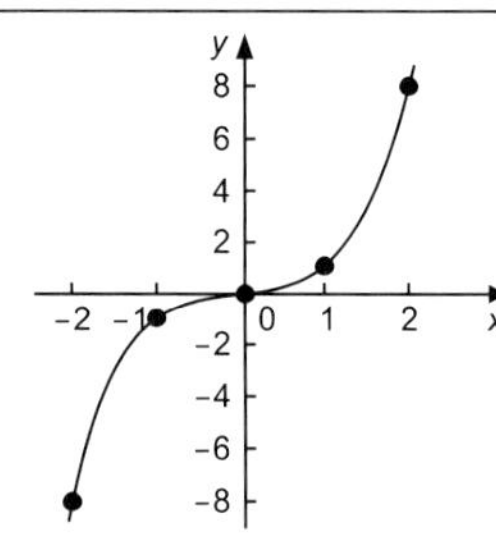

$y = \frac{1}{x}$

x	-1.5	-1	-0.5	0	0.5	1	1.5
y	-0.7	-1	-2	↑	2	1	0.7

$\frac{1}{0}$ cannot be calculated

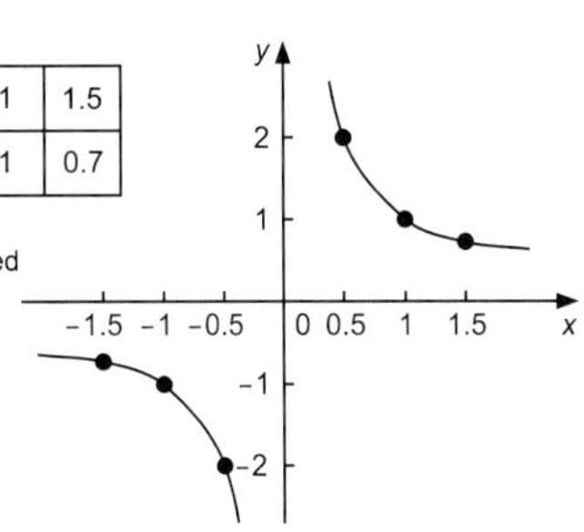

$y = \frac{1}{x^2}$

x	-1.5	-1	-0.5	0	0.5	1	1.5
y	0.4	1	4	↑	4	1	0.4

$\frac{1}{0^2}$ cannot be calculated

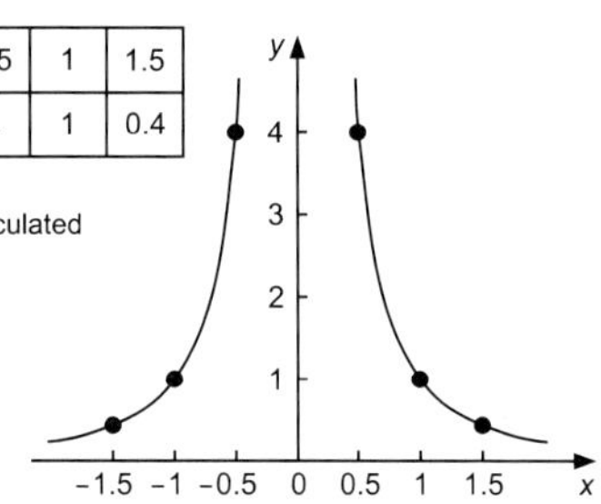

$y = \sqrt{x}$

x	0	1	4	9	16
y	0	1	2	3	4

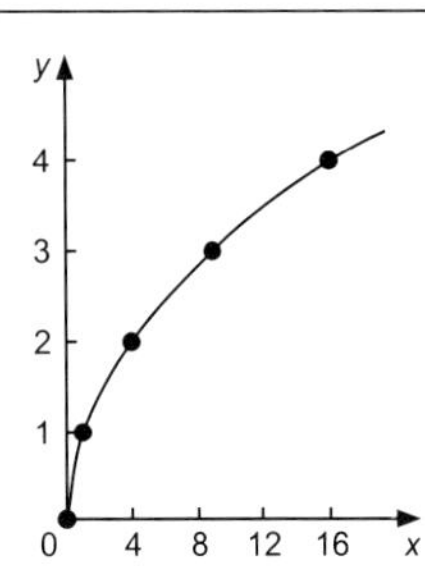

$y = x$

x	-2	-1	0	1	2
y	-2	-1	0	1	2

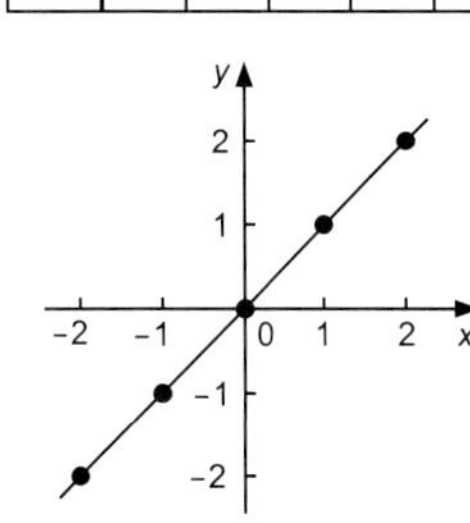

$y = -x$

x	-2	-1	0	1	2
y	2	1	0	-1	-2

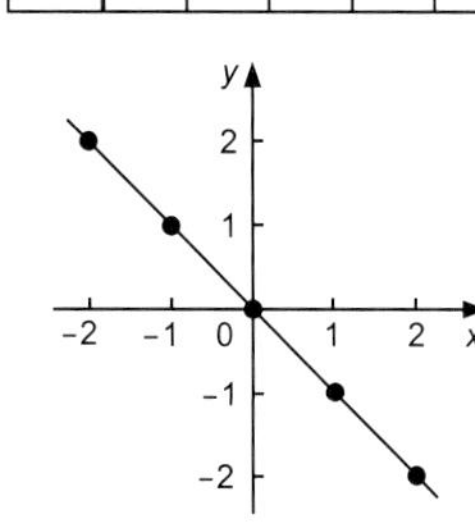

Equation of a Straight Line

1. Gradient of a Straight Line

The **gradient** of a straight line is a measure of how steep it is: the bigger the gradient, the steeper the line.

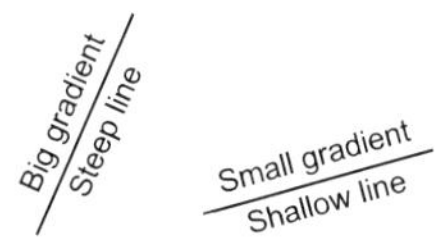

To find the gradient of a straight line, first choose any two points on the line (the further apart, the more accurate your answer).

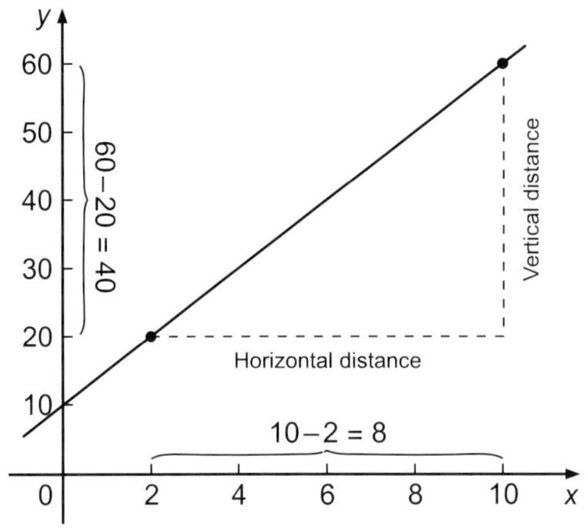

Find the vertical and horizontal distances between the two points:

Vertical distance = 60 − 20 = 40
Horizontal distance = 10 − 2 = 8

Then calculate the gradient using the formula:

$$\text{Gradient} = \frac{\text{Vertical distance}}{\text{Horizontal distance}}$$

For this line, we have:

$$\text{Gradient} = \frac{40}{8} = 5$$

A gradient of 5 means 5 up for every 1 along

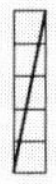

Sign of the Gradient

Gradients can be positive or negative (or 0), depending on the direction of the line.

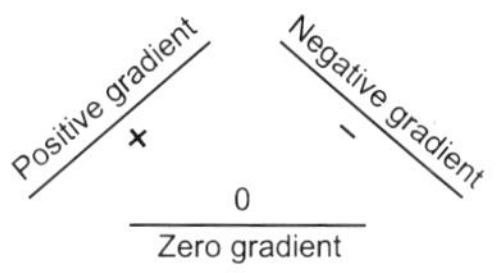

Example

Vertical distance
= 8 − 2 = 6
Horizontal distance
= 4 − 0 = 4

$$\text{Gradient} = -\frac{6}{4} = -1.5$$

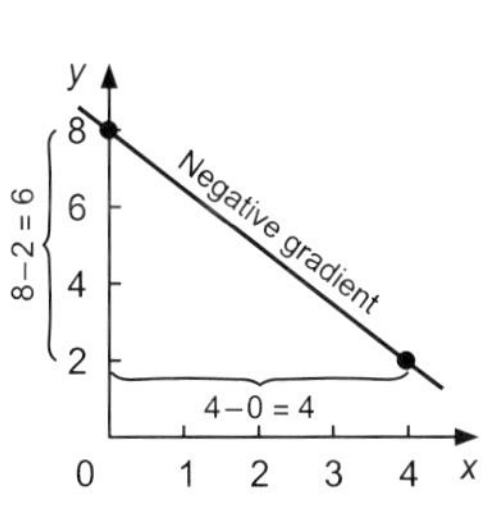

NOTE The horizontal and vertical distances are always *positive* numbers (6 and 4). Put the minus sign before the fraction.

Parallel lines have the same gradient.

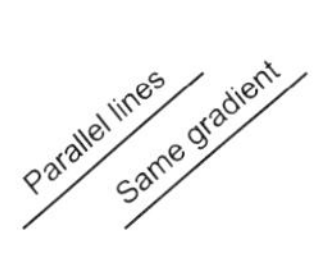

2. Equation of a Straight Line

The equation $y = 5x + 10$ was used to calculate the coordinates in the table below.

x	0	2	4	6	8	10
y	10	20	30	40	50	60

For example, when $x = 4$, $y = 5 \times 4 + 10 = 30$. Plotting these points gives a straight-line graph. This is the same straight line as in Box 1.

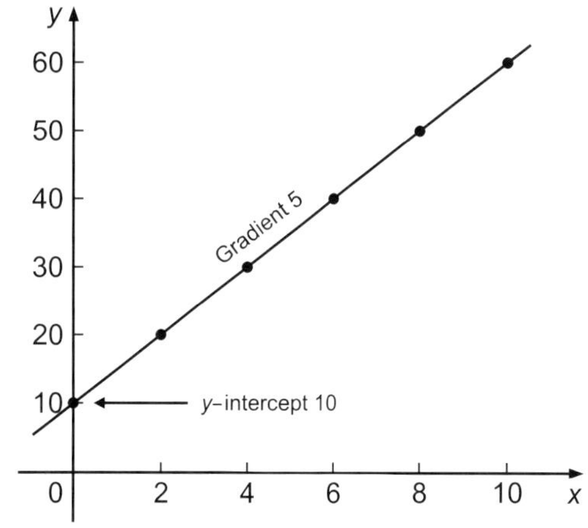

Notice that the line crosses the y-axis at $y = 10$. This is called the ***y*-intercept**.

Also, the gradient of the line is 5 (see Box 1).

These two numbers, 5 and 10, appear in the equation of the straight line (see top of page):

$$y = 5x + 10$$

(5: Gradient; 10: y-intercept)

So, the equation of a straight line can be written:

$$y = (\text{gradient})x + (y\text{-intercept})$$

This can also be written algebraically as:

$$y = mx + c$$

(m: Gradient; c: y-intercept)

For example, $y = -3x + 30$ is the equation of a straight line whose gradient is −3 and which crosses the y-axis at $y = 30$ (see diagram below).

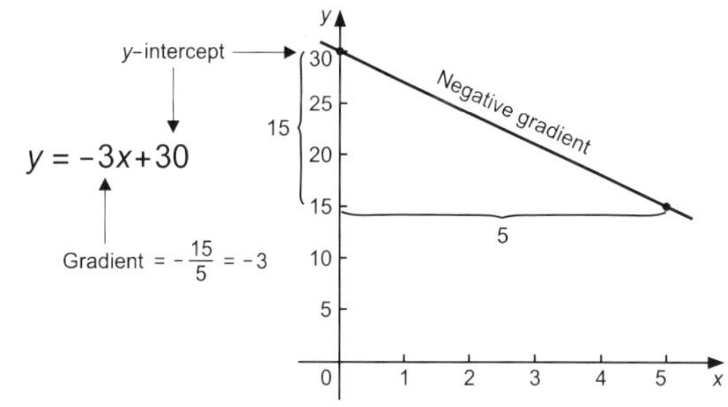

Comparing $y = -3x + 30$ with the general equation of a straight line $y = mx + c$, we can see that the gradient is $m = -3$ and the y-intercept is $c = 30$.

Equation of a Straight Line (Contd)

3. Straight–Line Graphs You Should Know

Here are some graphs whose equations do not look like $y = mx+c$. You should be able to recognise each graph and its equation. You should also be able to draw each graph, given its equation.

$y = 2$

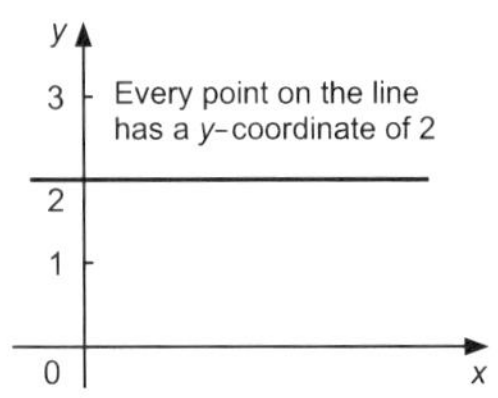

$x = 2$

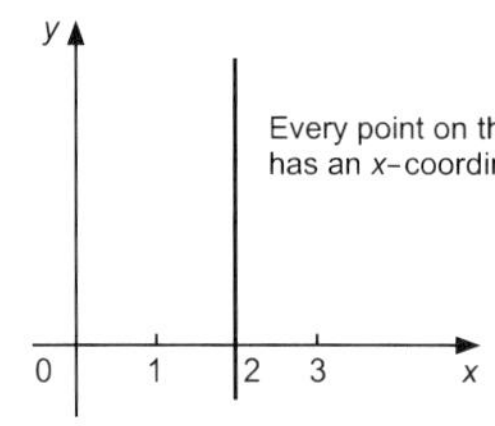

$y = x$

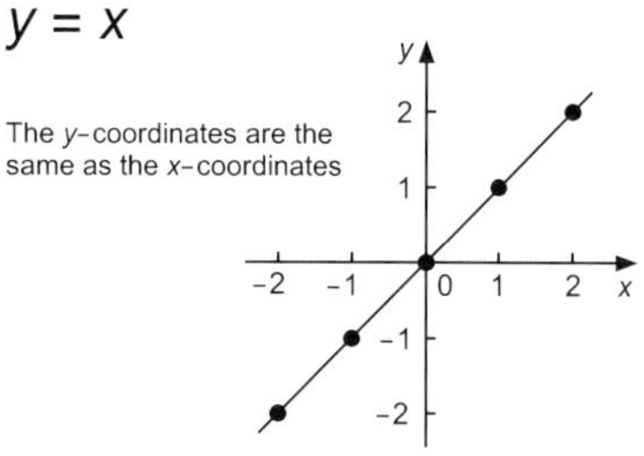

$y = 0$

The x-axis is a straight line with equation $y = 0$. Each point on the line has a y-coordinate of 0

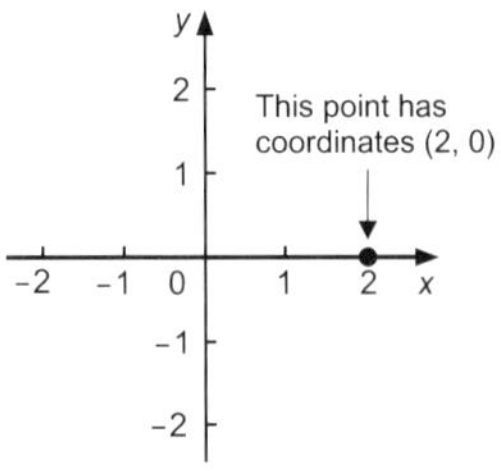

$x = 0$

The y-axis is a straight line with equation $x = 0$. Each point on the line has an x-coordinate of 0

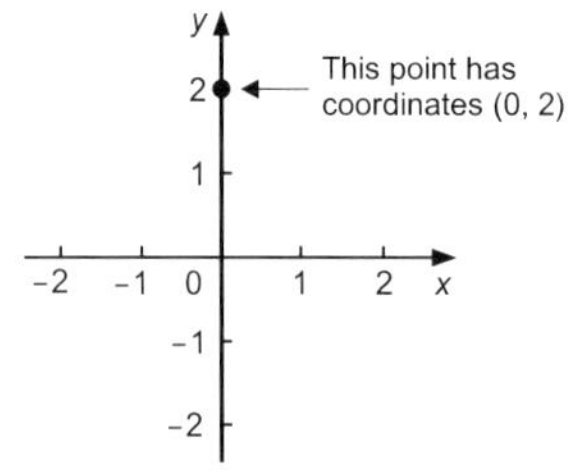

$y = -x$ *or* $x+y = 0$

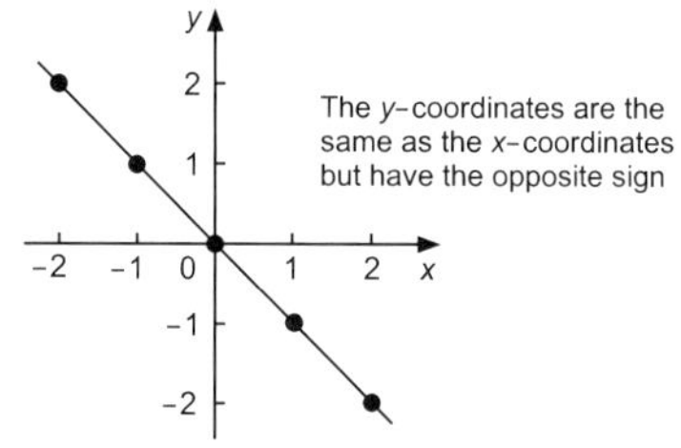

4. Rearranging Equations

Quite often the equation of a straight line is not in the form $y = mx+c$. Rearrange the equation into this form if you need to find the gradient or y-intercept.

Question *Find the gradient and y-intercept of the straight line with equation $x-2y = 6$. Make a sketch of the graph.*

Rearrange the equation into the form $y = mx+c$.

$x-2y = 6$

$\therefore\ x = 6+2y$ Move $-2y$ to RHS; change to $+2y$

$\therefore\ x-6 = 2y$ Move $+6$ to LHS; change to -6

$\therefore\ 2y = x-6$ Turn the equation around

$\therefore\ \frac{2y}{2} = \frac{x}{2} - \frac{6}{2}$ Divide both sides by 2 (cancelling: $\frac{6}{2}$ gives 3)

$\therefore\ y = \frac{1}{2}x-3$ NOTE $\frac{x}{2}$ is the same as $\frac{1}{2}x$

(Gradient ↑ $\frac{1}{2}$, y-intercept ↑ -3)

So, the gradient is $\frac{1}{2}$ and the y-intercept is -3.

To make a quick sketch, draw both axes (don't bother about scales). Mark the y-intercept. Since the gradient is positive, draw an upward sloping line through the y-intercept.

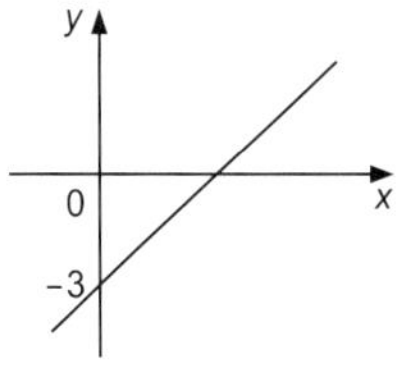

5. Quick Way of Drawing Straight Line Graphs

A quick way to draw a straight-line graph is to join the points where it crosses the x and y axes.

Question *Draw the graph of the straight line with equation $5x+3y = 12$.*

The point where the line crosses the x-axis has a y-coordinate of 0 (see Box 3). So, we can find this x-coordinate by putting $y = 0$ in the equation.

$5x+3\times0 = 12$

$\therefore\ 5x+0 = 12$

$\therefore\ 5x = 12$

$\therefore\ x = \frac{12}{5} = 2.4$

So, the straight line crosses the x-axis at $x = 2.4$ (see graph below).

Similarly, we can find the y-coordinate by putting $x = 0$ in the equation:

$5\times0+3y = 12$

$\therefore\ 0+3y = 12$

$\therefore\ 3y = 12$

$\therefore\ y = 4$

So, the straight line crosses the y-axis at $y = 4$.

Mark these two points on the axes and join them up.

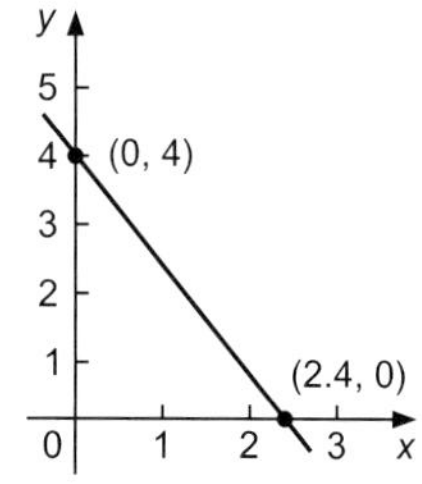

6. Finding the Equation of a Straight Line

Question *The table below shows some temperatures measured both in degrees Fahrenheit and Celsius.*

Celsius (C)	*-10*	*0*	*15*	*40*	*60*
Fahrenheit (F)	*14*	*32*	*59*	*104*	*140*

(a) Draw the graph of F against C.
(b) Find the F-intercept of the straight line.
(c) Find the gradient of the straight line.
(d) Write down the equation of the straight line.
(e) Use your equation to calculate the temperature which is the same measured in degrees Fahrenheit or Celsius.

(a) Carefully plot the points and join them with a straight line.

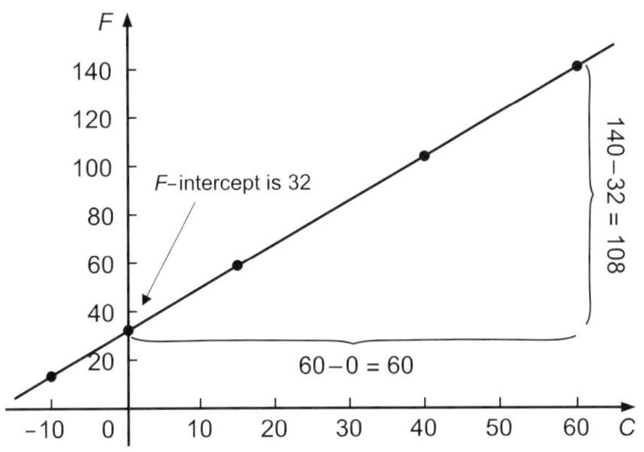

(b) The line crosses the F-axis at $F = 32$

(c) Gradient = $\frac{\text{Vertical distance}}{\text{Horizontal distance}} = \frac{108}{60} = 1.8$

(d) Equation of a straight line is (see Box 2, p57):

y = (gradient)x + (y-intercept)

Replace y by F and x by C

F = (gradient)C + (F-intercept)

$\therefore\ F = 1.8C+32$

(e) When F is the same as C the equation becomes:

$C = 1.8C+32$

$\therefore\ 0 = 1.8C-C+32$ Move C to RHS; change to $-C$

$\therefore\ 0 = 0.8C+32$ Simplify

$\therefore\ -32 = 0.8C$ Move +32 to LHS; change to −32

$\therefore\ 0.8C = -32$ Turn the equation around

$\therefore\ C = \frac{-32}{0.8} = -40$ Move $0.8\times$ to RHS; change to $\div 0.8$

So, −40 °C is the same as −40 °F.

Rate of Change

1. Rate of Change

Rate of change tells you how much one quantity changes as another, related quantity changes. For example, a car hire firm might hire out a Ford Fiesta at a rate of £40 per day, i.e. the hire charge increases by £40 for each extra day of hire.

A rate of change can be written as an equation. The above rate of £40 per day could be written:

Hire charge for 1 day = £40

This can help to solve problems.

Question *1 litre of water flows from a hosepipe every 12 seconds. What is the rate of flow in litres per minute?*

We must find how much water flows per minute, i.e. in 1 minute.

12 seconds of flow = 1 litre

$\therefore$ 1 second of flow = $\frac{1\text{ litre}}{12}$ = 0.0833 litres

$\therefore$ 60 seconds of flow = 60×0.0833 litres = 5 litres

$\therefore$ 1 minute of flow = 5 litres

So, the rate of flow is 5 litres per minute. This can be written briefly as 5ℓ per min or 5ℓ/min.

2. Rate of Change and Gradient

The rate of change of two quantities is the same as the gradient of their graph.

Question *Calculate the gradient of this currency conversion graph. Explain what your answer means?*

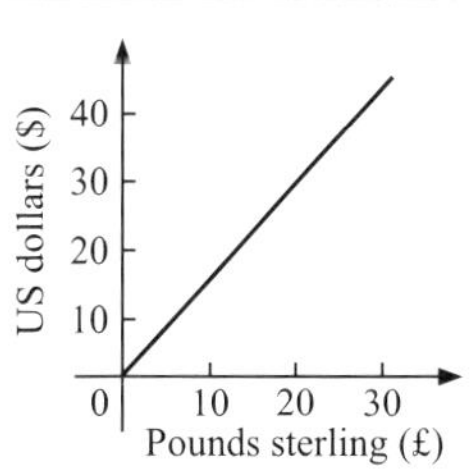

Gradient

$= \frac{\text{Vertical distance}}{\text{Horizontal distance}}$

$= \frac{\$30}{£20}$

$= \frac{\$1.5}{£1}$

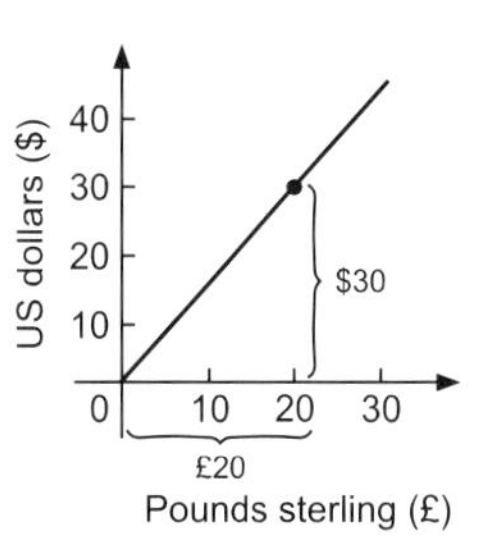

Now, $\frac{\$1.5}{£1}$ can be written as \$1.5/£1, or simply \$1.5/£, which is said '\$1.5 per £'. In other words, £1 is equivalent to \$1.5. So, this gradient is the rate of exchange from pounds sterling (£) to US dollars (\$).

Rate of Change (Contd)

3. Practical Examples of Rate of Change

You may be shown a graph and asked to comment on the rate of change. If the graph is a straight line, it has a constant gradient and so the rate of change is also constant.

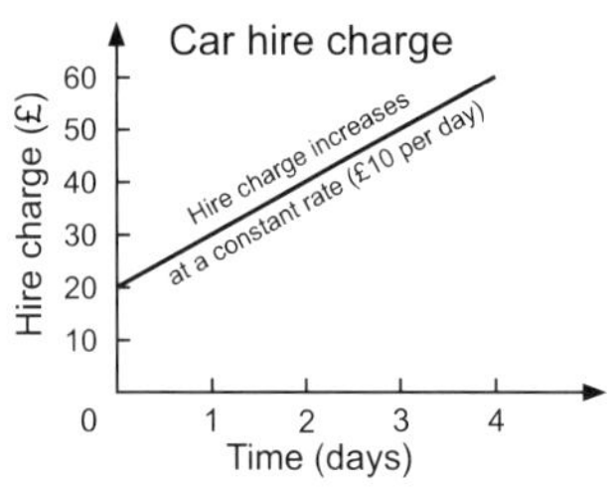

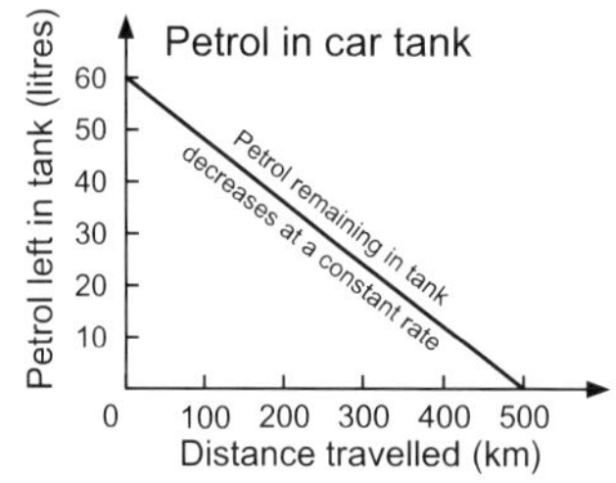

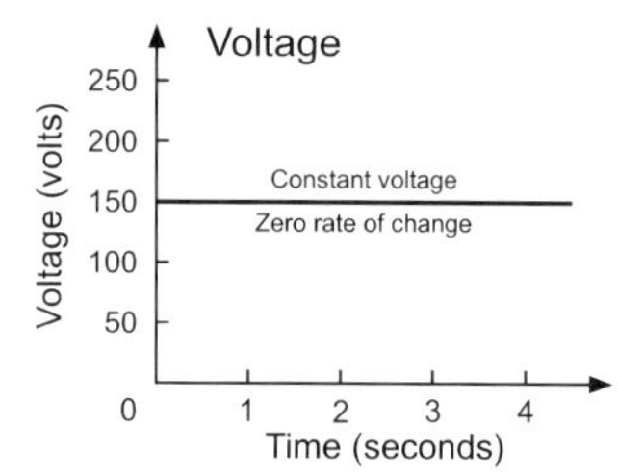

If the graph is curved, the gradient (steepness) changes and so the rate of change also changes: the steeper the line, the faster the rate of change.

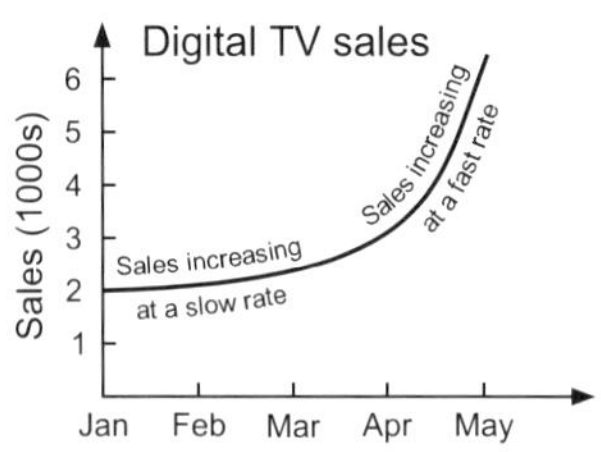

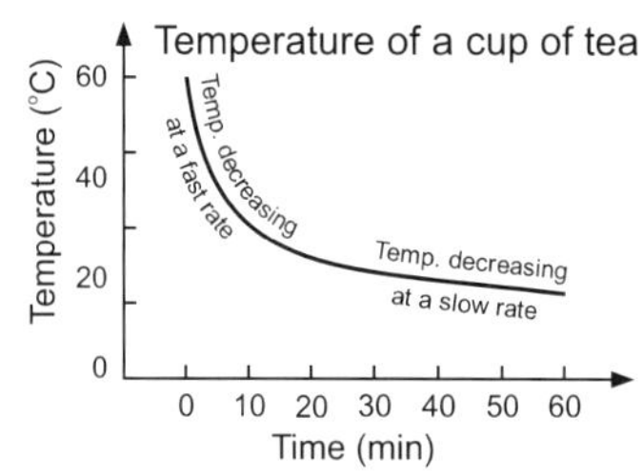

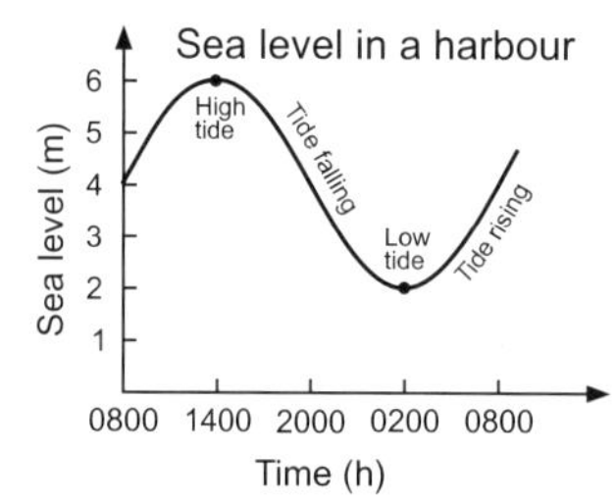

Speed, Distance and Time

1. Time

It is sometimes necessary to convert a period of time into different units. Use the conversion diagram below.

To convert a big unit to a small unit, **multiply**

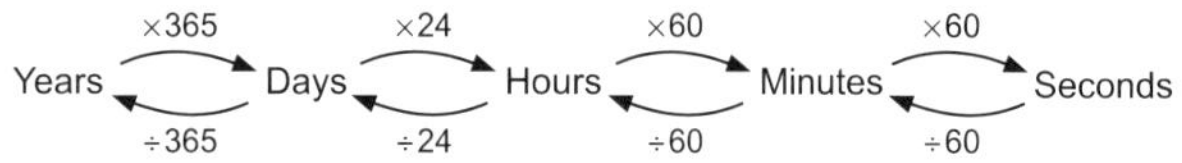

To convert a small unit to a big unit, **divide**

Question *How many minutes are there in a year?*

1 year = 365 days = 365×24 hours
= 365×24×60 minutes
= 525 600 minutes

Question *Convert 132 minutes to hours.*

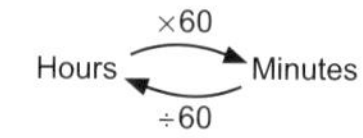

132 minutes = $\frac{132}{60}$ hours = 2.2 hours

0.2 h = 0.2×60 min = 12 min

So, 132 min = 2 h 12 min.

2. Speed

A speed of 5 km/h means that a distance of 5 km is travelled in 1 hour.

Speed (s), distance (d) and time (t) are related by the formula:

$$\text{Speed} = \frac{\text{Distance}}{\text{Time}} \quad \text{or} \quad s = \frac{d}{t}$$

Question *A racing cyclist travelled 65 km in 1 hour 48 minutes. Calculate her speed in km/h.*

First, convert 48 minutes to hours:
48 minutes = $\frac{48}{60}$ hours = 0.8 hours
So, 1 hour 48 minutes = 1.8 hours

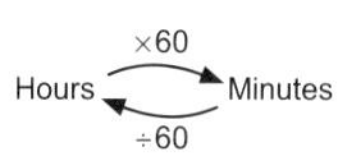

$$\text{Speed} = \frac{\text{Distance}}{\text{Time}} = \frac{65\text{ km}}{1.8\text{ h}} = 36.1\text{ km/h}$$

NOTE It is unlikely that the cyclist kept up a *constant* speed of exactly 36.1 km/h. So we say that his **average speed** was 36.1 km/h for the journey.

Question *Convert 25 km/h to m/s.*

First, convert 25 km to metres:

25 km = 25×1000 = 25 000 m

Then, convert 1 hour to seconds:

1 h = 60×60 = 3600 s

25 km/h means 25 km is travelled in 1 hour and so:

$$\text{Speed} = \frac{\text{Distance}}{\text{Time}} = \frac{25\text{ km}}{1\text{ h}} = \frac{25\,000\text{ m}}{3600\text{ s}} = 6.94\text{ m/s}$$

Question *John drives his car at a constant speed of 50 km/h for 18 minutes. How far does he travel?*

In 1 hour John travels 50 km
∴ In 60 minutes John travels 50 km
∴ In 1 minute John travels $\frac{50}{60} = \frac{5}{6}$ km
∴ In 18 minutes John travels $18 \times \frac{5}{6}$ km = 15 km

3. Accuracy When Measuring Speed

Question *Jill ran a distance of 100 m, measured correct to the nearest metre. Her stopwatch showed she took 14.8 seconds, correct to the nearest tenth of a second.*
(a) Write down the longest possible distance she ran.
(b) Write down the shortest possible time she took.
(c) Calculate her maximum possible average speed.

(a) Maximum possible distance run is 100.5 m

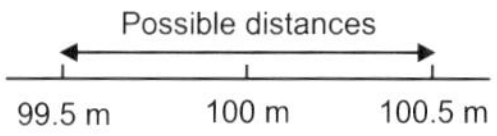

(b) Minimum possible time taken is 14.75 s

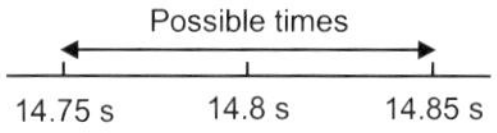

(c) Maximum possible speed $= \dfrac{\text{Maximum possible distance}}{\text{Minimum possible time taken}}$

$= \dfrac{100.5\text{ m}}{14.75\text{ s}} = 6.8\text{ m/s (1 dp)}$

4. Distance–Time Graph

A **distance–time graph** illustrates a journey.The y-axis represents the distance (route) travelled and the x-axis represents time.

Question *Dieter left home at 3.00 pm and cycled along the main road to the park 5 km away. He arrived at 3.30 pm and played football for 1 hour 30 minutes and then cycled home at an average speed of 6 km/h.*
(a) Draw a distance–time graph for his journey.
(b) Which part of the journey did he cycle the faster?
(c) What was his average speed on the outward journey?

(a) Dieter started from home at 3.00 pm, which is where the axes cross. At 3.30 pm he arrived at the park, 5 km away. Mark this point. At 5.00 pm he was still at the park. Mark this point. He cycled home over a distance of 5 km at a speed of 6 km/h. We need to find the time he arrived home.
A speed of 6 km/h means:
6 km = 1 hour = 60 minutes
$\therefore$ 1 km = 10 minutes
$\therefore$ 5 km = 50 minutes
So, Dieter arrived home at 5.50 pm. Mark this point. Join the points with straight lines.

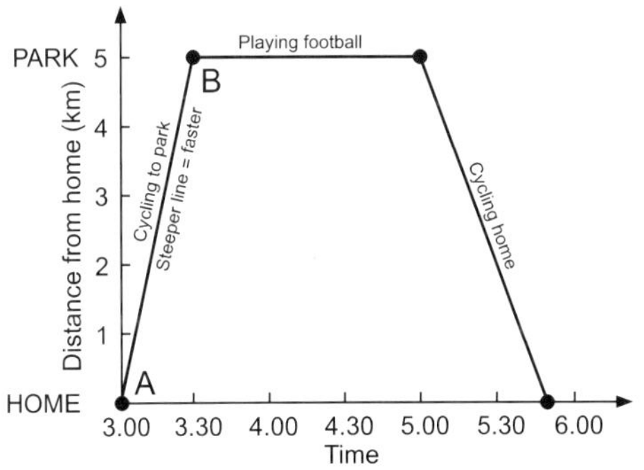

(b) The steeper the line, the greater the speed.
So, Dieter cycled faster from home to the park.

(c) Speed = Gradient of distance–time graph

Speed from home to park = Gradient of line AB

$= \dfrac{\text{Distance travelled}}{\text{Time taken}}$

$= \dfrac{5\text{ km}}{30\text{ min}} = \dfrac{5\text{ km}}{0.5\text{ h}} = 10\text{ km/h}$

5. Acceleration

Acceleration is the rate at which speed changes. For example, an acceleration of 5 m/s per second means that the speed increases by 5 m/s every second. An acceleration of 5 m/s per second is written briefly as 5 m/s^2.

Acceleration (a), speed (s) and time (t) are related by the formula:

$$\text{Acceleration} = \frac{\text{Change in speed}}{\text{Time taken}} \quad \text{or} \quad a = \frac{s}{t}$$

Question *A car increases its speed from 15 m/s to 20 m/s in 10 seconds. Calculate the car's acceleration.*

$$\text{Acceleration} = \frac{\text{Change in speed}}{\text{Time taken}} = \frac{20-15}{10}$$
$$= \frac{5}{10} = 0.5\text{ m/s}^2$$

6. Speed–Time Graph

Question *A tram travels between two stops. The speed–time graph for the journey is shown below.*
(a) What is the speed of the tram after 8 seconds?
(b) Calculate the acceleration of the tram during the first 10 seconds.
(c) Calculate the distance travelled by the tram when it is travelling at a constant speed.

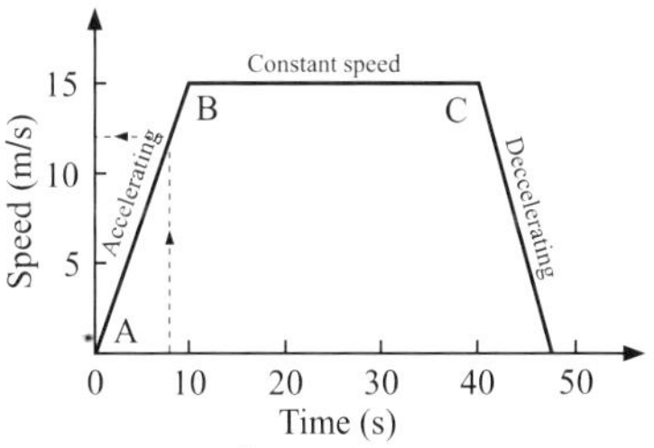

(a) Speed after 8 seconds is 12 m/s (see dashed line).

(b) Acceleration = Gradient of speed–time graph
The tram accelerates from 0 m/s to 15 m/s in 10 seconds. So:
Acceleration = Gradient of AB $= \frac{15-0}{10} = \frac{15}{10} = 1.5\text{ m/s}^2$

(c) Tram speed is 15 m/s for 30 seconds from B to C. 15 m/s means the tram travels 15 m in 1 second. So, in 30 seconds, tram travels 30×15 m = 450 m.

Tangents and Area Under a Curve H

1. Tangent to a Curve

When a straight line touches a curve at just one point, it is called a **tangent** to the curve.

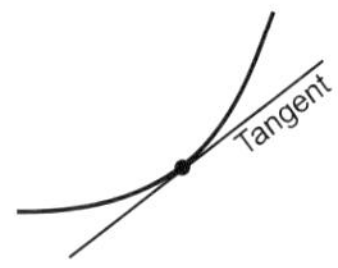

2. Drawing a Tangent to a Curve

To draw a tangent to a curve, first mark the point on the curve where you want the tangent. Gradually move your ruler towards the point, keeping an equal length of curve on either side of the point. When you can only see the point on the curve, draw the tangent.

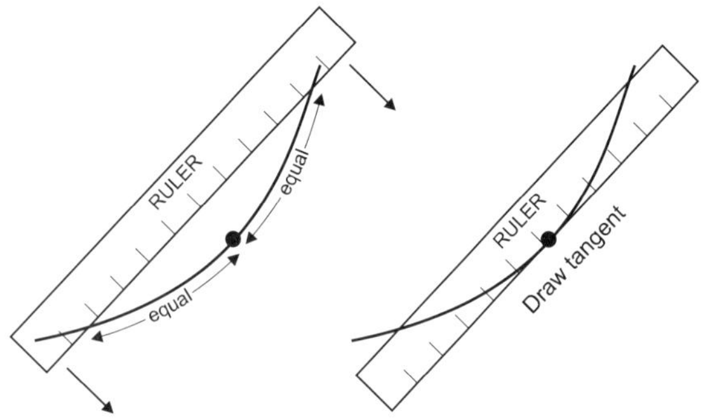

3. Gradient of a Curve

The gradient (steepness) of a curve varies from point to point. So we can only find the gradient at a particular point on a curve

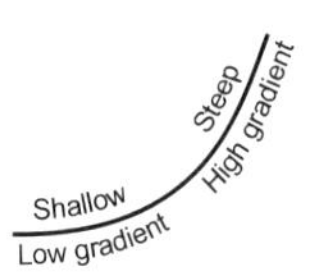

The gradient of a curve at a particular point is the gradient of the tangent to the curve at that point

Question *The distance–time graph shows the first 20 seconds of a car journey.*
(a) Draw the tangent to the curve where t = 15 seconds.
(b) Calculate the gradient of the curve at this point.
(c) What does this gradient represent?

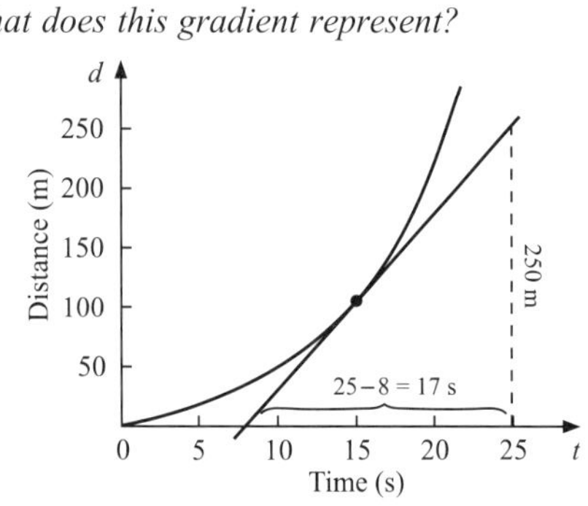

(a) Mark the point on the curve where $x = 15$ and draw the tangent to the curve at this point. Draw the tangent as long as possible.

(b) Calculate the gradient of the tangent. First, choose any two points on the tangent (the further apart, the more accurate your answer). Then:

$$\text{Gradient of curve} = \text{Gradient of tangent} = \frac{250\text{ m}}{17\text{ s}} = 14.7\text{ m/s}$$

(c) From part (b), the gradient represents the speed of the car after the first 15 seconds of its journey.

4. Estimating Areas on Graph Paper

Graph paper is divided into 1 cm squares. The area of a 1 cm square is written 1 cm^2. If you have to estimate the area of a shape drawn on graph paper, it is only necessary to estimate to the nearest quarter of a square.

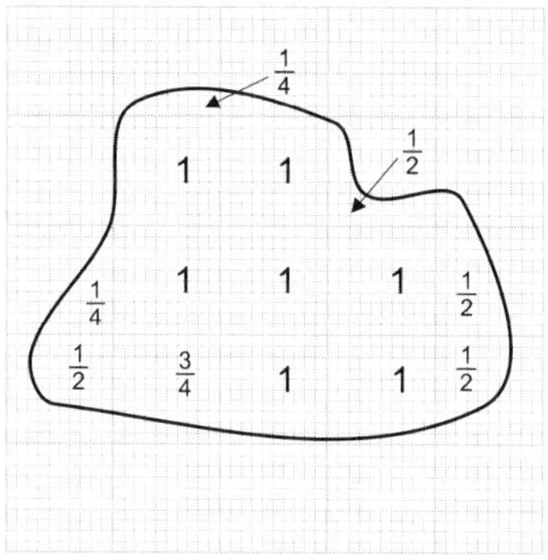

The area of the above shape is approximately:
$7 + \frac{1}{4} + \frac{1}{2} + \frac{1}{4} + \frac{3}{4} + \frac{1}{2} + \frac{1}{2} + \frac{1}{2}$
= 10.25 square centimetres = 10.25 cm^2

5. Estimating the Area Under a Curve

Question *Estimate the area under this curve.*

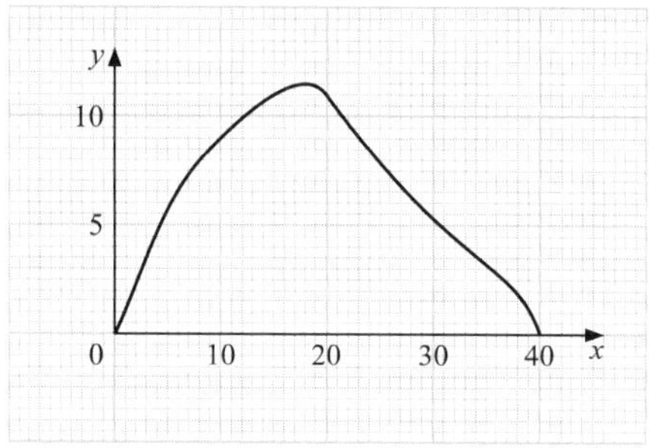

First, use the scales for the x and y axes to find what a 1 cm square *represents*.

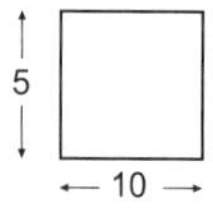

Area = Base × Height
= 10 × 5 = 50 square units

Now estimate the number of 1 cm squares under the curve (from left to right):

$\frac{3}{4} + \frac{1}{4} + 1 + 1 + \frac{1}{4} + \frac{1}{2} + 1 + \frac{1}{2} = 5.25$

So, area under curve $\approx 5.25 \times 50$ square units
≈ 260 square units

NOTE Sometimes we can calculate the *exact* area under a curve. The area under the curve below is the area of the trapezium ABCD.

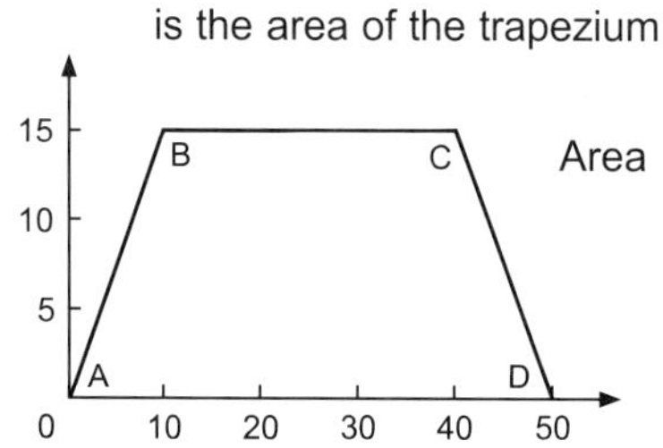

Area = $\frac{1}{2}(50 + 30) \times 15$
= 600 square units

6. Trapezium Rule

Another way of estimating the area under a curve is to use the **trapezium rule**.

Question *The diagram shows the cross-section of an excavated trench. The depth of the trench was measured at 1 metre intervals as shown.*

(a) Use the trapezium rule to estimate the area of the cross-section.

(b) Estimate the volume of earth removed in digging the trench, given that the trench is 20 m long.

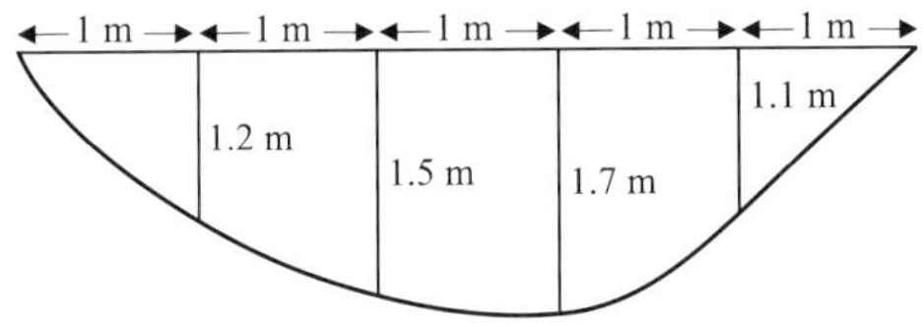

Trapezium Rule

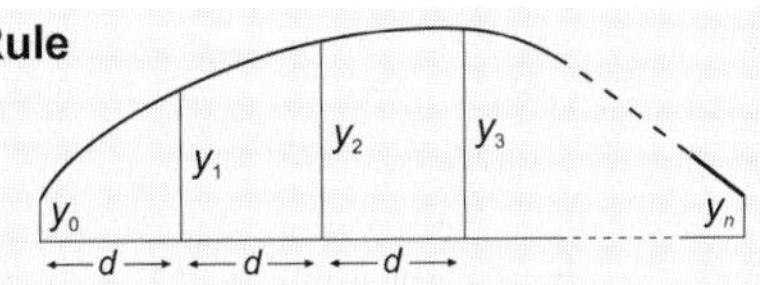

Area under curve $\approx \frac{d}{2}(y_0 + 2y_1 + 2y_2 + \ldots + y_n)$

(a) d is the interval between measurements. In this case, $d = 1$.

The measurements are labelled y_0, y_1, y_2, ... In this case, $y_0 = 0$, $y_1 = 1.2$, $y_2 = 1.5$, $y_3 = 1.7$, $y_4 = 1.1$ and $y_5 = 0$.

NOTE The left and right edges of the trench have a depth of 0, and so $y_0 = 0$ and $y_5 = 0$

Area $\approx \frac{1}{2}(0+2\times1.2+2\times1.5+2\times1.7+2\times1.1+0)$
$= 0.5\times11 = 5.5 \text{ m}^2$

(b) Volume of trench = Area of cross-section $\times$ Length
$= 5.5\times20 = 110 \text{ m}^3$

7. Area Under a Speed–Time Curve

The area under a speed-time curve is equal to the distance travelled.

Question *A motorcycle travels between two sets of traffic lights. The speed-time graph is shown below.*

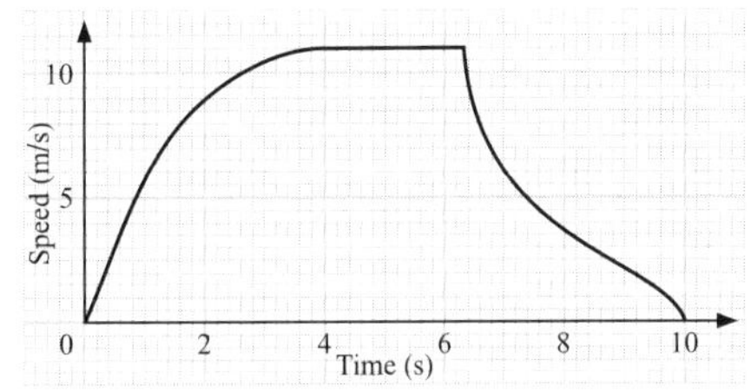

(a) Estimate the acceleration of the motorcycle after 2 seconds.

(b) Estimate the distance between the traffic lights.

(a) Acceleration = Gradient of curve after 2 s
So, draw a tangent to the curve at 2 s:

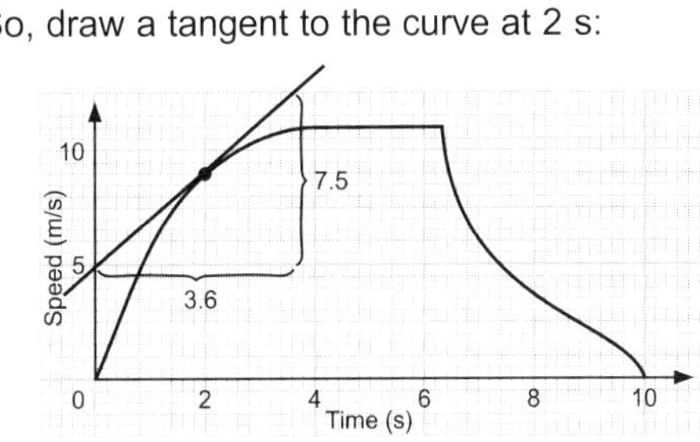

Acceleration = Gradient $\approx \frac{7.5}{3.6} = 2.1 \text{ m/s}^2$

(b) A 1 cm square represents a distance of $2\times5 = 10$ m

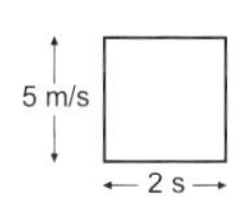

Estimate number of 1 cm squares under curve:

$5 + \frac{3}{4} + \frac{1}{4} + \frac{1}{4} + \frac{1}{2} + \frac{1}{2} = 7.25$

So, distance = $7.25\times10 = 72.5$ m

Graphical Solution of Equations

1. Solving Quadratic Equations Using Graphs

Question *(a) Complete the table below for the equation $y = x^2 - x - 2$.*

x	−3	−2	−1	0	1	2	3
y	10		0		−2	0	

(a) Use your graph to solve the equation $x^2 - x - 2 = 0$.

(b) Use your graph to solve the equation $x^2 - x - 5 = 0$.

(a) When $x = -2$ $y = (-2)^2-(-2)-2 = 4+2-2 = 4$
When $x = 0$ $y = 0^2-0-2 = 0-0-2 = -2$
When $x = 3$ $y = 3^2-3-2 = 9-3-2 = 4$

x	−3	−2	−1	0	1	2	3
y	10	4	0	−2	−2	0	4

Carefully plot these points and join them with a smooth curve.

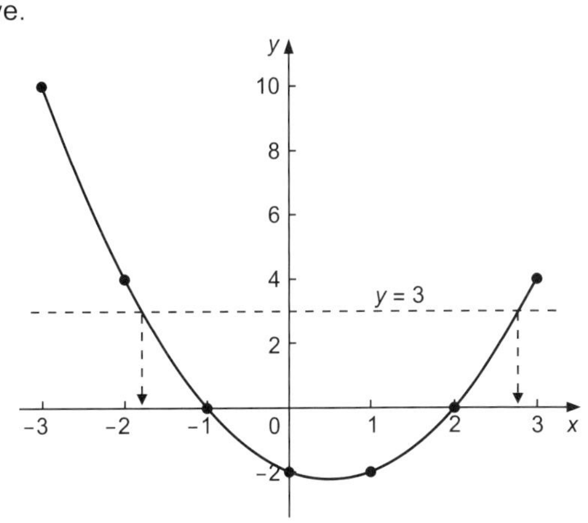

more

Graphical Solution of Equations (Contd)

1. Solving Quadratic Equations Using Graphs (Contd)

(b) The curve crosses the x-axis at two points, where $x = -1$ and $x = 2$. Every point on the x-axis has a y-coordinate of $y = 0$. So if we put $x = -1$ or $x = 2$ in the equation of the curve the answer should be 0. Check:
When $x = -1$, $y = (-1)^2-(-1)-2 = 1+1-2 = 0$ ✓
When $x = 2$, $y = (2)^2-(2)-2 = 4-2-2 = 0$ ✓
This means that $x = -1$ and $x = 2$ are the solutions of the equation $x^2-x-2 = 0$.

(c) Since we must use the graph of $y = x^2-x-2$ to solve the equation $x^2-x-5 = 0$, we need to change the LHS to x^2-x-2:

$x^2-x-5 = 0$
$\therefore\ x^2-x-2-3 = 0$ Split -5 into -2-3
$\therefore\ x^2-x-2 = 3$

So now we have to find the values of x which make $x^2-x-2 = 3$. On the graph, $y = x^2-x-2$ and so we are looking for the values of x where $y = 3$. Draw the line $y = 3$ on the graph and read off the two values of x where it crosses the curve (see dashed lines). From the graph, we can see that $x \approx -1.8$ and $x \approx 2.8$. These are the required solutions of the equation $x^2-x-5 = 0$.

2. Solving Simultaneous Equations

Question *Complete the tables of values below for the equations $y = x+3$ and $2y+x = 9$. Hence, solve the simultaneous equations $y = x+3$ and $2y+x = 9$.*

$y = x+3$

x	-2	0	3
y	1		

$2y+x = 9$

x	-2	0	3
y			3

For $y = x+3$:
When $x = 0$, $y = 0+3 = 3$
When $x = 3$, $y = 3+3 = 6$

x	-2	0	3
y	1	3	6

For $2y+x = 9$:
When $x = -2$ $2y-2 = 9$
$\therefore\ 2y = 11\ \therefore\ y = 5.5$
When $x = 3$ $2y+3 = 9$
$\therefore\ 2y = 6\ \therefore\ y = 3$

x	-2	0	3
y	5.5	4.5	3

Plot the points and join them with straight lines:

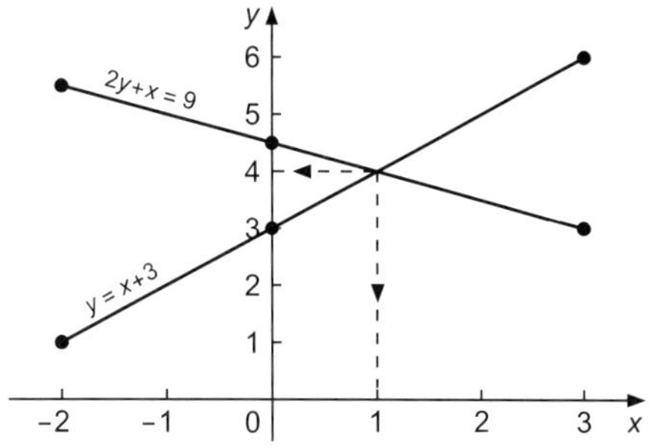

Read the x and y coordinates of the point of intersection of the lines: $x \approx 1$ and $y \approx 4$.

Check these values are the solution of the equations.

For $y = x+3$: $4 = 1+3$ ✓
For $2y+x = 9$: $2\times4+1 = 9$ ✓

3. More Difficult Simultaneous Equations

Question *The graph of $y = x^3-6x^2+9x+4$ has been drawn below.*

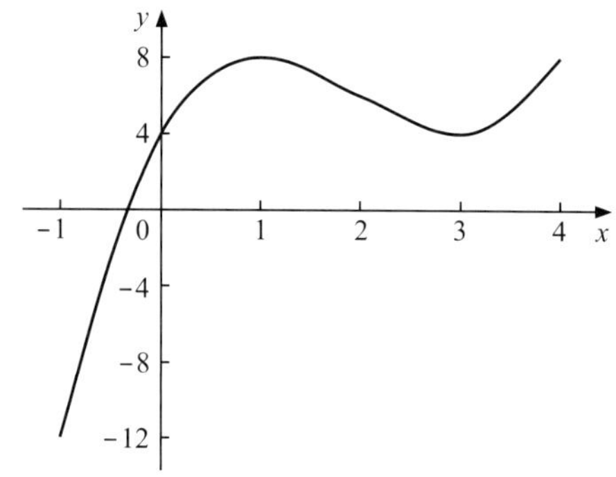

(a) Draw the graph of the straight line $y = 4x-4$ using the same axes.
(b) Find the x-coordinates of the points of intersection of the two graphs.
(c) Show that the answers to part (b) are the solutions of the equation $x^3-6x^2+5x+8 = 0$.

(a) Make a table of values for the equation $y = 4x-4$:
When $x = -1$ $y = 4\times(-1)-4 = -4-4 = -8$
When $x = 1$ $y = 4\times1-4 = 4-4 = 0$
When $x = 3$ $y = 4\times3-4 = 12-4 = 8$

x	-1	1	3
y	-8	0	8

Carefully plot the points using the same axes as for the graph above. Join the points with a straight line.

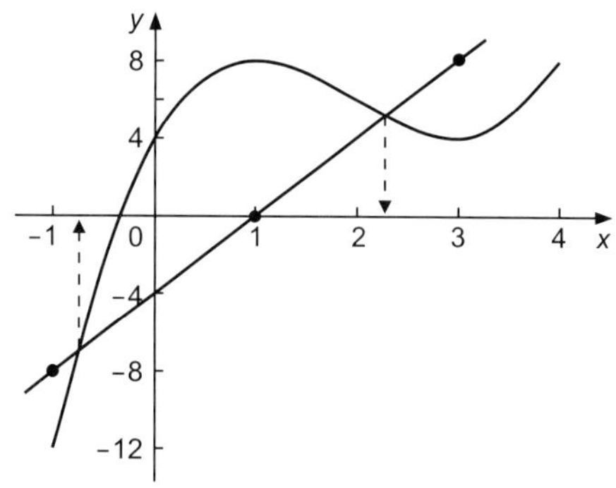

(b) Read off the x-coordinates of the points of intersection of the straight line and curve: $x \approx -0.7$ and $x \approx 2.3$.

(c) Since the points of intersection lie on *both* lines, either equation can be used to calculate their y-coordinates: $y = x^3-6x^2+9x+4$ or $y = 4x-4$. They both give the same y-coordinates for the x-coordinates found in part (b). So, for these particular x-coordinates:

$x^3-6x^2+9x+4 = 4x-4$
$\therefore\ x^3-6x^2+9x+4-4x+4 = 0$
$\therefore\ x^3-6x^2+9x-4x+4+4 = 0$
$\therefore\ x^3-6x^2+5x+8 = 0$

Thus, $x = -0.7$ and $x = 2.3$ are the approximate solutions of this equation.

Graphical Solution of Inequalities

1. Graphs of Simple Inequalities

Inequalities can be used to describe regions on graph paper. For example, the inequality $x < 4$ tells us that the x-coordinates of all points in the region are less than 4.

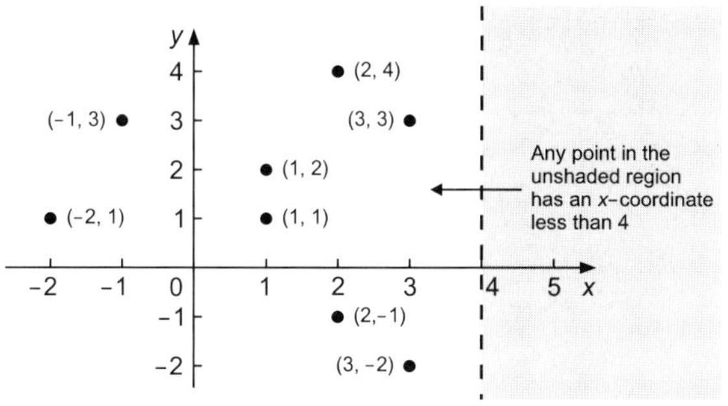

Points on the dashed line have x-coordinates equal to 4 and so are **not** included in the region $x < 4$. We say that the points in the unshaded region **satisfy** the inequality $x < 4$.

Similarly, $y \leqslant 3$ defines the region of points whose y-coordinates are less than or equal to 3. The points on the line $y = 3$ *are* included in the region.

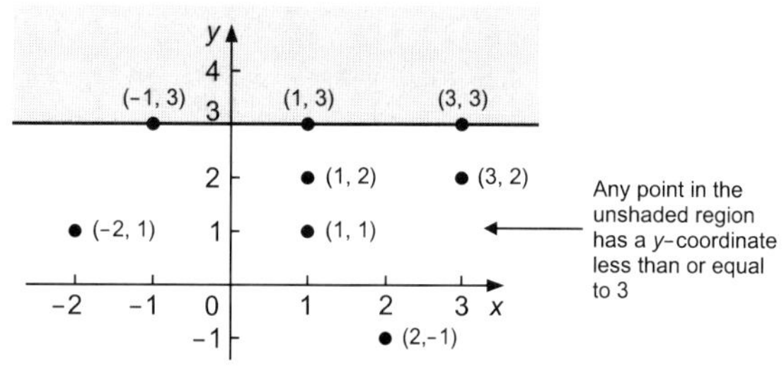

2. Straight Lines and Inequalities

A straight line divides the graph paper into two regions. For example, the straight line with equation $y = 2x+3$ divides the graph paper into regions A and B below:

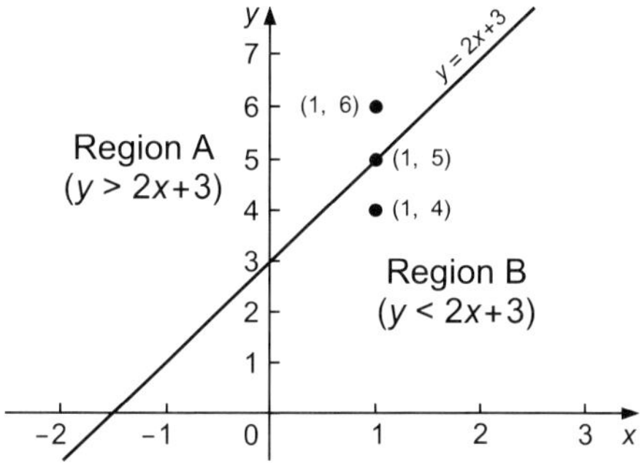

The coordinates of points lying *on the line* satisfy the equation $y = 2x+3$, e.g. for the point (1, 5) we have $x = 1$, $y = 2\times1+3 = 5$ ✓ See graph.

The coordinates of points lying in region A satisfy the inequality $y > 2x+3$, e.g. for the point (1, 6) we have $6 > 2\times1+3 = 5$ ✓ See graph.

The coordinates of points lying on the opposite side of the line in region B satisfy the inequality $y < 2x+3$, e.g. for the point (1, 4) we have $4 < 2\times1+3 = 5$ ✓ See graph.

Question *Mark the region R satisfying the inequality* $2x+y \geqslant 20$.

First, replace the $\geqslant$ with = and draw the straight line $2x+y = 20$. Make a table of values using three points.

When $x = 0$ $\quad 2\times(0)+y = 20 \quad \therefore 0+y = 20 \quad \therefore y = 20$
When $x = 6$ $\quad 2\times6+y = 20 \quad \therefore 12+y = 20 \quad \therefore y = 8$
When $x = 10$ $\quad 2\times10+y = 20 \quad \therefore 20+y = 20 \quad \therefore y = 0$

x	0	6	10
y	20	8	0

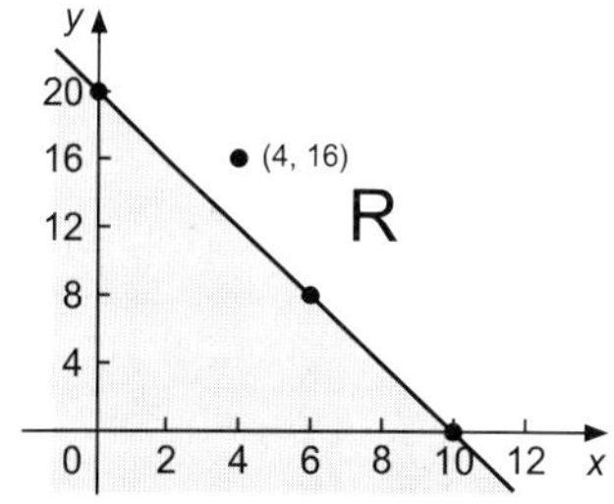

Choose any point that is ***not*** on the line, e.g. (4, 16) and check whether its coordinates satisfy the inequality:

$2x+y \geqslant 20$ becomes $2\times4+16 \geqslant 20$, which is true.

Label this side of the line R. Shade the unwanted side.

NOTE If we had chosen a point on the other side, say (4, 8), the inequality would not have been satisfied: $2x+y \geqslant 20$ becomes $2\times4+8 \geqslant 20$, which is not true. We would shade the other side.

3. Simultaneous Inequalities

Question *Use shading to indicate the region satisfying the inequalities*

$$y < 2 \qquad x+y > 4 \qquad x < 3$$

Each inequality describes a region of points. The points in the overlap of these regions satisfy all three inequalities at once (simultaneously).

Replace < and > with = to change the inequalities into equations and draw the lines $y = 2$, $x+y = 4$ and $x = 3$ on graph paper (use dashed lines).

To find the region described by the inequality $x+y > 4$, choose any point ***not*** on the line $x+y = 4$, e.g. (2, 1).

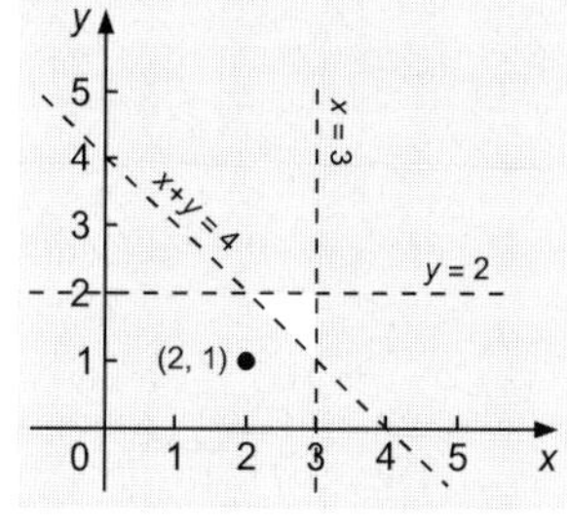

Check to see whether this point satisfies the inequality: $x+y > 4$ becomes $2+1 > 4$, which is not true. So, the other side of the line satisfies the inequality.

Shade the unwanted regions.

Graphical Solution of Inequalities

4. Solving Problems Using Inequalities

Question *At each performance of a school play the audience had to satisfy the following conditions:*

(i) no more than 250 children

(ii) a maximum size of audience of 300

(iii) at least twice as many children as adults

(a) If there were x children and y adults, write down three inequalities that x and y must satisfy.

(b) Use shading to indicate the region on graph paper that satisfies all three inequalities.

(c) Use your graph to find the maximum number of adults at a performance.

(d) If tickets cost £3 for children and £4 for adults, find the maximum possible income

(a) (i) Number of children (x) less than or equal to 250 gives the inequality $x \leqslant 250$.

(ii) Audience = Children + Adults = $x+y$
The audience is less than or equal to 300 and so $x+y \leqslant 300$.

(iii) The number of children (x) is greater than or equal to twice the number of adults ($2y$), which gives the inequality $x \geqslant 2y$.

(b) Draw the lines $x = 250$, $x+y = 300$ and $x = 2y$ and shade the unwanted regions. The points in the unshaded region satisfy all three inequalities.

(c) The number of adults is represented by the vertical axis and so the greatest possible number of adults corresponds to the highest point in the unshaded region, i.e. A (200, 100).

This point represents an audience of 200 children and 100 adults. So, the greatest possible number of adults is 100.

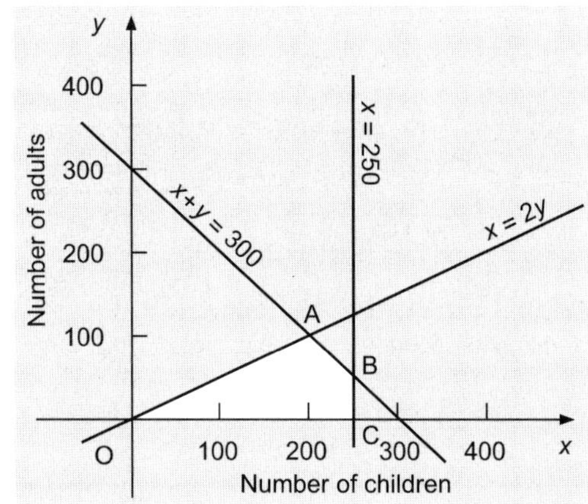

(d) The total income from x children's tickets at £3 and y adults' tickets at £4 is £$(3x+4y)$. We must find the point in the unshaded region which maximises this income. This will be one of the corner points.

For A (200, 100), $x = 200$ and $y = 100$, giving an income of £$(3\times200+4\times100)$ = £1000

For B (250, 50), $x = 250$ and $y = 50$, giving an income of £$(3\times250+4\times50)$ = £950

For C(250, 0), $x = 250$ and $y = 0$, giving an income of £$(3\times250+4\times0)$ = £750

So, the maximum possible income is £1000.

Graphs and Algebraic Laws

1. Algebraic Laws

An **algebraic law** is a formula connecting several quantities. For example, the law connecting voltage (V), resistance (R) and current (I) in an electrical circuit is given by the formula $V = IR$.

Simple algebraic laws are sometimes verified by drawing a graph of two quantities in the formula.

Question *A group of science students measured the time, t seconds, taken for a ball to fall to the ground from various heights, h metres, and recorded their results in the table below.*

Time, t (seconds)	0.65	0.85	1.00	1.10	1.20
Height, h (metres)	1	2	3	4	5
$\sqrt{h}$	1				

Their teacher told them that the relationship between the height, h, and time, t, is given by the formula $t = a\sqrt{h}+b$, *where a and b are constant numbers.*

(a) Complete the table by calculating the values of $\sqrt{h}$.

(b) Plot the graph of t against $\sqrt{h}$.

(c) Use your graph to estimate the values of a and b.

(a) Square root the values of h in the table.

Time, t (seconds)	0.65	0.85	1.0	1.1	1.2
Height, h (metres)	1	2	3	4	5
$\sqrt{h}$	1	1.4	1.7	2	2.2

(b) Carefully plot t against $\sqrt{h}$. This means that t is represented by the vertical axis.

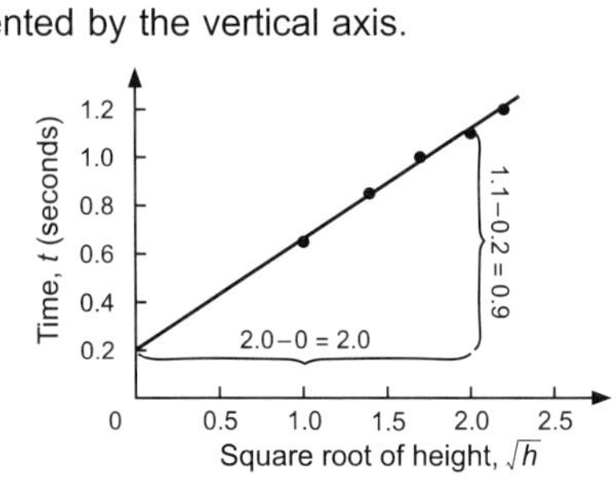

(c) Notice that you get a straight line. Compare the formula $t = a\sqrt{h}+b$ with the equation of a straight line:

$t = a\sqrt{h} + b$
$y = m\,x + c$

a corresponds to the gradient and b the y-intercept.

From the graph, a = gradient = $\frac{1.1-0.2}{2.0-0} = \frac{0.9}{2} = 0.45$

And b = y-intercept = 0.2

The formula connecting h and t is: $t = 0.45\sqrt{h}+0.2$

2. Exponential Growth **H**

A quantity **grows exponentially** if it is repeatedly multiplied by a number greater than 1.

Question *A single cell is placed in a dish. The number of cells in the dish doubles every 5 hours.*
(a) Complete the table below.
(b) Plot a graph of the number of cells, n, against time, t
(c) Write down an algebraic formula connecting n and t.

Time, t (hours)	0	5	10	15	20	25	30
Number of cells, n	1						

(a) After $t = 0$ hours, number of cells $= n = 1$
After $t = 5$ hours, number of cells $= n = 2\times1 = 2$
After $t = 10$ hours, number of cells $= n = 2\times2 = 4$
After $t = 15$ hours, number of cells $= n = 2\times4 = 8$
etc.

Time, t (hours)	0	5	10	15	20	25	30
Number of cells, n	1	2	4	8	16	32	64

(b) Carefully plot these points and join them with a smooth curve.

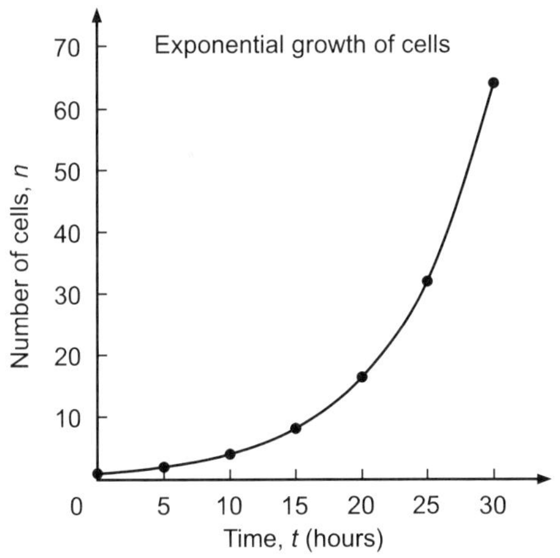

(c) The numbers of cells in the table are powers of 2.

Time (t hours)	0	5	10	15	20	25	30
Number of cells (n)	1	2	4	8	16	32	64
	2^0	2^1	2^2	2^2	2^3	2^4	2^5

Notice that the indices of these powers can be found by dividing the number of hours by 5. For example, $5\div5 = 1$, $10\div5 = 2$, $15\div5 = 3$.

So, the rule connecting the number of cells and time is:

$$\text{Number of cells} = 2^{\text{Time}\div5}$$

which can be written algebraically as:

$$n = 2^{\frac{t}{5}}$$

3. Exponential Decay **H**

A quantity **decays exponentially** if it is repeatedly multiplied by a number less than 1.

Question *The graph below shows how the value of a piece of machinery decreases over the years. The value, £V, of the machinery after t years is given by the formula $V = p(q^t)$, where p and q are constant numbers.*
(a) Use the graph to estimate the values of p and q.
(b) Estimate the value of the machinery when it is 10 years old.

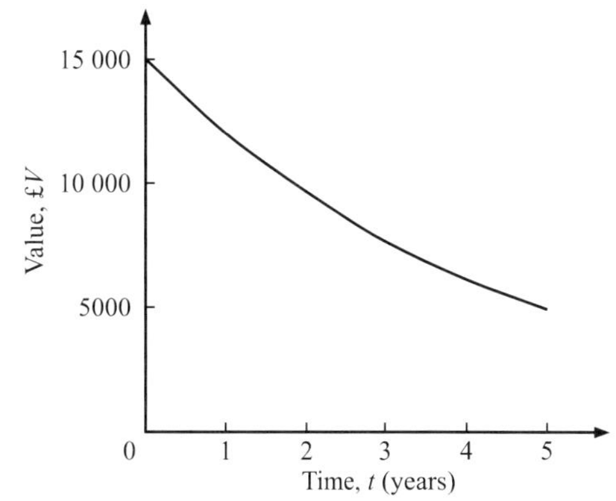

(a) We can use the graph to find pairs of values of V and t.

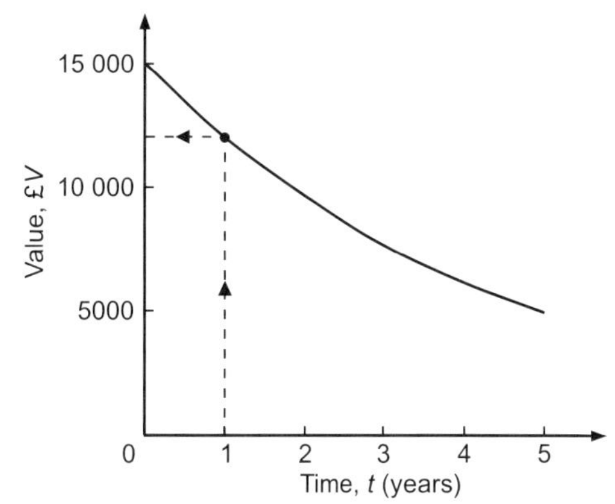

When $t = 0$, $V = 15\,000$
When $t = 1$, $V = 12\,000$

We can now substitute these pairs of values into the formula $V = p(q^t)$.
Substituting $t = 0$, $V = 15\,000$ gives:

$15\,000 = p(q^0)$
$\therefore 15\,000 = p\times1$ Because $q^0 = 1$
$\therefore p = 15\,000$

So we now know the formula is $V = 15\,000(q^t)$.

Substituting $t = 1$, $V = 12\,000$ gives:

$12\,000 = 15\,000(q^1)$
$\therefore 12\,000 = 15\,000\times q$
$\therefore 15\,000q = 12\,000$
$\therefore q = \frac{12\,000}{15\,000} = 0.8$

So the formula is $V = 15\,000(0.8^t)$.

(b) When $t = 10$, $V = 15\,000(0.8^{10}) \approx 1611$

1 5 0 0 0 × 0 . 8 x^y 1 0 = 1610.61274

The machinery will have a value of approximately £1611 when it is 10 years old.

Functions and Graphs

1. Functions

A **function** is a rule that changes one number into another. For example, the rule 'add 2' is a function. Functions are often represented by the letters f, g, F, and G. Let's represent the function 'add 2' by the letter f. Then, f is the rule that adds 2 to a given number:

f changes 3 into 3+2
f changes 4 into 4+2
f changes 5 into 5+2

Generally, f changes x into $x+2$, which can be written briefly as $f: x \to x+2$
or simply $f(x) = x+2$

Question *Given that* $f(x) = x^2 - 2x$, *calculate the values of (a)* $f(5)$ *and (b)* $f(-2)$.

(a) Replacing x by 5 in the function $f(x) = x^2-2x$ gives:

$f(5) = 5^2 - 2\times 5 = 25 - 10 = 15$

NOTE f(5) is called the **function value** when $x = 5$.

(b) $f(-2) = (-2)^2 - 2\times(-2) = 4+4 = 8$

The function $f(x) = x+2$ can also be illustrated using a **mapping diagram**:

$x \to x+2$
$1 \to 1+2$
$2 \to 2+2$
$3 \to 3+2$

2. Graph of a Function

You can use a function to calculate the y-coordinates of a graph.

Question *Plot the graph of* $y = f(x)$, *where* $f(x) = x^2+2x$ *for the values* $-3 \leqslant x \leqslant 3$.

$y = f(x)$ means that the y-coordinates are calculated using the function $f(x)$. In this case, $f(x)$ is the function x^2+2x. So this question is another way of asking you to plot the graph of the equation $y = x^2+2x$.

The inequality $-3 \leqslant x \leqslant 3$ tells you that the values of x range from $x = -3$ to $x = 3$.

Complete the table of values and carefully plot the points, joining them with a smooth curve.

$f(-3) = (-3)^2+2(-3) = 9-6 = 3$
$f(-2) = (-2)^2+2(-2) = 4-4 = 0$ etc.

x	−3	−2	−1	0	1	2	3
f(x)	3	0	−1	0	3	8	15

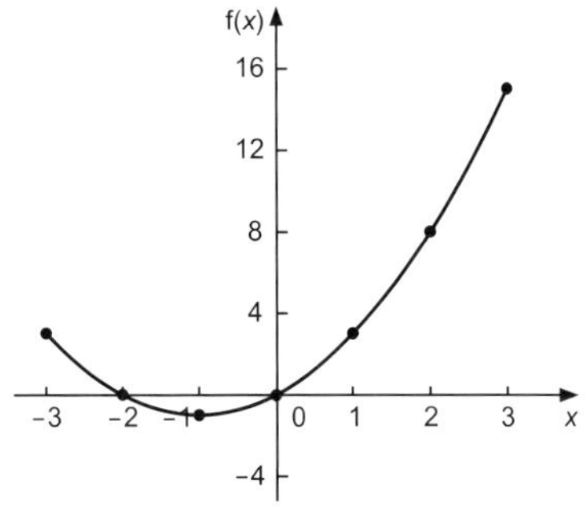

3. Related Functions

H

Consider the function $f(x) = \frac{1}{x}$.

If we replace x by $x+1$ in both sides of the equation, we get the **related function** $f(x+1) = \frac{1}{x+1}$.

Here are some more functions related to $f(x)$. Each of these functions is obtained by modifying both sides of the original function $f(x) = \frac{1}{x}$.

$f(x) = \frac{1}{x}$	Original function
$f(2x) = \frac{1}{2x}$	Replace x by $2x$
$f(x)+2 = \frac{1}{x}+2$	Add 2 to both sides
$f(-x) = \frac{1}{-x}$	Replace x by $-x$

4. Graphs of Related Functions

H

In Box 2 we drew the graph of the function $f(x) = x^2+2x$.

If we modify the function, the graph is also modified. Here are some of the ways the function $f(x) = x^2+2x$ and its graph can be modified. These changes can be applied to all functions.

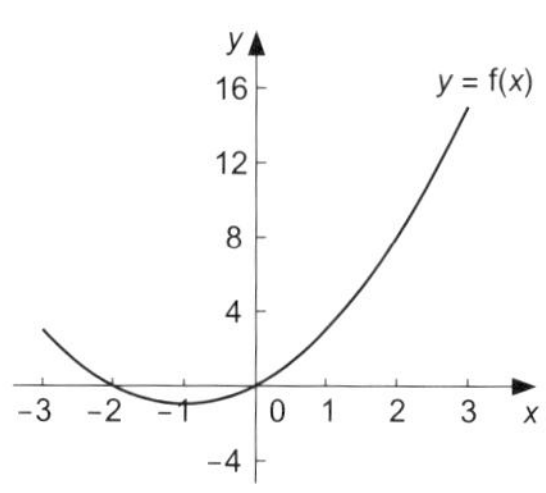

The original graph of $f(x)$ is shown as a dashed line.

Translation in the y-direction

Changing $f(x)$ to $f(x)+6$ moves the curve up 6. We say that the curve $y = f(x)$ has been translated +6 in the y-direction.

Subtracting 6 would move the curve 6 downwards.

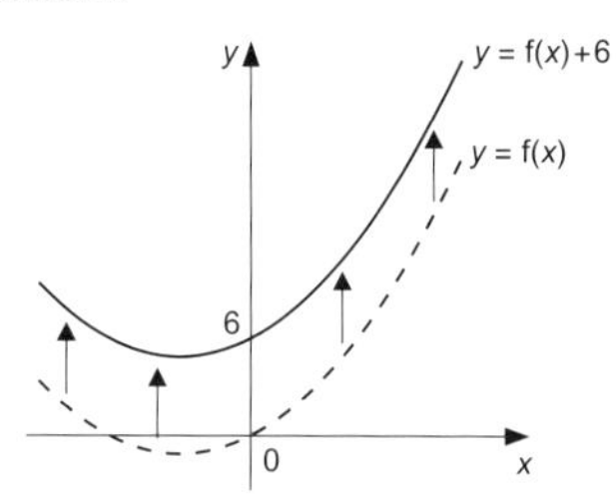

Translation in the x-direction

Changing $f(x)$ to $f(x-2)$ moves the curve 2 to the right in the x-direction. We say that the curve $y = f(x)$ has been translated +2 in the x-direction.

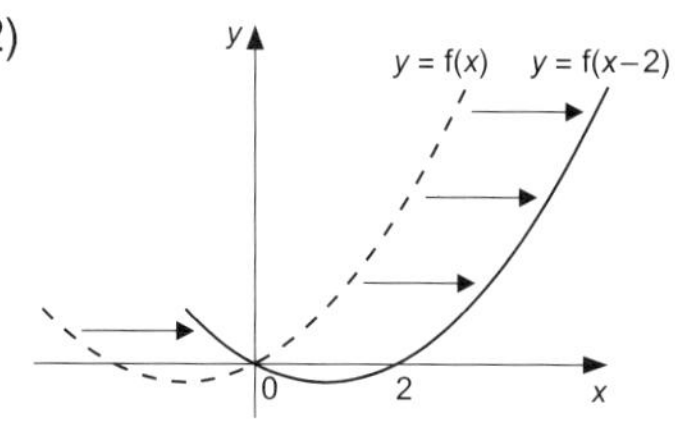

Changing $f(x)$ to $f(x+2)$ would move the curve 2 to the left. Don't get these mixed up

Reflection in the x-axis

Changing $f(x)$ to $-f(x)$ reflects the curve in the x-axis.

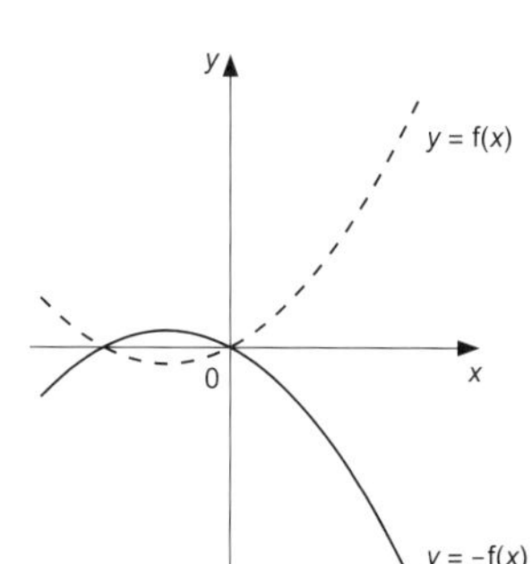

more

4. Graphs of Related Functions (Contd)

Reflection in the *y*–axis

Changing f(x) to f($-x$) reflects the curve in the y–axis.

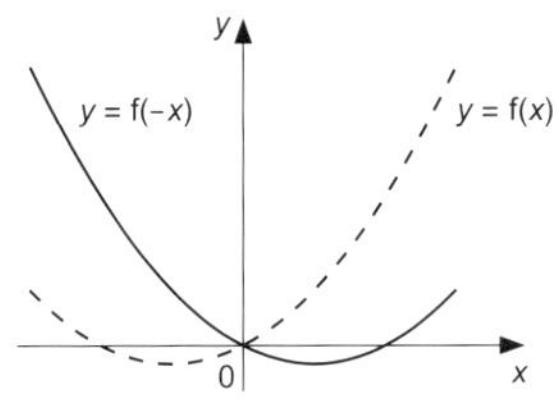

Stretching in the *y*–direction

Changing f(x) to 2f(x) stretches the curve by a factor of 2 in the y–direction: the distance from the x–axis of every point on the curve is doubled. This makes the curve steeper.

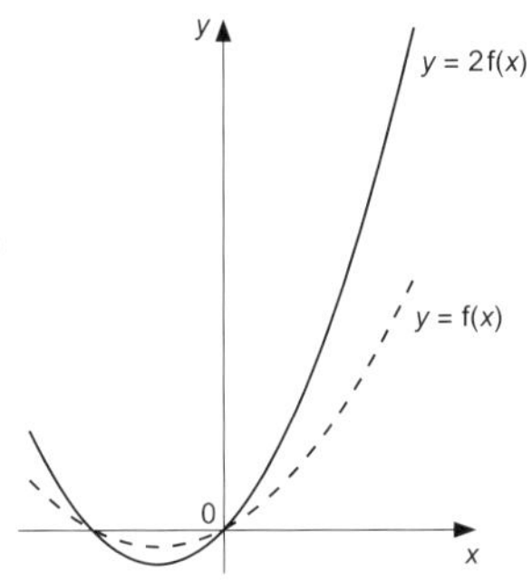

Stretching in the *x*–direction

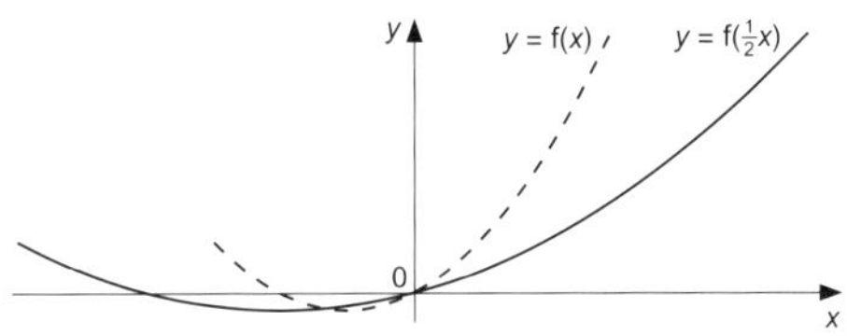

Changing f(x) to f($\frac{1}{2}x$) stretches the curve by a factor of 2 in the x–direction: the distance from the y–axis of every point on the curve is doubled. This makes the curve less steep.

Reciprocal of a function

Replacing f(x) by $\frac{1}{f(x)}$ changes large y–coordinates into small y–coordinates and vice–versa. If the original graph of f(x) crosses the x–axis, the graph of $\frac{1}{f(x)}$ breaks at this point. See example below for more details.

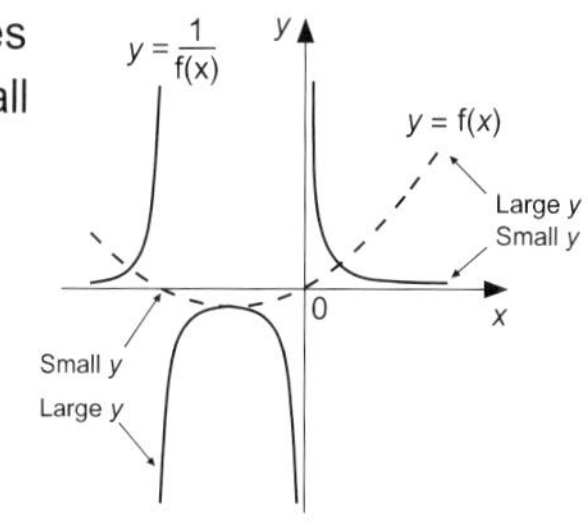

When you forget how to draw a related graph ....

Another way of sketching graphs of related functions is to calculate a few y–coordinates, though this is usually more time consuming.

Question *The graph of the function* f(x) = $\frac{1}{x}$ *is shown opposite. Sketch the graph of the function* f(x+0.5).

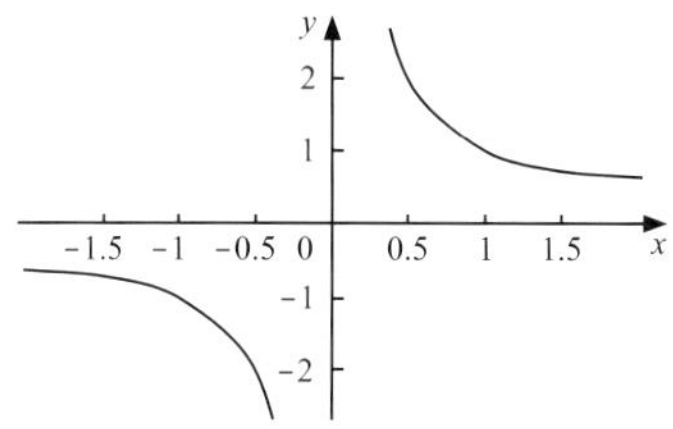

f(x+0.5) = $\frac{1}{x+0.5}$. Make a table of values.

When x = −1.5: f(x+0.5) = $\frac{1}{-1.5+0.5}$ = −1, etc.

x	-1.5	-1	-0.5	0	0.5	1	1.5
f(x+0.5)	-1	-2		2	1	0.7	0.5

f(x+0.5) cannot be calculated when x = −0.5. This is where the curve breaks. As x approaches −0.5, the curve gets closer to the dashed line (the **assymptote**), but never quite reaches it.

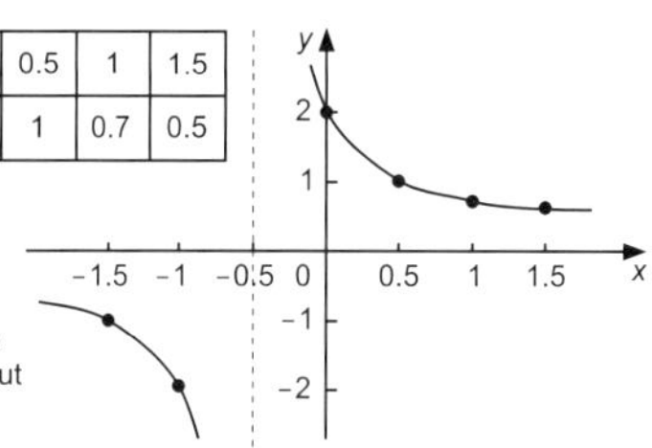

Three–dimensional Coordinates

The position of a point on a flat surface can be described using a pair of x and y coordinates, e.g. the point (3, 2). The x and y axes lie on the flat surface of the page.

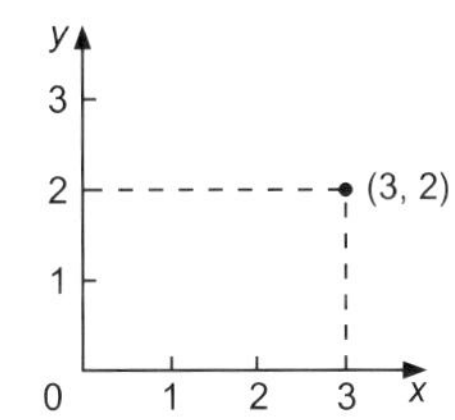

A point in 3–D space needs an extra z–coordinate to describe its position, e.g. the point (3, 2, 4). Think of the x and y axes as lying on the page with the z–axis coming towards you at right angles.

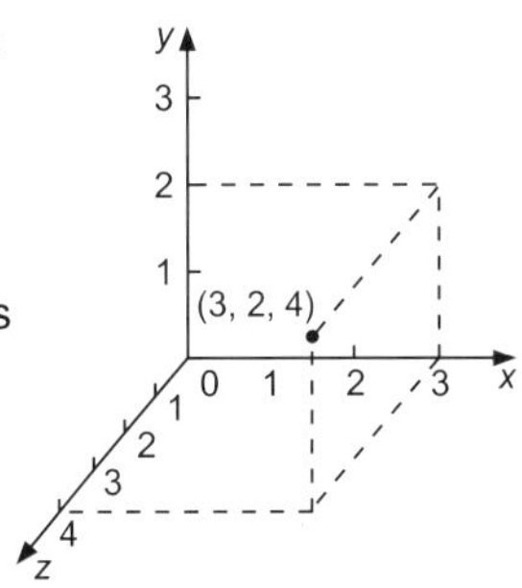

Question *The cuboid OABCDEFG is drawn using 3–D axes. Point P is the midpoint of edge DE. Find the coordinates of P.*

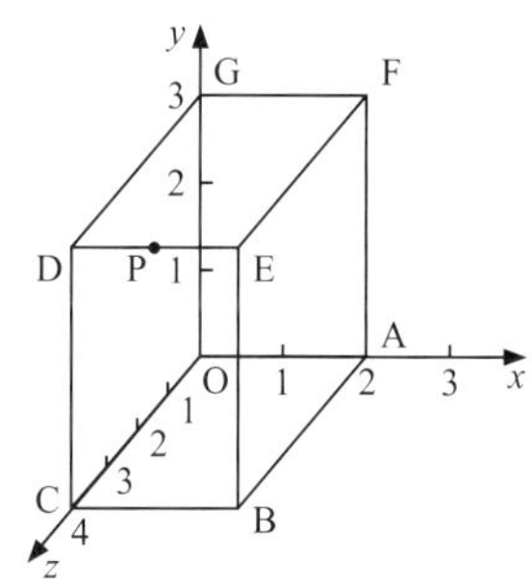

Draw lines parallel to the axes leading from 0 to P (see dashed lines). Starting from O, move +1 along the x–axis to M, then +3 parallel to the y–axis to N, then +4 parallel to the z–axis to point P. So, the coordinates of P are (1, 3, 4).

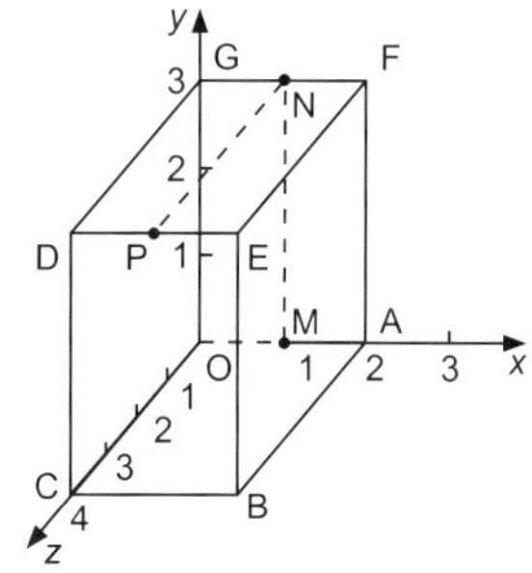

Perimeters

The **perimeter** of a shape is the total length of its outside edges.

Question *Calculate the perimeter of this rectangular door, giving your answer in metres.*

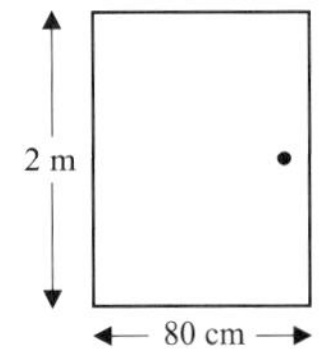

First, convert 80 cm to m:
80 cm = 80÷100 m = 0.8 m

Then perimeter of door
= 0.8+0.8+2+2 = 5.6 m
(opposite edges of a rectangle are equal)

Sometimes you have to calculate missing lengths of sides, as in the following question.

Question *Calculate the perimeter of this shape, giving your answer in centimetres.*

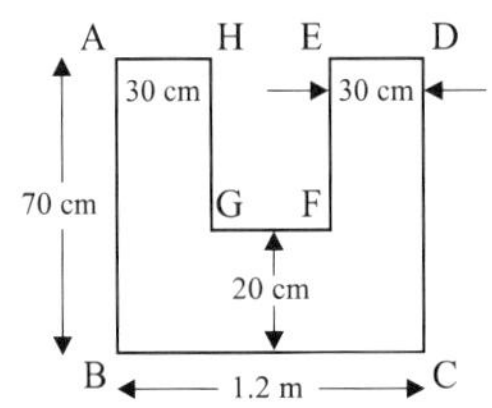

Convert 1.2 m to cm:
1.2 m = 1.2×100 cm = 120 cm

Then calculate missing sides:
EF = HG = 70−20 = 50 cm
GF = 120−30−30 = 60 cm

Perimeter = 70+120+70+30+50+60+50+30 = 480 cm

Question *The fence around the sheep pen is a tenth of the length of the fence around the field. Calculate the width of the sheep pen.*

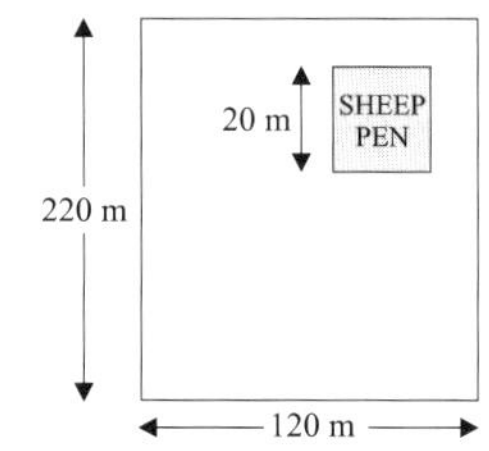

Perimeter of field = 120+220+120+220 = 680 m

So, perimeter of sheep pen = $\frac{1}{10}$ of 680 m = 68 m

The total length of the two given sides of the sheep pen is 20+20 = 40 m

The remaining two sides have a total length of 68−40 = 28 m

So, width of sheep pen = 28÷2 = 14 m

Area

1. Units of Area

The basic unit of area is the square metre (m^2). 1 m^2 is the area of a square with 1 m sides.

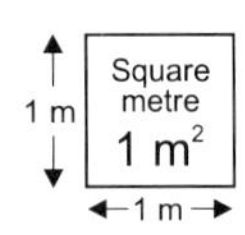

Similarly, 1 cm^2 is the area of a square with 1 cm sides.

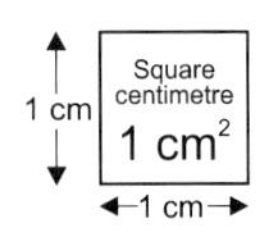

And 1 mm^2 is the area of a square with 1 mm sides.

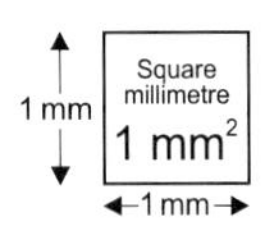

Question *Convert 3.2 m^2 to cm^2.*

For converting **length** we used the diagram below (see Box 6, p28):

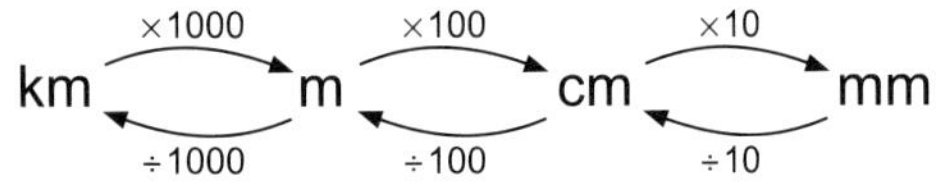

For converting **area**, square the numbers on the diagram:

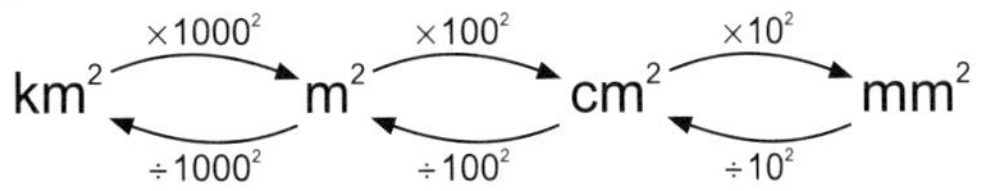

So, 3.2 m^2 = 3.2×100^2 cm^2 = 32 000 cm^2

Question *Convert 15mm^2 to cm^2.*

15 mm^2 = 15÷10^2 cm^2 = 0.15 cm^2

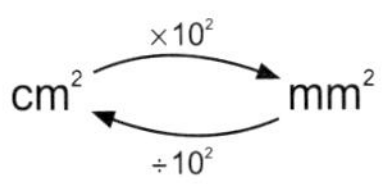

2. Area of a Rectangle

Area of a rectangle
= Base × Height

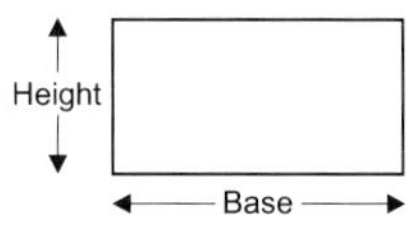

Question *Calculate the area of the photograph in cm^2.*

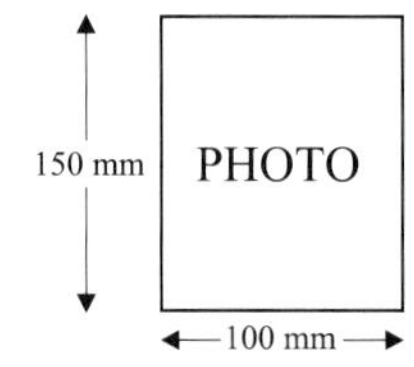

First, convert the lengths to cm. This is easier than converting the answer to cm^2 later on.

100 mm = 100÷10 cm = 10 cm
150 mm = 150÷10 cm = 15 cm

So, Area = Base × Height = 10 cm×15 cm = 150 cm^2

Question *Calculate the area of the shaded frame of the photograph.*

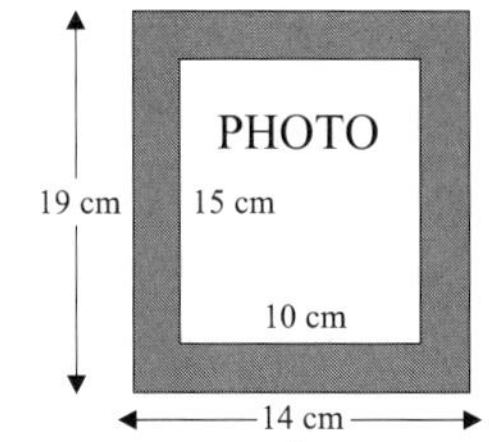

Area of frame = Area of outer rectangle − Area of inner rectangle
= 14×19 − 10×15
= 266−150
= 116 cm^2

3. Area of a Square

A **square** is a rectangle with equal sides.

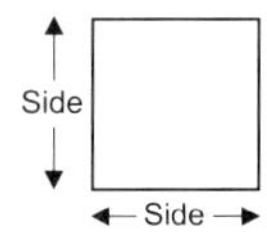

Area of a square = Side $\times$ Side
= $(\text{Side})^2$

Question *The square and rectangle have the same area. Calculate the side of the square.*

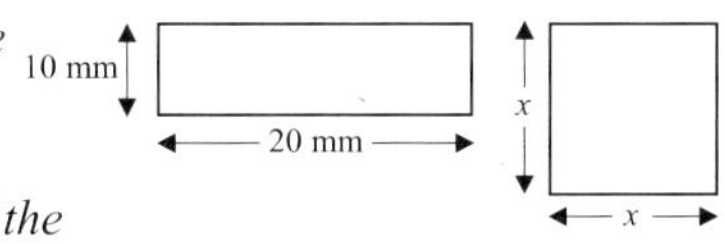

Let x be the length of the side of the square

Area of rectangle = Base$\times$Height = $20\times10 = 200\ \text{mm}^2$

Area of square = $(\text{Side})^2 = x^2$

These areas are equal, so: $x^2 = 200\ \text{mm}^2$

$\therefore\ x = \sqrt{200} = 14.1\ \text{mm}$

4. Tiling Problem

Question *Harry covered a wall using 15 cm square tiles, as shown in the diagram below. Some tiles needed cutting in half.*

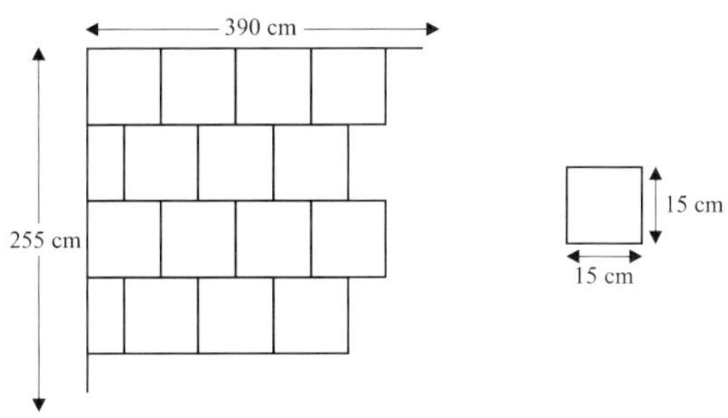

(a) How many tiles are there in the top row?
(b) How many rows are there?
(c) How many tiles did Harry use to cover the wall?
(d) How many tiles did he cut in half?
(e) If a box of tiles covers an area of 0.45 m², how many boxes did Harry need to buy?

(a) No. of tiles in top row = $390\div15 = 26$

(b) No. of rows = $255\div15 = 17$

(c) No. of tiles = No. of rows $\times$ No. of tiles per row
= $17\times26 = 442$

(d) Since 15 cm divides exactly into 390 cm, rows that begin with a whole tile must also end in one. Rows that begin with a half tile must also end in one; there are 8 such rows and so Harry needed to cut 8 tiles in half.

(e) Convert measurements of the wall to m:

390 cm = $390\div100 = 3.9$ m
255 cm = $255\div100 = 2.55$ m

Wall area = Base$\times$Height = $3.9\times2.55 = 9.945\ \text{m}^2$

No. of boxes needed = $9.945\div0.45 = 22.1 \approx 23$

5. Area of a Triangle

The area of a triangle is given by the formula:

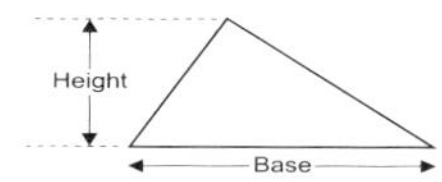

Area = $\frac{1}{2}$ Base $\times$ Height

The height used must be perpendicular to the base.

Question *Calculate the area of this triangle.*

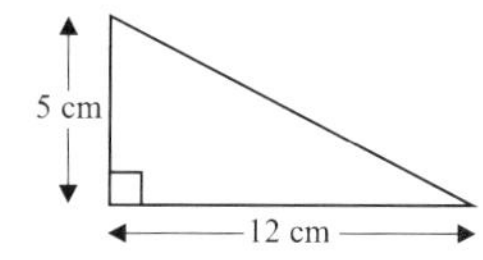

Since the 5 cm side is perpendicular to the 12 cm side we can take 12 cm as the base and 5cm as the perpendicular height

Area = $\frac{1}{2}$ Base $\times$ Height = $0.5\times12\ \text{cm}\times5\ \text{cm} = 30\ \text{cm}^2$

Question *The window shown in the diagram has square pane of side 30 cm and a triangular pane. The overall height of the window is 50 cm. Calculate the area, in cm², of the window. Convert your answer to m².*

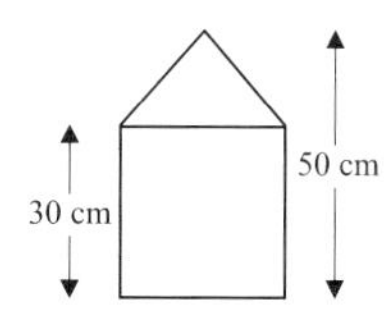

Height of triangle = 50 cm − 30 cm = 20 cm

Base of triangle = Base of square = 30 cm

Area of window = Area of square + Area of triangle
= $30\times30 + \frac{1}{2}\times30\times20$
= $900 + 0.5\times30\times20$
= $900+300$
= $1200\ \text{cm}^2$

$1200\ \text{cm}^2 = 1200\div100^2\ \text{m}^2$
$= 1200\div10\ 000\ \text{m}^2$
$= 0.12\ \text{m}^2$

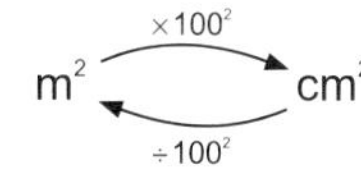

6. Area of a Trapezium

The area of a trapezium is given by the formula:

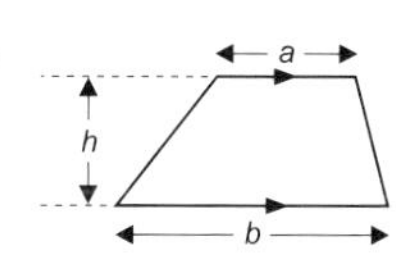

Area = $\frac{1}{2}(a+b)h$

In words, the area of a trapezium is half the sum of the parallel sides times the perpendicular distance between them.

Question *Calculate the area, in mm², of the scalpel blade shown in the diagram . Convert your answer to cm².*

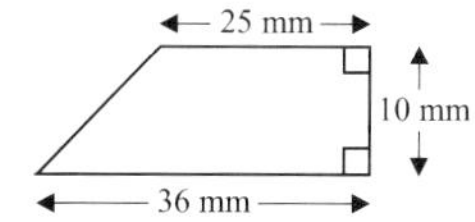

The 25 mm and 36 mm sides are parallel because they are both at right angles to the 10 mm side. The perpendicular distance between them is 10 mm.

Area = $\frac{1}{2}(a+b)h$
= $0.5(25+36)10 = 0.5\times61\times10 = 305\ \text{mm}^2$

$305\ \text{mm}^2 = 305\div10^2\ \text{cm}^2$
$= 305\div100\ \text{cm}^2$
$= 3.05\ \text{cm}^2$

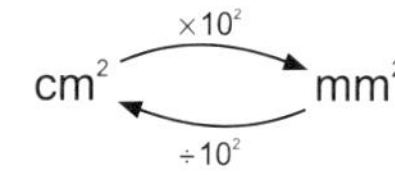

Area (Contd)

7. Area of a Parallelogram

The area of a parallelogram is given by the formula:

Area = Base × Height

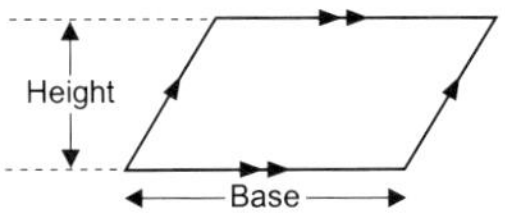

Question *Calculate the area of this tiled pattern.*

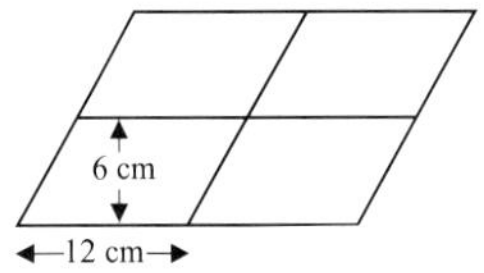

Area of 1 tile
= Base×Height
= 12 cm×6 cm = 72 cm^2

Area of 4 tiles = 4×72 = 288 cm^2

8. Area of a Circle

The area of a circle is given by the formula:

Area = $\pi\times(\text{Radius})^2$

or, using algebra:

Area = πr^2

Circumference = $2\pi r$

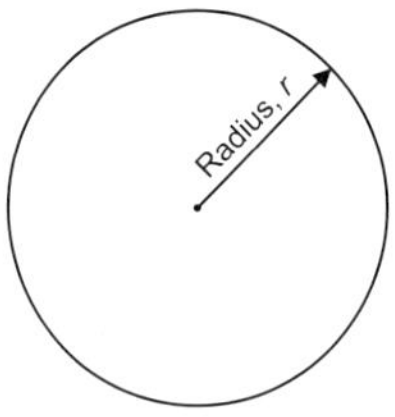

Question *Calculate the area of the top of this can of paté.*

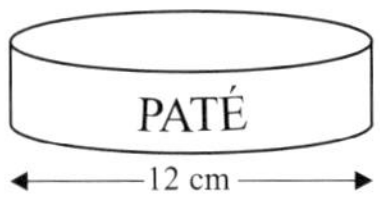

The radius of a circle is half its diameter.

Radius of circular can top = 12 cm÷2 = 6 cm

So, area of top = $\pi\times(\text{Radius})^2$ = 3.14×6^2 = 113.1 cm^2

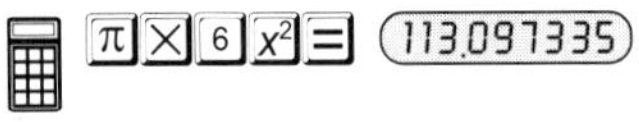

9. Sector of a Circle

H

A **sector** is a fraction of a circle, bounded by two radii and an arc.

The area of a sector and the length of arc are given by the formulas:

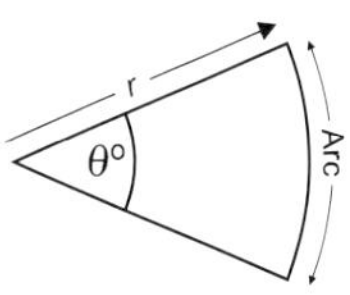

Area of sector = $\frac{\theta}{360}\times$ Area of circle = $\frac{\theta}{360}\times\pi r^2$

Length of arc = $\frac{\theta}{360}\times$ Circumference = $\frac{\theta}{360}\times2\pi r$

Question *The beam from a lighthouse sweeps through an angle of 40° and can be seen from a distance of 4 km. Calculate the area from which the lighthouse can be seen.*

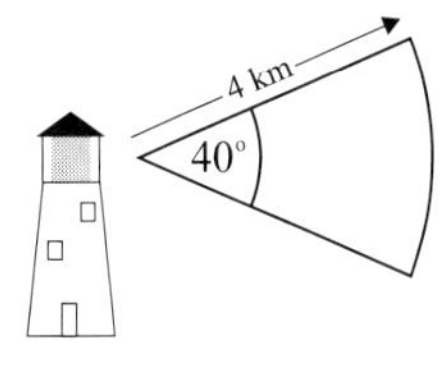

Area of sector = $\frac{\theta}{360}\times\pi r^2 = \frac{40}{360}\times\pi\times4^2$ = 5.59 km^2

10. Segment of a circle

H

The diagram shows a **segment** of a circle.

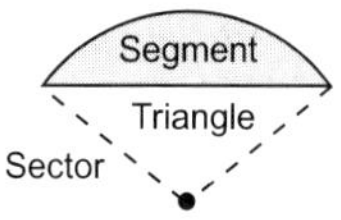

Area of segment = Area of sector − Area of triangle

Question *The diagram shows the end view of a barn. Calculate the area of this end.*

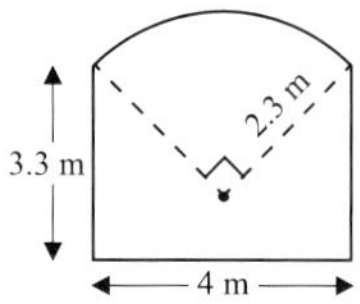

Draw a horizontal line to make a segment and a triangle.

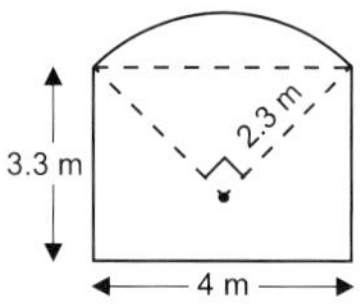

Area of barn = Area of rectangle + Area of segment

Area of rectangle = 3.3×4 = 13.2 m^2

Area of segment = Area of sector − Area of triangle

$= \frac{\theta}{360}\times\pi r^2 - \frac{1}{2}\times\text{Base}\times\text{Height}$

$= \frac{90}{360}\times\pi(2.3)^2 - \frac{1}{2}\times2.3\times2.3$

$= 90\div360\times\pi\times2.3^2 - 0.5\times2.3\times2.3$

= 4.155−2.645

= 1.51 m^2

So, area of barn end = 13.2+1.51 = 14.71 m^2

11. Concentric Circles

Concentric circles have the same centre.

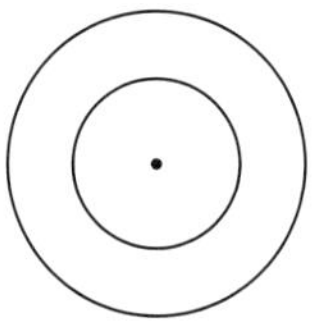

Question *Calculate the shaded area of this archery target. What percentage of the target is shaded?*

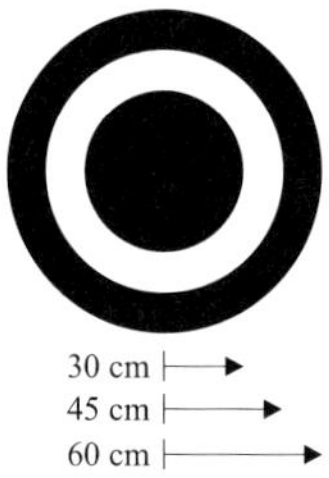

Shaded area = Inner circle + Outer ring

Area of inner circle = $\pi r^2 = \pi\times30^2$ = 2827.4cm^2

Area of ring = Area of 60 cm circle − Area of 45 cm circle

$= \pi\times60^2 - \pi\times45^2$

= 11 309.7−6361.7

= 4948.0 cm^2

So, shaded area = 2827.4+4948.0 = 7775.4 cm^2

Total area of target = Area of 60 cm circle

= 11309.7 cm^2 (from above)

So, % area shaded = $\frac{7775.4}{11309.7}\times100$ = 68.7%

Surface Area

1. Surface Area of a Cuboid

A **cuboid** is a block or box with six rectangular faces. Its surface area is the sum of the areas of its six rectangular faces.

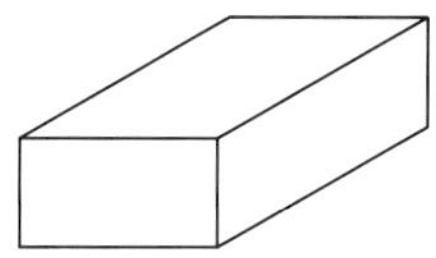

Question *(a) Calculate the surface area of this wooden block.*

(b) A can of red paint can cover an area of 2.2 m². How many blocks could be given a single coat of paint?

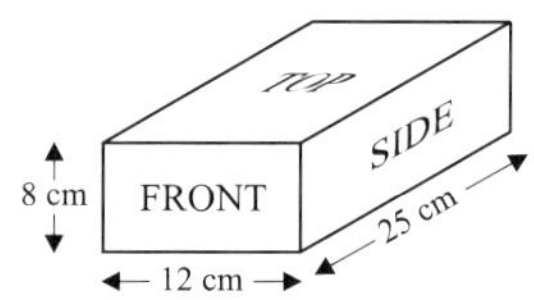

(a) Area of front face = $8\times12 = 96\text{ cm}^2$
Area of side face = $25\times8 = 200\text{ cm}^2$
Area of top face = $12\times25 = 300\text{ cm}^2$

Total surface area = $2\times96+2\times200+2\times300$
= 1192 cm^2 Opposite faces have equal area

(b) First, convert 2.2 m^2 to cm^2:
$2.2\text{ m}^2 = 2.2\times100^2\text{ cm}^2 = 22\,000\text{ cm}^2$
No. of bricks = $22\,000\div1192 = 18.46$
$\approx$ 18 complete blocks

2. Surface Area of a Sphere **H**

The surface area of a sphere is given by the formula.

Surface area = $4\pi r^2$

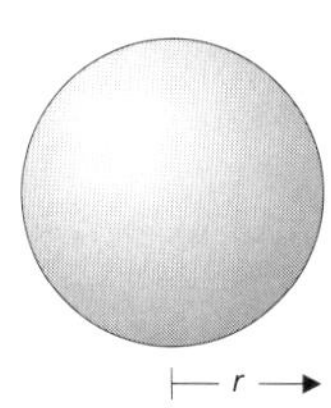

You will be given this formula in the question or on the formula sheet of the question paper.

A **hemisphere** is half a sphere, and so its surface area is $2\pi r^2$.

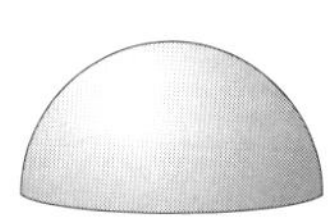

Question *A child made a party hat by gluing a plastic hemisphere onto a cardboard circular ring.*

(a) Calculate the surface area of the hemisphere.

(b) Calculate the surface area of the top of the hat.

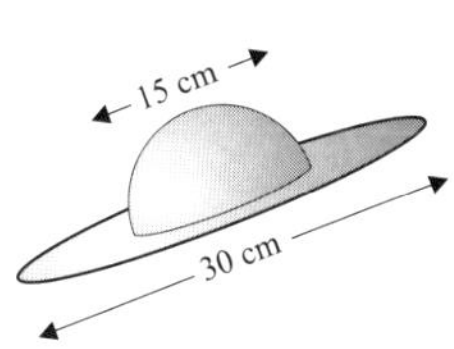

(a) Radius of hemisphere = 15 cm ÷ 2 = 7.5 cm
Surface area of hemisphere = $2\pi r^2 = 2\times\pi\times7.5^2$
= 353.4 cm^2

(b) Area of ring = Large circle − Small circle
= $\pi\times15^2 - \pi\times7.5^2$
= 530.1 cm^2

So, total surface area of hat
= $353.4+530.1 = 883.5\text{ cm}^2$

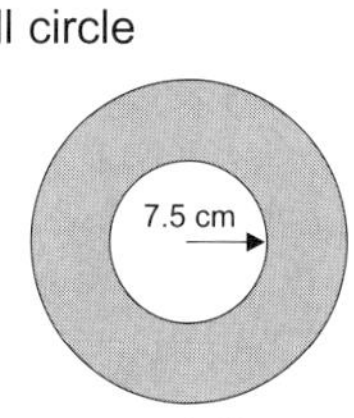

3. Surface Area of a Cylinder

A solid cylinder has two circular ends and a curved wall. The area of the curved wall is given by the formula:

Area of curved wall = $2\pi rh$

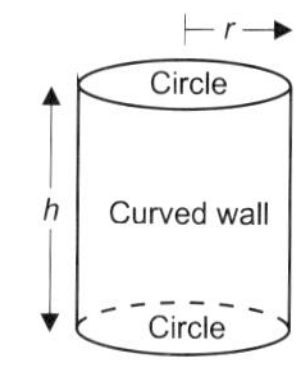

You will be given this formula in the question or on the formula sheet of the question paper.

Question *A rectangular sheet of paper is rolled up to form the curved wall of the cylinder shown in the diagram. You may assume that there are no overlapping edges.*

(a) Calculate the area of the wall of the cylinder.

(b) Calculate the length of the sheet of paper.

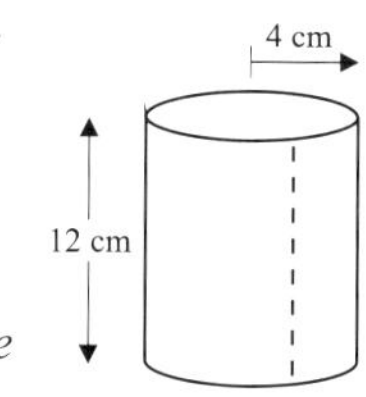

(a) Area of wall = $2\pi rh = 2\times\pi\times4\times12 = 301.6\text{ cm}^2$

(b) If the cylinder is unrolled, the top circle becomes the top edge of the sheet of paper.

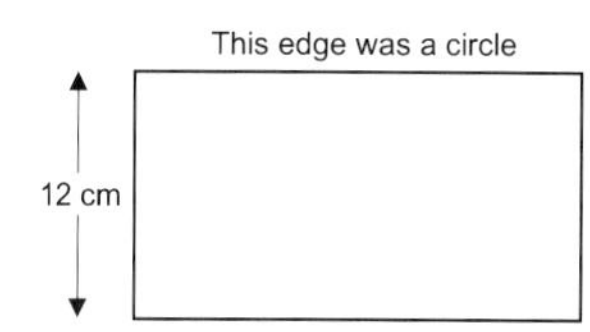

So, length of paper = Circumference of circle
= $2\pi r$
= $2\times\pi\times4 = 25.1$ cm

Question **H** *The diagram shows a wedge of cheese cut from a cylindrical block.*

(a) Calculate the area of the sector OAD.

(b) Calculate the total surface area of the wedge.

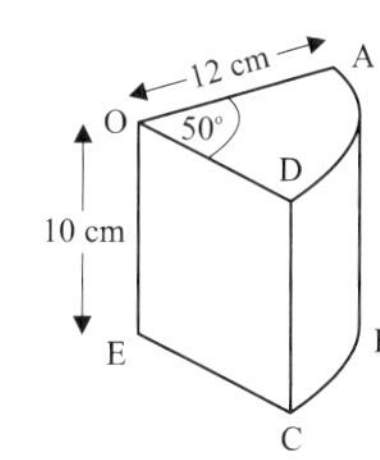

(a) Area of sector = $\frac{50}{360}$ of Area of circle (see Box 9, p72)
= $\frac{50}{360}\times\pi\times12^2$
= $50\div360\times\pi\times12^2$
= 62.83 cm^2

(b) Area of rectangular side ODCE = $10\times12 = 120\text{ cm}^2$
Area of curved side = $\frac{50}{360}$ of Wall of cylinder
= $\frac{50}{360}\times2\pi rh$
= $\frac{50}{360}\times2\times\pi\times12\times10$
= 104.7 cm^2

Total surface area
= 2 Rectangles + 2 Sectors + Curved side
= $2\times120+2\times62.83+104.7$
= 470.4 cm^2

Question *A pot of paint just covers 20 wooden spheres with a single coat. Wooden cylinders are manufactured whose radius and height are both equal to the radius of the spheres. How many cylinders can be coated with the pot of paint?*

Surface area of a sphere = $4\pi r^2$

Cylinders have radius r and height r. So:
Surface area of a cylinder = $2\pi r\times r + 2\pi r^2 = 4\pi r^2$

This is the same as the surface area of a sphere. So, the same number of cylinders can be painted, i.e. 20.

Surface Area (Contd)

4. Surface Area of a Cone

H

A cone has a circular base and a curved wall. The area of the curved wall is given by the formula:

Area of curved wall = πrl

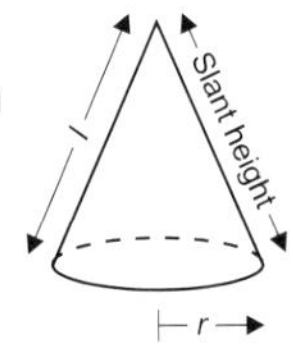

where l is the slant height of the cone (see diagram). You will be given this formula in the question or on the formula sheet of the question paper.

Question *A conical oil can has a base of radius 3 cm and a slant height of 12 cm. Calculate its total surface area.*

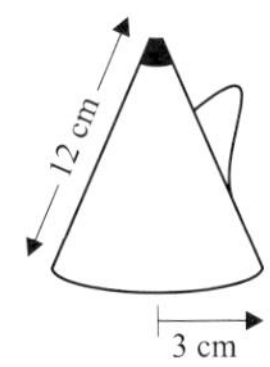

Area of circular base = $\pi r^2 = \pi \times 3^2 = 28.3$ cm^2

Area of wall = $\pi rl = \pi \times 3 \times 12 = 113.1$ cm^2

So, total surface area = 28.3+113.1 = 141.4 cm^2

5. Surface Area of a Pyramid

A square pyramid has a square base.

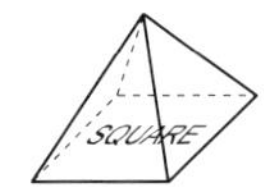

A triangular pyramid has a triangular base.

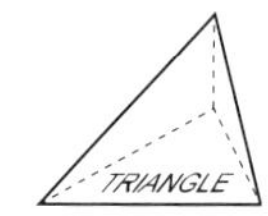

The surface area of a pyramid is the sum of the areas of its faces.

Question *A student cut out this cardboard shape and folded it up to make a square pyramid. Calculate the total surface area of the outside of the pyramid.*

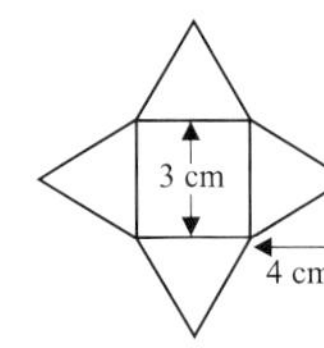

Area of square base = $3 \times 3 = 9$ cm^2

Area of triangular face = $\frac{1}{2}$ Base$\times$Height
= $0.5 \times 3 \times 4$
= 6 cm^2

Total surface area = Base + 4 Triangular faces
= $9 + 4 \times 6$
= 33 cm^2

Volume

1. Units of Volume

The basic unit of volume is the **cubic metre** (m^3).

1 m^3 is the volume of a cube with 1 m sides.

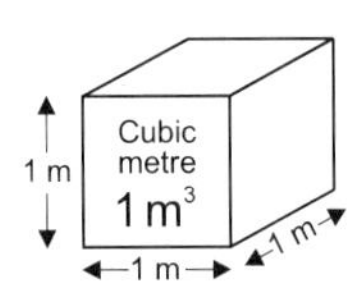

Similarly, 1 cm^3 is the volume of a cube with 1 cm sides.

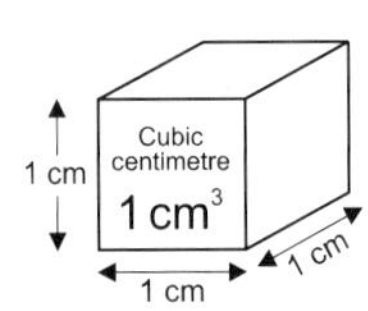

And 1 mm^3 is the volume of a cube with 1 mm sides.

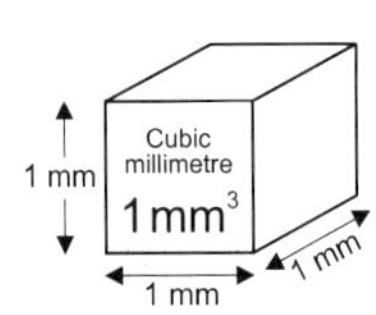

Question *Convert 3125 cm^3 to m^3.*

For converting **length** we used the diagram below (see Box 6, p28).

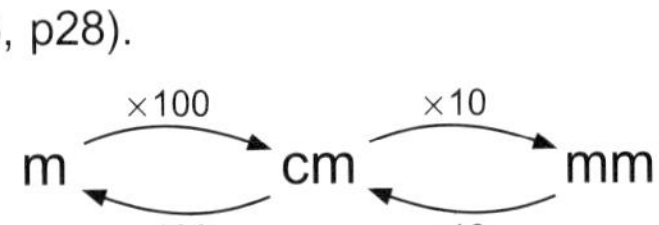

For converting **volume**, cube the numbers on the above diagram:

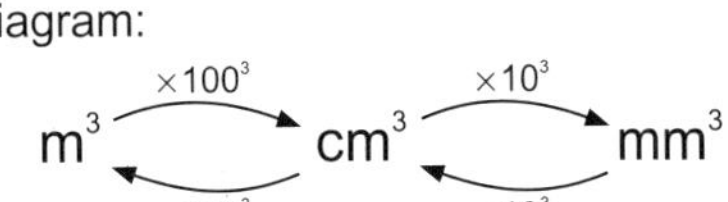

So, 3125 cm^3 = $3125 \div 100^3$ m^3 = $3125 \div 1\,000\,000$ m^3
= 0.003 125 m^3

2. Volume of a Cuboid

Volume of a cuboid
= Length$\times$Breadth$\times$Height
= lbh

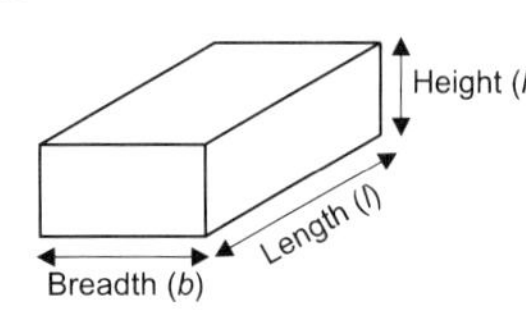

Question *The diagram below shows a concrete platform used for the presentation of medals.*
(a) Calculate the volume of concrete, in cm^3.
(b) What is the volume in m^3?

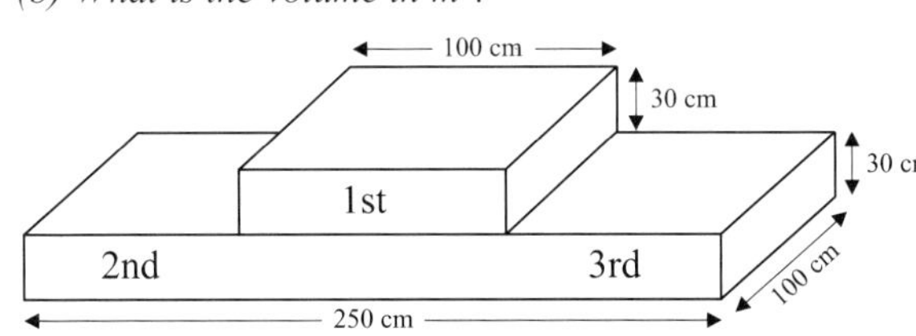

(a) The platform can be split into two cuboids.

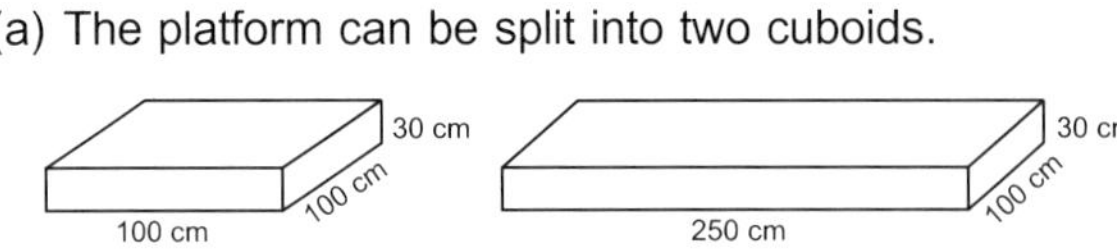

Volume
= lbh
= $100 \times 100 \times 30$
= 300 000 cm^3

Volume
= lbh
= $250 \times 100 \times 30$
= 750 000 cm^3

So, volume of platform = 300 000+ 750 000
= 1 050 000 cm^3

(b) 1 050 000 cm^3 = $1\,050\,000 \div 100^3$ m^3
= $1\,050\,000 \div 1\,000\,000$ m^3
= 1.05 m^3

Volume

3. Problems Involving Cuboids

Question *A cube of gold of side 2.5 cm is rolled into a thin rectangular sheet of width 10 cm and length 2 m.*

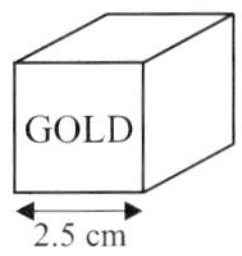

(a) Calculate the volume of gold, in mm³.

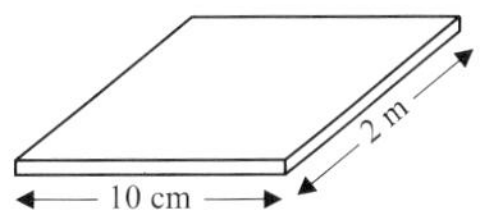

(b) Calculate the thickness of the rectangular sheet, correct to the nearest 0.01 mm.

(a) A **cube** is a cuboid whose sides are equal.

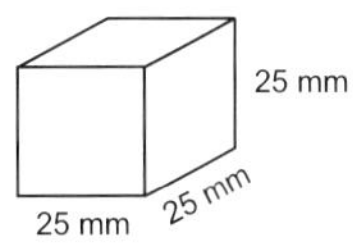

First, convert 2.5 cm to mm:
2.5 cm = 2.5×10 mm = 25 mm

Volume of cube = lbh = 25×25×25 = 15 625 mm^3

(b) The rectangular sheet is also a cuboid.

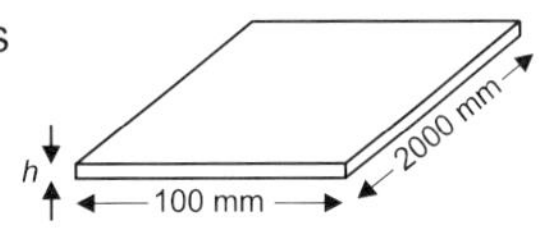

First, convert lengths to mm:
10 cm = 10×10 mm = 100 mm
2 m = 2×1000 mm = 2000 mm

Although we do not know the height, h, we can still substitute the length and breadth into the formula:

Volume of sheet = lbh = 100×2000×h
= 200 000h mm^3

Now, this is the same volume as the cube. So:

200 000h = 15 625

$\therefore\ h = \frac{15\,625}{200\,000} = 0.078\,125 \approx 0.08$ mm

4. Volume of a Prism

A **prism** is a solid whose cross-section is the same throughout its length.

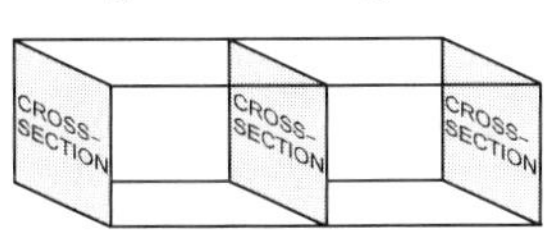

Square prism

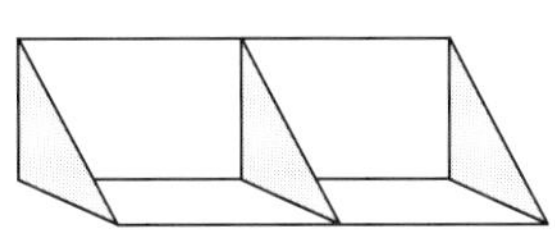

Triangular prism

Volume of a prism
= Area of cross-section × Length

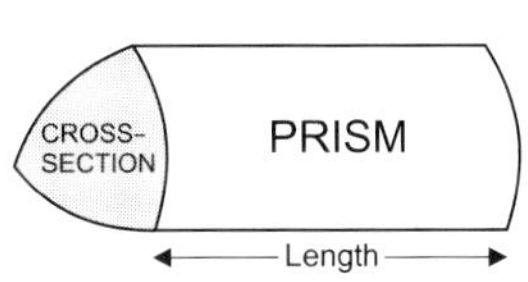

Question *The diagram shows a paint trough in the shape of a prism.*
(a) Calculate the area of the cross-section.
(b) Calculate the volume of the trough.

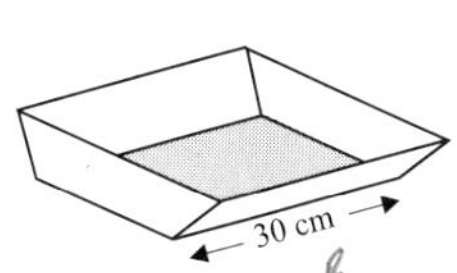

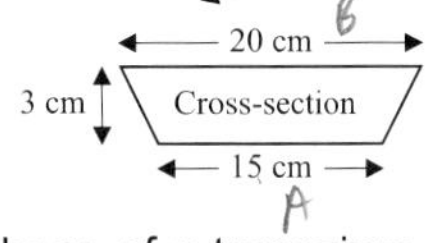

(a) The cross-section is in the shape of a trapezium.
Area of cross-section
= Area of trapezium
= $\frac{1}{2}(a+b)h = \frac{1}{2}\times(15+20)3 = 0.5\times35\times3 = 52.5$ cm^2

(b) Volume of trough = Area of cross-section×Length
= 52.5×30 = 1575 cm^3

5. Volume of a Cylinder

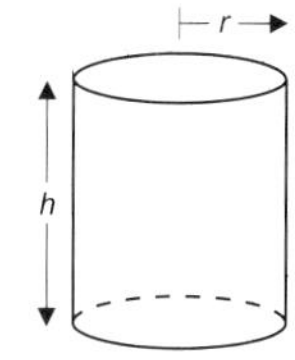

Volume of a cylinder
= Area of circular base × Height
= $\pi r^2 h$

Question *A company drinks machine accepts tokens made in the shape of a cylinder, as shown in the diagram.*

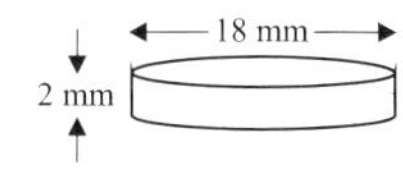

(a) Calculate the volume of a token, in mm³.
(b) A cylindrical hole is drilled in the middle of each, reducing its volume to 400 mm³. Calculate the radius of the hole.

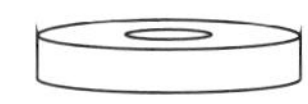

(a) Volume of cylinder = $\pi r^2 h = \pi\times9^2\times2 = 508.9$ mm^3

(b) If the new volume is to be 400 mm^3 then:

Volume of hole = 508.9−400 = 108.9 mm^3

The hole is a cylinder with height 2 mm and unknown radius, r. So, using the formula, we have:

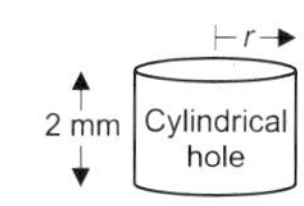

Volume of hole = $\pi r^2 h = \pi\times r^2\times2$
$= 2\pi r^2$

So, $2\pi r^2 = 108.9$

$\therefore\ r^2 = \frac{108.9}{2\pi} = 17.3$

$\therefore\ r = \sqrt{17.3} = 4.2$ mm (1 dp)

Question *Calculate the volume of this loaf of bread to the nearest 100 cm³. The cross-section has a semicircular top.*

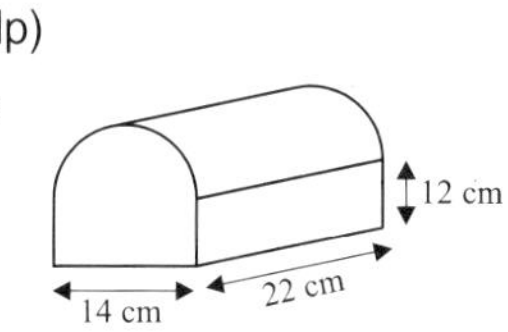

Prism method

Radius of circle = 7 cm

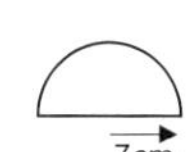

Cross-section
= Rectangle + Semi-circle
= $bh + \frac{1}{2}$of πr^2
= $14\times12 + 0.5\times\pi\times7^2$
= 245.0 cm^3

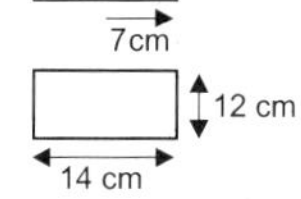

Volume of loaf = Area of cross-section × Length
= 245.0×22
≈ 5400 cm^3 (to nearest 100 cm^3)

Cylinder method

Radius of circle = 7 cm

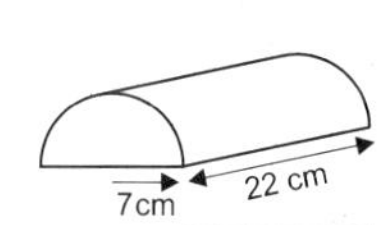

Volume of loaf
= Volume of cuboid + Volume of half cylinder
= $lbh + \frac{1}{2}$ of $\pi r^2 h$
= $22\times14\times12 + 0.5\times\pi\times7^2\times22$
= 5389.3 cm^3
≈ 5400 cm^3 (to nearest 100 cm^3)

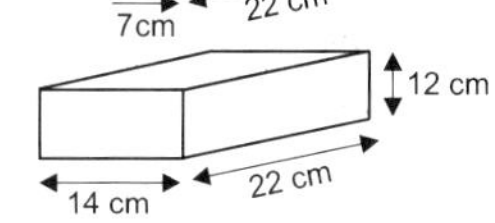

Volume (Contd)

6. Volume of a Sphere

H

Volume of a sphere = $\frac{4}{3}\pi r^3$

You will be given this formula in the question or on the formula sheet of the question paper.

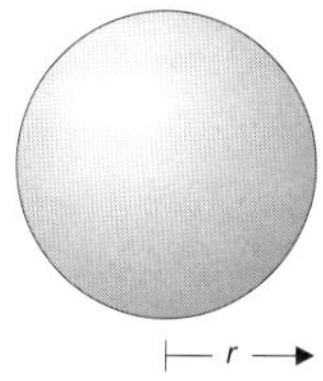

A **hemisphere** is half a sphere, and so its volume is $\frac{2}{3}\pi r^3$.

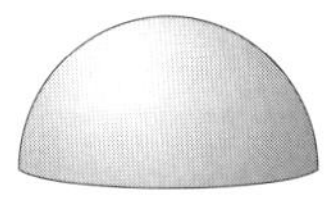

Question *A spherical glass ball of radius 5 cm is placed in a tank of water and is completely covered by the water.*

(a) Calculate the volume of the glass ball.

(b) Calculate the rise in water level.

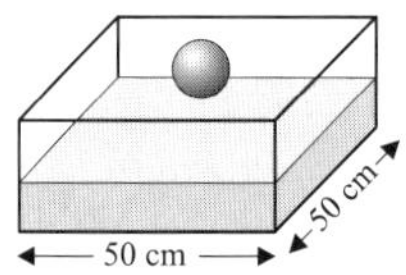

(a) Volume of sphere = $\frac{4}{3}\pi r^3 = \frac{4}{3}\times\pi\times5^3$

$= 4\div3\times\pi\times5^3$

$= 523.6\ \text{cm}^3$

Calculator: 4 ÷ 3 × π × 5 x^y 3 = → 523.598776

(b) When the ball is immersed, the water level rises. The water that rises is the dark shaded cuboid in the diagram and has the same volume as the sphere.

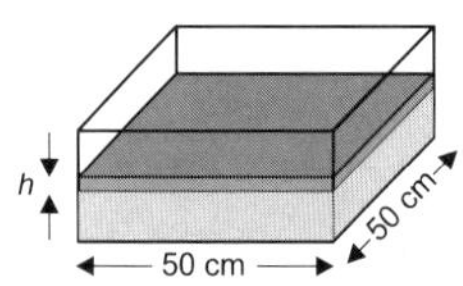

Volume of cuboid = $lbh = 50\times50\times h = 2500h\ \text{cm}^3$

Now, volume of cuboid = Volume of sphere

So, $2500h = 523.6$

$\therefore\ h = \frac{523.6}{2500} = 0.21$ cm

The water level rises 0.21 cm.

7. Volume of a Cone

H

Volume of cone = $\frac{1}{3}\pi r^2 h$

where h is the vertical height of the cone (see diagram). You will be given this formula in the question or on the formula sheet of the question paper.

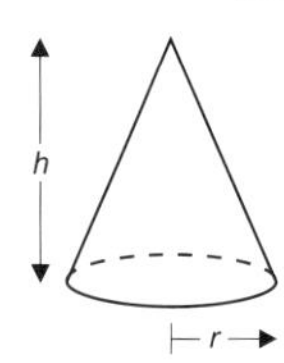

Question *The diagram shows a conical paper drinking cup.*

(a) Calculate the maximum volume of water the cup can hold.

(b) How many cups can be filled from a tank containing 25 litres?

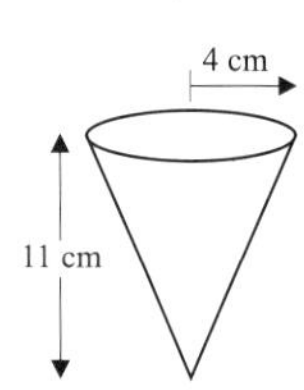

(a) Volume of cup = $\frac{1}{3}\pi r^2 h = \frac{1}{3}\times\pi\times4^2\times11 = 184.3\ \text{cm}^3$

Calculator: 1 ÷ 3 × π × 4 x^2 × 1 1 = → 184.306769

(a) 1 litre = $1000\ \text{cm}^3$

25 litres = $25\times1000\ \text{cm}^3 = 25\,000\ \text{cm}^3$

Number of cups = $\frac{25\,000}{184.3} = 135.6$

So, 135 cups can be filled from the tank.

8. Volume of a Pyramid

H

Volume of a pyramid

= $\frac{1}{3}\times$ Area of base $\times$ Height

NOTE The height must be perpendicular to the base

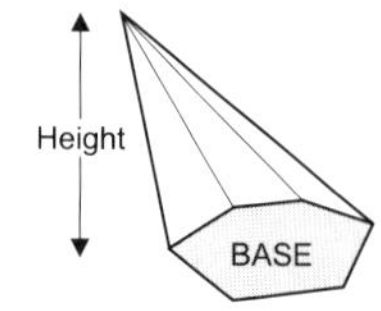

Question *A wax candle is made in the shape of a pyramid with a square base. The candle burns down a distance of 6 cm.*

(a) Calculate the volume of wax that has melted.

(b) Calculate the volume of wax remaining.

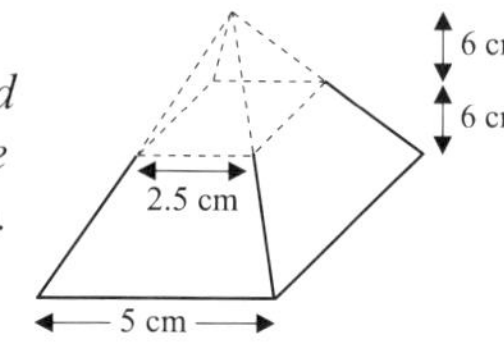

(a) Volume of melted wax

= Volume of small pyramid

$= \frac{1}{3}\times\text{Base}\times\text{Height}$

$= \frac{1}{3}\times2.5\times2.5\times6$

$= 12.5\ \text{cm}^3$

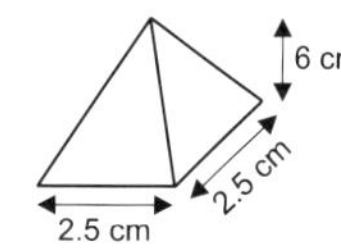

(b) Volume of large pyramid

$= \frac{1}{3}\times\text{Base}\times\text{Height}$

$= \frac{1}{3}\times5\times5\times12$

$= 100\ \text{cm}^3$

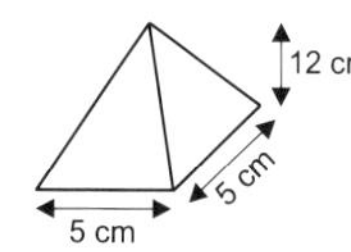

Wax remaining = $100-12.5 = 87.5\ \text{cm}^3$

9. Density

The **density** of a substance tells you its mass per unit volume. For example, cork has a density of $0.25\ \text{g/cm}^3$, which means that $1\ \text{cm}^3$ of cork has a mass of 0.25 g.

Question *An ice-making machine can produce cylindrical blocks of ice with radius 7 cm and of any length.*

(a) Calculate the volume of a block of ice 20 cm long.

(b) Given that the density of ice is $0.9\ \text{g/cm}^3$, calculate the weight this block.

(c) Calculate the length of a block of ice weighing 4 kg.

(a) Volume of block = $\pi r^2 h = \pi\times7^2\times20 = 3078.8\ \text{cm}^3$

(b) $1\ \text{cm}^3$ of ice weighs 0.9 g

So, $3078.8\ \text{cm}^3$ weighs 3078.8×0.9 g = 2771 g

(c) Weight of block of ice = 4 kg = 4×1000 g = 4000 g

Now, 0.9 g of ice has a volume of $1\ \text{cm}^3$

$\therefore$ 1 g has a volume of $\frac{1}{0.9} = 1.\dot{1}\ \text{cm}^3$

$\therefore$ Volume of 4000 g = $4000\times1.\dot{1}\ \text{cm}^3 = 4444.\dot{4}\ \text{cm}^3$

Store this in the memory of your calculator.

So, volume of block of ice = $4444.\dot{4}\ \text{cm}^3$.

Using the formula for the volume of a cylinder:

Volume of block of ice = $\pi r^2 h = \pi\times7^2\times h\ \text{cm}^3$

So, $\pi\times7^2\times h = 4444.\dot{4}$

$\therefore\ h = \frac{4444.4}{\pi\times7^2} = 28.9$ cm

Calculator: MR ÷ (π × 7 x^2) = → 28.871645

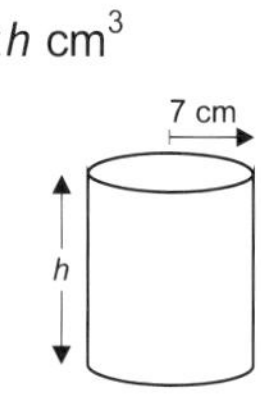

Dimensions and Formulas

Length (1 dimension)

When you use a formula to calculate a length, the answer is in metres, m (or in cm, mm, km).

For example, the perimeter of a rectangle is given by the formula:

Perimeter = $2(l+b)$

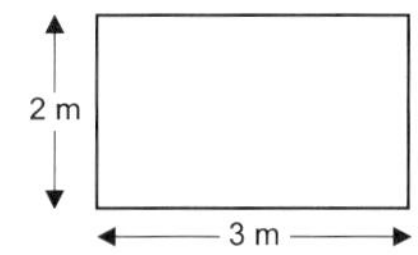

For this rectangle, l = 3 m and b = 2 m. So:

Perimeter = 2(3 m + 2 m) = 2×5 m = 10 m

Area (2 dimensions)

When you use a formula to calculate an area (length×length), the answer is in square metres, m^2 (or in cm^2, mm^2, km^2).

For example, the total surface area of a cylinder is given by the formula:

Area = $2\pi r(r+h)$

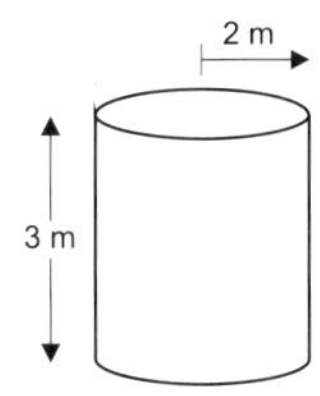

For this cylinder, r = 2 m and h = 3 m. So:

Area = 2×π×2 m×(2 m + 3 m)
= 2×π×2 m×5 m
= 2×π×2×5 m×m Rearranging letters
= 62.8 m^2

Volume (3 dimensions)

When you use a formula to calculate a volume (length×length×length), the answer is usually in cubic metres, m^3 (or in cm^3, mm^3).

For example, the volume of a cuboid is given by the formula:

Volume = lbh

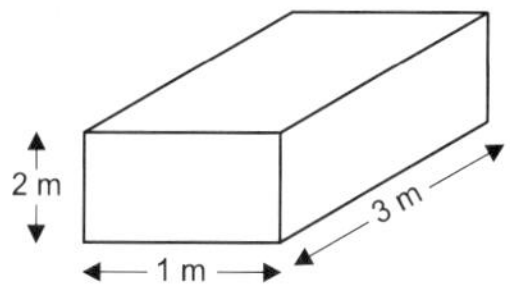

For this cuboid, l = 3 m, b = 1 m and h = 2 m. So:

Volume = 3 m × 1 m × 2 m = 3×1×2 m^3 = 6 m^3

Question *The diagram shows a deodorant stick in the shape of a prism. The following formulas can be used in connection with this prism:*

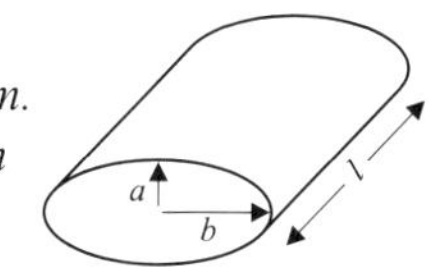

πab $\pi(a+b)$ πabl $\pi(a+b)l$

Which of these formulas represent area?

Replace each letter by m, the unit for length.

πab becomes πmm = π m^2	Area
$\pi(a+b)$ becomes π(m+m) = π×2 m = 2π m	Length
πabl becomes πmmm = π m^3	Volume
$\pi(a+b)l$ becomes π(m+m)m = π×2 m×m = 2π m^2	Area

So, the formulas πab and $\pi(a+b)l$ represent areas.

Similar Shapes and Solids

1. Enlargement and Scale Factors

Statue B is an enlarged replica of statue A. The lengths of statue A have been multiplied by 2 to obtain the lengths of statue B. We say that statue A has been enlarged by a **scale factor** of 2.

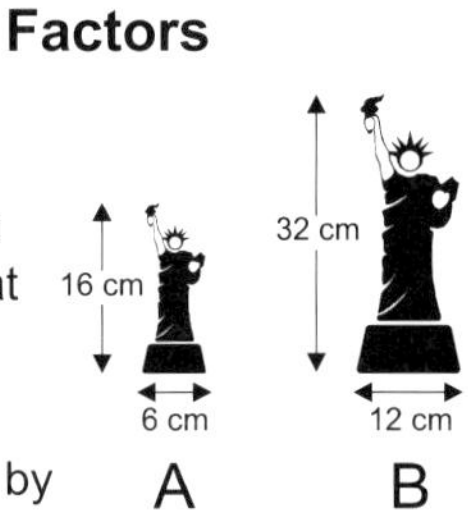

The scale factor can be calculated by dividing any two corresponding lengths of the statues. For example, using their widths gives: 12÷6 = 2.

Another way of describing the relationship between the lengths of the statues is to use the ratio of their lengths. The widths of the statues are in the ratio 6 : 12 or, equivalently, 1 : 2.

A	:	B
6 cm	:	12 cm
1	:	2

Alternatively, we could use the heights of the statues. The heights of the statues are in the ratio 16 : 32 or, equivalently, 1 : 2.

A	:	B
16 cm	:	32 cm
1	:	2

2. Similar Shapes and Solids

Two shapes or solids are said to be **similar** if one is an enlargement of the other. So, the statues in Box 1 are similar solids.

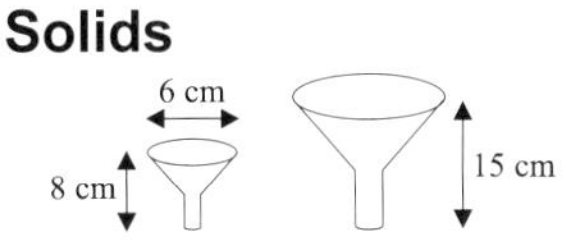

Question *These two funnels are similar in shape. Calculate the width of the larger funnel.*

The funnels are similar and so one is an enlargement of the other. The scale factor of the enlargement can be found by dividing the heights of the funnels:

Scale factor = 15÷8 = 1.875

Width of larger funnel = 1.875 ×Width of smaller funnel
= 1.875×6 = 11.25 cm

Question *A4 and A5 sizes of paper are similar in shape. Calculate the width of an A5 sheet.*

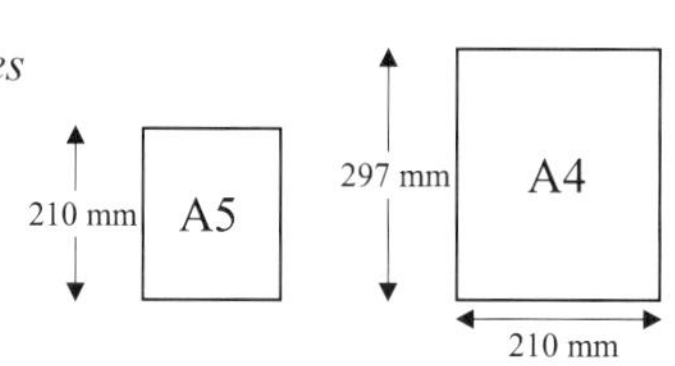

The A4 sheet is an enlargement of the A5 sheet. Use their heights to find the scale factor:

Scale factor = 297÷210 = 1.41

So, width of A4 sheet = 1.41 × Width of A5 sheet

We know that the width of an A4 sheet is 210 mm. So,

So, 210 = 1.41 × Width of A5 sheet

∴ Width of A5 sheet = 210÷1.41 ≈ 148 mm

Similar Shapes and Solids (Contd)

3. Areas of Similar Shapes and Solids

Rectangles A and B are similar. The lengths of B are 3 times the lengths of A. So, the lengths of A have been enlarged by a scale factor of 3.

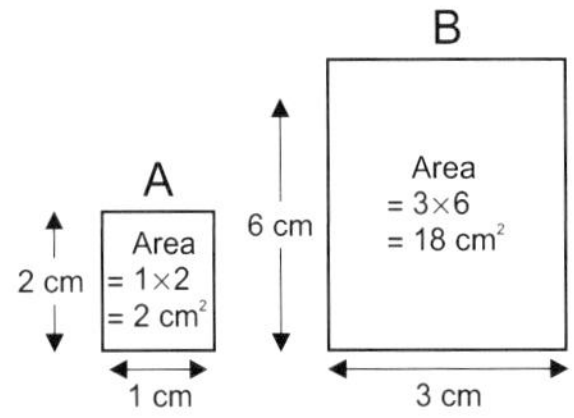

Notice that the **area** of B is 9 times the area of A.
So, the area of A has been enlarged by a scale factor of **9**, which is the square of the length scale factor 3.

Scale factor for area = (Scale factor for length)^2

Question *These two cans of soup have a simlar shape. The area of the smaller label is 85 cm^2. Calculate the area of the larger label.*

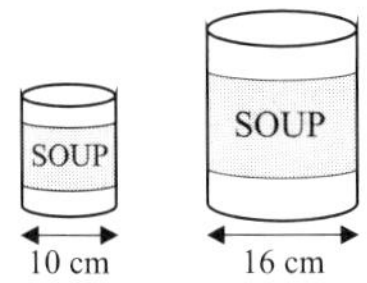

Scale factor for length = $\frac{16}{10}$ = 1.6

So, scale factor for area = 1.6^2 = 2.56

So, area of larger label = 2.56 × Area of smaller label
= 2.56×85 cm^2
= 217.6 cm^2

4. Volumes of Similar Shapes and Solids

Cuboids A and B are similar. The lengths of B are twice the lengths of A. So, the lengths of A have been enlarged by a scale factor of 2.

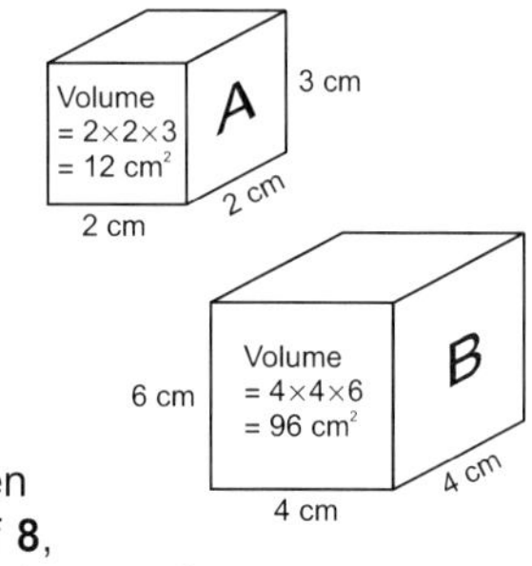

Notice that the **volume** of B is 8 times the volume of A.
So, the volume of A has been enlarged by a scale factor of **8**, which is the cube of the length scale factor 2.

Scale factor for volume = (Scale factor for length)^3

Question *Brian has built a scale model of an oil tanker, using a scale of 1:72. The real oil tanker can carry 25 000 m^3 of oil. How many cm^3 of oil can the model tanker carry?*

Scale factor for length = 72

So, scale factor for volume = 72^3 = 373 248

∴ Volume of tanker = 373 248 × Volume of model

∴ 25 000 m^3 = 373 248 × Volume of model tanker

∴ Volume of model tanker = $\frac{25\,000}{373\,248}$ = 0.067 m^3

And 0.067 m^3 = 0.067×100^3 cm^3 = 67 000 cm^3

5. Length, Area and Volume of Similar Shapes and Solids

Question *These two hot-air balloons have a similar shape. The surface area of Sky is 800 m^2 and the surface area of Ascender is 1000 m^2. When fully inflated, the volume of air in Ascender is 2700 m^3. Calculate the volume of air in Sky, correct to the nearest 100 m^3.*

Scale factor for areas = $\frac{1000}{800}$ = 1.25

So, scale factor for length = $\sqrt{1.25}$ = 1.1180

So, scale factor for volume = 1.1180^3 = 1.3975

So, volume of air in Sky = 2700÷1.3975 ≈ 1900 m^3

Angles and Parallel Lines

1. 360° in a Circle

There are 360° in a circle.

Question *Find the angle x.*

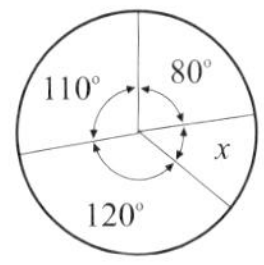

Total angle = 360°

$\therefore$ $x+80°+110°+120° = 360°$

$\therefore$ $x = 360°-80°-110°-120°$

$\therefore$ $x = 50°$

2. Names of Angles

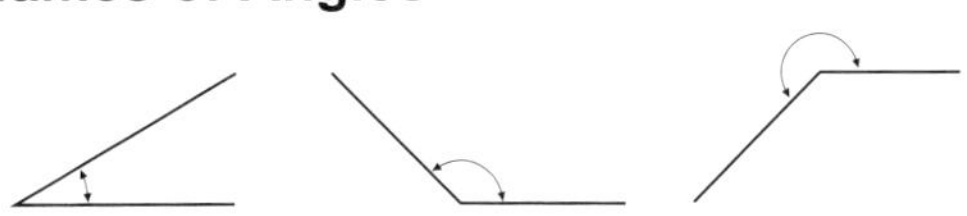

Acute angle (less than 90°) — Obtuse angle (between 90° and 180°) — Reflex angle (between 180° and 360°)

Question *Mark the acute angle(s) in this shape.*

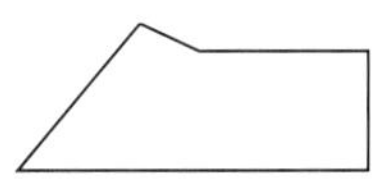

Acute angles are *less* than 90° and so the right angles have not been marked. There is only one acute angle in this shape.

Right Angle

An angle of 90° is called a **right angle** and is indicated by a small square (see diagram).

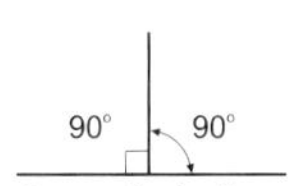

Perpendicular lines

Lines that meet at right angles are called **perpendicular** lines. For example, horizontal and vertical lines are perpendicular.

Question *Find the angle marked x.*

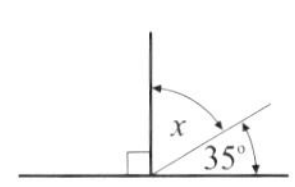

$x+35° = 90°$

$\therefore$ $x = 90°-35° = 55°$

3. Describing an Angle

There are various ways of referring to an angle in a diagram.

The 50° angle in this triangle is called angle ABC, because it lies between the sides AB and BC. Other ways of writing angle ABC are: $A\hat{B}C$, $\hat{B}$ or simply B.

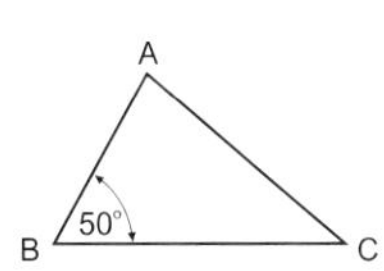

4. Angle of a Straight Line

The angle of a straight line is 180°.

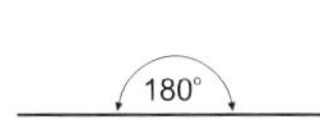

Question *Calculate the angle x.*

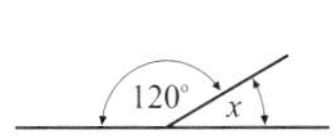

$\therefore$ $x+120° = 180°$

$\therefore$ $x = 180°-120° = 60°$

5. Opposite Angles

Opposite angles are equal.

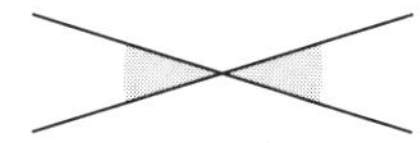

Question *Calculate the angles x and y.*

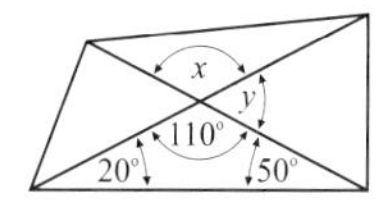

$x = 110°$ Opposite angles

$\therefore$ $y = 180°-110° = 70°$ Angle of a straight line

6. Parallel Lines

Parallel lines have the same direction, indicated by arrows.

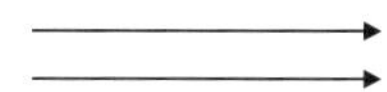

A straight line crossing parallel lines meets them at the same angle.

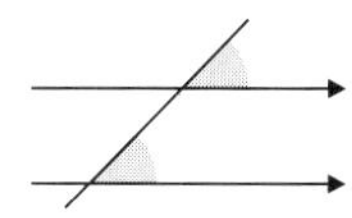

Question *Find angles a, b and c.*

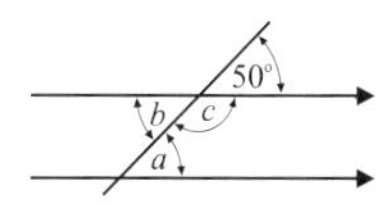

$a = 50°$ Parallel lines

$\therefore$ $b = 50°$ Opposite angles

$\therefore$ $c = 180°-50° = 130°$ Angle of a straight line

Alternate (Z) angles are equal.

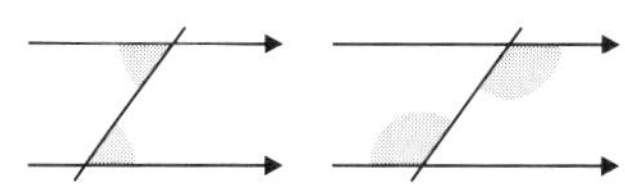

Question *Calculate angles x and y*

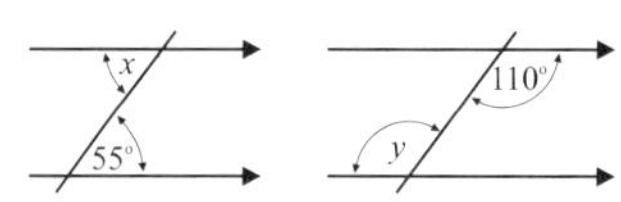

$x = 55°$ Alternate angles

$y = 110°$ Alternate angles

Interior angles add up to 180°.

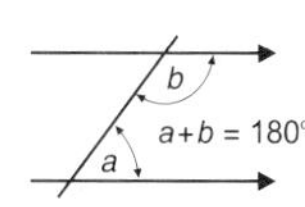

Question *Calculate the angle x.*

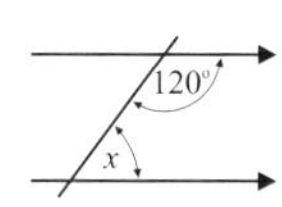

$x+120° = 180°$

$\therefore$ $x = 180°-120° = 60°$

Question *The diagram shows the side elevation of a lean-to shed. Find the angle marked x.*

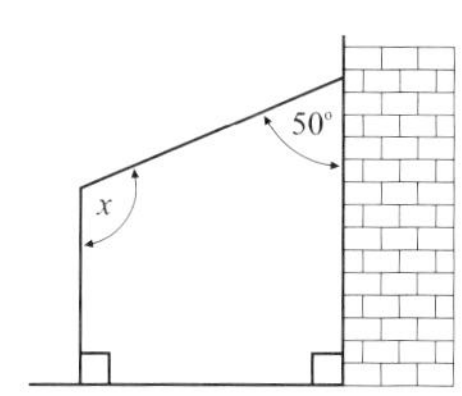

The walls are parallel because they are both vertical (at right angles to the ground). So, x and 50° are interior angles:

$\therefore$ $x+50° = 180°$

$\therefore$ $x = 180°-50° = 130°$

Triangles

1. Angles of a Triangle

The interior angles of a triangle add up to 180°.

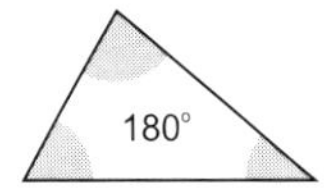

Question *Find the angle x.*

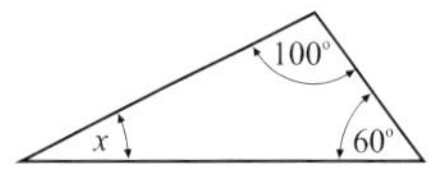

Sum of angles = 180°

$\therefore\ x+60°+100° = 180°$

$\therefore\ x = 180°-60°-100°$

$\therefore\ x = 20°$

2. Right–angled Triangles

A right-angled triangle contains a 90° angle, indicated by a small square.

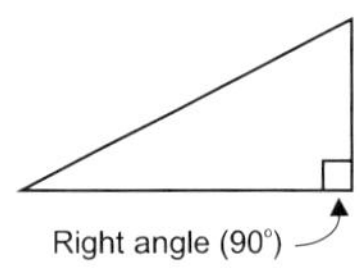

Question *Find the angle x.*

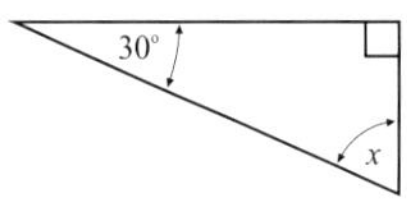

$x+30°+90° = 180°$

$\therefore\ x = 180°-90°-30°$

$= 90°-30°$ This shows that we only need to

$= 60°$ subtract 30° from 90° to find x

Question *Calculate angles a, b, c and d in this roof structure.*

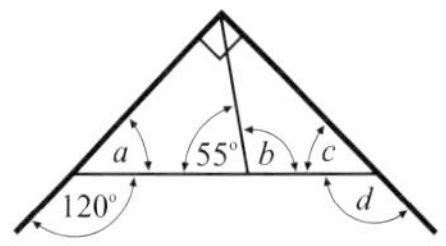

$a = 180°-120° = 60°$	Angle of a straight line
$b = 180°-55° = 125°$	Angle of a straight line
$c = 90°-a = 90°-60° = 30°$	Angles of a triangle
$d = 180°-c = 180°-30° = 150°$	Angle of a straight line

NOTE As soon as you find an angle, put it on the diagram and see how it can help you find the next angle.

3. Equilateral Triangle

An **equilateral triangle** has all its sides equal. The interior angles of an equilateral triangle are all 60°

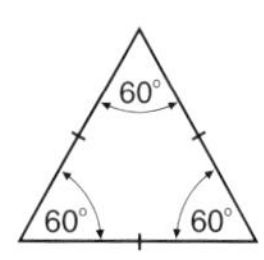

Question *12 identical matches were used to make this pattern.*
(a) State the size of angle GAB, giving a reason for your answer.
(b) Calculate angle ABC.

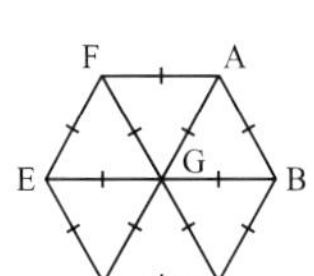

(a) GAB = 60°
Equilateral triangle

(b) ABC = ABG+GBC
= 60°+60°
= 120°

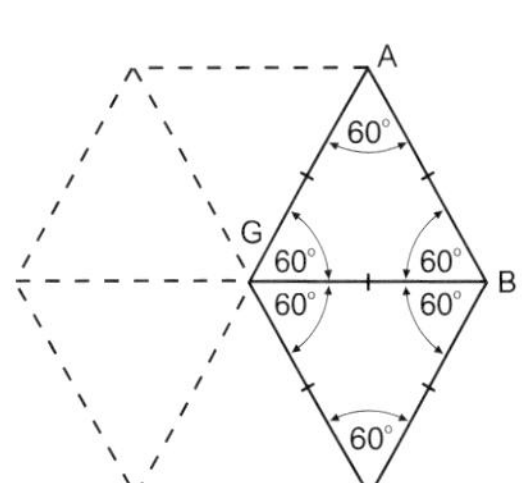

4. Isosceles Triangle

An **isosceles triangle** has two sides the same length (indicated here by dashes). Its two **base angles** are equal. Isosceles triangles are very common in questions involving angles.

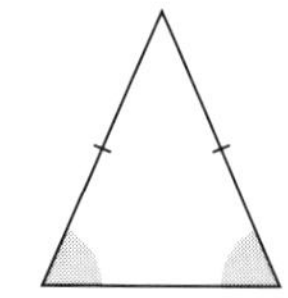

Question *ABC is an isosceles triangle with AB = AC and angle ABC = 50° Calculate angle BAC.*

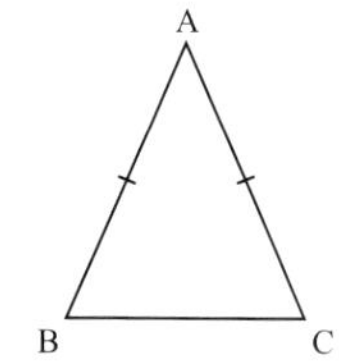

Mark angle ABC = 50° on the diagram.

Then, ACB = 50°

Base angles are equal

So, BAC+50°+50° = 180°

Angles of a triangle add up to 180°

$\therefore$ BAC = 180°−50°−50° = 80°

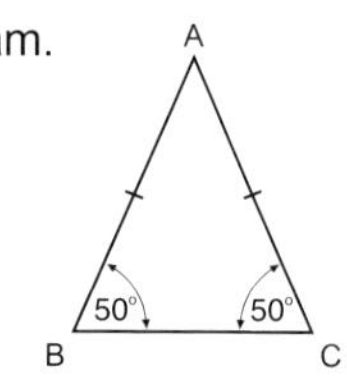

Question
(a) Calculate the size of angle ACD.
(b) If triangle ABC is also isosceles, what are the two possible values of angle ABC.

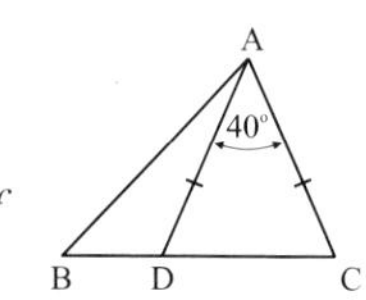

(a) The angles of triangle ADC add up to 180°. We know that angle DAC is 40°, which leaves 180°−40° = 140° for the two base angles ADC and ACD. Since the base angles are equal, then ACD = 140°÷2 = 70°.

(b) Triangle ABC can be isosceles in the two ways shown below.

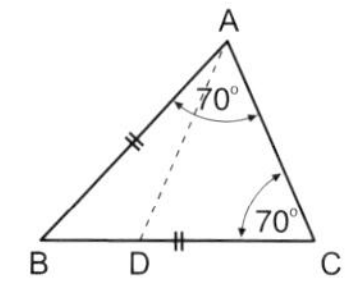

ACB and CAB are equal base angles and so CAB = 70°

ABC = 180°−70°−70°
= 40°

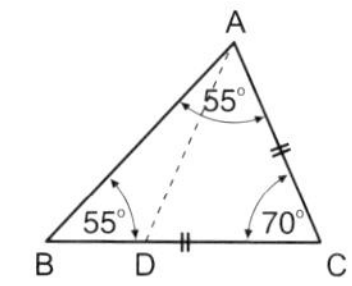

ABC and CAB are equal base angles that add up to 180°−70° = 110°

So, ABC = 110°÷2 = 55°

Question *A garden trellis is made in the form of a rectangle. Given that AB = BC, calculate the size of angle ABF.*

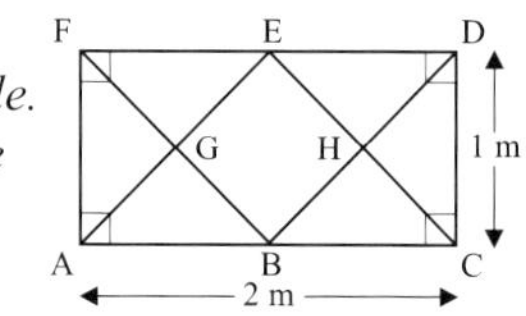

Since AB = BC and AC = 2 m, then AB = 1 m and AF = 1 m also. So, triangle FAB is isosceles (see diagrams).

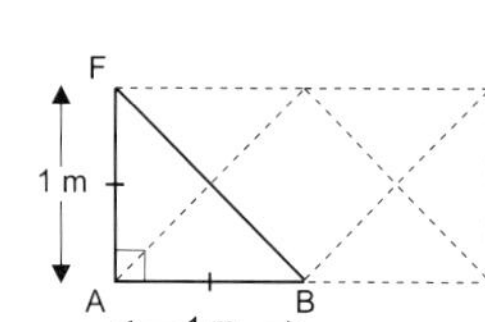

The base angles AFB and ABF add up to 90° (because 180°−90° = 90°).

So, ABF = 90°÷2 = 45°

5. Pythagoras' Theorem

The longest side of a right-angled triangle is opposite the right angle and is called the **hypotenuse**.

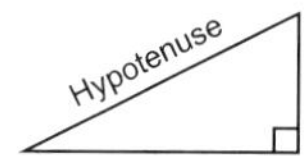

Pythagoras' theorem says that the square of the hypotenuse is equal to the sum of the squares of the other two sides:

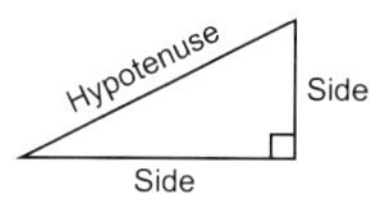

$(\text{Hypotenuse})^2 = (\text{Side})^2 + (\text{Side})^2$

This equation can be written algebraically as: $a^2 = b^2 + c^2$

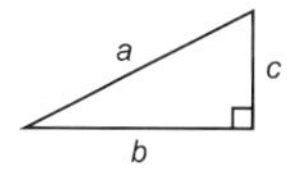

Question *A ladder reaches 3 m up a wall when its foot is 1 m from the wall. Calculate the length of the ladder.*

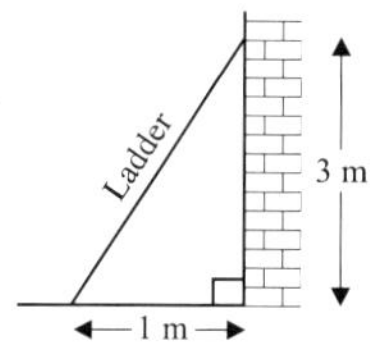

$(\text{Ladder})^2 = 1^2+3^2 = 1+9 = 10$

So, length of ladder = $\sqrt{10}$ = 3.16 m

Question *A 50 cm television screen means that it is 50 cm from corner to corner. How tall is this screen?*

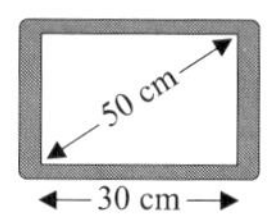

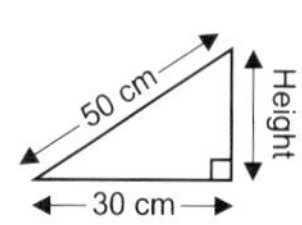

$50^2 = (\text{Height})^2+30^2$ Pythagoras

$\therefore (\text{Height})^2 + 30^2 = 50^2$

$\therefore (\text{Height})^2 = 50^2-30^2$

$= 2500-900 = 1600$

So, height of screen = $\sqrt{1600}$ = 40 cm

Question *Calculate the area of this isosceles triangle.*

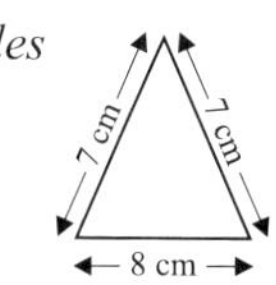

An isosceles triangle can be divided into two identical right-angled triangles

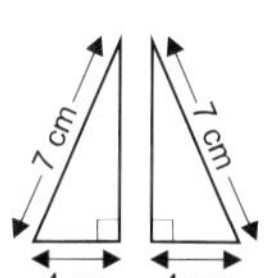

$7^2 = (\text{Height})^2+4^2$ Pythagoras

$\therefore (\text{Height})^2 + 4^2 = 7^2$

$\therefore (\text{Height})^2 = 7^2-4^2 = 49-16 = 33$

So, height of triangle = $\sqrt{33}$ = 5.74 cm

Area of triangle = $\frac{1}{2}\times$ Base $\times$ Height

$= 0.5\times8\times5.74 = 23.0 \text{ cm}^2$

6. Similar Triangles

Triangles are **similar** if they have the same angles. This means they have the same shape.

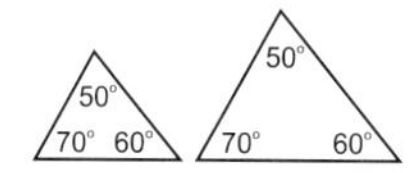

The third angle of each of these triangles must be the same and so the triangles are similar.

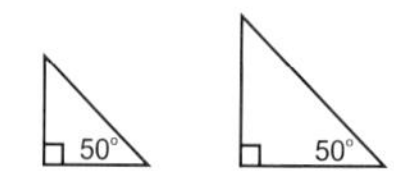

Question *Write down the two triangles that are similar.*

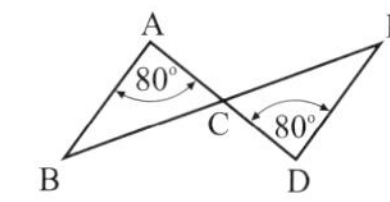

Angle ACB = Angle DCE   Opposite angles

So, angles ABC and CED must also be equal.

So, triangle CAB is similar to triangle CDE.

NOTE Make sure that the letters of one triangle correspond to the letters of the other, e.g. angle A is the same as angle D.

C A B
↕ ↕ ↕
C D E

If two triangles are similar, then one is an enlargement of the other, and vice-versa.

Question *The diagram shows a playground slide.*
(a) Which two triangles are similar?
(b) Calculate the height of the support AB.

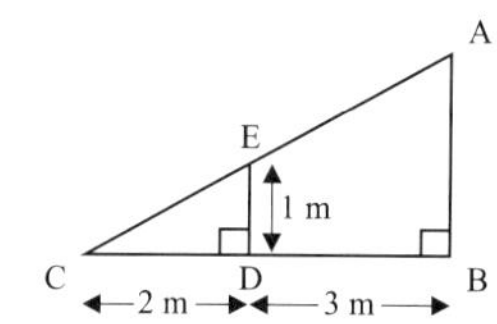

(a) Triangles ACB and ECD are similar because they have angles C and 90° in common.

(b) So, triangle ACB is an enlargement of triangle ECD.

Draw and label the two triangles separately so that you can see them more clearly.

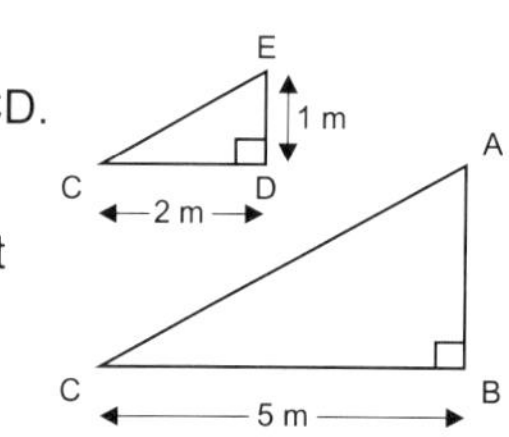

Scale factor of enlargement = $\frac{5}{2}$ = 2.5

So, all lengths of triangle ACB are 2.5 times the corresponding lengths of triangle ECD

So, AB = 2.5×ED = 2.5×1 m = 2.5 m

7. Congruent Triangles

Triangles are **congruent** if they are identical. This means they have the same sides and angles.

You may be asked to explain why two particular triangles are congruent. The three possibilities are shown below.

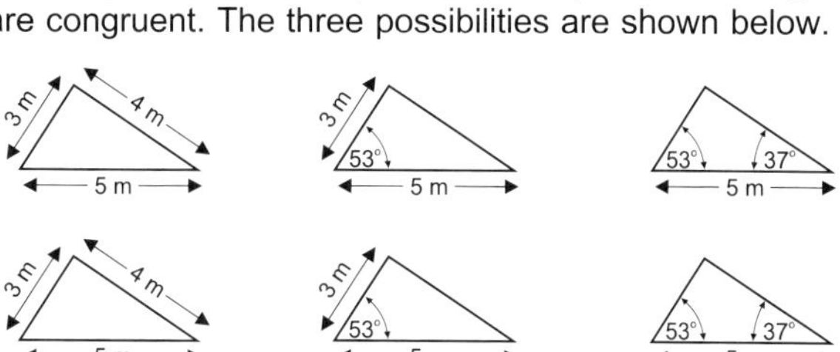

All sides equal   Two sides and included angle equal   One side and two matching angles equal

Question *Explain why triangles ABE and CDE are congruent.*

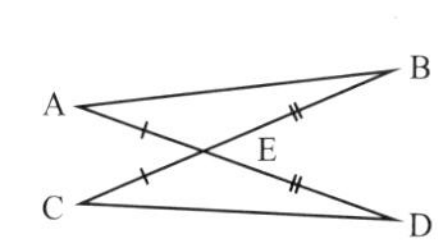

Angle AEB = Angle CED   Opposite angles

So, triangles ABE and CDE are congruent because two sides and their included angle are equal (CE = AE, ED = EB and angle AEB = angle CED).

Question *Explain why triangle ADB is **not** congruent to triangle DCB.*

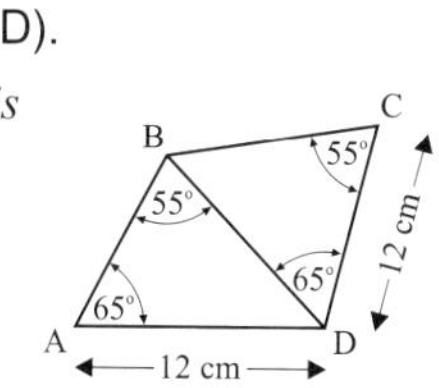

Angle ADB = 180°−65°−55° = 60° which is not the same as angle DCB. So, although the sides AD and DC are equal, the matching angles are not.

Quadrilaterals

1. Quadilaterals

A **quadrilateral** is a shape which has four sides.

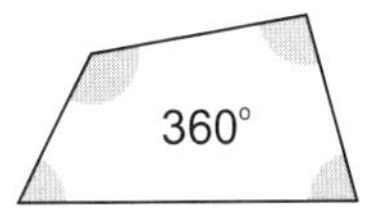

The angles of a quadrilateral add up to 360°.

Question *Find the angle marked x.*

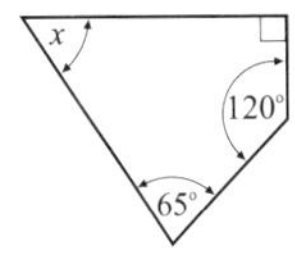

Sum of angles = 360°

$\therefore x+90°+120°+65° = 360°$

$\therefore x = 360°-90°-120°-65°$

$\therefore x = 85°$

2. Rectangle

A **rectangle** has four right angles and equal opposite sides.

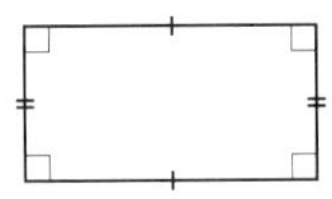

The diagonals are equal and cut each other in half (**bisect** each other).

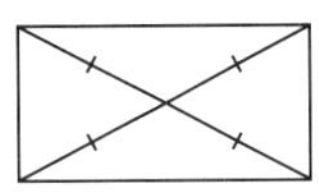

Opposite sides are parallel.

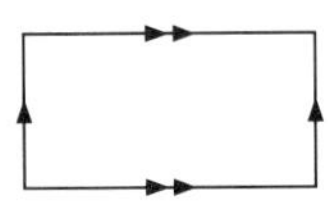

Question *Find the length of the diagonal bar of this rectangular gate.*

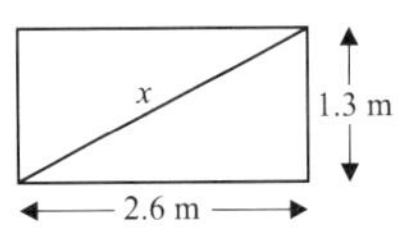

Using Pythagoras' theorem:

$x^2 = 1.3^2+2.6^2 = 8.45$

$\therefore x = \sqrt{8.45} = 2.9$ m

Question *ABCD is a rectangle. Find angle ACD, giving a reason for your answer.*

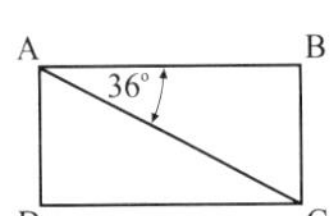

AB and CD are parallel. So:
ACD = 36° Alternate angles
(See Box 6, p79)

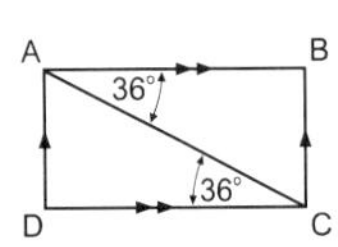

3. Square

A **square** is a rectangle with four equal sides.

The diagonals are equal and cross at right angles (90°). They cut the corner angles in half.

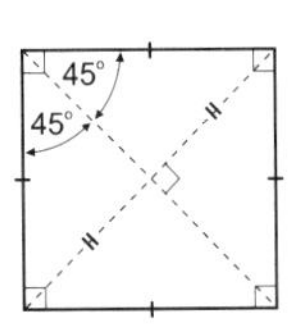

Question *The diagonal of a square boxing ring is 8.5 m. What is the length of its sides, to the nearest m?*

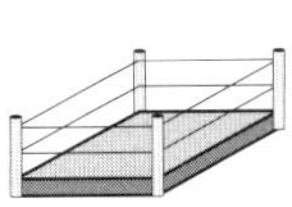

Let x be the length of its sides.

Using Pythagoras' theorem:

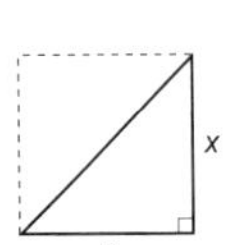

$x^2+x^2 = 8.5^2$

$\therefore 2x^2 = 72.25$

$\therefore x^2 = 72.25 \div 2$

$\therefore x^2 = 36.13$

$\therefore x = \sqrt{36.13} \approx 6$ m

4. Rhombus

A **rhombus** is like a square that has been pushed over.

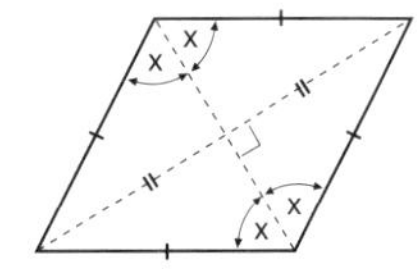

Opposite sides are parallel and equal.

Opposite angles are equal.

The diagonals bisect each other and cross at 90°.

Question *PQRS is a rhombus. Find angle PQS, giving a reason for your answer.*

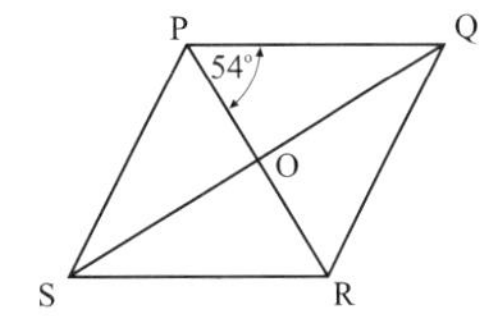

Diagonals cross at right angles and so angle POQ is 90°.

So, PQO = 180°−54°−90° = 36°

5. Parallelogram

A **parallelogram** is like a rectangle that has been pushed over.

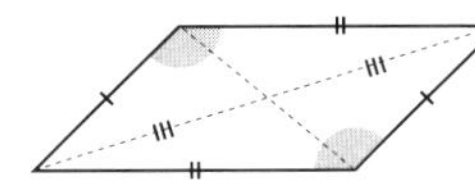

Opposite sides are parallel and equal.

Opposite angles are equal.

The diagonals bisect each other but do ***not*** cut the corner angles in half.

Question *ABCD is a parallelogram. Find angle x.*

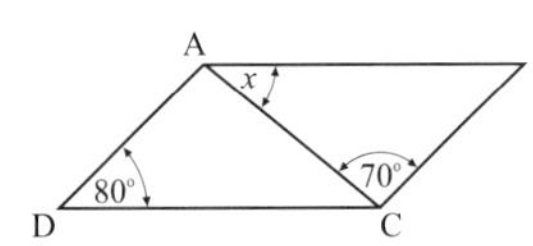

Opposite angles are equal and so angle ABC = 80°.

So, $x+70°+80° = 180°$

$\therefore x = 180°-70°-80 = 30°$

6. Trapezium

A **trapezium** has just two sides parallel. It has no other special properties.

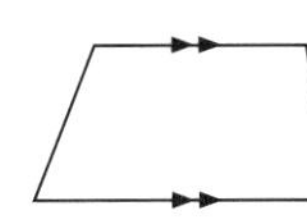

Question *Calculate angle a.*

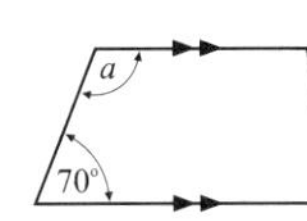

a and 70° are interior angles and so add up to 180° (see Box 6, p79)

$\therefore a = 180°-70° = 110°$

7. Kite

A **kite** has two pairs of equal sides and one pair of equal opposite angles.

The diagonals cross at right angles.

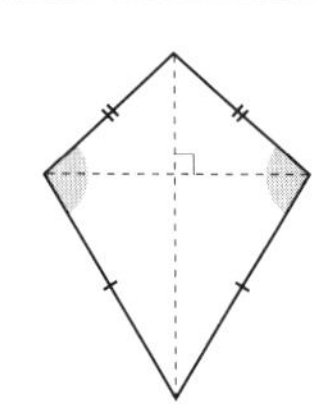

Question *PQRS is a kite. Find angle x.*

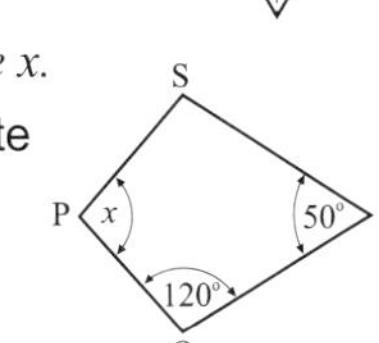

Angle PSR = 120° because opposite angles are equal.

The angles add up to 360° and so:

$x+50°+120°+120° = 360°$

$\therefore x = 360°-50°-120°-120° = 70°$

Polygons

1. Regular Polygons

Regular polygons are geometrical shapes with all sides and angles equal. You need to know the names of the regular polygons below.

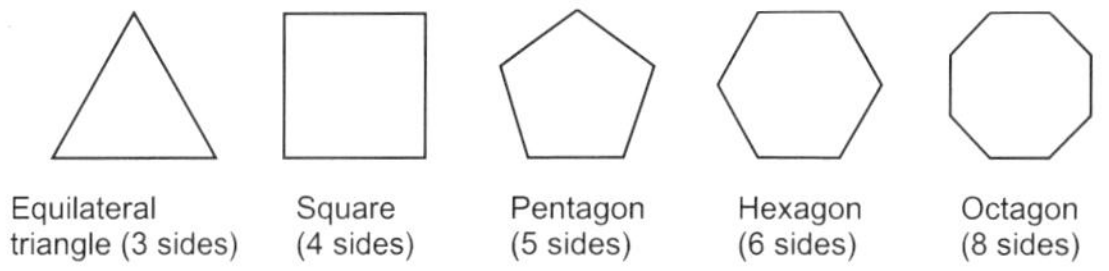

2. The Centre of a Regular Polygon

A circle can be drawn around a regular polygon. The centre of the circle is the centre of the polygon.

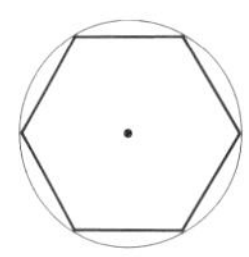

Joining the corners of the polygon to the centre creates identical isosceles triangles. The equal sides of these triangles are radii of the circle.

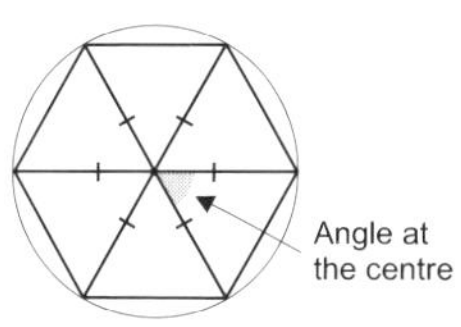

The angles at the centre of a regular polygon are equal and add up to 360°.

Question *Calculate the angle at the centre of a regular pentagon.*

A pentagon has 5 sides and so has 5 equal angles at the centre that add up to 360°.

So, each angle = 360°÷5 = 72°

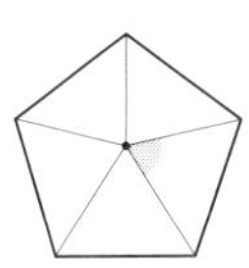

3. Interior and Exterior Angles

The interior angles of a regular polygon are all equal.

The exterior angles of a regular polygon are equal to the angle at its centre (see Box 2) and so:

The exterior angles of any polygon add up to 360°.

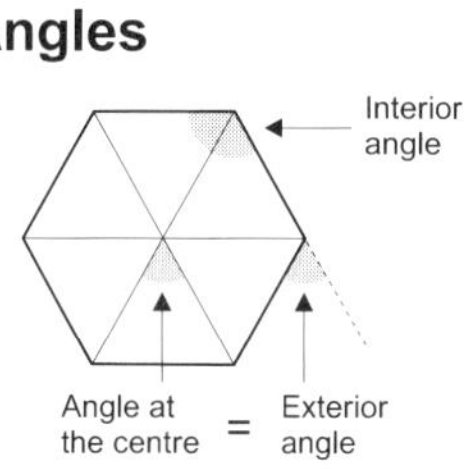

Question *Find the interior angle of a regular octagon. You must show your working.*

Make a sketch, including the angles at the centre.

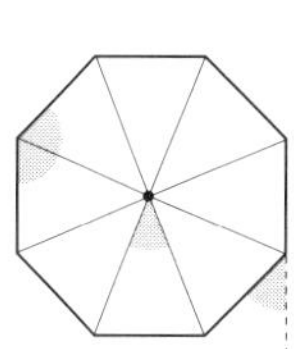

It is easiest to find the angle at the centre first.

An octagon has 8 sides and so 8 equal angles at the centre that add up to 360°. So, the angle at the centre is 360°÷8 = 45°, which is the same as the exterior angle.

Then, interior angle = 180°−45° = 135°
(Angle of a straight line is 180°)

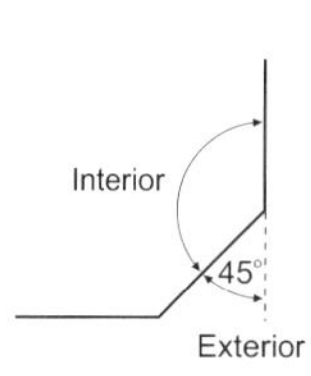

4. Finding the Number of Sides of a Polygon

For each side there is an exterior angle and so there are the same number of sides as exterior angles. It is easy to find the number of exterior angles.

Question *The interior angle of a polygon is 160°. How many sides does it have?*

Find the exterior angle first.

Exterior angle = 180°−160° = 20°

The exterior angles add up to 360°.

So, number of exterior angles = 360°÷20° = 18

So, the polygon has 18 sides.

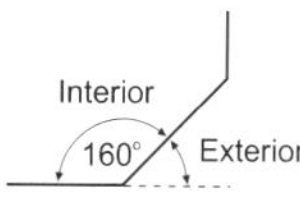

5. Problems Involving Hexagons

Regular hexagons are easy polygons to work with and often appear in examination questions.

Question *The diagram shows a new coin made in the shape of a regular hexagon. Calculate the smallest height of slot that the coin could fit into.*

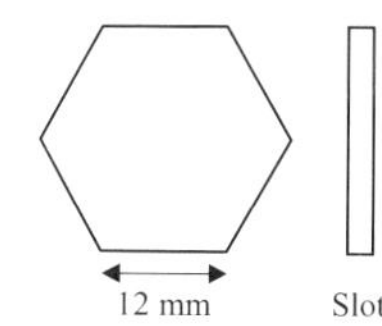

A hexagon can be divided into six equilateral triangles.

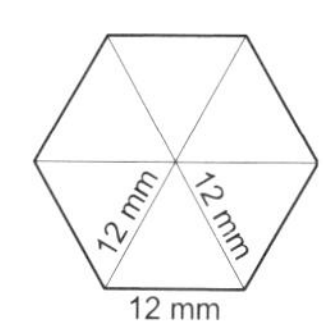

Split an equilateral triangle in half to make two identical right-angled triangles of height h, say. Then by Pythagoras' theorem:

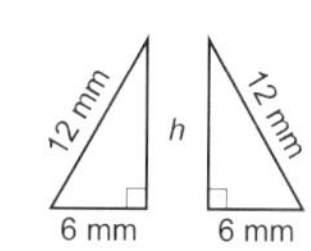

$h^2+6^2 = 12^2$

$\therefore\ h^2 = 12^2-6^2 = 144-36 = 108$

$\therefore\ h = \sqrt{108} = 10.4$ mm

So, the distance between the flat sides of the coin is 2×10.4 = 20.8 mm, the minimum possible slot height.

6. Sum of Interior Angles of Any Polygon

The following result is true for *all* polygons.

Sum of interior angles = (Number of sides − 2)×180°

For example, a pentagon has 5 sides, so:

Sum of interior angles = (5−2)×180°
= 3×180° = 540°

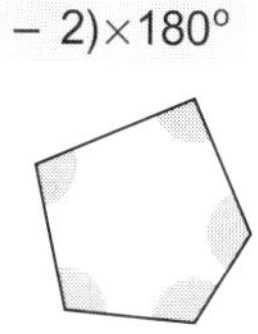

Question *The interior angles of a polygon add up to 900°. How many sides does it have?*

Let n be the number of sides, then:

$(n-2)\times 180 = 900$

$\therefore\ n-2 = 900\div 180 = 5$

$\therefore\ n = 5+2 = 7$

Circles H

You need to know the following results about circles. These results are called **theorems**.

A **chord** is a straight line that joins any two points on the circumference of a circle.

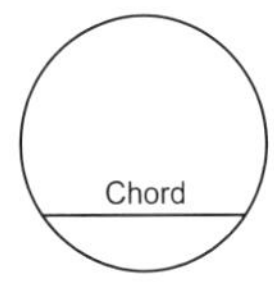

A radius that is perpendicular (at right angles) to a chord cuts the chord in half (bisects the chord). The opposite is also true: if a radius cuts a chord in half, it is perpendicular to the chord.

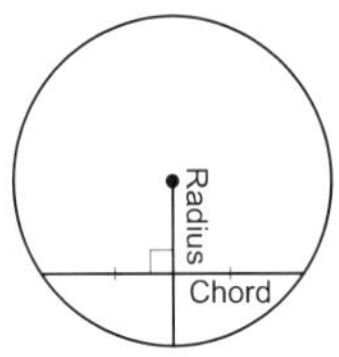

Question *Calculate the drop in water level in this fish bowl.*

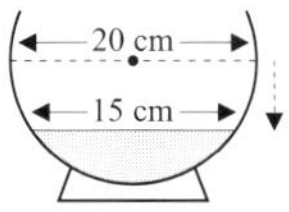

Radius of circle = 10 cm

Let h be the distance the water level has dropped. From Pythagoras' theorem:

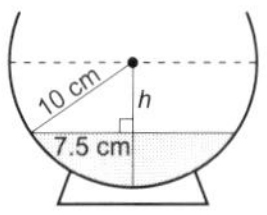

$h^2+7.5^2 = 10^2$

$\therefore\ h^2 = 10^2-7.5^2 = 100-56.25 = 43.75$

$\therefore\ h = \sqrt{43.75} = 6.6$ cm

If the corners of a triangle lie on the circumference of a circle and one of its sides is a diameter, the angle opposite the diameter is 90°. The opposite is also true: if you draw a circle around a right-angled triangle, the hypotenuse will be a diameter of the circle.

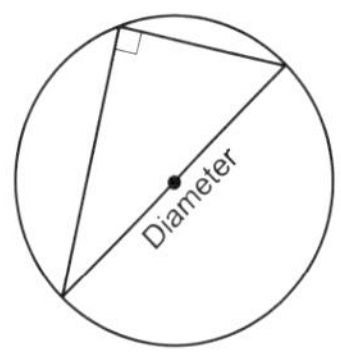

Question *Find angle BAC, given that O is the centre of the circle.*

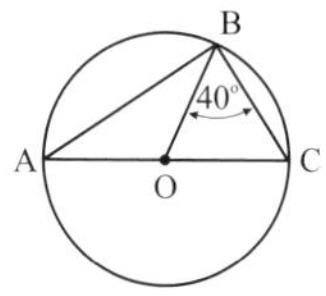

Mark the radii with dashes to show any isosceles triangles.

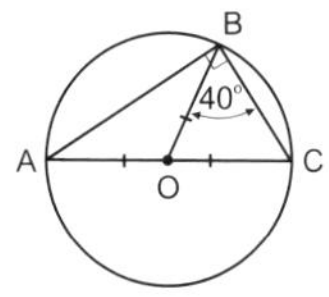

BCO = 40° Isosceles triangles
ABC = 90° 90° opposite diameter
So, BAC = 180°−90°−40° = 50°

When the ends of a chord are joined to a point on the circumference they make an angle, called the **subtended angle**. The subtended angles on the same side of the chord are all equal.

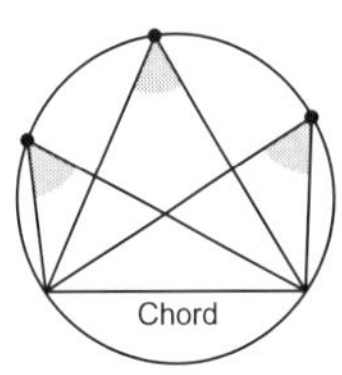

Question *Find angle x.*

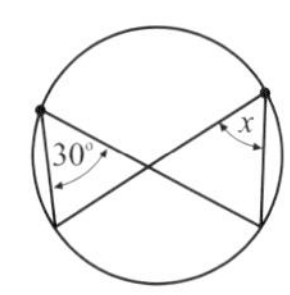

Draw in the missing chord. Then you can see that angles x and 30° are subtended by the same chord. So, x = 30°.

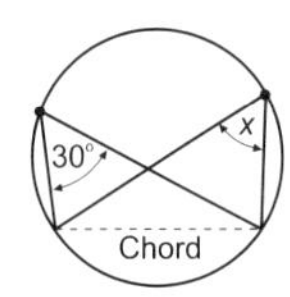

The angle subtended by a chord at the centre of a circle is twice the angle subtended at the circumference by the chord.

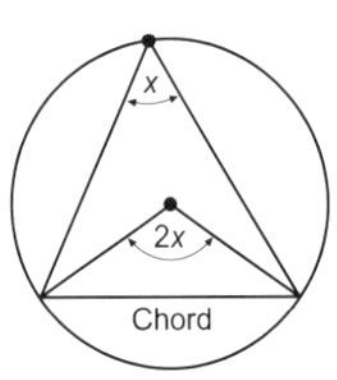

Question *Find angles x and y.*

Angle x is subtended by the chord PQ at the centre O. Angle 40° is subtended by the same chord at the circumference. So, x = 2×40° = 80°

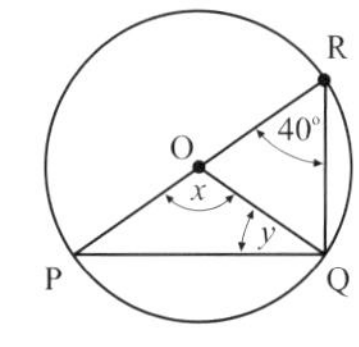

Triangle POQ is isosceles.
Base angles of triangle POQ add up to 180°−80° = 100°.
So, y = 100°÷2 = 50°

If all four corners of a quadrilateral lie on the circumference of a circle, the quadrilateral is said to be **cyclic**.

Opposite angles of a cyclic quadrilateral add up to 180°.

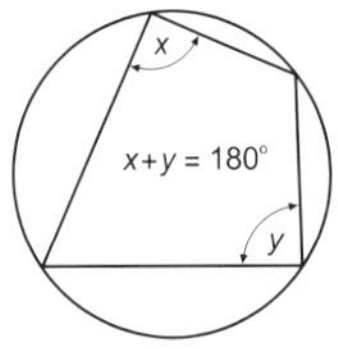

Question *Find angles x and y.*

x+90° = 180° $\therefore\ x$ = 180°−90° = 90°

y+115° = 180° $\therefore\ y$ = 180°−115° = 65°

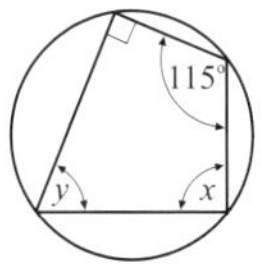

A **tangent** is a line that touches a circle at just one point.The two tangents drawn from point T to the circle are equal and meet radii OA and OB at right angles. Triangles AOT and BOT are congruent (identical).

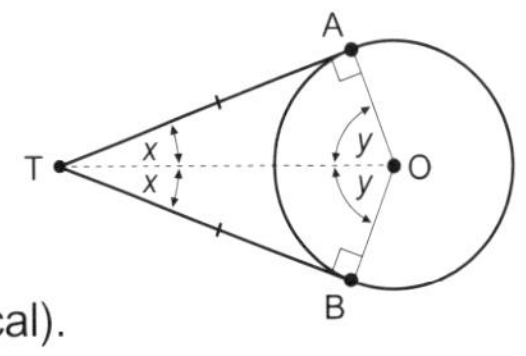

Question *Find angle AOB.*

The angles between the radii and tangents are 90°. The angles of a quadrilateral add up to 360°, so:

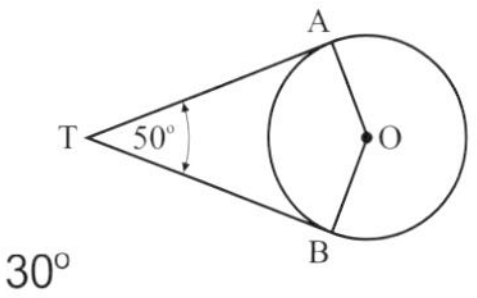

AOB+50°+90°+90° = 360°

$\therefore$ AOB = 360°−50°−90°−90° = 130°

Measuring. Scale Drawing. Loci

1. Measuring with a Ruler

Measure lengths as accurately as you can. You will lose marks if your answer is more than 1 mm out.

Question *Measure the length of the line AB.*

Align the 0 mark of your ruler with the end of the line.

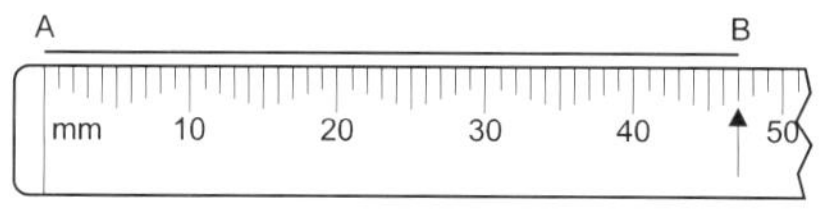

The line AB is approximately 47 mm long (or 4.7 cm).

2. Scale Drawing

Question *The diagram shows the plan of a tennis court, drawn to a scale of 1 cm to 7 yards. Find the length of the serve XY correct to the nearest yard.*

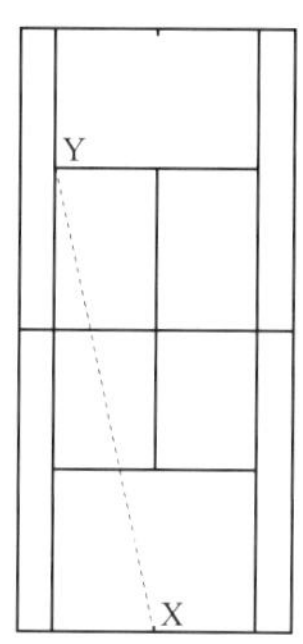

Use your ruler to accurately measure the length XY. You should find that XY $\approx$ 31 mm = 3.1 cm.

A scale of 1 cm to 7 yards means that 1 cm on the drawing represents a real distance of 7 yards. So,

$$\begin{aligned} XY = 3.1 \text{ cm} &= 3.1 \times 7 \text{ yards} \\ &= 21.7 \text{ yards} \\ &\approx 22 \text{ yards} \end{aligned}$$

3. Measuring Angles with a Protractor

Use a semicircular protractor numbered 1° to 180°, both clockwise and anticlockwise. A clear plastic protractor is best because you can see the diagram underneath it.

Measure angles as accurately as you can. You will lose marks if your answer is more than 1° or 2° out.

Question *Measure the size of angle ABC.*

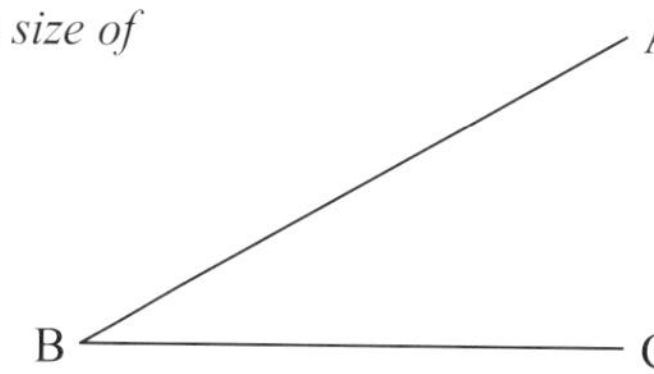

Accurately place the centre point of your protractor on point B with the 0° line over line BC.

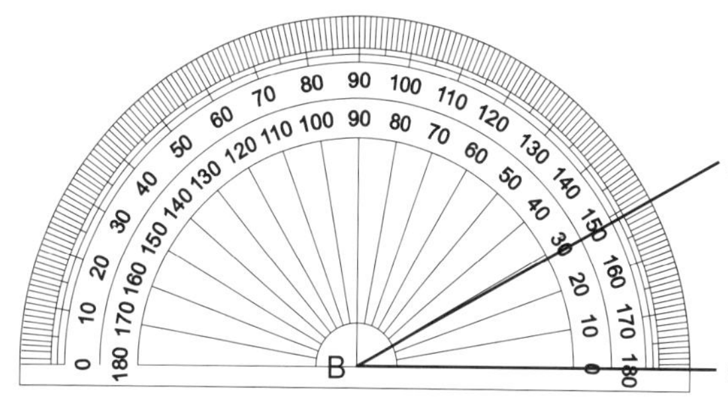

The angles on the protractor increase from 0°, on line BC, through 10°, 20° and up to angle 29° on line AB. So, the angle ABC is 29°.

4. Drawing Geometrical Shapes

Question *Accurately draw this triangle.*

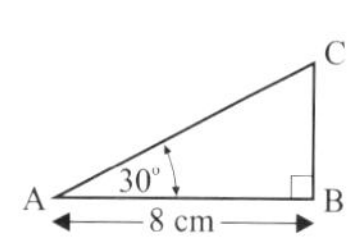

Use your ruler to draw side AB, 8 cm long.

Use your protractor to mark a 90° angle at point B.

NOTE Do not estimate right angles (90°) by eye.

Draw a feint line perpendicular to AB. Feint lines like this are called **construction lines**. Don't rub them out, as the examiner will want to see how you made your drawing.

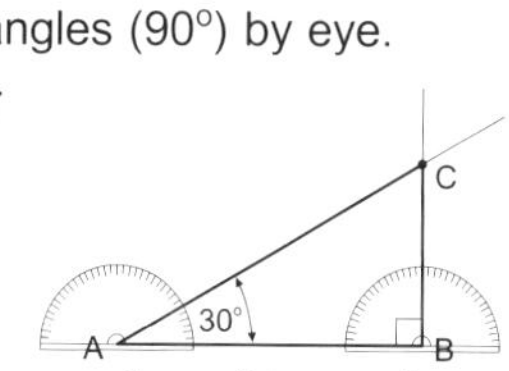

Use your protractor to mark a 30° angle at point A. Draw a feint line at 30° to AB, crossing the perpendicular line at point C.

Draw lines AC and BC to complete the triangle. Label the corners A, B and C.

Question *Make an accurate drawing of this shape. The curved lines are quarter circles.*

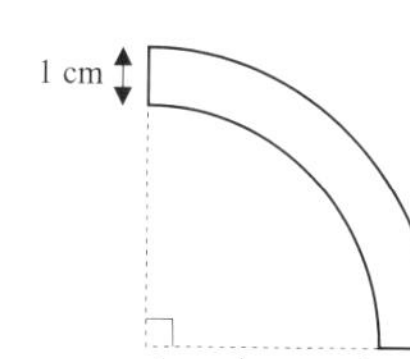

Think about the method you are going to use before you start drawing.

Draw perpendicular construction lines 5 cm long.

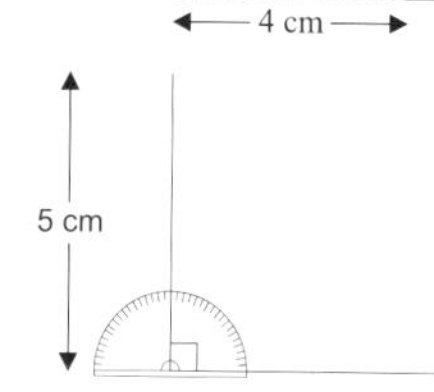

Open your compasses to 5 cm, place the point on A and draw a quarter circle. Reduce the compasses to 4 cm and draw another quarter circle.

Join up the ends of the quarter circles to complete the shape.

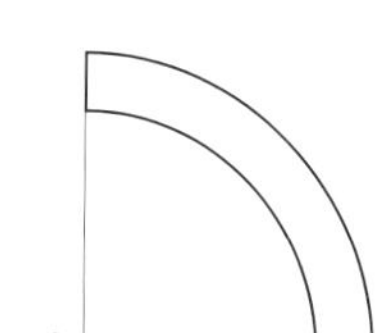

5. Locus

A **locus** is a line whose points obey a rule. The plural of the word locus is *loci*. Often, this line is the path traced out by a moving object. For example, the dashed line in the diagram shows the locus (path) of a playground swing. You need to know the loci described in the boxes overleaf. The title of each box describes the rule of the locus.

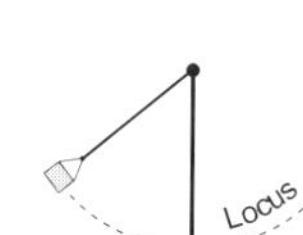

Measuring. Scale Drawing. Loci (Contd)

6. Equal Distance from a Point

Question *A and B represent two radio transmitters. Signals from A can be received 300 km away. Signals from B can be received 350 km away. Shade the region in which signals can be received from both transmitters.*

1 cm represents 100 km

Transmitter A sends signals 300 km in all directions. So the signals can be received within a circle of radius 300 km with centre A. Since 1 cm represents 100 km, draw a circle of radius 3 cm.

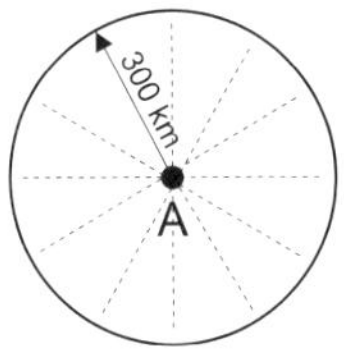

For transmitter B, 350 km is represented by 3.5 cm on the diagram. So draw a circle of radius 3.5 cm, centre B. Shade the intersection, where radio signals can be received from both transmitters.

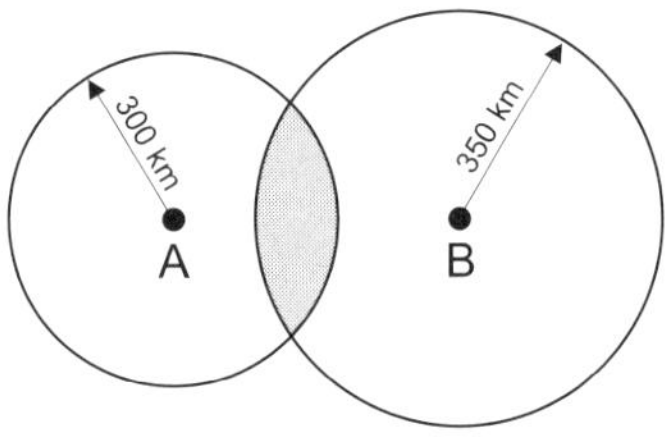

7. Equal Distance from Two Points

Question *A ship sails between the two buoys P and Q, keeping an equal distance from each. Accurately draw the course of the ship.*

Every point on the dashed line is an equal distance from P and Q.

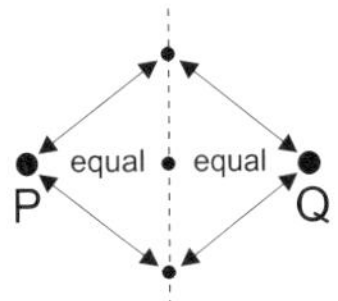

Place the point of your compasses on Q and open them to about $\frac{3}{4}$ of the distance PQ. Draw an arc as shown. Repeat for point P. Using a ruler, join the intersections of these arcs to obtain the course of the ship.

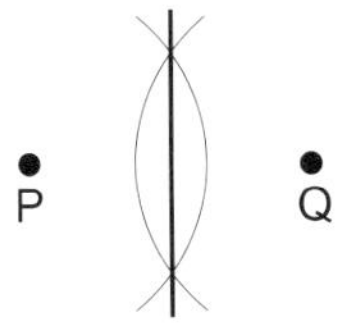

8. Equal Distance from a Line (Contd)

Question *The light, S, above a front door has a sensor which turns on the light when anything moves within a distance of 2 m. A man walks towards the door.*
(a) Draw the locus of point H.
(b) Draw the locus of points 2 m from S.
(c) How far is the point H from the door when the light is switched on?

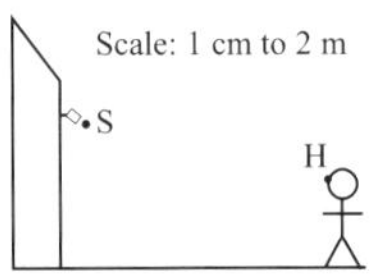

(a) Draw a line parallel to the ground through H.

(b) Draw a circle with centre S and radius 1 cm (representing 2m).

(c) Measure AB to be 0.5 cm, which represents a distance of 1 m.

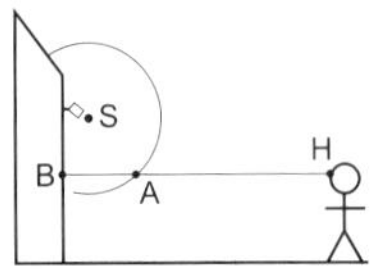

8. Equal Distance from a Straight Line

Question *In a certain type of cricket match, players must stand within 15 yards of the line AB joining the wickets.*

Use a scale of 1 cm to 5 yards to draw the area in which the players must stand.

All points on the dashed line are 15 yards from the line AB. The straight lines are parallel to AB and 15 yards from AB. The semicircles have radii 15 yards.

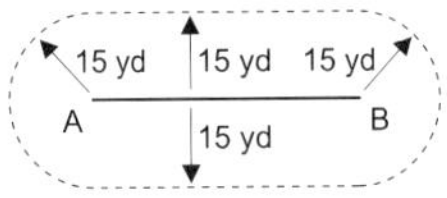

1 cm represents 5 yards and so 3 cm represents 15 yards. Use your protractor to draw construction lines perpendicular to AB through points A and B. Measure 3 cm along these lines away from AB and join up the points. Use your compasses to draw semicircles of radius 3 cm with centres at A and B.

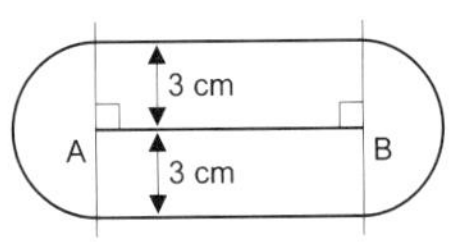

9. Equal Distance from Two Lines

Question *The diagram shows a car park. A guard patrols the park, entering at point B. He walks across the park, keeping the same distance from walls AB and BC until he is 5 m from wall CD. He then walks parallel to wall CD to door E in wall AD. Draw the path the guard takes.*

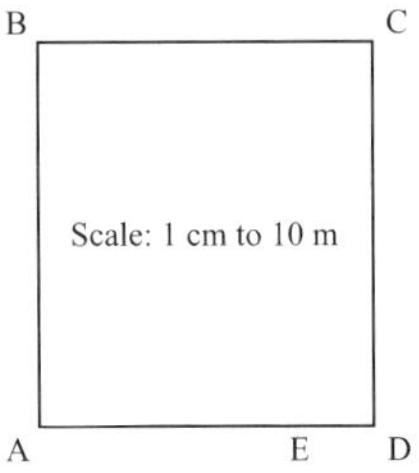

Place the point of your compasses on B and draw an arc cutting AB and BC at P and Q. Place the point of your compasses on P and draw an arc. Repeat for point Q, keeping your compasses the same distance apart. Join corner B to the intersection of these arcs.

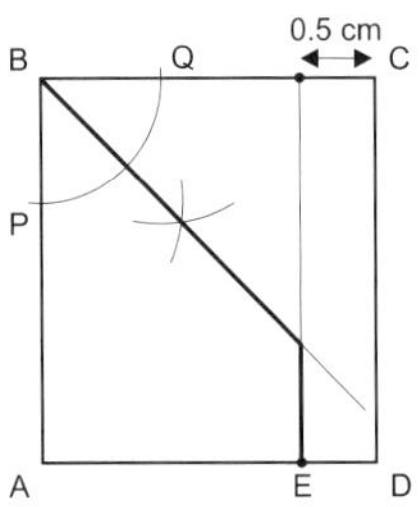

Points on this line are an equal distance from walls AB and BC. Now, 1 cm represents 10 m and so 0.5 cm represents 5 m. Mark points 0.5 cm along CB and DA, join them with a construction line and draw the rest of the guard's path.

more

Bearings

1. Bearings

A **bearing** is an angle measured clockwise from North, and is a way of describing a direction.

Bearings are always given as three digit numbers; so 45° is written 045°.

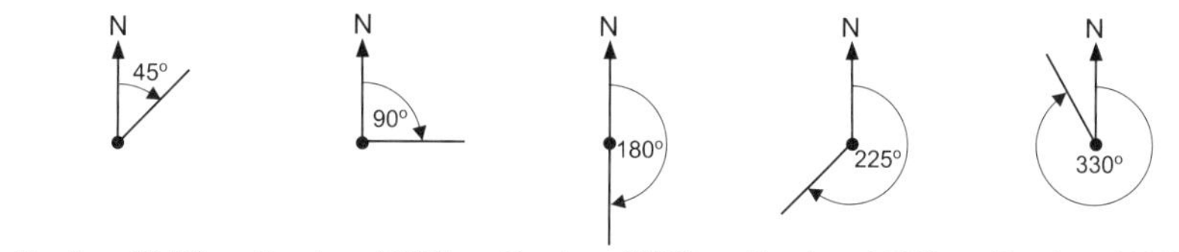

Bearing of 045° Bearing of 090° Bearing of 180° Bearing of 225° Bearing of 330°

Compass Points

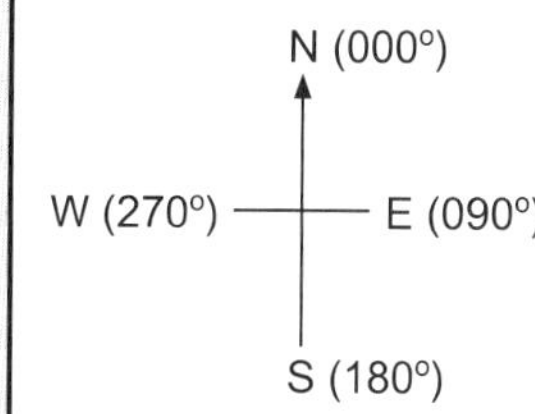

2. Measuring a Bearing

A bearing is always measured **from** a certain point.

Question *Measure the bearing of the buoy B* ***from*** *the ship S.*

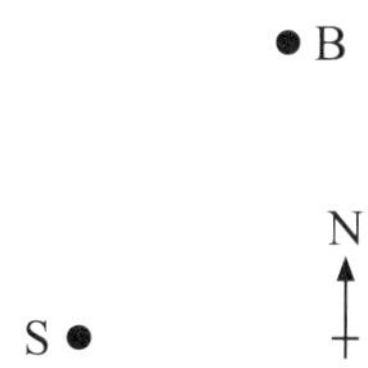

Join the two points with the straight line SB.

Draw a North arrow through point S, **from** which you are measuring the bearing. Make sure that your North line is parallel with the given North line.

Place the centre of your protractor on point S, aligning 0° with North. Read off the **clockwise** angle the line SB makes with North.

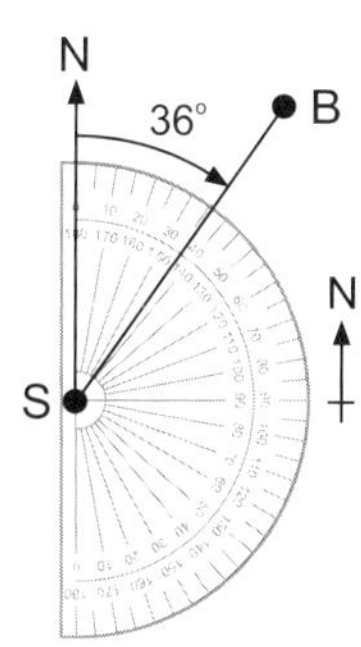

This gives a bearing of 036°.

When the bearing is greater than 180°, turn your protractor around and add 180° to your reading.

Question *The street map opposite is drawn to a scale of 1 cm to 200 m. Calculate:*

(a) the bearing of B from A

(b) the distance between the centres of these two buildings.

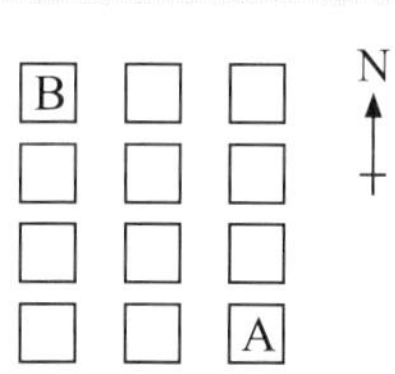

(a) Join the centres of the two buildings with a straight line. Draw a North line through A, from which the bearing is measured. Place the centre of your protractor on A, aligning 0° with North, as shown. Read the angle of 138° in a clockwise direction.

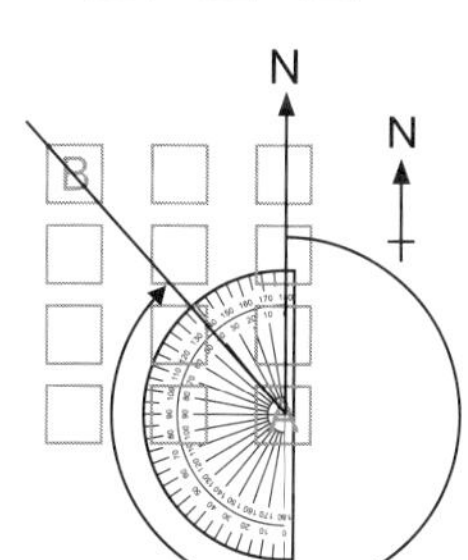

Bearing = 180°+138° = 318°

(b) Measure the distance between the centres of the two buildings:

Distance = 23 mm = 2.3 cm

1 cm represents a real distance of 200 m and so:

Actual distance = 2.3×200 = 460 m

3. Drawing a Bearing (Plotting a Course)

You may be asked to plot the course of, for example, a ship or an aircraft.

Question *A fishing boat sails 50 km from port P to a fishing ground G on a bearing of 240°. Plot the course of the boat and mark the fishing ground G. Use the scale 1 cm to 10 km.*

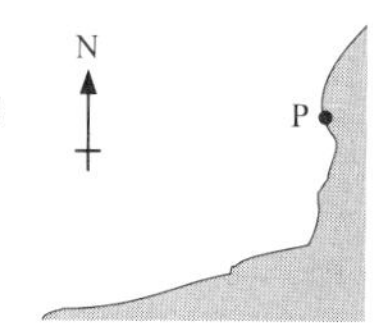

Draw a North line through point P. Place the centre of the protractor on point P, aligning 0° with North. The bearing of 240° can be split into 180°+60°. Mark a point at an angle of 60°, measured clockwise from 0° at the bottom of the protractor. Join point P to this point. This line is the course of the boat.

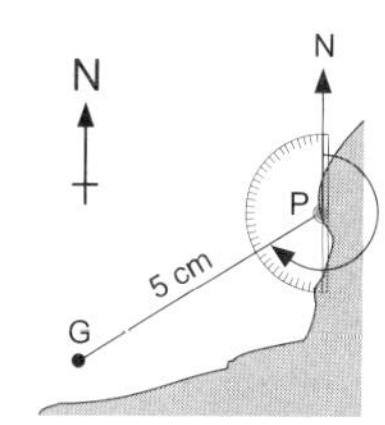

Since 1 cm represents 10 km, then 5 cm represents 50 km. So, mark fishing ground G 5 cm from port P.

4. Locating a Point

Question *A coastguard, C, picks up a distress signal from a yacht Y due East. The same signal is received by a lighthouse L, on a bearing of 050°.*

(a) Plot the position of the yacht Y.

(b) Measure CY and calculate the distance of the yacht from the coastguard.

(c) How long would a helicopter take to reach the yacht from the coastguard, travelling at an average speed of 180 km/h? Give your answer to the nearest minute.

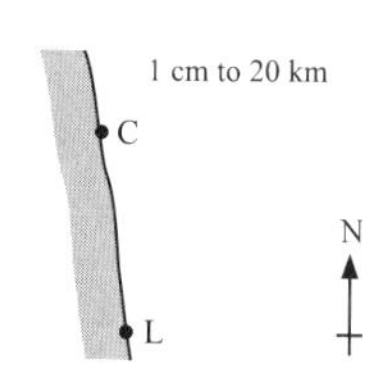

(a) Draw North lines through C and L. Use your protractor to draw bearings of East (090°) from C and 050° from L. Mark the position of yacht Y where these bearings intersect.

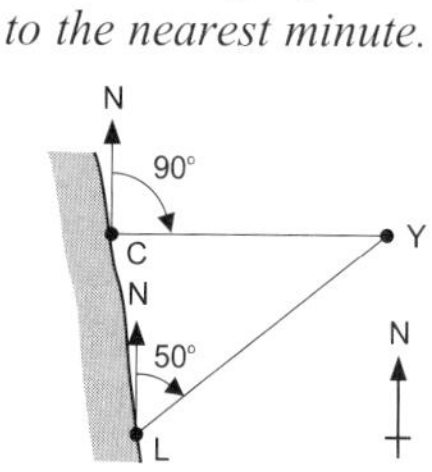

(b) Carefully measure CY to be 1.9 cm. The scale in the diagram tells us that 1 cm represents 20 km and so CY = 1.9×20 = 38 km.

(c) 180 km/h means 180 km travelled in 60 min

So, 1 km travelled in $\dfrac{60 \text{ min}}{180} = 0.\dot{3}$ min

So, 38km travelled in $38 \times 0.\dot{3}$min ≈ 13 min

Vectors H

1. Vectors

A **vector** quantity has both size (magnitude) and direction. For example, a car travelling 50 km/h on a bearing of 045°.

A vector can be represented by a straight line with an arrowhead. The length of the line represents the size of the vector and the arrowhead indicates its direction.

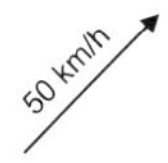

Vectors can be labelled using bold letters.

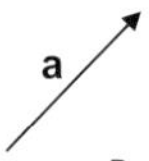

Or they can be labelled using capital letters.

This vector is written $\overrightarrow{AB}$.

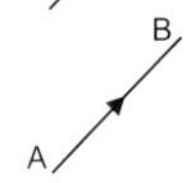

Column vectors

Vectors can be described using a pair of numbers like coordi–nates, only written vertically. The top number represents the horizontal direction, the bottom number the vertical direction.

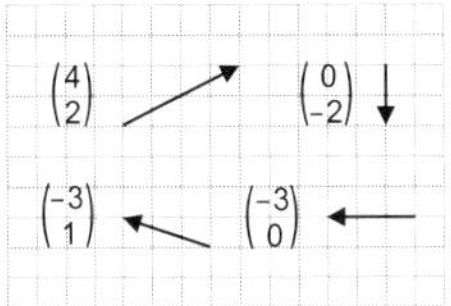

Position vectors

The **position vector** of a point is the vector joining the origin to the point. The position vector of P is described using a column vector $\overrightarrow{OP} = \binom{3}{2}$.

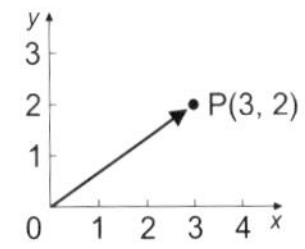

2. Comparing Vectors

Parallel vectors

Parallel vectors have the same (or opposite) direction, but may have different sizes.

Vectors with the same gradient are parallel.

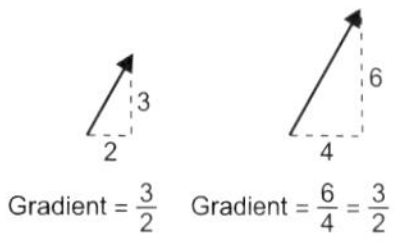

Negative vectors

Vector −**a** has the same size as vector **a** but opposite direction

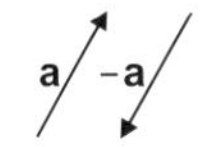

If $\overrightarrow{AB}$ is the vector from A to B, then $-\overrightarrow{AB}$ is the opposite vector from B to A, i.e. $\overrightarrow{BA}$

So, $-\overrightarrow{AB} = \overrightarrow{BA}$

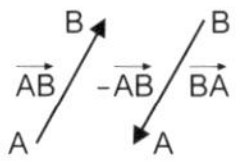

Multiplying vectors by a number

Multiplying a vector by a positive number changes its size but not its direction. Multiplying by a negative number reverses the direction.

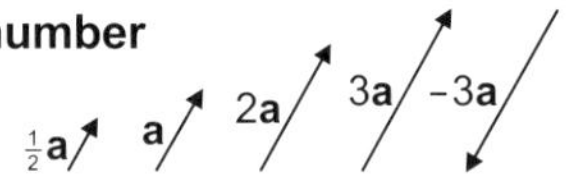

3. Vector Arithmetic

Vectors can be added: $\binom{3}{2} + \binom{5}{4} = \binom{3+5}{2+4} = \binom{8}{6}$

Vectors can be multiplied by a number. Remember to multiply *both* numbers. $2\times\binom{1}{3} = \binom{2\times1}{2\times3} = \binom{2}{6}$

4. Vector Diagrams

In this diagram, $\overrightarrow{OA} = \binom{2}{3}$ and $\overrightarrow{AB} = \binom{4}{2}$

Adding these vectors, we get:

$$\binom{2}{3} + \binom{4}{2} = \binom{2+4}{3+2} = \binom{6}{5}$$

which is the same as vector $\overrightarrow{OB}$. So we have the result: $\overrightarrow{OA} + \overrightarrow{AB} = \overrightarrow{OB}$

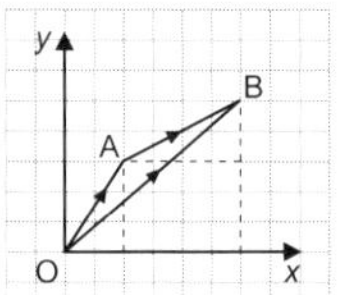

In other words, travelling from O to A and then from A to B is the same as travelling directly from O to B

Question *M is the midpoint of OQ. Given that* $\overrightarrow{OM} = \binom{2}{3}$ *and* $\overrightarrow{OP} = \binom{5}{7}$ *find the vectors* $\overrightarrow{OQ}$ *and* $\overrightarrow{PQ}$.

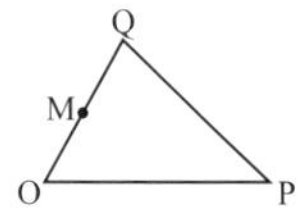

Since M is the mid–point of OQ then $\overrightarrow{OQ} = 2\times\overrightarrow{OM}$

So, $\overrightarrow{OQ} = 2\times\overrightarrow{OM} = 2\times\binom{2}{3} = \binom{4}{6}$

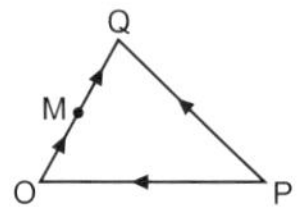

Travelling from P to Q is the same as travelling from P to O and then from O to Q. So, we can write: $\overrightarrow{PQ} = \overrightarrow{PO} + \overrightarrow{OQ} = -\overrightarrow{OP} + \overrightarrow{OQ}$

Thus, $\overrightarrow{PQ} = -\binom{5}{7} + \binom{4}{6} = \binom{-5+4}{-7+6} = \binom{-1}{-1}$

Question *In this diagram,* $\overrightarrow{OR} = \mathbf{a}$ *and* $\overrightarrow{OP} = \mathbf{b}$. *Given that* $\overrightarrow{PQ} = 2\overrightarrow{OR}$ *and* $\overrightarrow{OS} = \frac{1}{3}\overrightarrow{OQ}$, *express in terms of* **a** *and* **b** *the vectors* $\overrightarrow{PQ}$, $\overrightarrow{OQ}$ *and* $\overrightarrow{OS}$. *Use your answers to express* $\overrightarrow{PS}$ *and* $\overrightarrow{RS}$ *in terms of* **a** *and* **b**. *What can you say about points P, S and R?*

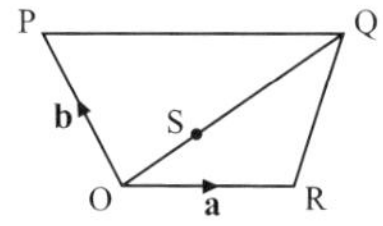

$\overrightarrow{PQ} = 2\times\overrightarrow{OR} = 2\times\mathbf{a} = 2\mathbf{a}$

$\overrightarrow{OQ} = \overrightarrow{OP} + \overrightarrow{PQ} = \mathbf{b}+2\mathbf{a}$

$\overrightarrow{OS} = \frac{1}{3}\times\overrightarrow{OQ} = \frac{1}{3}(\mathbf{b}+2\mathbf{a}) = \frac{1}{3}\mathbf{b}+\frac{2}{3}\mathbf{a}$

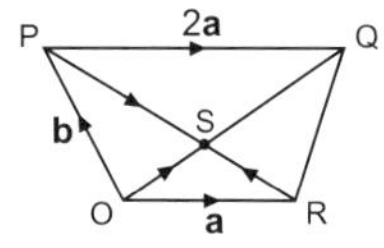

$\overrightarrow{PS} = \overrightarrow{PO} + \overrightarrow{OS} = -\overrightarrow{OP} + \overrightarrow{OS}$

$= -\mathbf{b}+\frac{1}{3}\mathbf{b}+\frac{2}{3}\mathbf{a} = -\frac{2}{3}\mathbf{b}+\frac{2}{3}\mathbf{a}$

$\overrightarrow{RS} = \overrightarrow{RO} + \overrightarrow{OS} = -\overrightarrow{OR} + \overrightarrow{OS} = -\mathbf{a}+\frac{1}{3}\mathbf{b}+\frac{2}{3}\mathbf{a} = \frac{1}{3}\mathbf{b} - \frac{1}{3}\mathbf{a}$

Notice that $\overrightarrow{PS} = -2\times\overrightarrow{RS}$ and so $\overrightarrow{PS}$ must be parallel to $\overrightarrow{RS}$ (see Box 2). But the lines PS and RS are joined at S, and so PSR must be a straight line.

5. Solving Problems Using Vectors

Question *A man wants to swim across a river. He swims through the water at a speed of 4 km/h, represented by the vector in the diagram. A current of 2 km/h drags him along the river.*

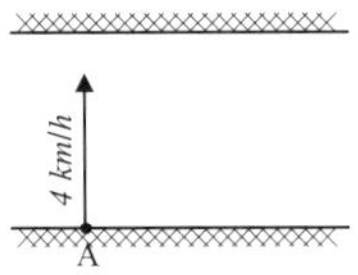

(a) Using a scale of 1 cm to 4 km/h, draw speed vectors for the current and the swimmer crossing the river.

(b) Find the speed of the swimmer by measurement and mark the point B where he reaches the opposite bank.

(a) Draw vector $\overrightarrow{PQ}$ 0.5 cm long to represent the speed of the current, 2 km/h. The actual speed of the swimmer is the combination of his speed through the water, $\overrightarrow{AP}$, and the speed of the current, $\overrightarrow{PQ}$. So, $\overrightarrow{AQ}$ represents the actual speed and course of the swimmer ($\overrightarrow{AQ} = \overrightarrow{AP} + \overrightarrow{PQ}$). Draw this vector.

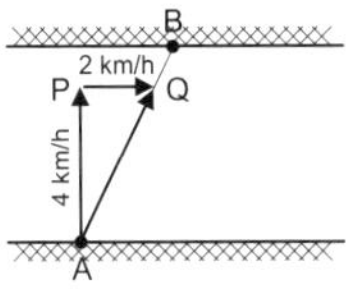

(b) Carefully measure $\overrightarrow{AQ}$ to be 1.1 cm. The scale is 1 cm = 4 km/h and so the speed of the swimmer is 1.1×4 km/h = 4.4 km/h. Continue the line AQ to B, where the swimmer reaches the opposite bank.

Transformations

1. Transformation

A **transformation** is a way of moving the points of a shape from one position to another.

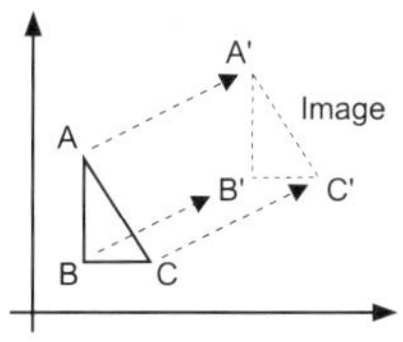

The corner points A, B and C have been transformed into the points A', B' and C'. We say that triangle A'B'C' is the **image** of triangle ABC under this transformation.

2. Translation

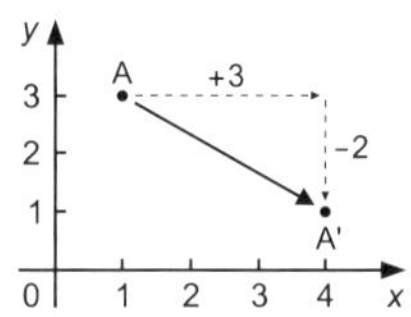

The point A has moved 3 units right and 2 units down. We say that the point A has been **translated** to point A'. This translation can be summarised using a column vector called a **translation vector**:

$\begin{pmatrix}3\\-2\end{pmatrix}$ ← Movement in the x-direction ← Movement in the y-direction

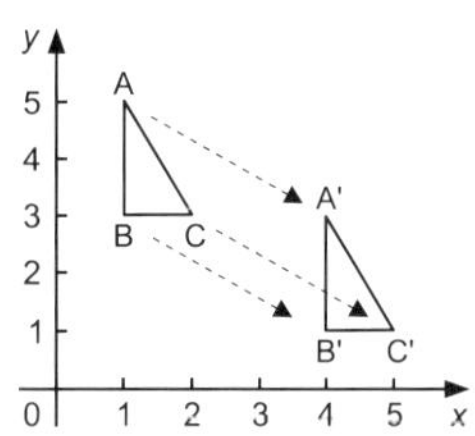

Each corner of triangle ABC has been translated using the vector $\begin{pmatrix}3\\-2\end{pmatrix}$.

3. Reflection

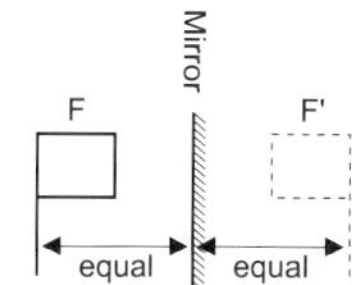

Flag F' is the **reflection** of flag F in the mirror. Each point of the flag F is the same distance from the mirror as the corresponding point of its reflection F'.

Question *Draw the reflection of triangle ABC in the y-axis and label the image A'B'C'.*

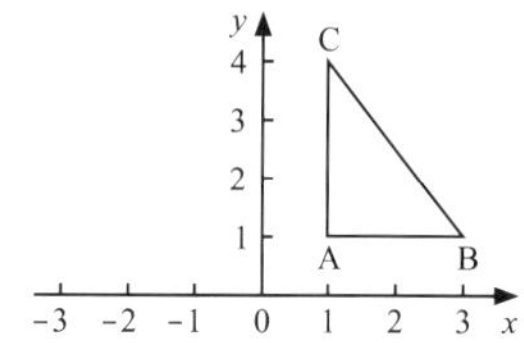

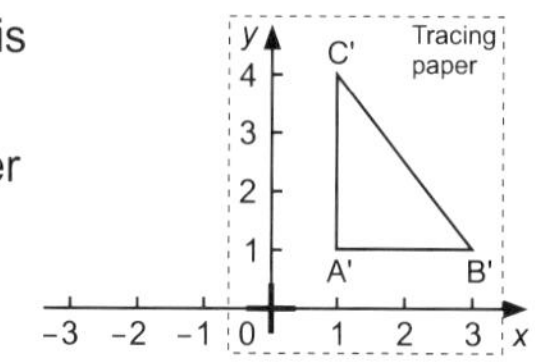

The mirror is the y-axis and is called the **line of reflection**.

Place a piece of tracing paper over the triangle and line of reflection. Trace the triangle and the y-axis. Label the corners A'B'C'. Draw a cross over the origin.

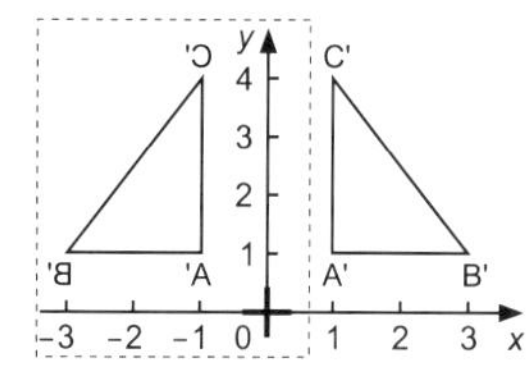

Turn over the tracing paper and align the cross over the origin. Use a pencil or the point of your compasses to mark points A', B' and C' and draw the new triangle A'B'C' (the image).

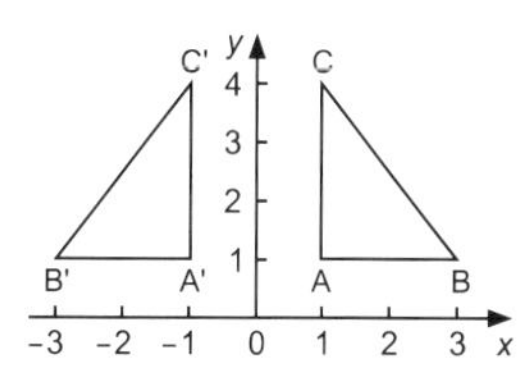

Carefully label the corners of the image A'B'C'. Corner A' should be opposite corner A and the same distance from the line of reflection; and the same for points B and C.

4. Rotation

To **rotate** a shape, you need to know the centre about which it is rotated and the angle of rotation.

Question *Rotate the shape PQRS 90° clockwise about the point (1, 0) and label the image P'Q'R'S'.*

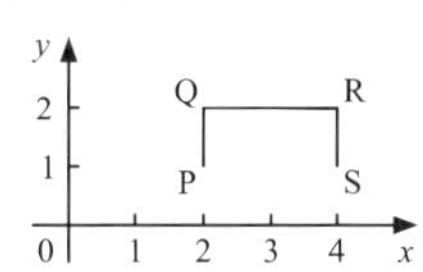

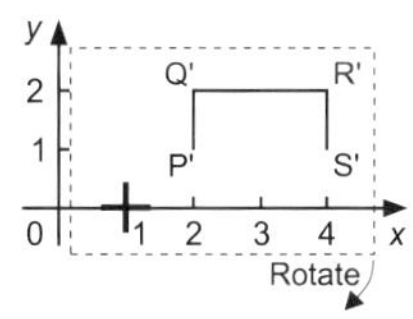

Place a piece of tracing paper over the shape and the point (1, 0). Trace the shape, labelling the corners P'Q'R'S'. Draw a cross over the centre of rotation (1, 0), as shown, and trace part of the x-axis. This will help you to position the image accurately.

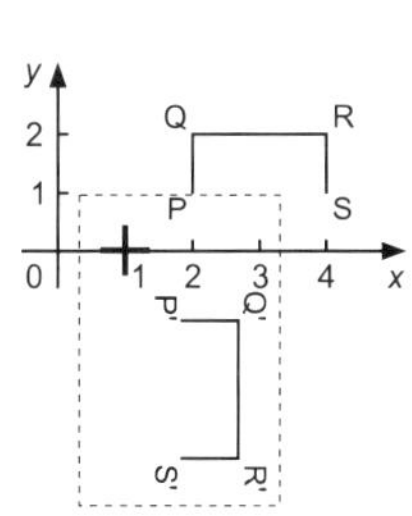

Push the point of your compasses through the centre of rotation and rotate the tracing paper 90° clockwise, aligning the cross over the centre of rotation once again.

Mark the points P'Q'R'S' and draw the image.

How to find the centre of rotation

Question *The triangle ABC has been rotated to obtain the image A'B'C'. Find the centre and angle of rotation*

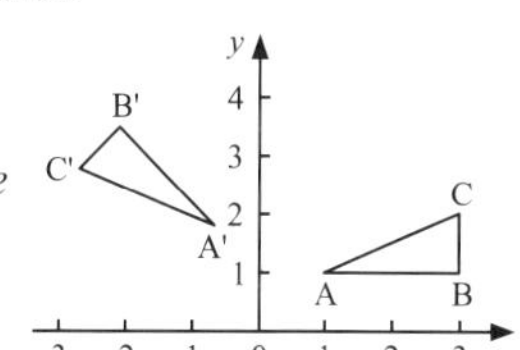

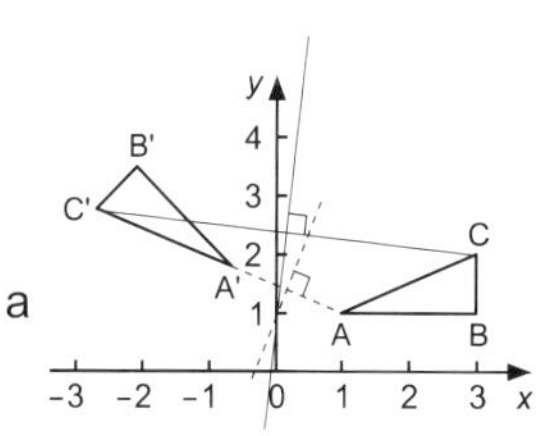

Use feint construction lines to join two pairs of corresponding points, e.g. AA' and CC'.

Use your ruler to find the midpoint of each line and draw a perpendicular line through it.

The centre of rotation is where these two lines cross. In this example, the centre of rotation is the point (0, 1).

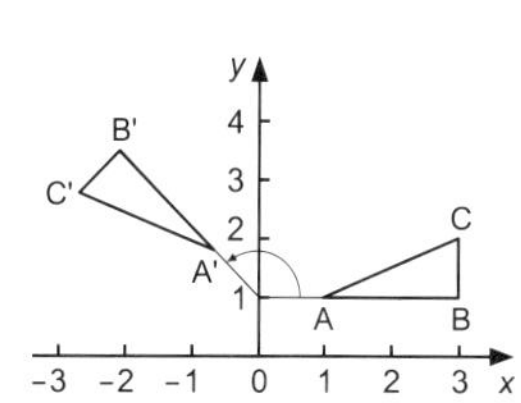

To find the angle of rotation, join two corresponding points to the centre of rotation, e.g. A and A'. Measure the angle from A to A', 130° anticlockwise.

Transformations (Contd)

5. Enlargement

To **enlarge** a shape, you need to know the centre from which to enlarge it and the scale factor.

Question *Enlarge the rectangle PQRS by a scale factor of 3, using the centre (0, 0).*

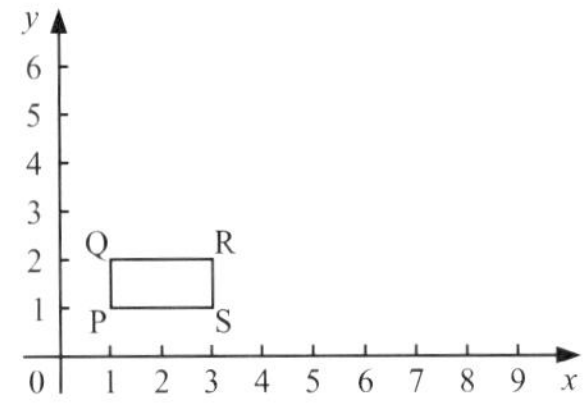

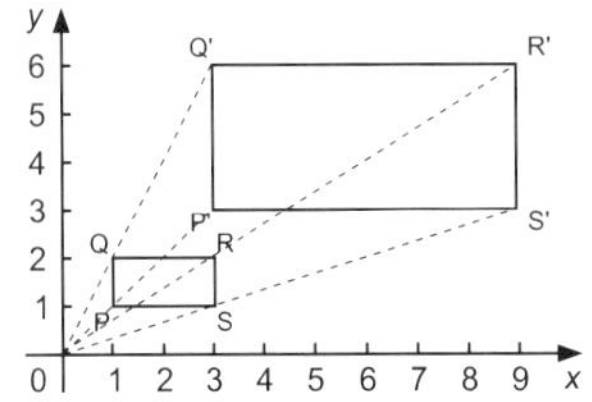

Take each point in turn. Let's start with P.

From the centre to P is 1 unit right, 1 unit up. Multiplying these distances by the scale factor 3 gives the position of the image P', 3 units right and 3 units up from the centre.

Similarly, from the centre to Q is 1 unit right, 2 units up. After multiplying by 3, these distances become 3 units right and 6 units up, giving the position of the image Q'. And so on. Join up the points P'Q'R'S' to give the enlarged image. Notice that the length and breadth of the rectangle P'Q'R'S' are 3 times those of PQRS.

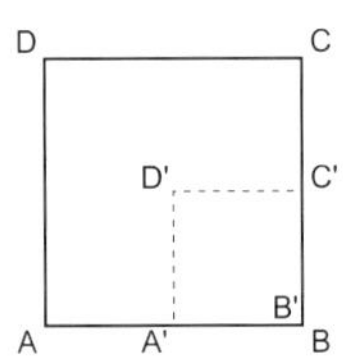

Here, square ABCD has been enlarged by a scale factor of $\frac{1}{2}$, using point B as centre.

Because the scale factor is a fraction (less than 1), this enlargement has *reduced* the size of square ABCD.

How to find the centre and scale factor of an enlargement

Question *Triangle A'B'C' is an enlargement of triangle ABC. Find the centre and scale factor of the enlargement.*

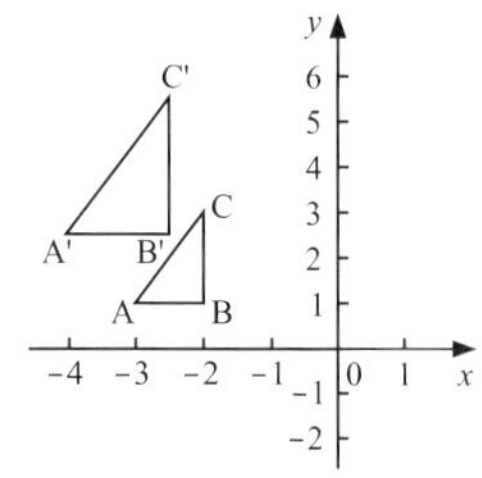

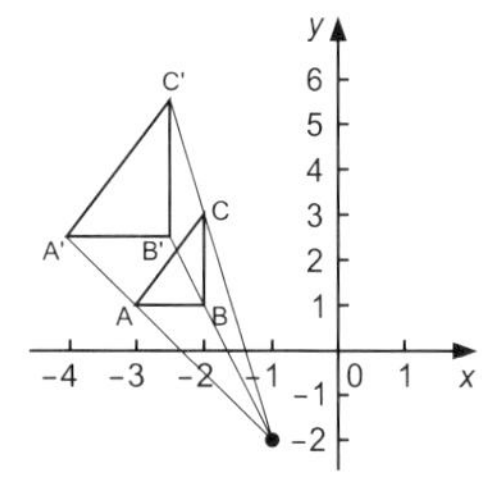

Draw straight lines through pairs of corresponding points, e.g. A to A'. The centre of enlargement is where these lines meet, i.e. (–1, –2).

To find the scale factor of enlargement, divide a length of the image A'B'C' by the corresponding length of triangle ABC. For example, A'B' = 1.5 and AB = 1. So, the scale factor = 1.5÷1 = 1.5.

Negative scale factor

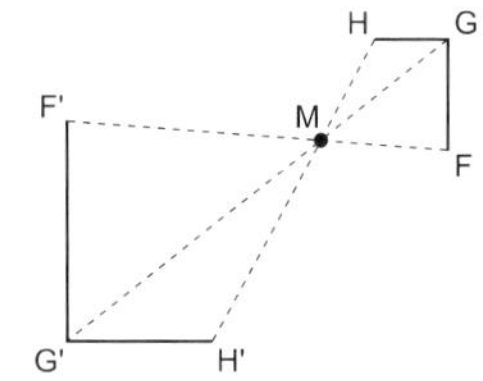

A negative scale factor, e.g. –2, means that the enlarged image of a shape will be on the opposite side of the centre.

The shape FGH has been enlarged by a scale factor of –2 using M as the centre. The distance from M to F' is 2 times the distance from M to F, but in the opposite direction.

6. Identifying Transformations

Question *P, Q, R and S are the images of different transformations of the shape T. Fully describe these transformations.*

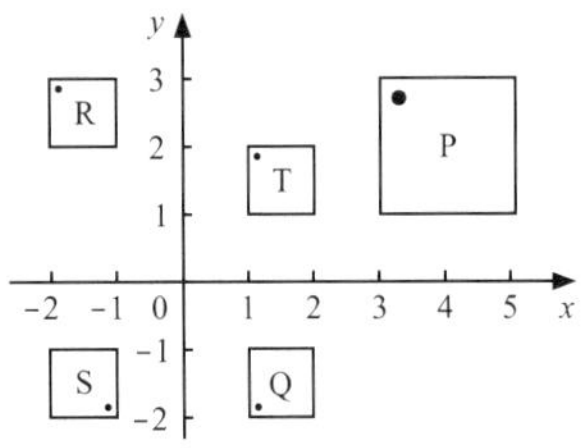

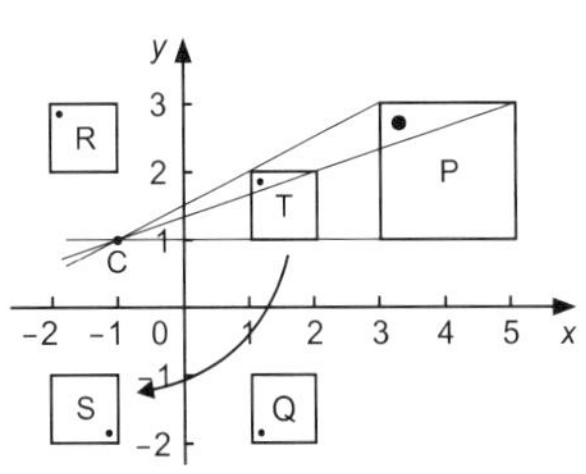

P is an enlargement of T, with scale factor 2. Joining up the corresponding points shows the centre to be (–1, 1).

Q is a reflection of T in the *x*–axis.

R is a translation with vector $\binom{-3}{1}$.

You can see from the position of the dot that S is a 180° rotation. The centre is the origin (0, 0).

7. Combining Transformations

Question *Triangle ABC has been reflected in the x–axis to give image A'B'C'. This image has then been reflected in the y–axis to give image A''B''C''. Fully describe the single transformation that changes ABC into A''B''C''.*

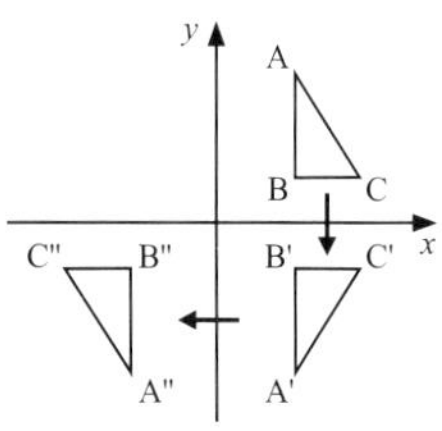

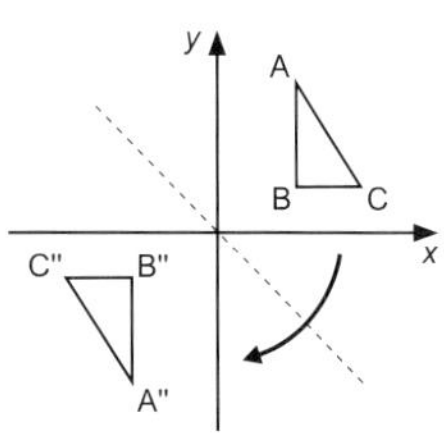

It cannot be a reflection in the dashed line because the reflection of A is not A''. So it must be a rotation of 180° with centre (0, 0). Check this deduction using tracing paper (see Box 4, p89).

8. Transformation Matrix

H

A **transformation matrix** is used to transform the points of a shape by multiplying their coordinates in a particular way. Transformation matrices are labelled using capital letters. For example, T = $\begin{pmatrix}-1 & 0\\0 & 1\end{pmatrix}$.

Question *Draw the image of the trapezium PQRS under the transformation*

$$T = \begin{pmatrix}-1 & 0\\0 & 1\end{pmatrix}$$

Fully describe the transformation.

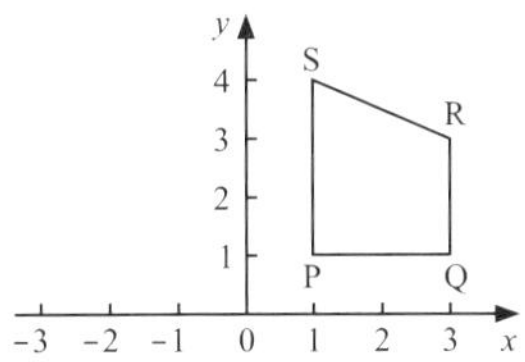

① Write the coordinates of each corner of the trapezium as a vector.

P(1, 1) becomes $\begin{pmatrix}1\\1\end{pmatrix}$ Q(3, 1) becomes $\begin{pmatrix}3\\1\end{pmatrix}$

R(3, 3) becomes $\begin{pmatrix}3\\3\end{pmatrix}$ S(1, 4) becomes $\begin{pmatrix}1\\4\end{pmatrix}$

② Multiply each vector by the matrix T to give the coordinates of the corners of the image.

For corner P: $\begin{pmatrix}-1 & 0\\0 & 1\end{pmatrix} \times \begin{pmatrix}1\\1\end{pmatrix} = \begin{pmatrix}-1\times1 + 0\times1\\0\times1 + 1\times1\end{pmatrix} = \begin{pmatrix}-1+0\\0+1\end{pmatrix} = \begin{pmatrix}-1\\1\end{pmatrix}$

So, the image P' of P has coordinates (−1, 1).

For corner Q: $\begin{pmatrix}-1 & 0\\0 & 1\end{pmatrix} \times \begin{pmatrix}3\\1\end{pmatrix} = \begin{pmatrix}-1\times3 + 0\times1\\0\times3 + 1\times1\end{pmatrix} = \begin{pmatrix}-3+0\\0+1\end{pmatrix} = \begin{pmatrix}-3\\1\end{pmatrix}$

So, the image Q' of Q has coordinates (−3, 1).

Similarly, R' has coordinates (−3, 3) and S' has coordinates (−1, 4).

③ Plot the coordinates of points P'Q'R'S'. Join the points to give the image P'Q'R'S'.

The diagram shows that the transformation T is a reflection in the *y*-axis.

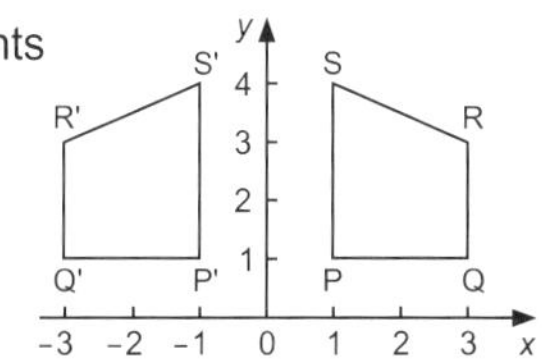

9. Identifying a Transformation Matrix

H

You may be given a transformation matrix and asked to describe the transformation it represents.

Question *Fully describe the transformation represented by the matrix:*

$$T = \begin{pmatrix}0 & -1\\1 & 0\end{pmatrix}$$

Make a sketch of the **unit square** OABC.

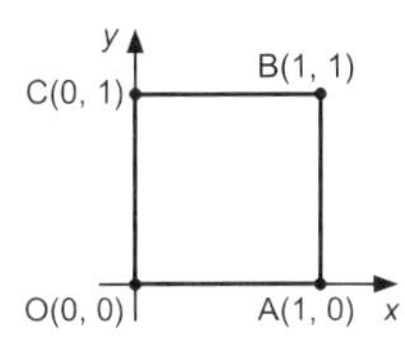

Use matrix T to transform points A, B and C.

For A(1, 0): $\begin{pmatrix}0 & -1\\1 & 0\end{pmatrix} \times \begin{pmatrix}1\\0\end{pmatrix} = \begin{pmatrix}0\\1\end{pmatrix}$

For B(1, 1): $\begin{pmatrix}0 & -1\\1 & 0\end{pmatrix} \times \begin{pmatrix}1\\1\end{pmatrix} = \begin{pmatrix}-1\\1\end{pmatrix}$

For C(0, 1): $\begin{pmatrix}0 & -1\\1 & 0\end{pmatrix} \times \begin{pmatrix}0\\1\end{pmatrix} = \begin{pmatrix}-1\\0\end{pmatrix}$

Sketch the image OA'B'C'. This shows that T represents a 90° anticlockwise rotation, centre (0, 0).

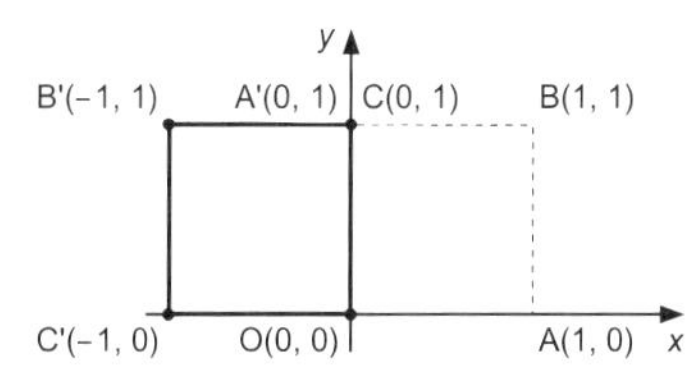

10. Combinations of Transformation Matrices

H

Question *(a) Fully describe the transformations represented by the matrices* $A = \begin{pmatrix}-1 & 0\\0 & 1\end{pmatrix}$ *and* $B = \begin{pmatrix}0 & -1\\1 & 0\end{pmatrix}$

(b) The shape T is transformed using the combination BA. Draw the image of T.

(c) Fully describe the single transformation that BA represents.

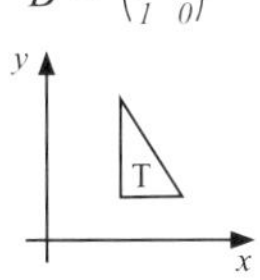

(a) From Boxes 8 and 9, A represents a reflection in the *y*-axis and B an anticlockwise rotation of 90° with centre (0, 0).

(b) The combination BA means 'apply A first, then B'.

Reflecting T in the *y*-axis and then rotating the image 90° anticlockwise about (0, 0) gives the image T'.

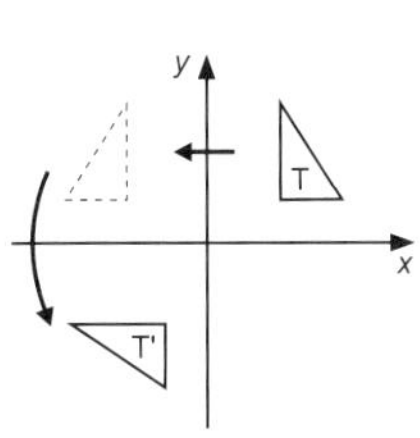

(c) T' is a reflection of T in the line *y* = −*x*.

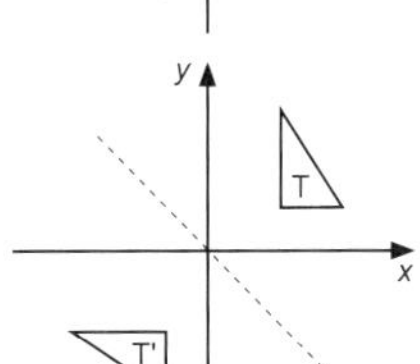

11. Inverse of a Transformation

H

The **inverse** of a transformation has the opposite effect. Here, T is a 90° anticlockwise rotation and so the inverse of T is a 90° clockwise rotation. The inverse of T is written T^{-1}. Applying T followed by its inverse T^{-1} leaves the shape unchanged. The inverse of T can be found using the formula:

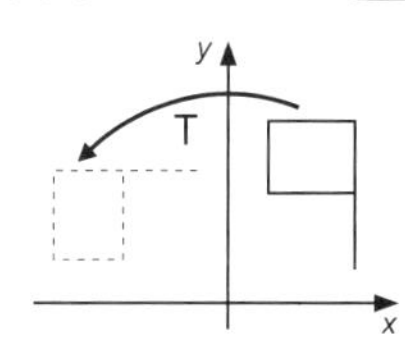

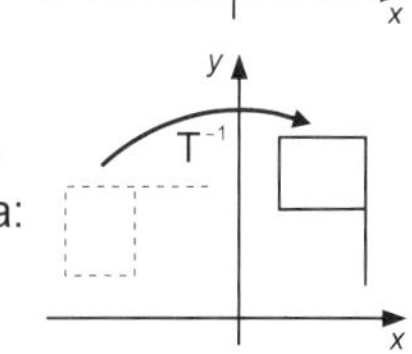

$$T = \begin{pmatrix}a & b\\c & d\end{pmatrix} \qquad T^{-1} = \frac{1}{ad-bc}\begin{pmatrix}d & -b\\-c & a\end{pmatrix}$$

Question *T is the transformation represented by the matrix* $T = \begin{pmatrix}0 & -1\\1 & 0\end{pmatrix}$.

(a) Describe the transformation T.

(b) Calculate the inverse T^{-1}.

(c) Describe the transformation represented by T^{-1}.

(a) From Box 9, we can see that T represents an anticlockwise rotation of 90° with centre (0, 0).

(b) For T = $\begin{pmatrix}0 & -1\\1 & 0\end{pmatrix}$, *a* = 0, *b* = −1, *c* = 1, *d* = 0 and so

$$T^{-1} = \frac{1}{0\times0--1\times1}\begin{pmatrix}0 & --1\\-1 & 0\end{pmatrix} = \frac{1}{1}\begin{pmatrix}0 & 1\\-1 & 0\end{pmatrix} = \begin{pmatrix}0 & 1\\-1 & 0\end{pmatrix}$$

(c) T^{-1} represents a clockwise rotation of 90° with centre (0, 0).

Symmetry. Nets. Tessellations

1. Line Symmetry

A **line of symmetry** divides a shape in half. It is like a mirror: each half is a reflection of the other.

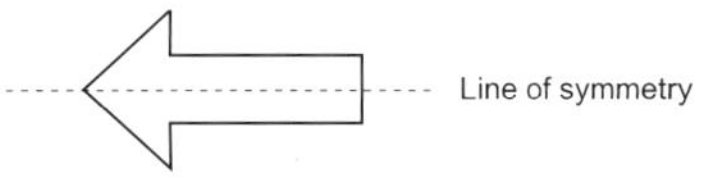

A shape can have several lines of symmetry:

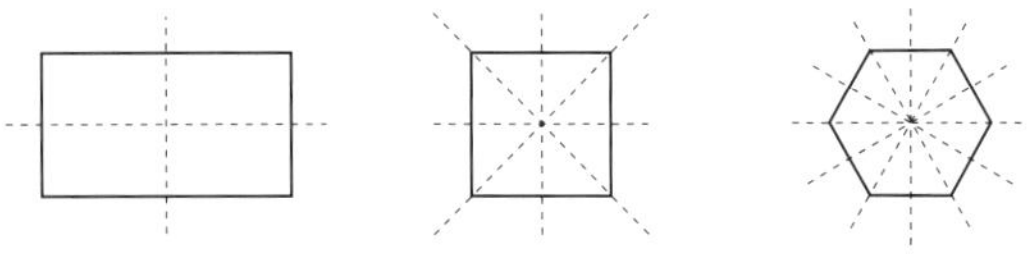

A rectangle has 2 lines of symmetry

A square has 4 lines of symmetry

A regular polygon has the same number of lines of symmetry as sides

Some shapes do not have line symmetry.

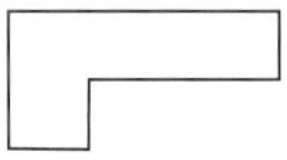

2. Rotational Symmetry

If the left-hand shape is rotated half a turn (180°), it becomes the right-hand shape. Ignoring the dot, these shapes appear the same. We say the shape has **rotational symmetry**. There are two positions in which the shape appears the same, so we say that the **order** of rotational symmetry is 2.

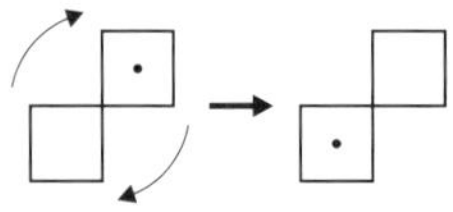

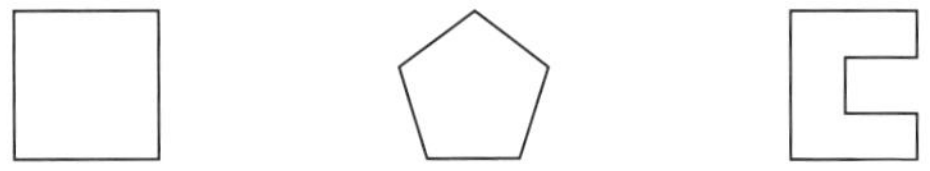

A square has rotational symmetry of order 4

A regular pentagon has rotational symmetry of order 5

This shape does not have rotational symmetry

Question *Which of the shapes below has*
(a) line symmetry but not rotational symmetry
(b) rotational symmetry but not line symmetry
(c) both line and rotational symmetry?

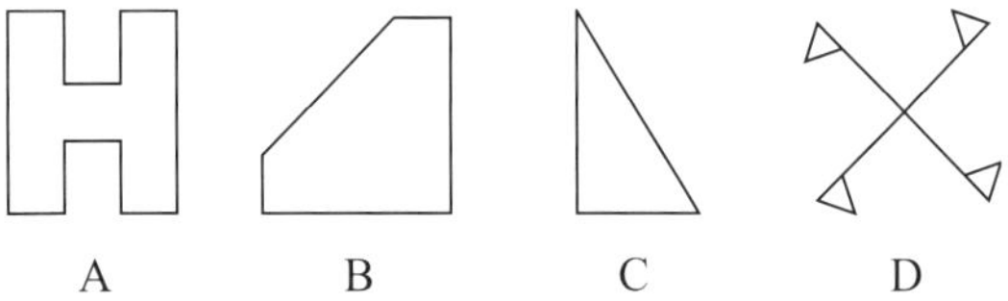

(a) B has line symmetry but not rotational symmetry.

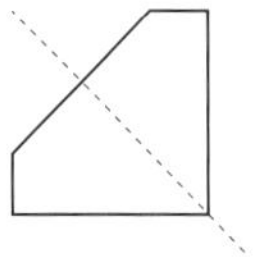

(b) D has rotational symmetry of order 4 but no lines of symmetry.

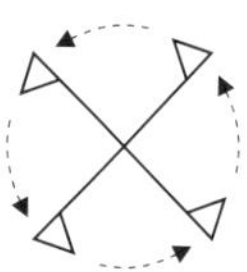

(c) A has both line symmetry and rotational symmetry of order 2.

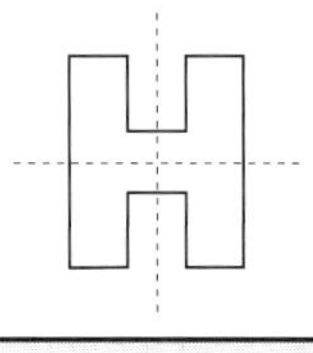

3. Nets

A **net** is a flat pattern that can be cut out and folded into a 3-D shape. The net opposite can be folded up to make a cube.

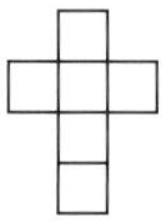

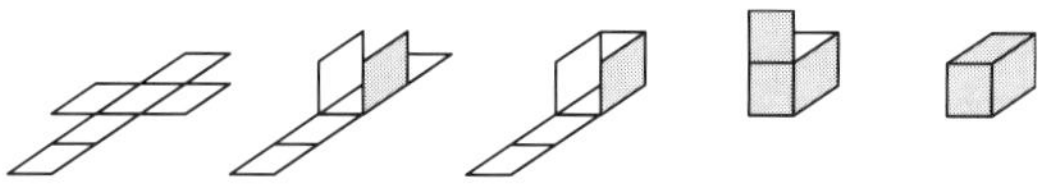

There are several different nets for each 3-D shape. Some of the nets for a cube are shown below, together with a pattern that is not a net.

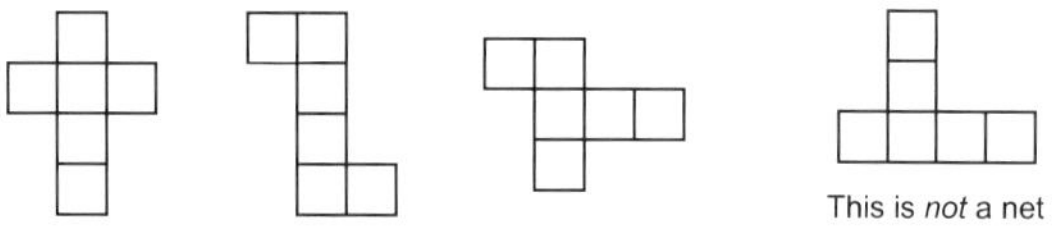

This is *not* a net

Question *Accurately draw a net of this prism.*

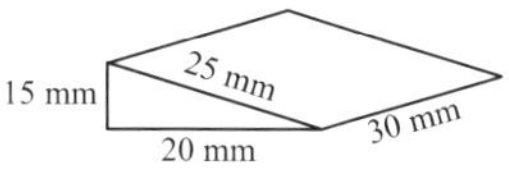

Start by drawing one of the central faces in the middle of the paper (see the bold lines in the diagram).

Add adjacent faces, mentally checking that they fold up correctly.

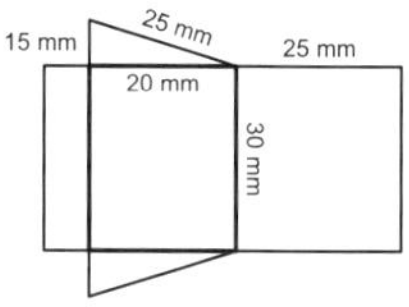

4. Tessellation

A tessellation is a repeating pattern made from one or more shapes. There are no gaps between the shapes.

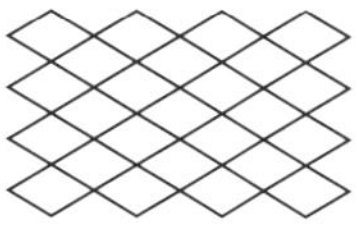

This tessellation is made from two shapes: a regular octagon and a square.

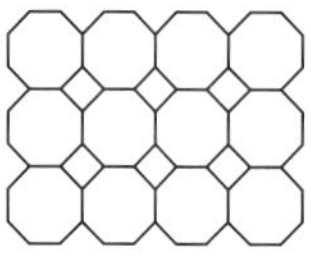

Any regular polygon whose interior angle divides exactly into 360° can be used to make a tessellation. This tessellation uses a hexagon, with an interior angle of 120° (see Box 3, p83).

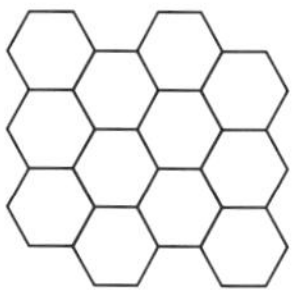

Any triangle can be used to make a tessellation. This is because the corner angles add up to 180° (half a circle) when placed next to each other.

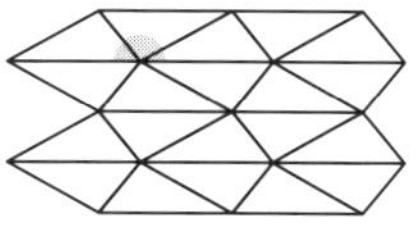

Any quadrilateral can be used to make a tessellation. This is because the corner angles add up to 360° (a circle) when placed next to each other.

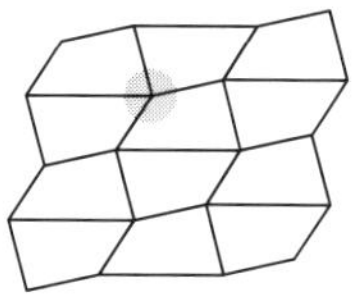

Trigonometry and Right-angled Triangles

1. Sides of a Right-angled Triangle

The side opposite the right angle is called the **hypotenuse**. It is the longest side.

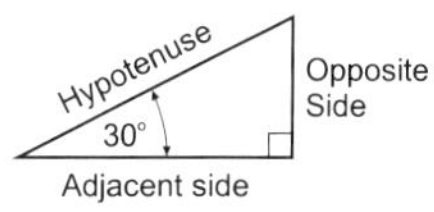

The side opposite the 30° angle is called the **opposite** side.

The side next to the 30° angle is called the **adjacent** side (because adjacent means 'next to').

2. Sin, Cos and Tan

Sine, cosine and tangent are the names of three fractions (called **trigonometric ratios**) made using the sides of a right-angled triangle. They are written briefly as sin, cos and tan.

$$\sin\theta = \frac{\text{Opposite}}{\text{Hypotenuse}}$$

$$\cos\theta = \frac{\text{Adjacent}}{\text{Hypotenuse}}$$

$$\tan\theta = \frac{\text{Opposite}}{\text{Adjacent}}$$

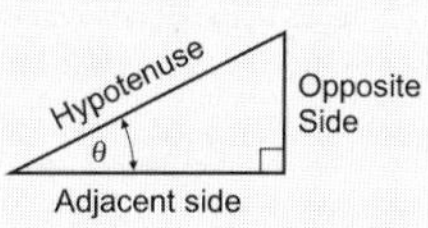

NOTE sin θ is written instead of just sin because the ratio depends on the size of the angle θ.

Question *Write down sin 30° as a fraction, using the sides of this triangle.*

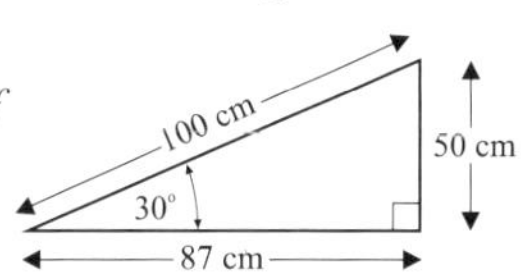

Label the sides as in Box 1

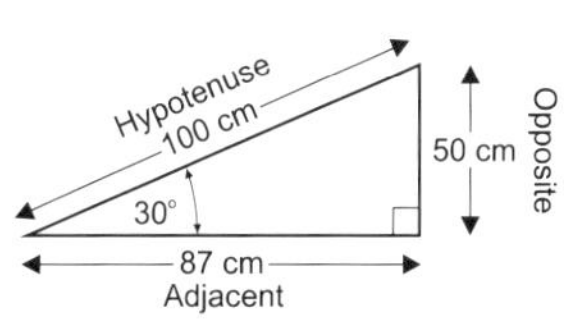

$$\sin 30° = \frac{\text{Opposite}}{\text{Hypotenuse}} = \frac{50}{100}$$

3. Using a Calculator to find Sin, Cos and Tan

Instead of drawing a triangle to find sin 30°, we can use a calculator. Make sure that your calculator is in **degrees mode**.

[3] [0] [sin] (0.5)

So, sin 30° = 0.5

NOTE More sophisticated calculators let you press the [sin] button first: [sin] [3] [0] [=]

Question *Calculate the value of tan 20° correct to 3 decimal places.*

[2] [0] [tan] (0.36397023)

So, tan 20° = 0.364 (3 dp)

4. Finding an Unknown Side

Question *Find the length of side AB, correct to 2 decimal places.*

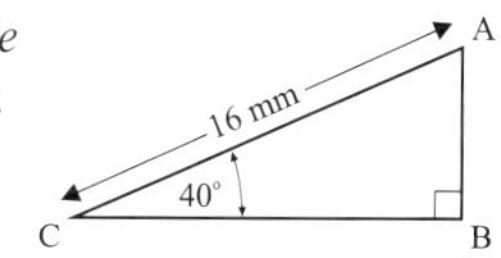

① Label the side you know (AC = 16 cm) and the side you want to find (AB).

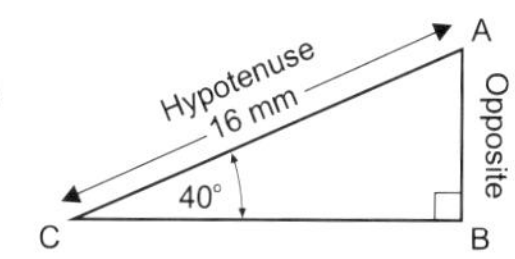

② Write down the equation in Box 2 that contains these two sides:

$$\sin\theta = \frac{\text{Opposite}}{\text{Hypotenuse}}$$

③ Fill in the angle and sides for this triangle:

$$\sin 40° = \frac{AB}{16}$$

④ Use your calculator to find the value of sin 40°

[4] [0] [sin] (0.64278761)

Keep this value displayed in your calculator, or store it in the memory. Write down an approximate value, e.g. sin 40° = 0.643.

⑤ Rewrite the equation and solve it:

$$0.643 = \frac{AB}{16}$$

$\therefore$ 0.643×16 = AB

$\therefore$ AB = 0.643×16 = 10.28 mm (2 dp)

All the calculator steps are shown below.

[4] [0] [sin] [×] [1] [6] [=] (10.2846018)

Question *A ladder leans against a wall and makes an angle of 72° with the ground. The foot of the ladder is 1.2 m from the wall. Calculate the length of the ladder.*

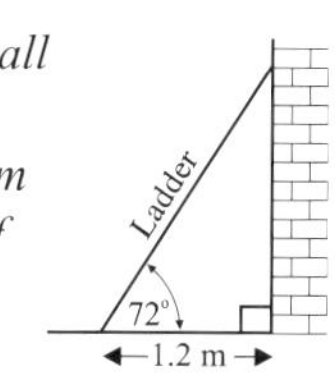

Let y be the length of the ladder

The ladder is the hypotenuse and the length 1.2 m is adjacent to the angle. From Box 2, use the equation:

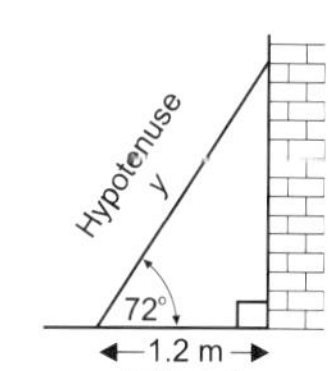

$$\cos\theta = \frac{\text{Adjacent}}{\text{Hypotenuse}}$$

$$\therefore \cos 72° = \frac{1.2}{y}$$

[7] [2] [cos] (0.30901699)

$\therefore 0.309 = \frac{1.2}{y}$ Move $\div y$ to LHS, and change to $\times y$

$\therefore 0.309y = 1.2$ Divide both sides by 0.309

$\therefore y = \frac{1.2}{0.309} = 3.88$ m

All the calculator steps are shown below:

[7] [2] [cos] [Min] [1] [·] [2] [÷] [MR] [=] (3.88328157)

Store the value of cos 72° in the memory

Trigonometry and Right–angled Triangles (Contd)

5. Sin^{-1}, Cos^{-1} and Tan^{-1}

If you enter an angle into your calculator and press the [sin] button you get a number. For example, pressing [3][0][sin] changes the angle 30° into the number 0.5.

The opposite (inverse) of sin is written $\sin^{-1}$. So, if you enter a number into your calculator and press [INV][sin] you get an angle. For example, pressing [0][·][5][INV][sin] changes the number 0.5 into the angle 30°.

Question *If* $\tan x = 1.3$ *find the angle* x.

When tan moves to the RHS it becomes the opposite (inverse) of tan, written $\tan^{-1}$.

$\tan x = 1.3$

$\therefore\ x = \tan^{-1} 1.3$

$= 52.43°$

[1][·][3][INV][tan] (52.4314080)

6. Finding an Unknown Angle

Question *Calculate the size of angle x, correct to the nearest 0.1°.*

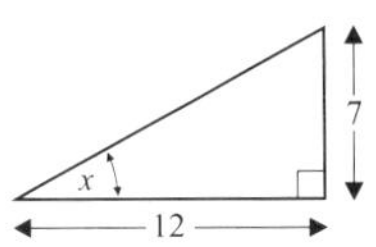

① Label the sides you know (12 and 7).

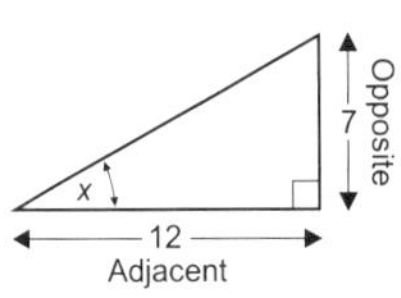

② Write down the equation in Box 2 that contains these two sides:

$$\tan x = \frac{\text{Opposite}}{\text{Adjacent}}$$

③ Fill in the sides for this triangle:

$$\tan x = \frac{7}{12}$$

④ Calculate the RHS using a calculator:

[7][÷][1][2][=] (0.58333333)

So, $\tan x = 0.58\dot{3}$

Keep this value displayed in your calculator.

⑤ Move tan to the RHS and change it to $\tan^{-1}$:

$\tan x = 0.58\dot{3}$

$\therefore\ x = \tan^{-1} 0.58\dot{3}$

⑥ Use your calculator to find the value of x.

You should still have 0.58333333 displayed on your calculator, from step 4. If not, calculate 7÷12 again. Here are all the calculator steps:

[7][÷][1][2][=][INV][tan] (30.2564372)

So, $x = 30.3°$

more

6. Finding an Unknown Angle (Contd)

Question *Find the angle between the blades of these scissors.*

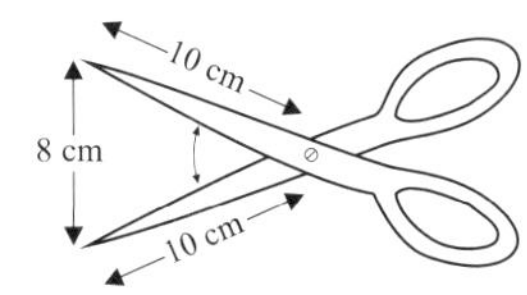

Draw the isosceles triangle made by the blades of the scissors:

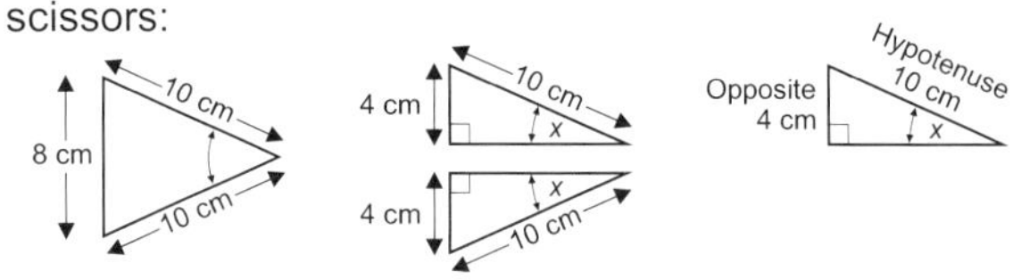

Split this into two identical right–angled triangles, each with angle x, say. Use one of the triangles to find x:

$\sin x = \frac{4}{10} = 0.4$

$\therefore\ x = \sin^{-1} 0.4 = 23.6°$

So, the angle between the blades = 2×23.6 = 47.2°.

7. Angles of Elevation and Depression

A person, P, looking straight ahead would have to look up (elevate) through an angle of 40° to see the bird and look down (depress) through an angle of 30° to see the ship.

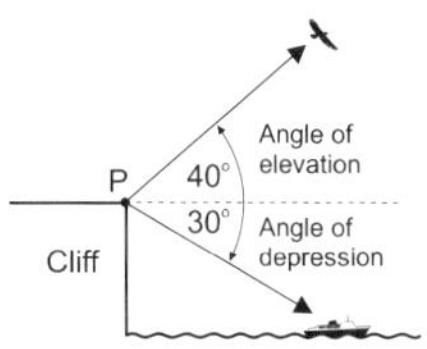

Question *A person looks out of a window 6 m above ground at a vertical flagpole. The angle of elevation of the top of the flagpole is 20°. The angle of depression of the foot of the flagpole is 40°.*

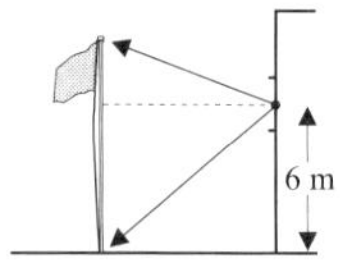

(a) Mark these angles on the diagram.
(b) Calculate the horizontal distance between the person and the flagpole.
(c) Calculate the height of the flagpole.

(a) See diagram.

(b) The diagram contains two right–angled triangles.

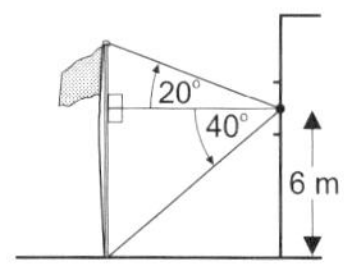

For this triangle:

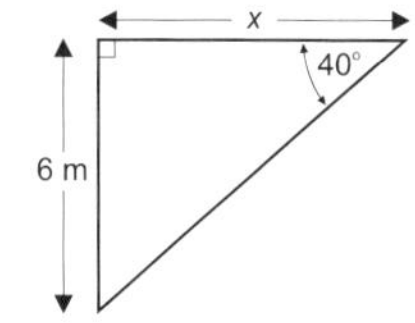

$\tan 40° = \frac{6}{x}$

$\therefore\ 0.839 = \frac{6}{x}$

$\therefore\ 0.839x = 6$

$\therefore\ x = \frac{6}{0.839} = 7.15\text{ m}$

Mark this length on the diagram to help you below.

For this triangle:

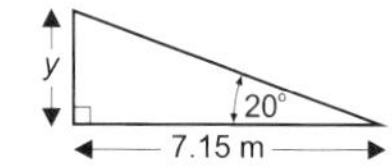

$\tan 20° = \frac{y}{7.15}$

$\therefore\ 0.364 = \frac{y}{7.15}$

$\therefore\ 0.364 \times 7.15 = y$

$\therefore\ y = 2.60\text{ m}$

So, height of flagpole = 6+2.60 = 8.60 m.

8. Trigonometry and Bearings

Question *Reading is 85 km due East of Bristol and 52 km due North of Portsmouth. Calculate*
(a) the bearing of Portsmouth from Bristol
(b) the distance between Portsmouth and Bristol.

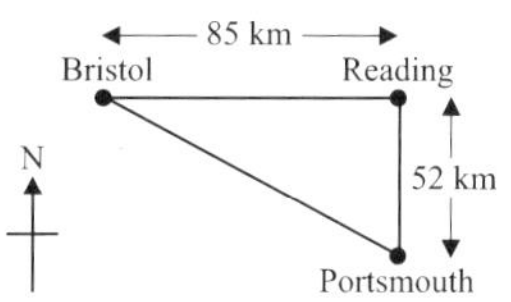

(a) The diagram shows that the bearing of Portsmouth from Bristol is 90° plus angle x. So, we must find x

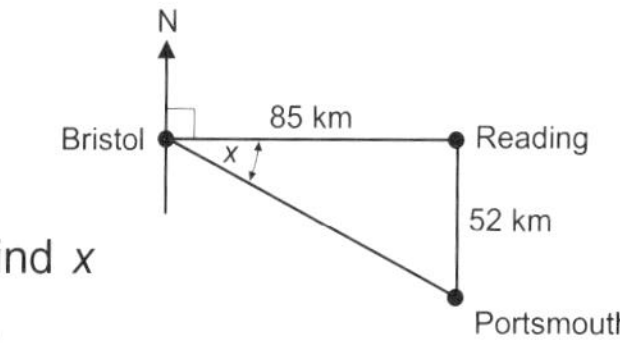

$$\tan x = \frac{52}{85} = 0.612$$

$\therefore\ x = \tan^{-1} 0.612 = 31.5°$

So, the bearing is $90°+31.5° = 121.5°$

(b) From Pythagoras' theorem, we have:

$(\text{Distance})^2 = 85^2+52^2 = 9929$

$\therefore$ Distance $= \sqrt{9929} = 99.6 \approx 100$ km

Sine and Cosine Rules H

1. Sine and Cosine Rules

The sine and cosine rules can be used to find the unknown sides and angles of any kind of triangle. They are normally used with triangles that do not have a right angle.

It is easy to decide which rule to use:

Use the sine rule if you know a side and the angle opposite.
Use the cosine rule if you cannot use the sine rule.

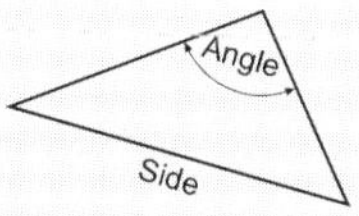

2. Labelling a Triangle

The capital letters A, B and C represent the corner angles.

The lower case letters a, b and c represent the sides opposite the angles A, B and C.

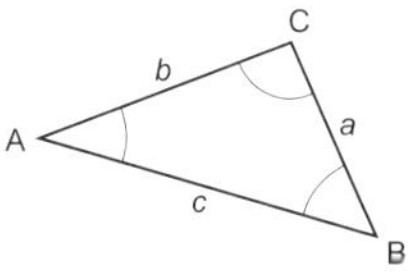

3. Sine Rule

The sine rule says that the fraction

$$\frac{\text{Side}}{\text{Sine of opposite angle}}$$

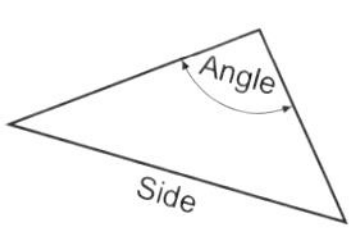

is the same for all three sides of the triangle.

The sine rule can be expressed algebraically:

$$\frac{a}{\sin A} = \frac{b}{\sin B} = \frac{c}{\sin C}$$

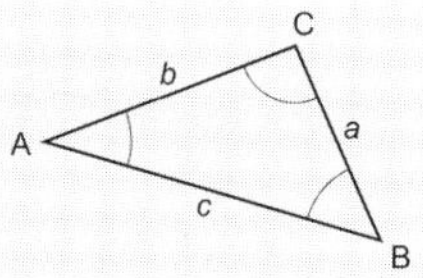

more

3. Sine Rule (Contd)

Question *Area ABC is a fenced area of woodland. Find the length of BC.*

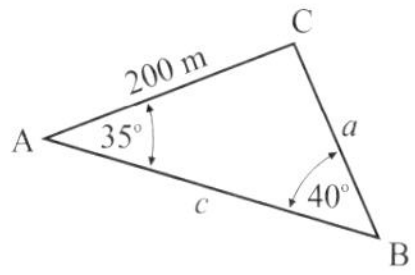

Here, $b = 200$, $A = 35°$ and $B = 40°$.

We want to find a, the length of side BC.

Substitute these values in the sine rule:

$$\frac{a}{\sin 35°} = \frac{200}{\sin 40°} \left(= \frac{c}{\sin C} \right)$$ Discard this unwanted part

$$\therefore\ \frac{a}{0.574} = \frac{200}{0.643}$$

$$\therefore\ \frac{a}{0.574} = 311.14$$

$\therefore\ a = 311.14 \times 0.574 = 178.5$ m

Question *Q is 25 km from P on a bearing of 075° and 40 km from R. P is due North of R. Calculate the bearing of Q from R, correct to the nearest degree.*

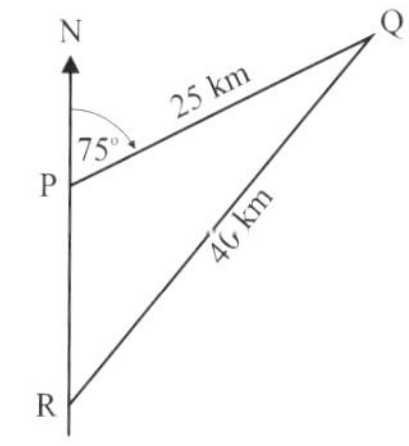

Angle $P = 180°-75° = 105°$

We want to find angle R.

Applying the sine rule, we have:

$$\frac{25}{\sin R} = \frac{40}{\sin 105°}$$

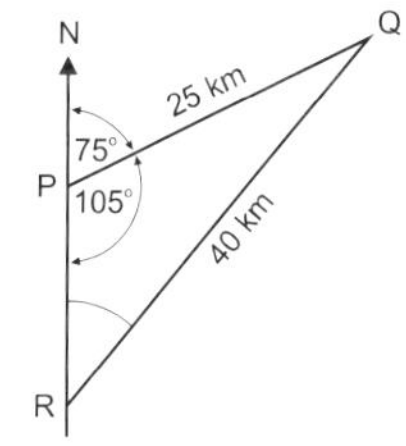

Turn both sides of the equation upside down, to get R on top:

$$\therefore\ \frac{\sin R}{25} = \frac{\sin 105°}{40}$$

$$\therefore\ \frac{\sin R}{25} = 0.024$$

$\therefore\ \sin R = 25 \times 0.024 = 0.604$

$\therefore\ R = \sin^{-1} 0.604 = 37.1° \approx 37°$

The bearing of Q from R is 037°.

Sine and Cosine Rules (Contd)

4. Cosine Rule

There are two formulae for the cosine rule: one to find lengths, the other to find angles.

To find the length a, use:

$a^2 = b^2 + c^2 - 2bc \cos A$

To find the angle A, use:

$\cos A = \dfrac{b^2 + c^2 - a^2}{2bc}$

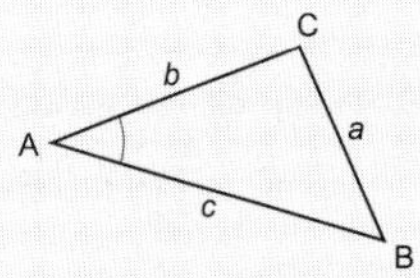

Question *Find the length BC.*

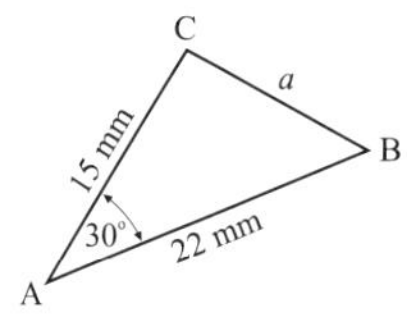

We do not know a side and its opposite angle and so cannot use the sine rule. So, we must use the cosine rule.

Here, $b = 15$, $c = 22$ and $A = 30^\circ$

We want to find a:

$a^2 = b^2 + c^2 - 2bc \cos A$

$= 15^2 + 22^2 - 2 \times 15 \times 22 \times \cos 30^\circ$

$= 225 + 484 - 571.6$

$= 137.4$

So, $a = \sqrt{137.4} = 11.7$ mm

Question *The diagram shows the cross-section of a roof. Calculate angles x, y and z.*

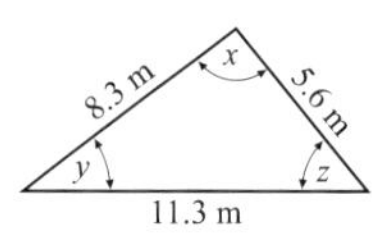

If we consider angle x to be angle A in the cosine rule, then $a = 11.3$, $b = 5.6$ and $c = 8.3$

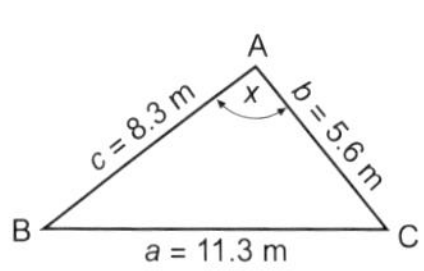

$\cos x = \dfrac{b^2 + c^2 - a^2}{2bc}$

$= \dfrac{5.6^2 + 8.3^2 - 11.3^2}{2 \times 5.6 \times 8.3}$

$= \dfrac{-27.44}{92.96}$

$= -0.295$

So, $x = \cos^{-1} -0.295 = 107.2^\circ$

Mark this angle on the diagram; it may help you to find other sides and angles.

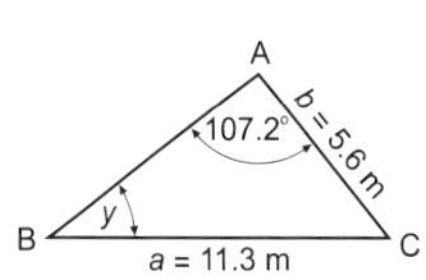

We now know a side (11.3 m) and the opposite angle (107.2°) and so can use the easier sine rule to find angle y:

$\dfrac{5.6}{\sin y} = \dfrac{11.3}{\sin 107.2^\circ}$

$\therefore \dfrac{\sin y}{5.6} = \dfrac{\sin 107.2^\circ}{11.3}$ Turn both sides of the equation upside down

$\therefore \dfrac{\sin y}{5.6} = 0.085$

$\therefore \sin y = 5.6 \times 0.085 = 0.473$

$\therefore y = \sin^{-1} 0.473 = 28.3^\circ$

Since the angles of a triangle add up to 180° we have:

$z = 180^\circ - 107.2^\circ - 28.3^\circ = 44.5^\circ$

4. Cosine Rule (Contd)

Question *Calculate the length of the ramp, BC.*

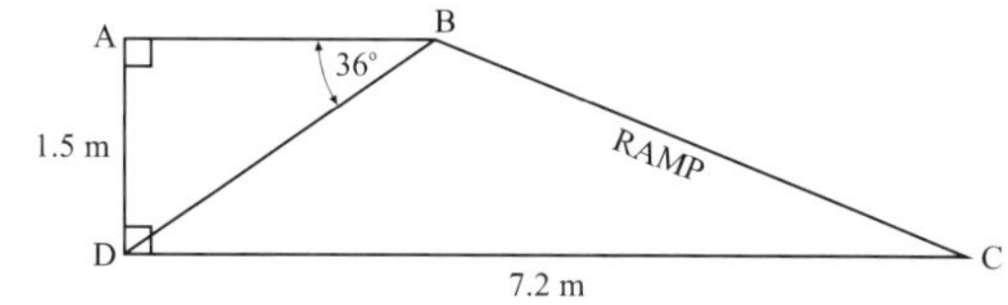

Triangle BCD contains the side BC we want.

To use the cosine rule with triangle BCD, we need to know sides BD and DC and angle BDC (opposite BC).

We know side DC = 7.2 m.

We can also find angle BDC. Sides AB and DC are parallel because they are both perpendicular to side AD. So, angle BDC = 36° (alternate angles).

We can find side BD from the right-triangled ADB using trigonometry:

$\sin 36^\circ = \dfrac{1.5}{BD}$

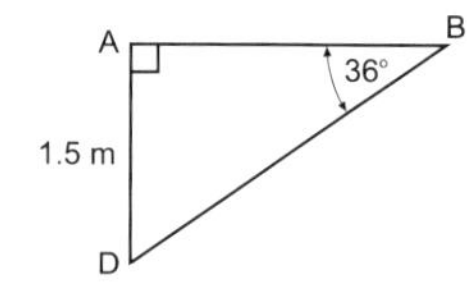

$\therefore BD \sin 36^\circ = 1.5$

$\therefore BD = \dfrac{1.5}{\sin 36^\circ} = 2.552$ m

Now we can apply the cosine rule to triangle BCD:

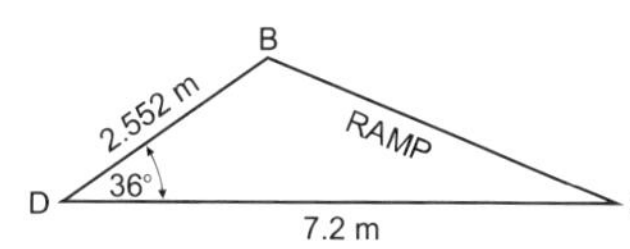

$BC^2 = BD^2 + DC^2 - 2 \times BD \times DC \times \cos D$

$= 2.552^2 + 7.2^2 - 2 \times 2.552 \times 7.2 \times \cos 36^\circ$

$= 6.513 + 51.84 - 29.73$

$= 28.62$

So, $BC = \sqrt{28.62} \approx 5.4$ m (1 dp)

5. Area of a Triangle = $\frac{1}{2} ab \sin C$

When you do not know the height of a triangle, use the following formula to calculate its area:

Area $= \frac{1}{2} ab \sin C$

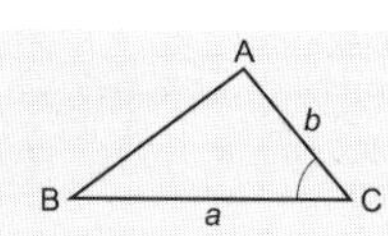

Question *The diagram shows a postage stamp in the shape of an equilateral triangle. Calculate its area.*

You could split this triangle into two right-angled triangles and calculate its height (see Box 6, p94).

But it is quicker to use the $\frac{1}{2} ab \sin C$ formula.

All the sides of this triangle are 3 cm and so $a = 3$ and $b = 3$. The angles are all 60° and so $C = 60^\circ$.

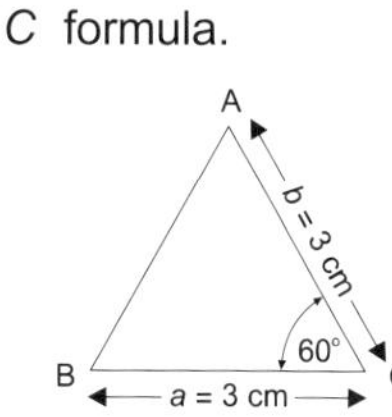

Area $= \frac{1}{2} ab \sin C$

$= \frac{1}{2} \times 3 \times 3 \times \sin 60^\circ$

$= 0.5 \times 3 \times 3 \times 0.866$

$= 3.9 \text{ cm}^2$

more

Three–dimensional Trigonometry

Triangles can be used to find unknown sides and angles of three–dimensional shapes.

The method is to find a triangle that contains the side or angle you want. Then draw the triangle, marking on it the lengths and angles you know. Look out for isosceles triangles and right–angled triangles.

Question *A road sign is made from an equilateral triangle of side 45 cm and is supported by two struts, each 30 cm long.*

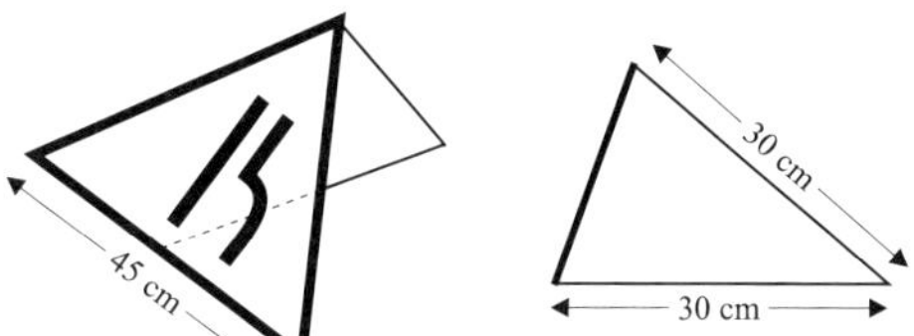

SIDE VIEW

(a) Calculate the height of the equilateral triangle.
(b) Calculate the angle the sign makes with the ground.

(a) The angles of an equilateral triangle are all 60°. Let h be the height of the triangle, then:

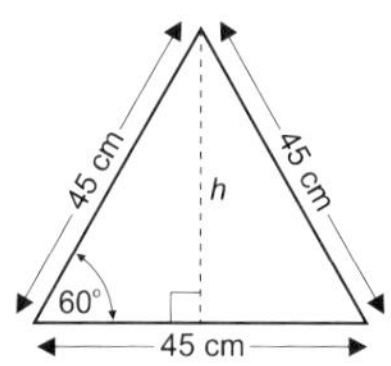

$$\sin 60^\circ = \frac{h}{45}$$
$$\therefore\ 0.866 = \frac{h}{45}$$
$$\therefore\ 0.866 \times 45 = h$$
$$\therefore\ h = 39.0 \text{ cm}$$

(b) Split the isosceles triangle into two identical right–angled triangles and let x be the angle between the sign and the ground. Then

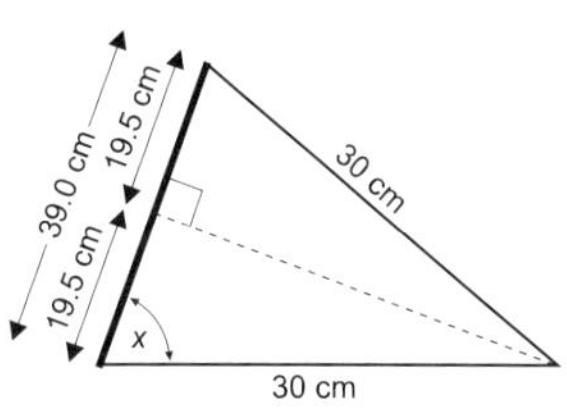

$$\cos x = \frac{19.5}{30} = 0.65$$
$$\therefore\ x = \cos^{-1} 0.65 = 49.5^\circ$$

NOTE You could also use the cosine rule since you know all three sides of the triangle.

Question **H** *The angle of elevation of the top of the Eiffel tower from Notre–Dame, 4.1 km away, is 4.5°. The Arc de Triomphe is 1.6 km from the Eiffel tower.*

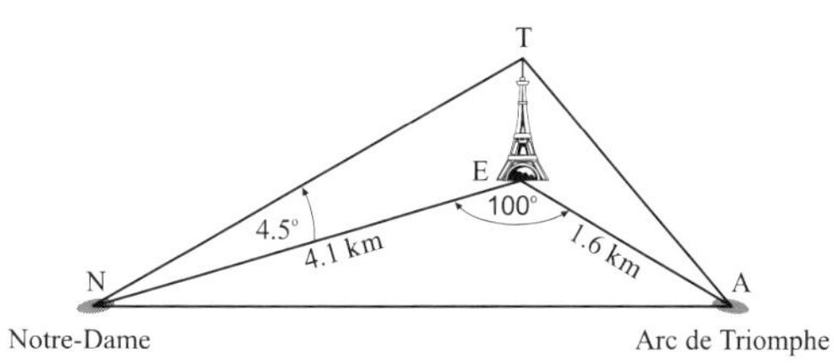

Calculate:
(a) the height of the Eiffel tower, correct to the nearest 10 m
(b) the angle of elevation of the top of the Eiffel tower from the Arc de Triomphe
(c) the distance between Notre–Dame and the Arc de Triomphe.

(a) Use the right–angled triangle NET to find the height of the Eiffel tower ET:

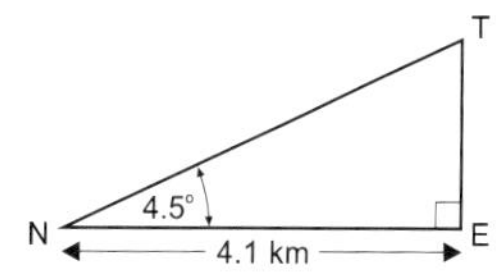

$$\tan 4.5^\circ = \frac{\text{ET}}{4.1}$$
$$\therefore\ 4.1 \times \tan 4.5^\circ = \text{ET}$$
$$\therefore\ \text{ET} = 0.3227 \text{ km} = 322.7 \text{ m} \approx 320 \text{ m}$$

(b) Use the right–angled triangle EAT to find the angle of elevation EAT of the top of the Eiffel tower from the Arc de Triomphe:

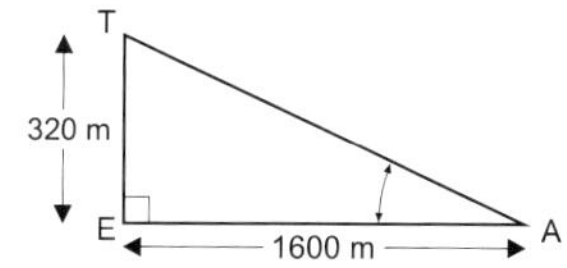

$$\tan A = \frac{322.7}{1600} = 0.2017$$
$$\therefore\ A = \tan^{-1} 0.2017 \approx 11.4^\circ \text{ (1 dp)}$$

(c) Apply the cosine rule to triangle NAE to find the distance NA between Notre–Dame and the Arc de Triomphe:

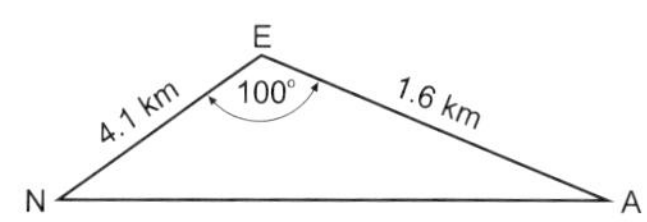

$$\text{NA}^2 = \text{NE}^2 + \text{EA}^2 - 2 \times \text{NE} \times \text{EA} \times \cos E$$
$$= 4.1^2 + 1.6^2 - 2 \times 4.1 \times 1.6 \times \cos 100^\circ$$
$$= 16.81 + 2.56 - -2.278$$
$$= 21.648$$

So, $\text{BC} = \sqrt{21.648} \approx 4.7$ km (1 dp)

Question **H** *The diagram shows a rectangular block of butter.*

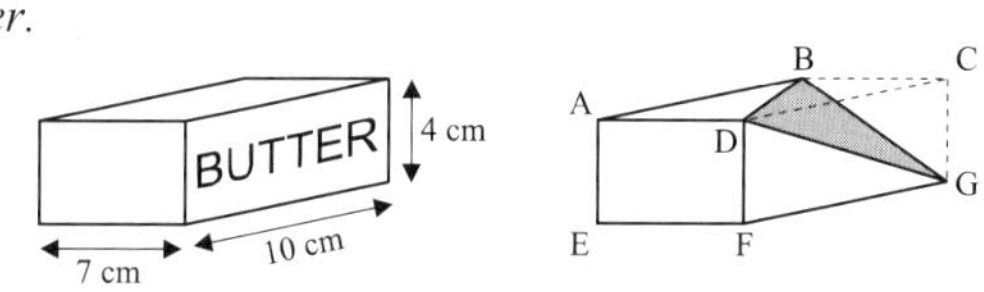

A corner of the butter is removed with a single knife cut. Calculate the area of the cut face of the remaining block.

We must first find the sides of triangular cut face BDG.

Using Pythagoras' theorem, we have:

$\text{DG}^2 = 4^2 + 10^2 = 116$ $\quad\therefore\ \text{DG} = \sqrt{116} = 10.77$ cm

$\text{BD}^2 = 10^2 + 7^2 = 149$ $\quad\therefore\ \text{BD} = \sqrt{149} = 12.21$ cm

$\text{BG}^2 = 4^2 + 7^2 = 65$ $\quad\therefore\ \text{BG} = \sqrt{65} = 8.06$ cm

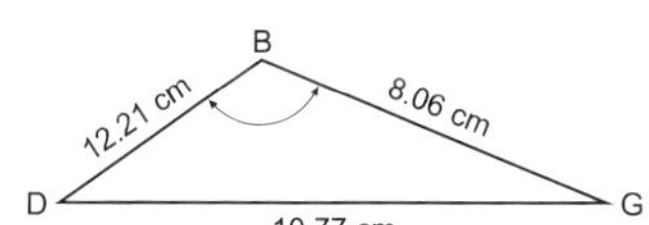

We can use $\frac{1}{2}ab \sin C$ to find the area of triangle BDG if we can find an angle. Using the cosine rule, we can find angle DBG:

$$\cos B = \frac{12.21^2 + 8.06^2 - 10.77^2}{2 \times 12.21 \times 8.06} = 0.4982$$
$$\therefore\ B = \cos^{-1} 0.4982 = 60.12^\circ$$

Area of triangle BDG $= \frac{1}{2} \times 12.21 \times 8.06 \times \sin 60.12^\circ$
$= 42.7 \text{ cm}^2$ (1 dp)

Graphs of Trigonometric Functions H

1. Graphs of Trigonometric Functions

Sine is a function that changes angles into numbers. If we calculate the value of sin x for different angles x, we will obtain a table of values like the one below

x	0°	30°	60°	90°	120°	150°	180°	210°	240°	270°	300°	330°	360°
$y = \sin x$	0	0.5	0.87	1	0.87	0.5	0	−0.5	−0.87	−1	−0.87	−0.5	0

Using these values, we can plot the graph of $y = \sin x$:

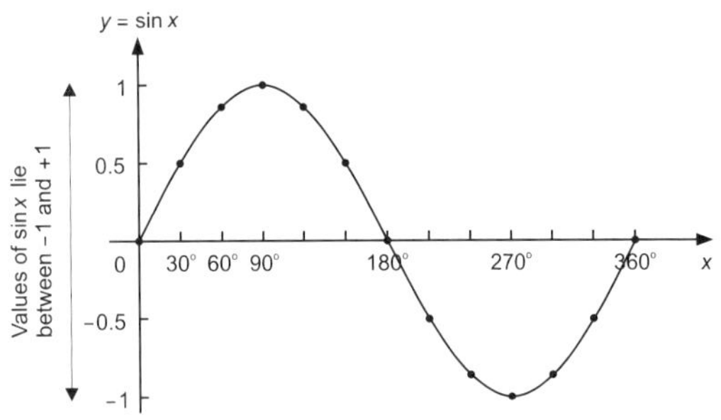

From the graph you can see that the values of sin x lie between −1 and +1. Also, the graph repeats itself every 360°. Each repetition is called a **cycle**.

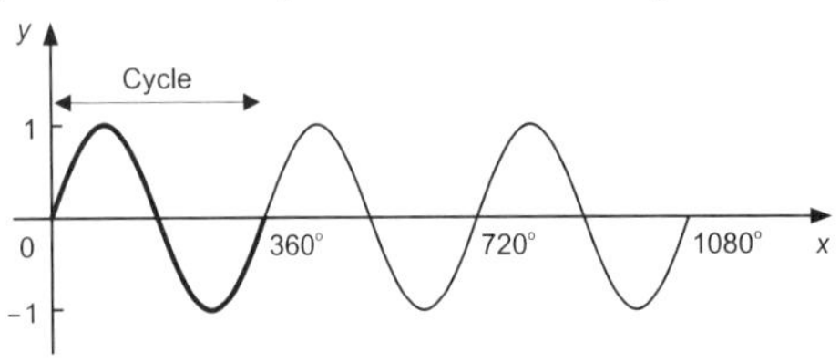

The graph of $y = \cos x$ is the same as the graph of $y = \sin x$ only shifted 90° left.

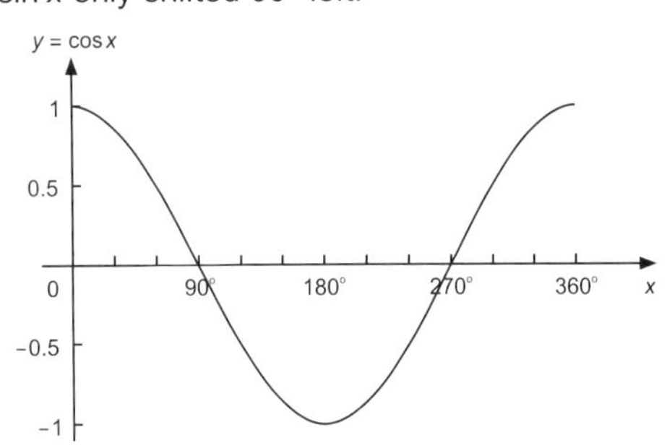

The graph of $y = \tan x$ is completely different:

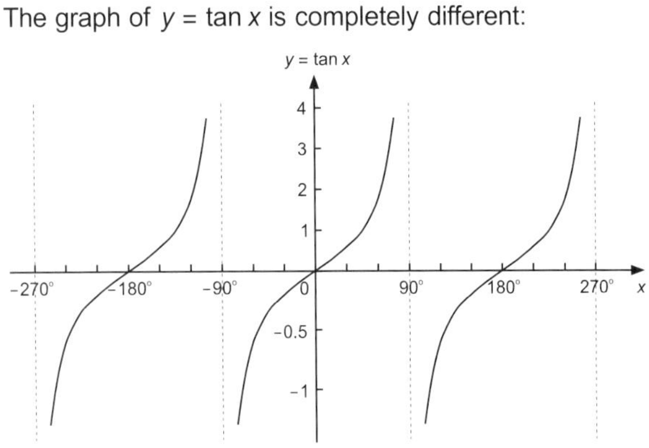

The graph is broken into separate sections (called **branches**). As x gets closer to 90° from the left, the y-values become very large. When x reaches 90°, y is infinitely large and cannot be calculated. Try calculating tan 90°: you will get a calculator error.

2. Graphs of Related Trigonometric Functions

Here is a cycle of the graph of $y = \sin x$.

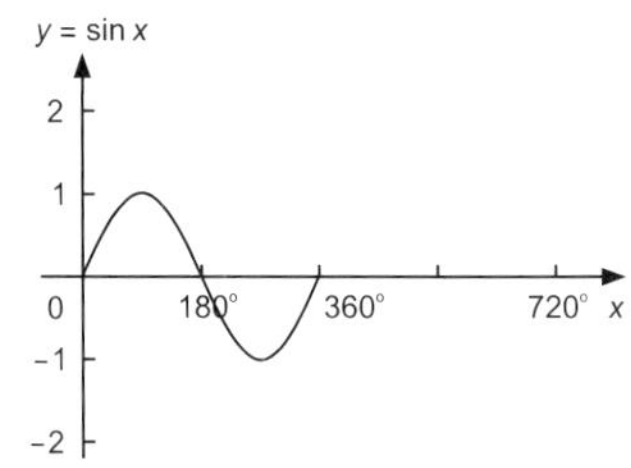

Here is a cycle of the graph of $y = 2 \sin x$. The y-values of sin x have been multiplied by 2 and so now lie between −2 and 2.

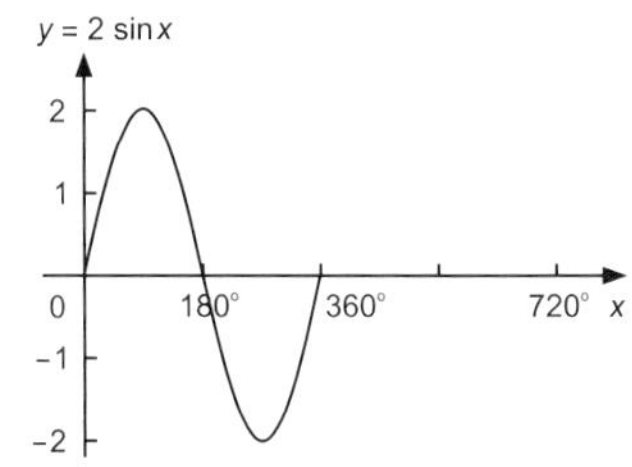

Here is a cycle of the graph of $y = \sin \frac{1}{2}x$. It is sin x stretched by a factor of 2 in the x-direction and so the cycle now repeats itself every 720° (360°×2).

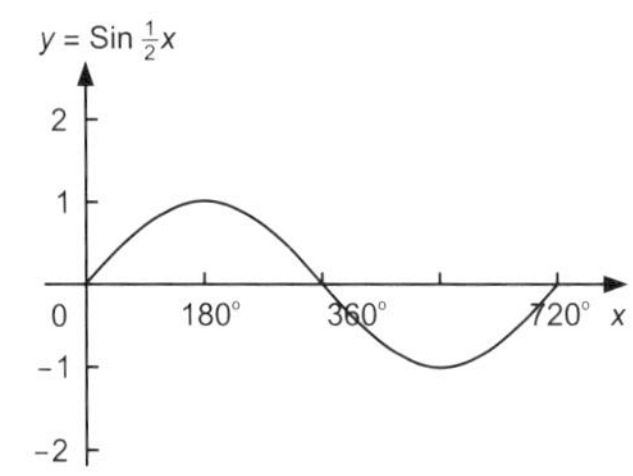

Here is a cycle of the graph of $y = \sin 2x$. It is sin x stretched by a factor of $\frac{1}{2}$ in the x-direction and so the cycle now repeats itself every 180° (360°÷2).

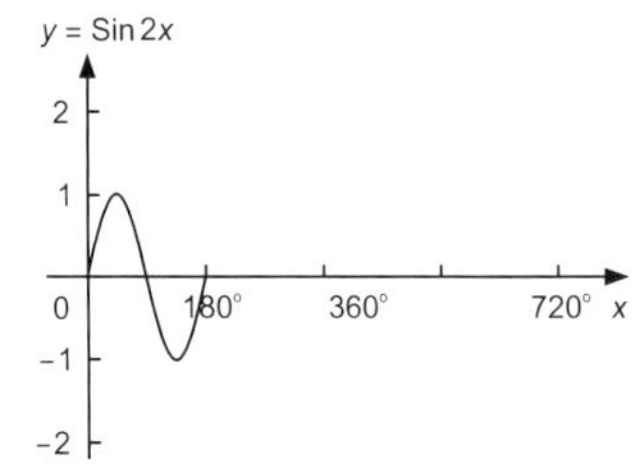

3. Solving Practical Problems

Question *The graph shows the tidal flow in a harbour over a 24-hour period.*

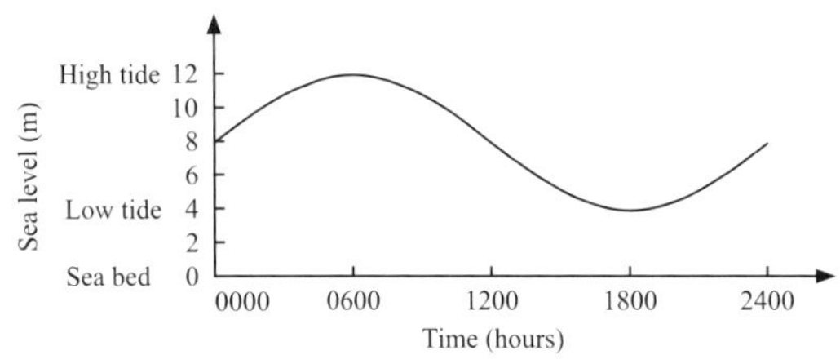

A ship needs 6 m of water to enter the harbour. Between which times is the ship unable to enter the harbour?

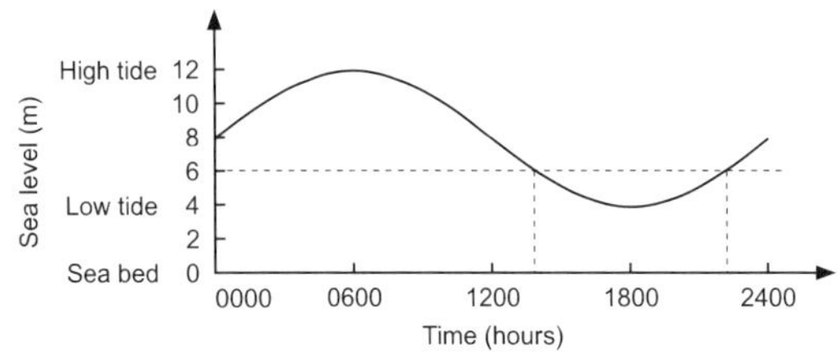

The dashed line shows that the ship cannot enter the harbour between approximately 1330 and 2230 hours.

4. Using Graphs to Solve Trigonometric Equations

Question *Complete the table below and plot the graph of $y = \cos \frac{1}{2}x$.*

x	0°	60°	120°	180°	240°	300°	360°
$y = \cos \frac{1}{2}x$		0.87		0		−0.87	

(a) Use your graph to solve the equation $\cos \frac{1}{2}x = -0.75$.

(b) By drawing a suitable straight line, solve the equation $\cos \frac{1}{2}x = \frac{x}{180}$

(a) When $x = 0$, $y = \cos \frac{1}{2}x = \cos \frac{1}{2} \times 0 = \cos 0 = 1$, etc., giving the table below:

x	0°	60°	120°	180°	240°	300°	360°
$y = \cos \frac{1}{2}x$	1	0.87	0.5	0	−0.5	−0.87	−1

Plot these points and join them with a smooth line.

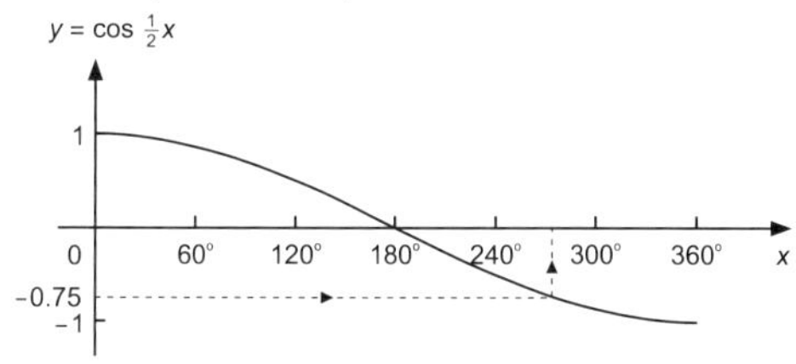

(b) We must solve the equation $\cos \frac{1}{2}x = -0.75$. We have drawn the graph of $y = \cos \frac{1}{2}x$ and so need to find the value of x when $y = -0.75$. Read off this value (see dashed line), giving $x \approx 277°$.

(c) Draw the straight line with equation $y = \frac{x}{180}$.

First, make a table of values:

When $x = 0$ $y = \frac{0}{180} = 0$

When $x = 180$ $y = \frac{180}{180} = 1$

When $x = 360$ $y = \frac{360}{180} = 2$

x	0°	180°	360°
$y = \frac{x}{180}$	0	1	2

Draw the straight line and read off the values where it intersects the curve.

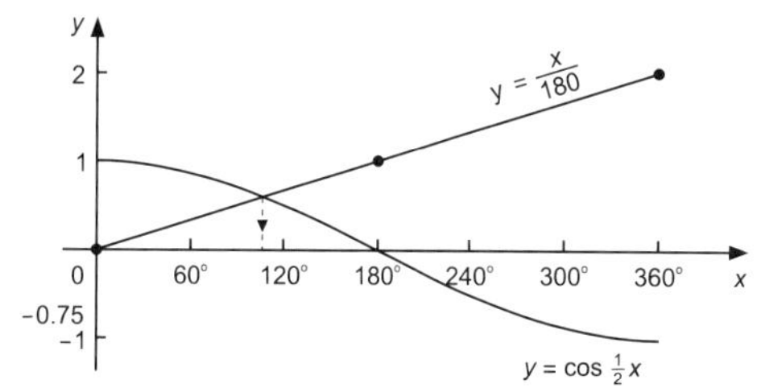

The solution of $\cos \frac{1}{2}x = \frac{x}{180}$ is $x \approx 107°$.

Collecting Data

1. Data

Data is information collected from the real world. There are several ways of collecting data:

Counting. For example, counting the number of tomatoes on a tomato plant.

Measuring. For example, measuring the height of children in a classroom.

Asking questions. For example, a questionnaire designed to find out what people think about mobile telephones.

Some data is **numerical**. For example, the numbers of tomatoes on 7 tomato plants: 15 28 7 14 32 21 20

The possible values of **discrete** data increase in steps, e.g. numbers of tomatoes increase by steps of 1. Each dot represents a possible number of tomatoes.

The possible values of **continuous** data increase smoothly, e.g. lengths of leaves. Each dot on the line represents a possible length of a leaf.

Non-numerical data can often be organised numerically. For example, 50 people were asked 'Do mobile telephones annoy you? YES/NO'. Although 'YES' and 'NO' are not numbers, it was recorded that 20 said 'YES' and 30 said 'NO'.

2. Populations and Samples

The collection of all objects being studied in an investigation is called the **population**.

For example, Katrina investigated the use of contact lenses and spectacles amongst the students of a college. The population under investigation was 'all students studying at the college'.

As the college population was very large, it would have been difficult for Katrina to get information about *every* student. Instead, she chose a **sample** of 60 students from the college population.

Provided her sample was *representative* of the college population and was *large enough*, she could then generalise her findings to the whole college.

Choosing a representative sample from a population

There were twice as many women as men studying at the college. Katrina felt that this aspect of the population should be reflected in the sample of 60 students (perhaps women prefer to wear contact lenses). So she made sure that the sample contained the same proportion of men and women, i.e. 40 women and 20 men. This is called **proportional sampling**.

Katrina enlisted the help of four research assistants to interview the 60 students. She gave each assistant a **quota** of 10 women and 5 men to interview.

It is important that the interviewees are chosen in an *unbiased* way. So each assistant was instructed to stand at a busy part of the college on a Monday morning and approach the first student that walked past. Each assistant was to continue in this way until they had filled their quota of 10 women and 5 men.

Random sample

Someone pointed out to Katrina that many students in the college did not have a chance of being included in her sample. A **random sample** is where *every* member of the population has an *equal chance* of being included.

Katrina thought up a way of choosing a random sample of 40 women and 20 men. Each student in the college had a three digit enrolment number, e.g. 378. She borrowed a **table of random numbers** from the library. Here are the first few numbers:

77803 61872 86245 68220 66267 01379 11304 01658
82404 46728 35228 49673 53552 51215 45611 83927

She read the numbers from left to right in groups of three digits: 778 036 187 286 245 682 etc.

She then wrote down the names of the students with these enrolment numbers, making a list of 40 female students and a list of 20 male students for interviewing. If one of the three digit numbers was not an enrolment number, she ignored it and moved on to the next one. The listed students were then interviewed.

3. Questionnaires

Use these guidelines when designing a questionnaire:

- **Not too many questions (could get boring)**
- **Make it clear how each question should be answered, e.g. tick one of the boxes, circle your preference**
- **Force the person to choose from a limited number of answers. Make sure that you include all possible answers**

  'Do you enjoy studying mathematics?'
  Invites too many possible answers.

  'Do you enjoy studying mathematics?' YES/NO
  Better. Forces a YES or NO answer.

  'Do you enjoy studying mathematics?'
  Not at all   A little   Quite a lot   Very much
  Better still. Covers all degrees of enjoyment.

- **Make sure the question has only one meaning**

  'Do you and your family go to the beach?' YES/NO
  Should you answer YES if you go to the beach on your own, or with your family, or some of your family?

- **The question should be unbiased**

  'Do you agree that TV violence should be banned because it causes violence in teenagers?' YES/NO
  This question is strongly suggesting that violence on TV does in fact cause violence in teenagers.

Frequency Distributions

1. Tally Charts and Frequency Distributions

Tally charts are a convenient way of recording data.

Peter investigated the hair colour of fellow students in a college. He prepared a record sheet like the one below.

Hair colour	
Brown	
Black	
Fair	
Other	

As students entered the college, he made a **tally mark** next to their hair colour. He recorded the hair colour of 60 students, as shown in the table below.

Hair colour	Tally
Brown	𝍸 𝍸 𝍸 𝍸 II
Black	𝍸 𝍸 IIII
Fair	𝍸 𝍸 𝍸 I
Other	𝍸 III

NOTE 𝍸 represents a group of 5 students

Peter observed that brown was the most common hair colour.

He replaced the tally marks by numbers, called **frequencies**, which are represented by the letter ***f***. The table of frequencies below is called a **frequency distribution**.

Hair colour	Number of students (Frequency, *f*)
Brown	22
Black	14
Fair	16
Other	8
Total	60

The table shows that 22 students had brown hair, 14 had black hair, 16 had fair hair and 8 students had some other hair colour. A total of 60 students.

This total can be found by adding together the frequencies in the table. The total frequency is represented by the letter ***n***. In this example, $n = 60$.

Question *Eugenia investigated the number of occupants of cars passing her school. She prepared the record sheet below and recorded a tally mark for each car passing the school.*

Number of occupants	Tally
1	𝍸 𝍸 𝍸 III
2	𝍸 𝍸 𝍸
3	𝍸 𝍸 I
4	𝍸 I

(a) Make a frequency distribution table for the data.
(b) How many cars did she record altogether?
(c) Suggest a way in which the record sheet could be improved.
(d) Eugenia noticed that the most common number of occupants was 1. She concluded that most cars are travelling around almost empty. Give a reason why this might not be true.

(a) Replace the tally marks by numbers (frequencies).

Number of occupants	Number of cars (Frequency, *f*)
1	18
2	15
3	11
4	6
Total	50

(b) The total number of cars Eugenia recorded is 50, found by adding up the frequencies.

(c) A large car could have 5 or more occupants and so Eugenia should have included this category in her record sheet.

(d) Some of the drivers may be parents who have just taken their children to school. Another way of describing the data is to say that most of the cars had 2 or more occupants. This gives a different impression.

2. Grouped Frequency Distributions

Ms French recorded the number of minutes it took her to drive to work over a period of 30 days. The results are shown below.

31 37 49 25 35 33 39 29 25 43
41 45 32 33 33 27 43 20 35 39
37 38 44 35 32 38 37 30 31 44

The values range from 20 to 49 minutes. This range could be divided into smaller **classes** (or **intervals**) of values. The class 20–24 contains the values 20, 21, 22, 23 and 24. We say that the values have been **grouped** into classes.

Classes
20–24
25–29
30–34
35–39
40–44
45–49

Each recorded time can be placed into one of these classes using a tally mark.

Time (min)	Tally
20–24	I
25–29	IIII
30–34	𝍸 III
35–39	𝍸 𝍸
40–44	𝍸
45–49	II

Replacing the tally marks by frequency numbers gives the **grouped frequency distribution**, as shown in this table.

Time (min)	No. of journeys
20–24	1
25–29	4
30–34	8
35–39	10
40–44	5
45–49	2
Total	30

more

Frequency Distributions (Contd)

2. Grouped Frequency Distributions (Cont)

Question *A fisherman's catch is recorded in the table below.*

Weight, w (kg)	Number of fish
$0.5 \leqslant w < 1.0$	75
$1.0 \leqslant w < 2.0$	150
$2.0 \leqslant w < 3.0$	90
$3.0 \leqslant w < 4.0$	45
$4.0 \leqslant w < 5.0$	25

(a) How many fish did he catch altogether?
(b) How many fish weighed between 1.0 kg and 3.0 kg?
(c) One fish weighed exactly 3.0 kg. Which class contains this weight?

(a) Add up the frequencies (numbers of fish) in the right column:

Total no. of fish = 75+150+90+45+25 = 385

(b) 150 fish weighed between 1.0 kg and 2.0 kg.
90 fish weighed between 2.0 kg and 3.0 kg.
So, 150+90 = 240 fish weighed between 1.0 kg and 3.0 kg.

(c) A weight of 3.0 kg is contained in the class $3.0 \leqslant w < 4.0$, because the sign $\leqslant$ means 'less than or *equal* to'. It is ***not*** contained in the class $2.0 \leqslant w < 3.0$, because the sign < means 'less than'.

Illustrating Data

1. Illustrating Data

Data can be illustrated using a variety of diagrams: pictograms, bar charts, line charts, pie charts, graphs, histograms, cumulative frequency diagrams. You will usually be asked to draw one of these, but sometimes you may have to choose an appropriate diagram for the given data.

2. Pictograms

A **pictogram** uses a symbol to represent a unit of data, e.g. 10 men. The symbol is usually related to the data, e.g. the symbol 웃 might represent 10 men

Question *The table shows the methods of transport used by the employees of a company. Illustrate this data using a pictogram.*

Mode of transport	Number of employees
Car	40
Train	15
Bus	7
Bicycle	2

Using the symbol 웃 to represent 5 employees, the pictogram for the data is:

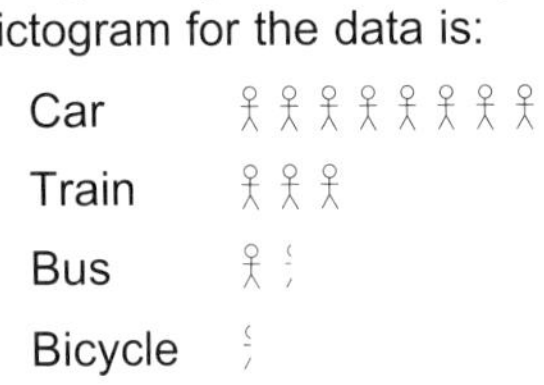

Some of the symbols are incomplete, representing less than 5 employees.

3. Bar Charts and Line Charts

A vertical (or horizontal) bar can be used to represent the amount of data in a category: the taller the bar, the more data it represents. The bars are of equal width and are separated from each other by narrow gaps. Lines can be used instead of bars.

Question *The table below shows the distribution of accommodation in Greater London during 1995.*

Owner occupied	Council housing	Housing association	Privately rented
56%	22%	7%	15%

Illustrate this data using a suitable chart.

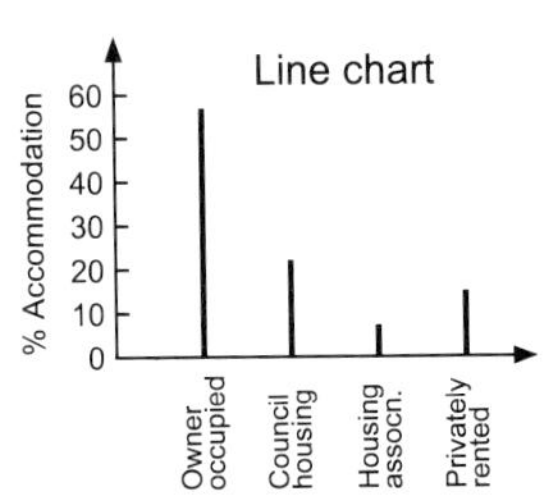

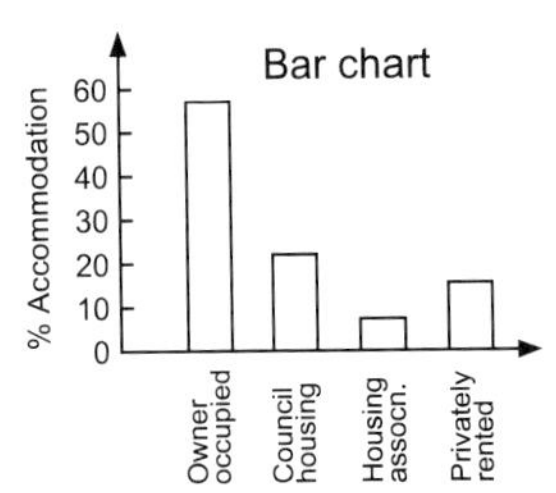

Question *100 students each took two exam papers in science. The bar chart shows the distribution of marks in both papers.*

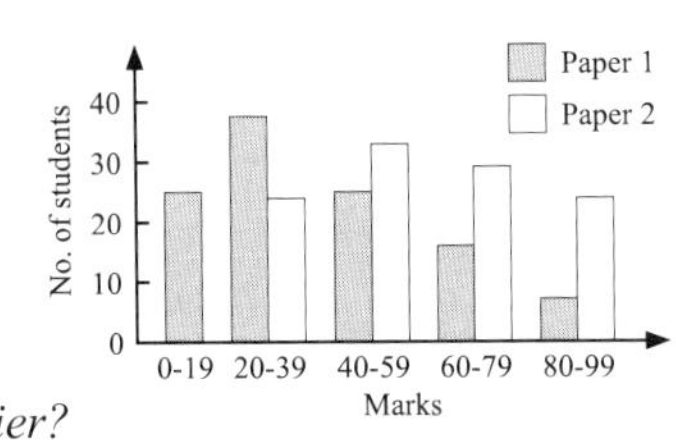

(a) Which paper was easier?
(b) In which exam was there a greater spread of marks?
(c) How many students scored 40 to 59 marks in Paper 2?

(a) Paper 2 appears to have been easier because more students got high marks (40 or more) than in Paper 1 (the white bars are higher than the shaded bars).

(b) The shaded bars show that students scored between 0 and 99 in Paper 1. But the white bars show they scored between 20 and 99 in Paper 2. So, the spread of marks is higher for Paper 1.

(c) The white bar for class 40–59 has a height of 33, so 33 students scored between 40 and 59 in Paper 2.

Illustrating Data

4. Pie Charts

A **pie chart** uses a circle to represent the total data. Each category of data is represented by a sector of the circle.

Question *Chantal's weekly expenditure is illustrated in the pie chart below. She spends £12 per week altogether.*

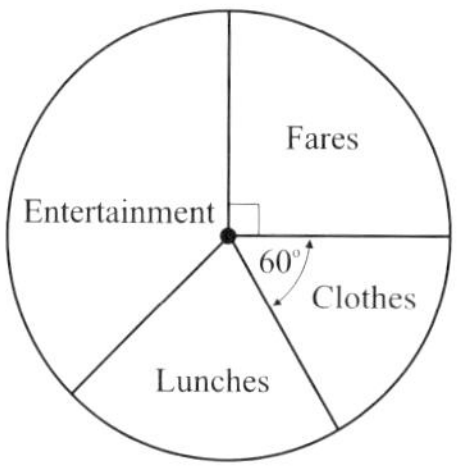

(a) How much does she spend on fares?

(b) How much does she spend on clothes?

(c) Measure the angle of the Entertainment sector.

(d) How much does she spend on entertainment?

(a) The whole circle represents £12.
There are 360° in a whole circle.
The angle of the Fares sector is 90°, which is $\frac{1}{4}$ of the whole circle ($\frac{90}{360} = \frac{1}{4}$).
So, money spent on fares = $\frac{1}{4}$ of £12 = £3

(b) The angle for the Clothes sector is 60°.
So, money spent on clothes = $\frac{60}{360}$ of £12
= 60÷360×£12 = £2

(c) Carefully measure the angle for the Entertainment sector to be 135°.

(d) Money spent on entertainment
= $\frac{135}{360}$ of £12 = 135÷360×£12 = £4.50

Question *Complete the pie chart for the distribution of British people living abroad.*

Location	Percentage
Asia & Australia	30
Africa	15
Middle East	15
Contintental Europe	20
Americas	20

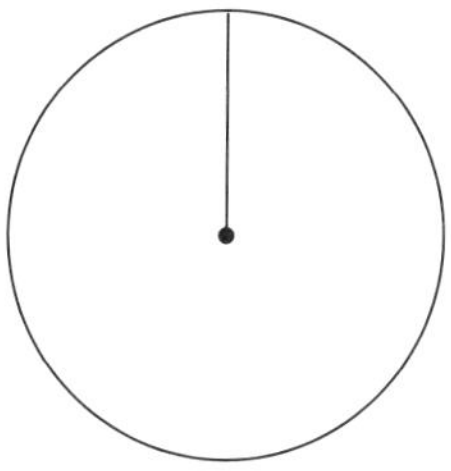

There are 360° in a circle. So:

Asia & Australia = 30% of a circle
= 30% of 360° = 30÷100×360° = 108°

Repeating this calculation for the other locations, we get:

Location	%	Angle
Asia & Australia	30	108°
Africa	15	54°
Middle East	15	54°
Contintental Europe	20	72°
Americas	20	72°

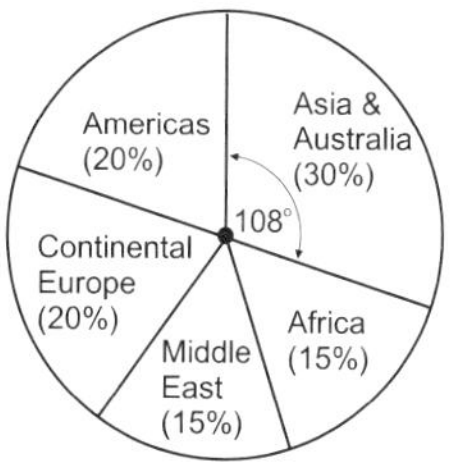

Place the centre of your protractor on the centre of the circle, aligning 0° with the given radius. Mark an angle of 108° and draw a new radius, labelling the sector 'Asia & Australia (30%)'. Repeat for the other locations.

Histograms

1. Histograms

A **histogram** is like a bar chart. The main differences are:

- There are no gaps between the bars.
- The horizontal axis has a continuous scale, just like a graph, e.g. 1 cm = 20 kg.
- The **areas** of the bars represent the frequencies (see Box 2).

Question *A student investigated the lengths of worms in the topsoil of an allotment. She measured the lengths of 115 worms and recorded them in the frequency distribution table. Draw a histogram to illustrate this data.*

Length (cm)	No. of worms
0–5	20
5–10	45
10–15	20
15–20	15
20–25	10
25–30	5
Total	115

The horizontal axis represents the recorded values (length of worms in cm). Mark the axis from 0 to 30 cm.

The vertical axis represents the frequencies (numbers of worms). Mark the axis from 0 to 45, the greatest number of worms in the table.

All the classes have the *same width*, 5 cm. For example, the width of the class 10–15 cm is 5 cm, as shown in the diagram below:

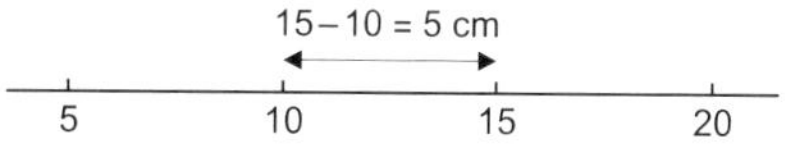

When all of the classes have the same width, make the height of each bar equal to the frequency of the class it represents.

Draw a bar above each class, with height equal to the frequency of the class. For example, the bar for the class 10–15 cm should have a height of 20.

NOTE The side of the first bar is part of the y-axis and there should be no gaps between the bars.

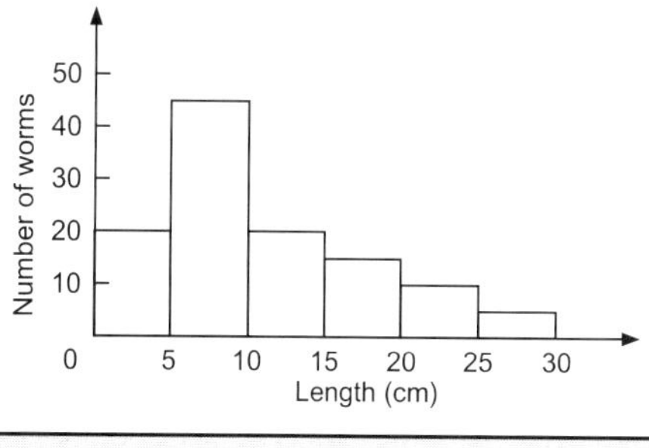

Histograms (Contd)

H

2. Histograms and Unequal Classes

When the class widths are unequal, the *areas* of the bars must be used to represent the frequencies.

Question *The weights of 60 rice portions from a Chinese take-away were measured and recorded in the table. Draw a histogram to illustrate this distribution.*

Weight, w (g)	No. of portions
$0 \leqslant w < 30$	6
$30 \leqslant w < 50$	14
$50 \leqslant w < 60$	20
$60 \leqslant w < 70$	12
$70 \leqslant w < 90$	8
Total	60

The class $0 \leqslant w < 30$ has a width of 30 g, whereas the class $30 \leqslant w < 50$ has a width of 20 g. So, the classes have different widths and we cannot simply make the heights of the bars equal to the frequencies.

Instead, we must make sure that the *area* of a bar represents the frequency of its class:

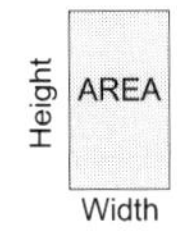

Area of a bar = Width × Height

But the area of a bar represents the frequency of the class and so the formula becomes:

Frequency = Class width × Bar height

Rearranging this formula, we get:

$$\text{Bar Height} = \frac{\text{Frequency}}{\text{Class width}}$$

The heights of the bars calculated in this way are called **frequency densities**.

Use this formula to calculate the bar height for each of the classes in the table of values.

Weight w (g)	Class width	No. of portions (Frequency)	Bar height (Frequency density)
$0 \leqslant w < 30$	30	6	0.2
$30 \leqslant w < 50$	20	14	0.7
$50 \leqslant w < 60$	10	20	2.0
$60 \leqslant w < 70$	10	12	1.2
$70 \leqslant w < 90$	20	8	0.4

Draw the histogram using these bar heights, choosing an easy scale for the y-axis.

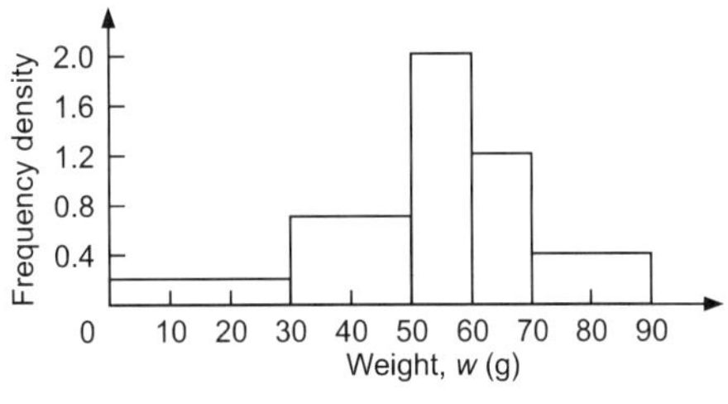

Question *The histogram below shows the distribution of spectators at professional football matches in England on a Saturday in 1993.*

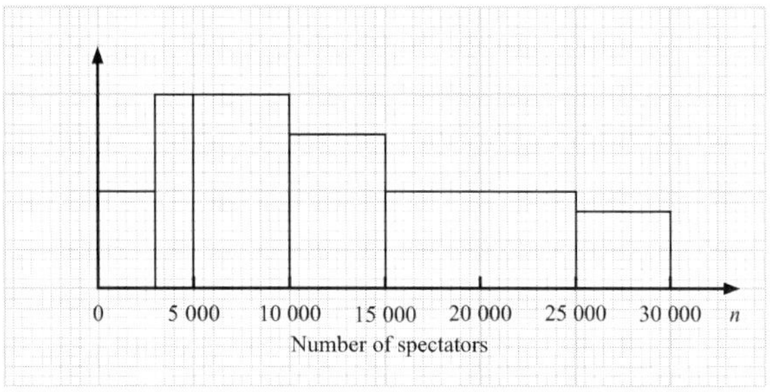

Use the information in the histogram to complete the frequency table.

No. of spectators	No. of matches
$0 \leqslant n < 3000$	
$3000 \leqslant n < 5000$	
$5000 \leqslant n < 10\,000$	10
$10\,000 \leqslant n < 15\,000$	
$15\,000 \leqslant n < 25\,000$	
$25\,000 \leqslant n < 30\,000$	

Remember that the area of a bar represents the frequency of its class. The area of the bar for the class $5000 \leqslant n < 10\,000$ is 1 cm × 2 cm = 2 cm^2 and the frequency for this class is 10. So:

2 cm^2 represents a frequency of 10

$\therefore$ 1 cm^2 represents a frequency of $10 \div 2 = 5$

Now we can calculate the areas of the other bars and convert them to frequencies.

For the class $0 \leqslant n < 3000$:

Area of bar = $0.6 \times 1 = 0.6$ cm^2 which represents a frequency of $0.6 \times 5 = 3$

Repeating this calculation for the other classes gives the remaining frequencies.

No. of Spectators	No. of Matches
$0 \leqslant n < 3000$	3
$3000 \leqslant n < 5000$	4
$5000 \leqslant n < 10\,000$	10
$10\,000 \leqslant n < 15\,000$	8
$15\,000 \leqslant n < 25\,000$	10
$25\,000 \leqslant n < 30\,000$	4

3. Frequency Polygons

Joining the mid-points of the tops of the bars of a histogram with straight lines makes a **frequency polygon**. It gives a good impression of the shape of the histogram.

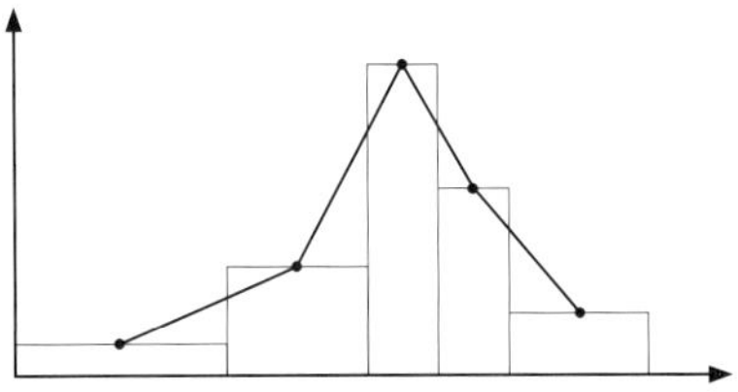

4. Shape of a Frequency Distribution

The shape of a frequency distribution is the shape of its frequency polygon. The diagram below shows three shapes you should know.

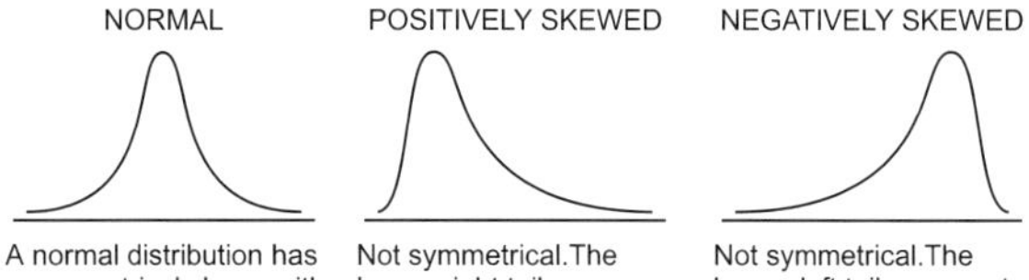

A normal distribution has a symmetrical shape with most values clustered around the centre

Not symmetrical. The longer right tail represents a few extremely large values, compared to the centre

Not symmetrical. The longer left tail represents a few extremely small values, compared to the centre

Averages

1. Averages

There are three ways of finding the average of a set of numbers: mean, median and mode. You need to be able to calculate all three.

2. Mode and Modal Class

The **mode** is the most frequently occurring value.

The mode of this set of numbers is 5 because it occurs more frequently than the others.

2 3 3 5 5 5 7 9 (5 5 5: Mode)

A student recorded the numbers of occupants travelling in 50 cars. The most frequent number of occupants is 1. We say that the mode is 1 occupant per car (or the modal number of occupants is 1).

No. of occupants	No. of cars
Mode 1	20
2	15
3	9
4	6

The **modal class** (or **modal interval**) is the class with the highest frequency. The numbers of words on the pages of a book are recorded in this table. The modal class is 360–379 words because this class has the highest frequency, 65.

No. of words	No. of pages
300–319	25
320–339	39
340–359	48
360–379	65
380–399	58

The mode does ***not*** tell you anything about the other values. Both these sets of numbers have a mode of 5 and yet their other values are very different.

2 3 3 5 5 5 7 9

5 5 5 10 50 100 200

3. Median

The **median** of a set of values is the middle value after all the values have been put in ascending order of size.

Janet recorded the number of dental fillings in 5 of her classmates, as shown below:

Number of fillings 7 4 8 3 6

She rearranged the values from smallest to largest:

3 4 6 7 8

(6: Median)

She could then see that the middle number is 6. We say that the median is 6 fillings per child.

A sixth classmate had 16 fillings! The values are now:

3 4 6 7 8 16

(Median: between 6 and 7)

The median is now halfway between the middle two values 6 and 7, and so is 6.5.

Notice that half (50%) of the values lie below the median and half (50%) lie above it.

Notice also that adding a large value, like 16, to the data has only slightly affected the median: it was 6 and is now 6.5. Very large and very small values are called **extreme values**.

4. Median of a Frequency Distribution

Question *In an experiment, 50 people were asked to estimate the length of a rod, to the nearest centimetre. The results are recorded in the table below:*

Length (cm)	8	9	10	11	12	13	14	15	16
No. of people	3	4	6	6	9	13	7	2	0

Calculate the median for the data.

Imagine the 50 estimated lengths in the table spread out from smallest to largest

8	8	8	9	9	9	....	?	?	....	14	14	14	14	15	15
1st	2nd	3rd	4th	5th	6th		25th	26th		45th	46th	47th	48th	49th	50th

(Median: between 25th and 26th)

The median lies halfway between the 25th and 26th values.

Add up the frequencies from left to right until you reach the 25th and 26th values:

3rd value is 8

3+4 = 7th value is 9

3+4+6 = 13th value is 10

3+4+6+6 = 19th value is 11

3+4+6+6+9 = 28th value is 12

So, the 25th and 26th values must both be 12. The median is therefore halfway between 12 and 12, which is 12.

5. Mean

The **mean** of a set of values is found by adding them up and dividing the total by the number of values.

$$\text{Mean} = \frac{\text{Total of values}}{\text{Number of values}}$$

Question *Rachna took a typing speed test. She typed 5 pages as fast as possible. The numbers of errors on the 5 pages are shown below:*

3 2 0 1 5

Calculate the mean number of errors per page.

$$\text{Mean} = \frac{\text{Total of values}}{\text{Number of values}}$$

$$= \frac{3+2+0+1+5}{5} = \frac{11}{5} = 2.2 \text{ errors per page}$$

NOTE The mean may not be a possible value. In this example, it is impossible to make 2.2 typing errors.

Rachna was then asked to type a sixth page and made 25 mistakes (due to nerves). Her results are now:

3 2 0 1 5 25

We can now recalculate the mean:

$$\text{Mean} = \frac{3+2+0+1+5+25}{6} = \frac{36}{6} = 6 \text{ errors per page}$$

Adding this single, extremely large value has had a considerable effect on the mean, dragging it up from 2.2 to 6 errors per page.

more

Averages (Contd)

5. Mean (Contd)

Question *A greengrocer counted the numbers of potatoes in 5 bags and calculated their mean to be 46 potatoes per bag.*

(a) Calculate the total number of potatoes contained in the 5 bags.

(b) He counted the potatoes in another 2 bags and found them both to contain 52 potatoes. Calculate the mean number of potatoes for all 7 bags.

(a) $\text{Mean} = \dfrac{\text{Total number of potatoes}}{\text{Number of bags}}$

$\therefore\ 46 = \dfrac{\text{Total number of potatoes}}{5}$

$\therefore\ 46\times5 = \text{Total number of potatoes}$

$\therefore\ \text{Total number of potatoes} = 230$

(b) Total potatoes in all 7 bags = 230+52+52 = 334
So, mean = 334÷7 = 47.7 potatoes per bag

6. Mean of a Frequency Distribution

Question *The frequency distribution table below shows the numbers of occupants of houses living in a certain area. Calculate the mean number of occupants per house.*

No. of occupants	0	1	2	3	4	5	6	7
No. of houses	2	8	15	34	48	41	22	16

① First, make sure you know what the numbers in the table mean.

The numbers in the top row are the **possible values** (the possible numbers of occupants living in a house).

The numbers in the bottom row are the **frequencies** (how often each value occurs).

② Calculate the total number of values (the total frequency).

There is a value (number of occupants) for each house. So:

Total no. of values = Total no.of houses
= Total frequency
= 2+8+15+34+48+41+22+16
= 186

③ Calculate the total of the values.

There are 2 values of 0; these total 2×0 = 0
There are 8 values of 1; these total 8×1 = 8
There are 15 values of 2; these total 15×2 = 30
etc.

Total of values
= 2×0+8×1+15×2+34×3+48×4
+41×5+22×6+16×7
= 781

④ Calculate the mean.

$\text{Mean} = \dfrac{\text{Total of values}}{\text{Number of values}}$

$= \dfrac{781}{186} = 4.2$ occupants per house

7. Mean of a Grouped Frequency Distribution

Question *833 people took a new kind of IQ test. Their results are recorded in the table. Calculate the mean IQ per person.*

IQ score	No. of people
60–79	52
80–99	145
100–119	392
120–139	165
140–159	58
160–179	21

The problem with calculating the mean here is that we do not know any values precisely. For example, we know that 52 people scored between 60 and 79 but we do not know their exact IQ scores. To get around this problem, we replace the class of scores 60–79 by its middle value, called the **mid–class value**. This can be found by adding the upper and lower class limits together and dividing by 2. For the class 60–79:

$$\text{Mid-class value} = \frac{60+79}{2} = \frac{139}{2} = 69.5$$

The table shows all of the mid–class values.

We can now calculate the mean using the method described in Box 6.

Mid–class value	No. of people (Frequency)
69.5	52
89.5	145
109.5	392
129.5	165
149.5	58
169.5	21

Number of values = Total frequency
= 52+145+392+165+58+21
= 833 Also given in question

Total of values = 52×69.5+145×89.5+....+21×169.5
= 93 113.5

$\therefore\ \text{Mean IQ} = \dfrac{\text{Total of values}}{\text{Number of values}}$

$= \dfrac{93\,113.5}{833} \approx 111.8$ per person

8. Using a Calculator

You can use the statistics mode of your calculator to find the mean of a frequency distribution and are less likely to make mistakes. However, you must still show the examiner how the mean is calculated.

Let's recalculate the mean of the distribution in Box 6, using a calculator.

First, put your calculator into **statistics mode**. See your instruction manual.

Now enter the first value, 0, and the frequency with which it occurs, 2. See your instruction manual.

Repeat this for all the values.

Press [n] to find the number of values entered (the total frequency). In this example, this is the total number of houses. You should get 186.

Press [Σx] to find the total of the values. You should get 781.

Press [$\bar{x}$] to find the mean. You should get 4.199.

Write down the working shown in Box 6, using these values obtained from your calculator.

NOTE If you do not show your working and calculate the wrong answer, you will get no marks.

9. Comparing Mean, Median and Mode

Mean and median

The mean and median have been calculated for the values below:

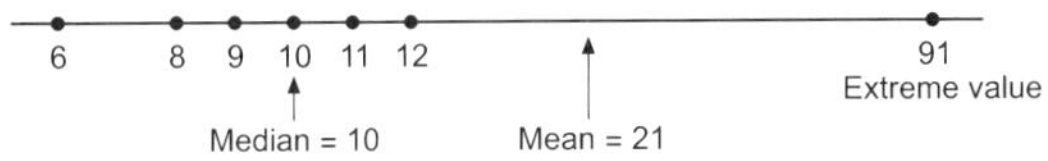

The median is the middle value, 10. The mean is greater than the median because of the extremely large value of 91 being included in its calculation.

The mean is greater than the median when the distribution is positively skewed (see Box 4, p104). The mode lies beneath the highest point of the curve.

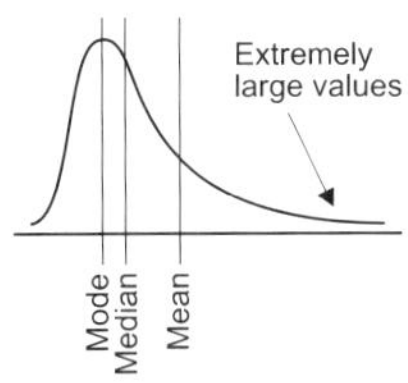

Similarly, if there are extremely small values (negatively skewed distribution), the mean will be less than the median

Mode

The mode only tells you which value occurs the most often. It tells you nothing about the other values (except that they occur less often).

You ***cannot*** calculate the mode by knowing the mean and median. You have to look at the values themselves. The two sets of values below have the same mean and median but completely different modes.

1 1 5 6 8 10 11 Mode = 1

Mean & Median

1 2 5 6 8 10 10 Mode = 10

Question *The table shows the mean and median heights of Jane's and Bill's friends. Jane says that Bill's friends are taller than her friends. Explain why this may not be true.*

	Mean height (cm)	Median height (cm)
Jane's friends	175	175
Bill's friends	180	175

The median height of Bill's friends is 175 cm and so half of his friends are shorter than 175 cm.

The median height of Jane's friends is also 175 cm and so half of her friends are taller than 175 cm.

So, half of Jane's friends are taller than half of Bill's friends. So, not all of Bill's friends are taller than Jane's.

Measures of Spread

1. Spread of Values

In an experiment at a health clinic, two methods of weight reduction were tried by different groups of people.

Group A were simply given a diet sheet.

Group B were given the same diet sheet and met once a week to give each other support.

The weight losses, in lb, for both groups are shown below:

Group A 0 1 2 4 6 8 10 11 12

Group B 4 5 5 6 6 6 7 7 8

Staff at the clinic calculated the mean weight loss for both groups and found them both to be 6 lb. So it appeared that the two methods had the same effect.

However, they noticed that the weight losses in Group A were more widely spread than those in Group B.

The clinic staff reasoned that the people in Group B ate in a more similar way than those in Group A because of their influence on each other at the weekly meetings.

There are three ways of measuring how spread out the values are: **range**, **standard deviation** (and **variance**) and **interquartile range**.

2. Range

The **range** of a set of values is simply the difference between the smallest and largest values. Here are the values from Groups A and B in Box 1:

Group A 0 1 2 4 6 8 10 11 12

Group B 4 5 5 6 6 6 7 7 8

Range for Group A = 12−0 = 12 lb

Range for Group B = 8−4 = 4 lb

The greater range for Group A reflects the greater spread of values.

3. Standard Deviation and Variance H

The standard deviation is based on the differences (**deviations**) between the values and their mean.

Here are the values for Groups A and B in Box 1. Each dot represents a value:

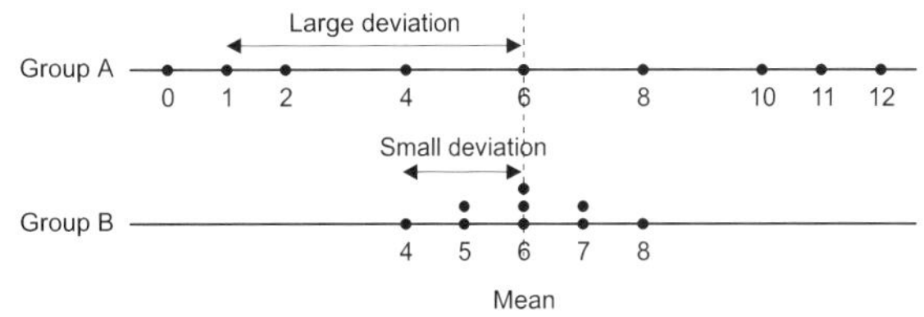

You can see that the deviations from the mean in Group A are greater, overall, than those in Group B.

more

Measures of Spread (Contd)

3. Standard Deviation & Variance (Contd)

The following steps show how to calculate the standard deviation for Group A.

Group A 0 1 2 4 6 8 10 11 12

Let x represent any value.

① Calculate the mean of the values and call it $\bar{x}$.

$$\text{Mean} = \bar{x} = \frac{0+1+2+4+6+8+10+11+12}{9} = \frac{54}{9} = 6$$

② Square the mean, giving $\bar{x}^2$.

$\bar{x} = 6$ and so $\bar{x}^2 = 6^2 = 36$

③ Square each value of x.

x	0	1	2	4	6	8	10	11	12
x^2	0	1	4	16	36	64	100	121	144

④ Add up the squares and call the sum $\sum x^2$.

Sum of squares = $\sum x^2 = 0+1+4+....+144 = 486$

⑤ Divide the sum of the squares, $\sum x^2$, by the number of values, n.

Number of values = $n = 9$

So, $\frac{\sum x^2}{n} = \frac{486}{9} = 54$

⑥ Subtract the square of the mean, $\bar{x}^2$, from $\frac{\sum x^2}{n}$. The result is called the **variance,** represented by s^2.

$$\text{Variance} = s^2 = \frac{\sum x^2}{n} - \bar{x}^2 = 54-36 = 18$$

⑦ Square root the variance to give the **standard deviation**, represented by s.

$$\text{Standard deviation} = s = \sqrt{\text{Variance}} = \sqrt{18} = 4.2 \text{ lb}$$

The standard deviation for group B was calculated in the same way and found to be 1.2 lb, which shows that the values of Group A are more spread out than those of Group B.

4. Standard Deviation of a Frequency Distribution **H**

Question *The table below shows the numbers and costs of surgical operations at a hospital during a one month period. The mean cost of an operation was calculated to be £4810. Calculate the standard deviation, correct to the nearest £100.*

Type of operation	Number of operations	Cost of operation (£)
Plastic surgery	3	2000
Varicose veins	3	4500
Knee replacement	1	5200
Hip replacement	3	7800

① Mean = $\bar{x}$ = £4810 Given in the question

② Square of mean = $\bar{x}^2 = 4810^2 = 23\,136\,100$

③ Square each value, x, giving x^2.

Type of operation	Number of operations, f	Cost of operation, x	x^2
Plastic surgery	3	2000	4 000 000
Varicose veins	3	4500	20 250 000
Knee replacement	1	5200	27 040 000
Hip replacement	3	7800	60 840 000

more

4. Standard Deviation of a Frequency Distribution (Contd)

④ Multiply each square, x^2, by its frequency, f, giving fx^2.

Type of operation	Number of operations, f	Cost of operation, x	x^2	fx^2
Plastic surgery	3	2000	4 000 000	12 000 000
Varicose veins	3	4500	20 250 000	60 750 000
Knee replacement	1	5200	27 040 000	27 040 000
Hip replacement	3	7800	60 840 000	182 520 000

Sum of squares = $\sum fx^2 = 282\,310\,000$

⑤ Calculate the sum of squares, $\sum fx^2$.

Sum of squares = $\sum fx^2 = 282\,310\,000$
(See table above)

⑥ Divide the sum of the squares, $\sum fx^2$, by the number of values, n (total frequency).

Number of values = $n = 3+3+1+3 = 10$

So, $\frac{\sum fx^2}{n} = \frac{282\,310\,000}{10} = 28\,231\,000$

⑦ Subtract the square of the mean, $\bar{x}^2$, from $\frac{\sum fx^2}{n}$ to give the variance, s^2.

$$\text{Variance} = s^2 = \frac{\sum fx^2}{n} - \bar{x}^2$$
$$= 28\,231\,000 - 23\,136\,100$$
$$= 5\,094\,900$$

⑧ Square root the variance to give the standard deviation, s.

$$\text{Standard deviation} = s = \sqrt{\text{Variance}}$$
$$= \sqrt{5\,094\,900}$$
$$= 2257$$
$$\approx £2300$$

5. Standard Deviation of a Grouped Frequency Distribution **H**

Question *Calculate the standard deviation of the IQ scores recorded in this table.*

IQ score	No. of people
60–79	52
80–99	145
100–119	392
120–139	165
140–159	58
160–179	21
Total	833

① Calculate the mid-class values (see Box 7, p106).

② Calculate the mean, $\bar{x}$, of these mid-class values.

The mean IQ was found to be $\bar{x} = 111.8$ in Box 7 on page 105

Mid-class value	No. of people (Frequency, f)
69.5	52
89.5	145
109.5	392
129.5	165
149.5	58
169.5	21
Total	833

③ Square the mean.

$\bar{x}^2 = 111.8^2 = 12\,499.2$

④ Square each mid-class value, x, giving x^2.

Mid-class value, x	x^2	No. of people (Frequency, f)
69.5	4830.25	52
89.5	8010.25	145
109.5	11 990.25	392
129.5	16 770.25	165
149.5	22 350.25	58
169.5	28 730.25	21

more

5. Standard Deviation of a Grouped Frequency Distribution (Contd)

⑤ Multiply each square, x^2, by its frequency, f, giving fx^2.

Mid-class value, x	x^2	No. of people (Frequency, f)	fx^2
69.5	4830.25	52	251 173
89.5	8010.25	145	1 161 486.3
109.5	11 990.25	392	4 700 178
129.5	16 770.25	165	2 767 091.3
149.5	22 350.25	58	1 296 314.5
169.5	28 730.25	21	603 335.25
		$\sum fx^2 =$	10 779 578

⑥ Calculate the sum of squares, $\sum fx^2$.

$\sum fx^2 = 10\,779\,578$ (See table above)

⑦ Divide the sum of the squares, $\sum fx^2$, by the number of values, n (total frequency).

Number of values = n = 833 (See table in question)

So, $\dfrac{\sum fx^2}{n} = \dfrac{10\,779\,578}{833} = 12\,940.7$

⑧ Subtract the square of the mean, $\bar{x}^2$, from $\frac{\sum fx^2}{n}$ to give the variance, s^2.

$$\text{Variance} = s^2 = \frac{\sum fx^2}{n} - \bar{x}^2$$
$$= 12\,940.7 - 12\,499.2$$
$$= 441.5$$

⑨ Square root the variance to give the standard deviation, s.

$$\text{Standard deviation} = s = \sqrt{\text{Variance}}$$
$$= \sqrt{441.5}$$
$$= 21.0 \text{ (1 dp)}$$

6. Special Problem

H

You may be asked how changing the values affects their mean and standard deviation.

Question *Five people invested £1000 in different bank accounts. At the end of 1 year, the bank accounts contained the following amounts:*

£1080 £1100 £1120 £1125 £1130

The mean amount was calculated to be £1111 and the standard deviation was £18.55. The sum of £1000 was subtracted from each amount to calculate the interest earned during the year. Calculate

(a) the mean interest earned

(b) the standard deviation of the interest earned.

(a) Increasing/decreasing every value by the same amount has the same effect on their mean.

Each amount has been reduced by £1000 and so the mean £1111 will also be reduced by £1000

So, mean interest earned = £1111−£1000 = £111

(b) Increasing/decreasing every value by the same amount has no effect on the standard deviation

The standard deviation of the amounts is £18.55

Reducing the amounts by £1000 has no effect on the standard deviation.

So, the standard deviation of the interest earned is also £18.55.

7. Comparing Distributions

Question *Two machines, A and B, were set to cut broom handles to a length of 180 cm. As part of a quality control check, 40 broom handles were taken from each machine and measured. The results are shown below together with their histograms. Comment on the performance of the machines.*

Machine A

Length (cm)	No. of handles
179.4–179.6	2
179.6–179.8	5
179.8–180.0	12
180.0–180.2	13
180.2–180.4	6
180.4–180.6	2

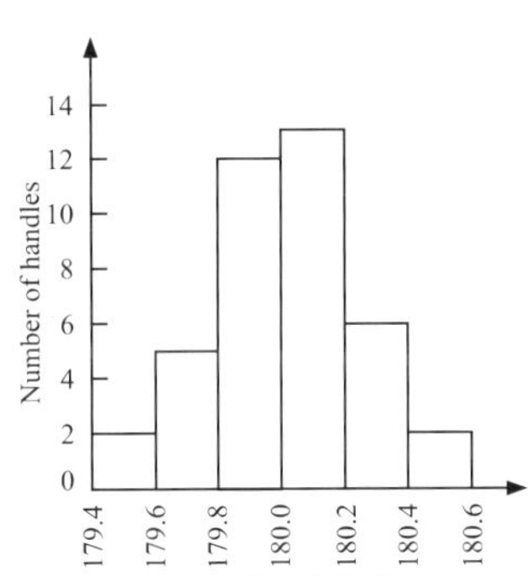

Machine B

Length (cm)	No. of handles
177.5–178.5	4
178.5–179.5	7
179.5–180.5	9
180.5–181.5	8
181.5–182.5	4
182.5–183.5	3
183.5–184.5	3
184.5–185.5	2

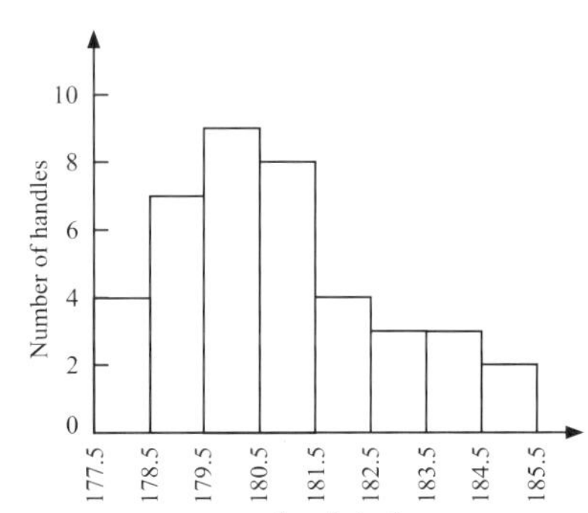

Machine A is working well, producing handles with lengths close to 180 cm. The lengths are not spread very much; the difference between the lowest and highest class limits is 180.6−179.4 cm = 1.2 cm (this is the maximum possible range of the values). The shape of the distribution is symmetrical and so the machine is producing a similar number of handles above and below 180 cm.

Machine B needs overhauling. The machine is producing a wide range of handle lengths; the difference between the lowest and highest class limits is 185.5−177.5 = 8 cm. The modal class (see tallest bar) is 179.5-180.5 cm and contains the machine setting of 180 cm. However, the shape of the distribution is positively skewed and so the mean and median handle lengths are above 180 cm (see Box 9. p107). This shows that the machine is producing more handles above 180 cm than below 180 cm.

Cumulative Frequency

1. Cumulative Frequency

The speeds of cars (mph) travelling along a main road are recorded in this table.

Speed (mph)	No. of cars (Frequency)
0–20	4
20–30	13
30–40	52
40–50	75
50–60	40
60–70	16
Total	200

This table shows that:

4 cars travelled at speeds between 0 and 20 mph
13 cars travelled at speeds between 20 and 30 mph
52 cars travelled at speeds between 30 and 40 mph
and so on

From this table we can deduce that:

4 cars travelled at speeds less than 20 mph
4+13 = 17 cars travelled at less than 30 mph
4+13+52= 69 cars travelled at less than 40 mph
and so on

Adding up the frequencies like this gives the so-called **cumulative frequencies** 4, 17, 69, ... These are summarised in the table below:

Speed (mph)	No. of cars (Cumulative frequency)
Less than 20	4
Less than 30	17
Less than 40	69
Less than 50	144
Less than 60	184
Less than 70	200

This table tells us, for example, that 144 cars travelled at speeds less than 50 mph.

2. Cumulative Frequency Curve

We can plot a graph of the cumulative frequencies in Box 1. Make sure that:

- The graph starts where the axes cross
- The points are joined with a smooth curve

This graph is called a **cumulative frequency curve** (or **ogive**) and has an S shape.

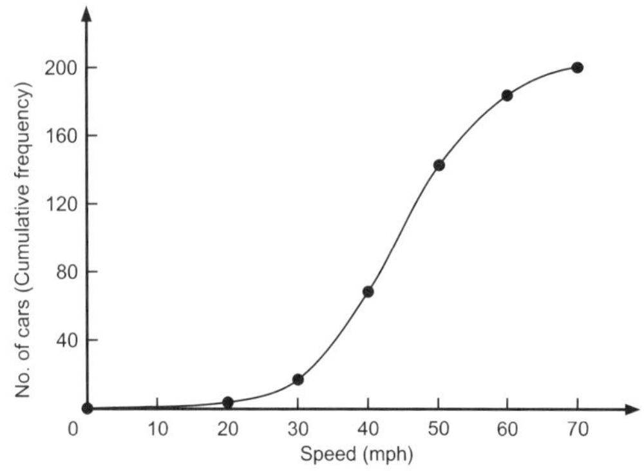

Question *Use the above cumulative frequency curve to answer the following questions.*

(a) How many cars had a speed less than 45 mph?
(b) How many cars had a speed above 45 mph?
(c) How many cars had a speed between 30 mph and 45 mph?
(d) 23 cars were travelling above the speed limit for this road. Use the graph to estimate the speed limit.

more

2. Cumulative Frequency Curve (Contd)

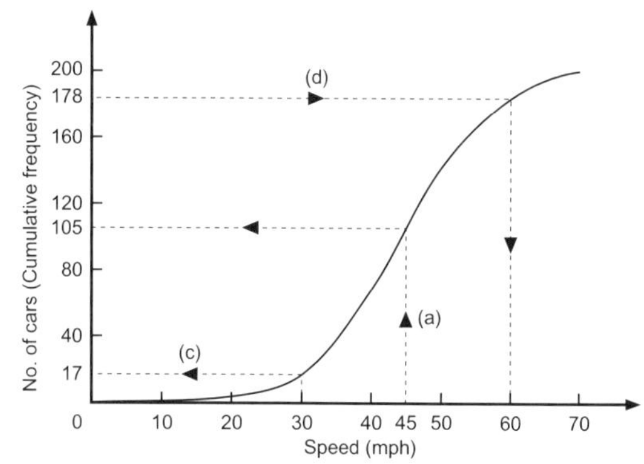

(a) From the graph, the cumulative frequency corresponding to 45 mph is 105. This tells us that 105 cars had a speed below 45 mph.

(b) There are 200 cars altogether (this is the highest cumulative frequency on the graph). So, if 105 cars had a speed below 45 mph, then, 200 − 105 = 95 cars had a speed above 45 mph.

(c) The graph shows that 105 cars had a speed below 45 mph and, of these, 17 cars had a speed below 30 mph. So, 105 − 17 = 88 cars had a speed between 30 mph and 45 mph.

(d) The 23 fastest cars correspond to the last 23 cumulative frequencies 178 to 200. The speed corresponding to the cumulative frequency 178 is 60 mph. So, the fastest 23 cars exceeded this speed limit.

3. Finding the Median

The median is the middle value. This means that half (50%) of the values are less than the median (see Box 3, p105). We can estimate the median from a cumulative frequency diagram.

Question *30 students took a memory test. The numbers of words they recalled are given in the table below.*

No. of words recalled	No. of students
16–20	2
21–25	5
26–30	12
31–35	8
36–40	3
Total	30

No. of words recalled	Cumulative frequency
20 or less	2
25 or less	
30 or less	
35 or less	
40 or less	

(a) Complete the cumulative frequency table.
(b) Draw the cumulative frequency curve.
(c) Use your graph to estimate the median number of words recalled per student.

(a) Add up the frequencies to give the cumulative frequencies.

No. of words recalled	Cumulative frequency
20 or less	2
25 or less	7
30 or less	19
35 or less	27
40 or less	30

more

3. Finding the Median (Contd)

(b) Carefully plot a cumulative frequency curve.

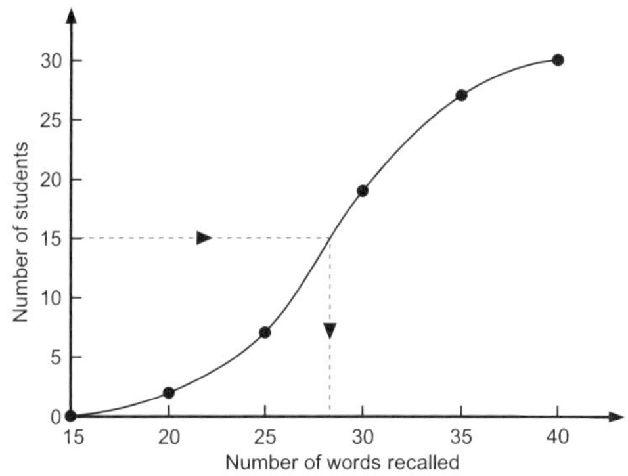

(c) There are 30 values altogether (see question). Half of these values are less than the median; so, 15 values are less than the median. From the graph, 15 students recalled less than 28 words. So, the median number of words recalled is 28 per student.

4. Interquartile Range

The yields (kg) from 12 dwarf apple trees are recorded below, arranged from smallest to largest:

6 8 8 12 15 16 18 19 22 24 25 26

The median yield is 17 kg. Half of the 12 values lie below this median. Now, half of 12 is 6 and so the first 6 trees each yielded less than 17 kg of apples:

Half (50%) of the values

6 8 8 12 15 16 ↑ 18 19 22 24 25 26

Median = 17

Similarly, a quarter of the 12 values lie below the **lower quartile**, 10 kg. Now, a quarter of 12 is 3 and so the first 3 trees each yielded less than 10 kg of apples:

Quarter (25%) of the values

6 8 8 ↑ 12 15 16 18 19 22 24 25 26

Lower quartile = 10

And three quarters of the 12 values lie below the **upper quartile**, 23 kg. Now, three quarters of 12 is 9 and so the first 9 trees each yielded less than 23 kg of apples:

Three quarters (75%) of the values

6 8 8 12 15 16 18 19 22 ↑ 24 25 26

Upper quartile = 23

The **interquartile range** is the difference between the two quartiles. In this example:

Interquartile range = Upper quartile − Lower quartile

= 23 − 10 = 13 kg

The middle half (50%) of the values lie within the interquartile range:

Middle half of the values

6 8 8 ↑ 12 15 16 18 19 22 ↑ 24 25 26

Lower quartile ← Interquartile range → Upper quartile

more

4. The Interquartile Range (Contd)

The quartiles can be estimated using a cumulative frequency curve, as shown in the following example.

Question *The cumulative frequency table shows the prices of houses for sale in an estate agent's window.*
(a) Draw a cumulative frequency curve.
(b) Use your graph to estimate
(i) the median, (ii) the interquartile range.

House price (£1000s)	No. of houses
$0 \leqslant p < 25$	5
$0 \leqslant p < 50$	23
$0 \leqslant p < 75$	39
$0 \leqslant p < 100$	51
$0 \leqslant p < 125$	58
$0 \leqslant p < 150$	60

(a) Carefully draw the cumulative frequency curve:

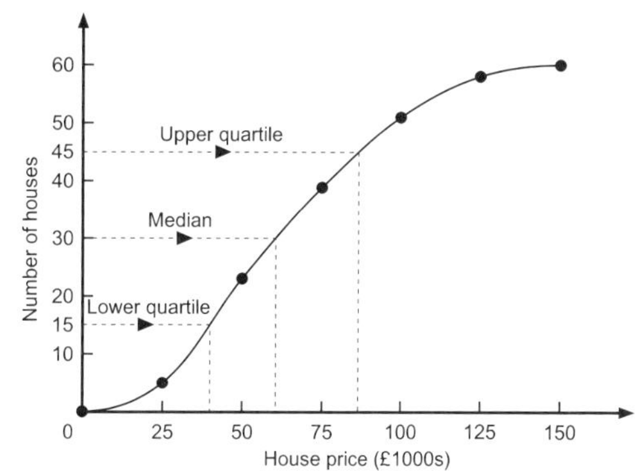

(b) (i) There are 60 values altogether (the highest cumulative frequency). The median corresponds to half of these, i.e. 30 values. From the graph, the median is £60 000.

(ii) First, find the lower and upper quartiles. The lower quartile corresponds to a quarter of the 60 values, i.e. 15 values. From the graph, the lower quartile is £40 000.
The upper quartile corresponds to three quarters of the 60 values, i.e. 45 values. From the graph, the upper quartile is £87 000.

So, interquartile range = £87 000 − £40 000
= £47 000

Question *25% of the employees of a company earn more than £23 000 per annum. The interquartile range of employees' incomes is £9000. 25% of the employees earn below a certain income. Calculate this income.*

The diagram shows that the lowest paid 25% of employees earn below £23 000 − £9000 = £14 000

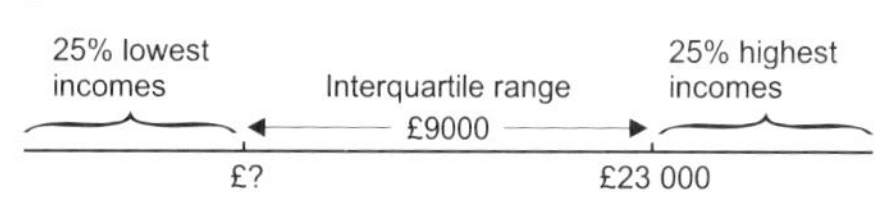

Scatter Graphs and Correlation

1. Scatter Graphs

A **scatter graph** shows the relationship between two quantities.

The heights and weights of 9 children are recorded in the table below.

Height (cm)	121	128	140	150	155	150	160	162	180
Weight (kg)	34	39	42	43	45	48	47	49	56

These values can be plotted as points to make the scatter graph shown below.

NOTE Do not join the points.

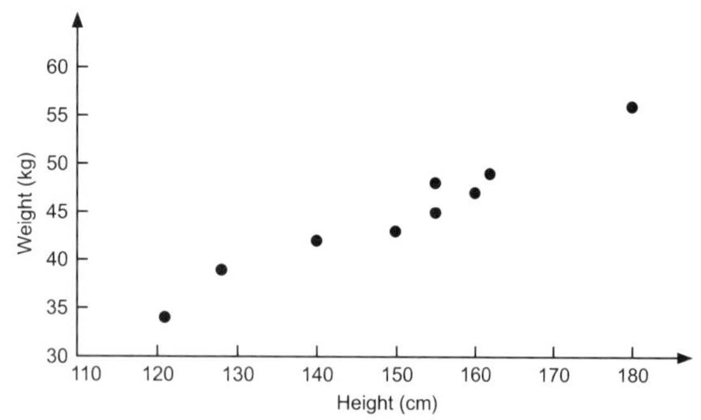

The upward slope of the points shows that, in general, the greater a child's height, the greater their weight.

2. Line of Best Fit

The points in the scatter graph of Box 1 lie roughly along a straight line, called the **line of best fit**.

The line of best fit is drawn so that there are roughly the same number of points on each side of it.

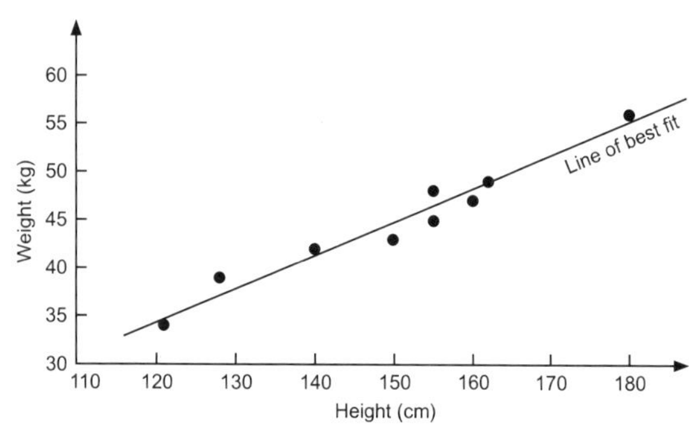

A more accurate line of best fit can be drawn by finding the 'central' point, as follows. Calculate the mean of the x-values (heights) and the mean of the y-values (weights):

Mean height $= \bar{x} = \frac{121+128+....+180}{9} = \frac{1346}{9} = 149.6$ cm

Mean weight $= \bar{y} = \frac{34+39+....+56}{9} = \frac{403}{9} = 44.8$ kg

Now plot the central point $(\bar{x}, \bar{y}) = (149.6, 44.8)$ and draw the line of best fit through it (see top right box).

more

2. Line of Best Fit (Contd)

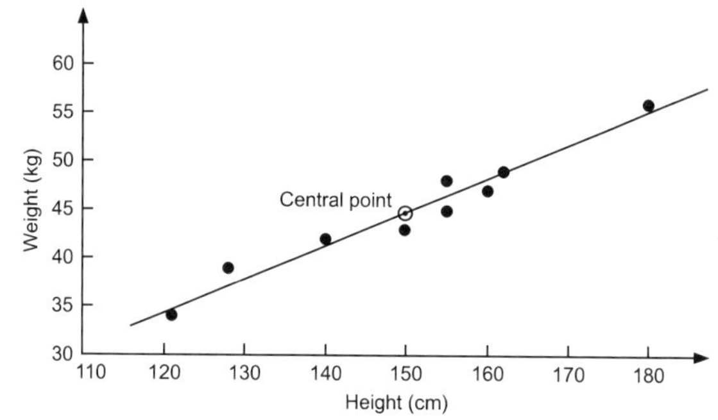

We can use the line of best fit to estimate the heights and weights of children. The graph below shows that a child with height 170 cm is likely to have a weight of about 52 kg. And a child whose weight is 40 kg is likely to be about 136 cm tall.

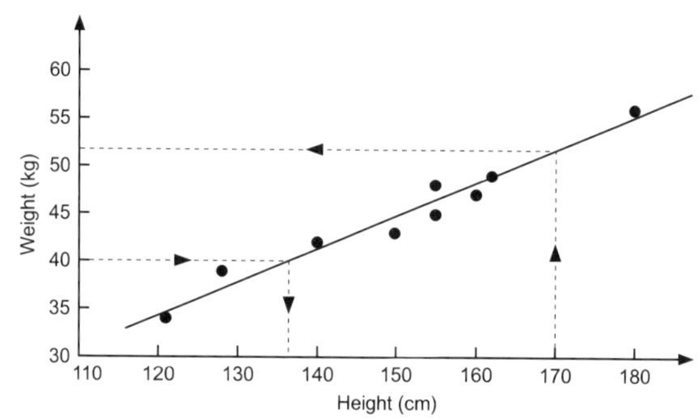

3. Correlation

The word **correlation** means mutual relationship. Two quantities are said to be **correlated** if changes in one are accompanied by changes in the other.

If two quantities are strongly correlated (closely related), the points of their scatter graph will be close to a straight line.

You need to be able to describe the correlation of a scatter graph. The different possibilities are shown below.

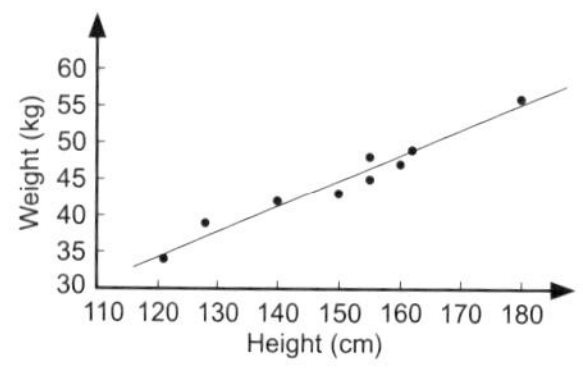

Strong, **positive** correlation because the line of best fit has a positive gradient. As a child's height increases, so does their weight.

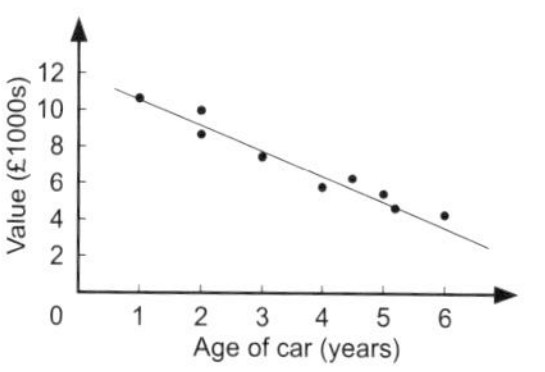

Strong, **negative** correlation because the line of best fit has a negative gradient. The older a car, the less its value.

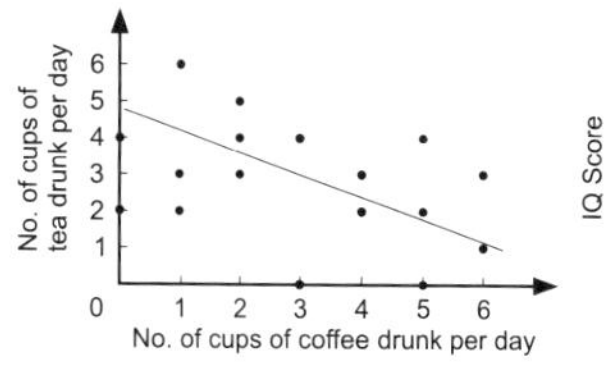

Weak, negative correlation; the points are not very close to the line of best fit but do have a downward slope. There is a weak relationship between tea and coffee drinking habits.

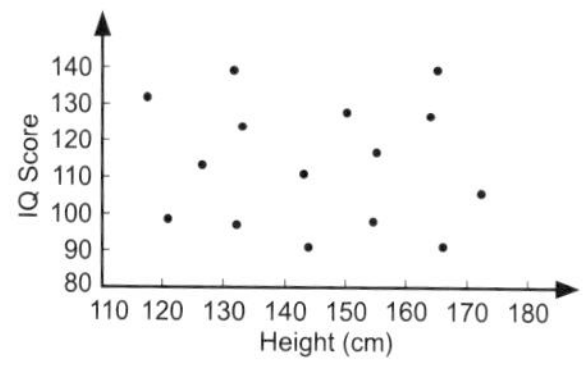

No correlation; because the points have no pattern. There is no relationship between a person's IQ and their height.

Probability

1. Definition of Probability

When a die is thrown, the six **possible outcomes** are 1, 2, 3, 4, 5, 6. If the die is perfectly balanced (**fair**), each of these outcomes is equally likely to occur.

An **event** is something that can happen. For example, when we throw a die an even number may turn up; so, 'an even number' is an event. There are three ways of throwing an even number: 2, 4 or 6; so, there are three ways the event 'an even number' can happen.

The **probability** of an event happening is a measure of how likely it will occur and can be calculated using the formula:

$$\text{Probability} = \frac{\text{Number of ways the event can happen}}{\text{Number of possible outcomes}}$$

NOTE This formula can be used only if all of the outcomes are **equally likely** to occur.

We can use this formula to find the probability of throwing an even number.

$$\text{Probability} = \frac{\text{No. of ways of throwing an even number (2, 4, 6)}}{\text{No. of possible outcomes (1, 2, 3, 4, 5, 6)}} = \frac{3}{6}$$

Probabilities can be written as fractions, decimals or percentages. In the above example:

Probability of throwing an even number $= \frac{3}{6} = 0.5 = 50\%$

Question *10 identical cards are numbered as below.*

2 4 4 4 6 8 8 8 8 8

The cards are shuffled and one chosen at random. Calculate the probability that the number on the card is (a) 2, (b) more than 2, (c) 8, (d) an even number, (e) an odd number.

(a) There are 10 possible outcomes when a card is chosen. The cards are identical and so each outcome is equally likely.

There is only 1 way of choosing a 2 and so:

Probability of choosing a 2 $= \frac{1}{10} = 0.1$

(b) There are 9 cards higher than 2 and so:

Probability of choosing above 2 $= \frac{9}{10} = 0.9$

(c) There are 5 ways of choosing an 8 (because there are 5 cards with an 8) and so:

Probability of choosing an 8 $= \frac{5}{10} = 0.5 = 50\%$

(This is commonly described as a **fifty–fifty** chance, since there is a 50% chance of choosing an 8 and a 50% chance of not choosing an 8.)

(d) All the cards are even and so choosing an even number is an **absolute certainty**. There are 10 ways of choosing an even number and so:

Probability of an even number $= \frac{10}{10} = 1$

(e) There are no cards with odd numbers and so choosing an odd number is **impossible**:

Probability of an odd number $= \frac{0}{10} = 0$

Notice that all the probabilities in this question are numbers between 0 and 1. They are illustrated below:

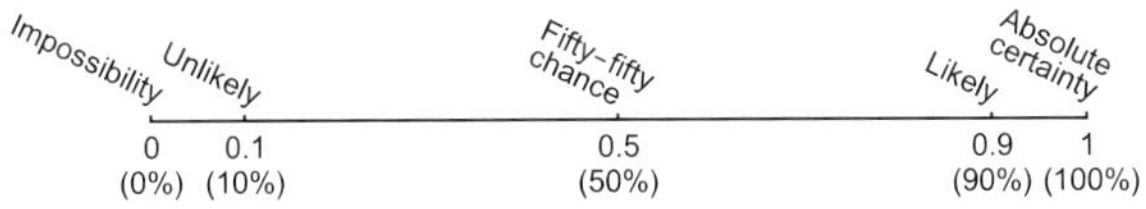

All probabilities lie between 0 and 1.

2. Probability of an Event Not Happening

Question *The numbers 1, 2, 3, 4 and 5 are equally likely when this spinner is used. The spinner is spun once. Calculate the probability of*
(a) obtaining a 3
(b) not obtaining a 3.

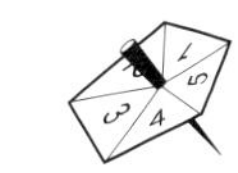

(a) There are 5 possible outcomes: 1, 2, 3, 4, 5.

There is only 1 way of spinning a 3 and so:

Probability of obtaining a 3 $= \frac{1}{5} = 0.2$

(b) There are 4 ways of ***not*** spinning a 3, namely 1, 2, 4 and 5. So:

Probability of ***not*** obtaining a 3 $= \frac{4}{5} = 0.8$

Notice that this probability can also be calculated by subtracting the probability in part (a) from 1: $1 - 0.2 = 0.8$. So:

Probability of ***not*** obtaining a 3 = 1 − Probability of obtaining a 3

Generally, we have the result:

Probability of an event ***not*** happening = 1 − Probability of event happening

Question *The probability of rain on a day in June is 0.3. What is the probability of no rain on 21 June?*

Probability of no rain = 1 − Probability of rain
$= 1 - 0.3 = 0.7$

3. Total Probability

If we add up the probability of every possible outcome, the answer is always 1. For example, if an ordinary fair die is thrown, the 6 possible outcomes are 1, 2, 3, 4, 5 and 6, each occuring with a probability of $\frac{1}{6}$.

So, total probability $= \frac{1}{6} + \frac{1}{6} + \frac{1}{6} + \frac{1}{6} + \frac{1}{6} + \frac{1}{6} = \frac{6}{6} = 1$

4. Adding Probabilities

Question *A bag contains 2 red, 3 green and 4 white counters. One counter is chosen from the bag at random. Calculate the probability that it is (a) red, (b) green, (c) red or green.*

(a) There are 9 possible outcomes (any one of 9 counters could be chosen).

There are 2 ways of choosing a red counter. So:

Probability of choosing a red counter $= \frac{2}{9}$

We can write this briefly as: P(red) $= \frac{2}{9}$

(b) Similarly, there are 3 ways of choosing a green counter and so:

P(green) $= \frac{3}{9} = \frac{1}{9}$

more

Probability (Contd)

4. Adding Probabilities (Contd)

(c) There are 5 ways of choosing a red or green counter (2 red + 3 green) and so:

P(red *or* green) = $\frac{5}{9}$

We can also calculate this probability by adding together the probabilities P(red) and P(green) found in parts (a) and (b):

P(red) + P(green) = $\frac{2}{9}+\frac{3}{9}=\frac{5}{9}$

and so we can say

P(red *or* green) = P(red) + P(green)

NOTE It looks as though the word *or* has been replaced by a plus sign

For any two events A and B, we have the general result:

P(A *or* B) = P(A) + P(B) **or means +**

NOTE You can only use this rule if the two events cannot happen together. This is the case in the above example because it is impossible to choose a counter that is both red and green. Here is a question where the rule **cannot** be used.

Question *One of these counters is chosen at random.*

[1] [2] (3) (4) △5

The probability of choosing a square counter is $\frac{2}{5}$ *and the probability of choosing an even number is* $\frac{2}{5}$*. Explain why the probability of choosing a square or an even number is not* $\frac{2}{5}+\frac{2}{5}$*.*

One of the counters is both square and even. This means that the events 'choosing a square' and 'choosing an even number' can happen both at once.

The correct answer is $\frac{3}{5}$, because choosing a square or an even number means choosing one of the 3 counters [1] [2] (4) out of a possible 5 counters.

5. Multiplying Probabilities

Question *An ordinary fair die has faces numbered 1, 2, 3, 4, 5, 6. An ordinary fair coin has two sides, head (H) and tail (T).*

(a) The die is thrown. Calculate the probability of obtaining a 3.

(b) The coin is flipped. Calculate the probability of obtaining a head (H).

(c) The die is thrown and the coin is flipped. Complete the list of possible outcomes below.

(1, H) (2, H) ( ) ( ) ( ) ( )
(1, T) ( ) ( ) ( ) ( ) ( )

(d) Use your list to calculate the probability of throwing a 3 and flipping a head.

(a) There are 6 possible outcomes (1, 2, 3, 4, 5, 6) but only 1 way of flipping a 3. So:

P(3) = $\frac{1}{6}$

(b) There two possible outcomes, head (H) or tail (T), but only 1 way of flipping a head. So:

P(H) = $\frac{1}{2}$

more

5. Multiplying Probabilities (Contd)

(c) Here are the possible outcomes:

(1, H) (2, H) (3, H) (4, H) (5, H) (6, H)
(1, T) (2, T) (3, T) (4, T) (5, T) (6, T)

There are 12 possible outcomes:
6 numbers × 2 sides = 12 combinations

(d) There are 12 possible outcomes but only 1 way of throwing a 3 and flipping a head, namely (3, H). So:

P(3 *and* H) = $\frac{1}{12}$

We can also calculate this probability by multiplying the probabilities P(3) and P(H) found in parts (a) and (b): $\frac{1}{6}\times\frac{1}{2}=\frac{1}{12}$ and so we can say

P(3 *and* H) = P(3) × P(H)

NOTE It looks as though the word *and* has been replaced by a multiplication sign

For any two events A and B, we have the general result:

P(A *and* B) = P(A) × P(B) **and means ×**

NOTE You can only use this rule if the two events are independent of each another, i.e. the occurence of one does not affect the occurence of the other. In this example, throwing a die obviously has no effect on flipping a coin, and vice-versa.

6. Tree Diagrams

Tree diagrams can help you to calculate the probability of two or more events happening.

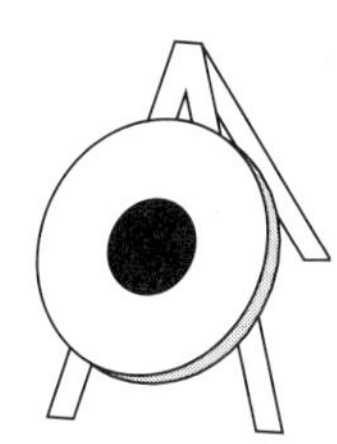

Question *The probability of an archer hitting the centre of this target is 0.2. The archer always hits the target.*

(a) Calculate the probability of the archer hitting the outer ring of the target with one arrow.

The archer shoots two arrows at the target. Her accuracy is unaffected by previous shots.

(b) Draw a tree diagram to illustrate the two shots.

(c) Use your diagram to calculate the probability of the archer hitting

(i) the centre of the target with both shots

(ii) the centre of the target with only one shot.

(a) P(hitting the outer ring) = P(not hitting the centre)
= 1 − P(hitting the centre)
= 1−0.2
= 0.8

(b) There are two possible outcomes for the first shot: hitting the centre and hitting the outer ring.

Each possible outcome is represented by a straight line labelled with its probability. Notice that the probabilities add up to 1.

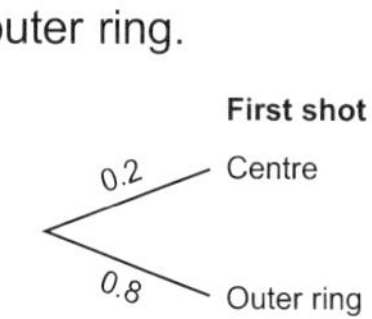

more

6. Tree Diagrams (Contd)

The second shot has the same two outcomes with the same probabilities (because the accuracy of the archer is unaffected by the first shot). The completed tree diagram is shown below:

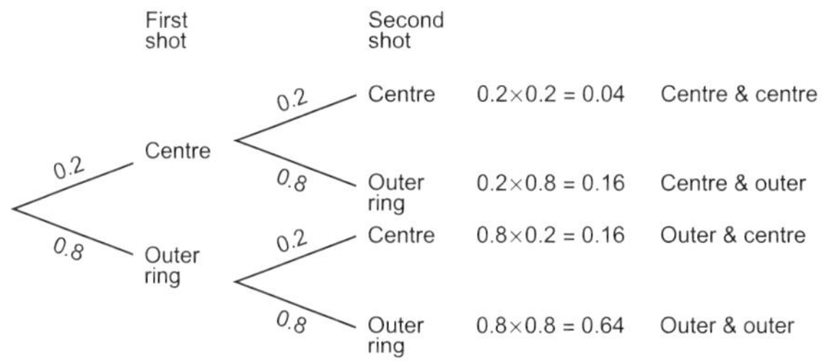

(c) (i) P(hitting centre with both shots)
= P(hitting centre with first shot *and* second shot)
= P(hitting centre with first shot) × P(hitting centre with second shot)
= 0.2×0.2 = 0.04

Remember: **and** means ×

This calculation is shown on the tree diagram above, together with the calculations for the other combinations of outcomes.

(ii) P(hitting centre with only one shot)
= P(hitting centre with first shot *and* then the outer ring with the second shot *or* hitting outer ring with first shot *and* then the centre with the second shot)
= 0.2×0.8 + 0.8×0.2
= 0.16+0.16
= 0.32

Remember: **and** means × and **or** means +

7. Conditional Probability

H

When the occurence of one event affects the probability of another event occuring, use a tree diagram to calculate probabilities.

Question *A bag contains 3 red and 5 blue marbles. Two marbles are taken from the bag. What is the probability that*
(a) they are both red
(b) only one of them is red?

It does not matter whether the two marbles are removed from the bag at the same time or one after the other: the result is the same. It is easier to assume they were removed one after the other, then we can draw a tree diagram.

First marble

There are 8 possible outcomes (3 red and 5 blue marbles) and 3 ways of choosing a red marble.

So, P(red marble first) = $\frac{3}{8}$
and P(blue marble first) = $\frac{5}{8}$

We can now draw the first branches of a tree diagram.

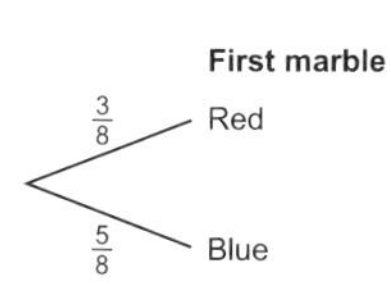

more

7. Conditional Probability (Contd)

Second marble

The choice of the second marble depends on (is conditional on) which one was chosen first.
If the *first marble* was *red*, the bag now contains 2 red and 5 blue marbles (7 marbles altogether) and so:

P(red second) = $\frac{2}{7}$ and P(blue second) = $\frac{5}{7}$

If the *first marble* was *blue*, the bag now contains 3 red and 4 blue marbles (7 marbles altogether) and so:

P(red second) = $\frac{3}{7}$ and P(blue second) = $\frac{4}{7}$

We can now complete the tree diagram:

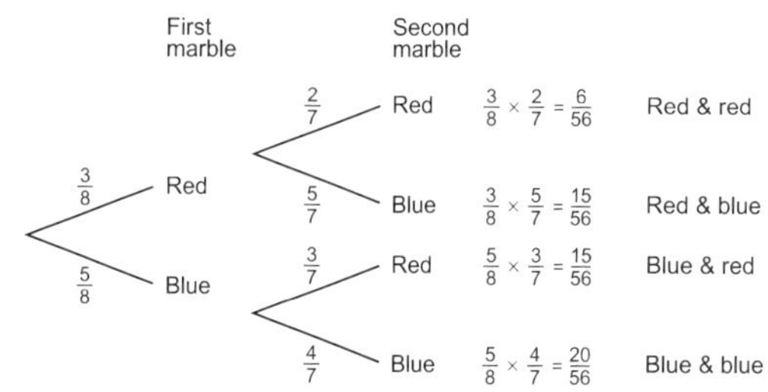

Now that the tree diagram is complete, we can use the same methods as in Box 6 to calculate the combined probabilities:

(a) P(both red) = P(red first *and* red second)
= P(red first) × P(red second)
$= \frac{3}{8} \times \frac{2}{7}$
$= \frac{6}{56}$

(b) P(only one red) = P(red first *and* blue second *or* blue first *and* red second)
$= \frac{3}{8} \times \frac{5}{7} + \frac{5}{8} \times \frac{3}{7}$
$= \frac{15}{56} + \frac{15}{56}$
$= \frac{30}{56}$
$= \frac{15}{28}$

8. Estimating Probability

Chin Lin did an experiment in which she flipped a 2p coin 50 times. She obtained 22 heads and 28 tails.

A **trial** is a repeatable act. In this experiment, each flip of the coin is a trial.

Chin Lin used her results to estimate the probability of obtaining heads or tails in the future:

$$\text{Probability of heads} = \frac{\text{Number of heads}}{\text{Number of flips}} = \frac{22}{50} = 0.44$$

$$\text{Probability of tails} = \frac{\text{Number of tails}}{\text{Number of flips}} = \frac{28}{50} = 0.56$$

These are called **experimental probabilities**.

She knew that, if the coin was perfectly balanced (fair), heads and tails would be equally likely and their probabilities would be:

P(head) = 0.5 and P(tail) = 0.5

Her estimates were fairly close to 0.5 and so she concluded that her 2p coin was probably fair.

more

Probability (Contd)

8. Estimating Probability (Contd)

To be more certain, Chin Lin repeated the experiment, this time flipping the 2p coin 100 times. She obtained 48 heads and 52 tails, giving the experimental probabilities below:

$$P(\text{head}) = \frac{48}{100} = 0.48 \quad \text{and} \quad P(\text{tail}) = \frac{52}{100} = 0.52$$

These estimates are even closer to 0.5 and so she was more certain that the 2p coin was fair.

Her teacher asked her how many heads she would **expect** if the coin were flipped another 200 times.

Chin Lin said that, if the coin were fair, the probability of a head would be 0.5, which is the same as 50%. So, she would expect 50% of flips to be heads:

Expected number of heads = 50% of 200
$= 0.5 \times 200 = 100$

You can see that the probability 0.5 has been multiplied by the number of flips (trials). So:

Expected number of times an event occurs = Probability of event × Number of trials

Critical Path Analysis H

1. Critical Path Analysis

Many tasks involve several steps before they can be completed. For example, making a loaf of bread involves weighing the ingredients, mixing the dry ingredients (flour, salt and yeast), adding water, kneading the dough, leaving the dough to rise, etc.

The aim of critical path analysis is to arrange the steps of a task in a logical order and find the shortest time in which the task can be completed.

2. Networks

A **network** is a diagram that helps you to organise the steps of a task.

Each step is represented by a straight line labelled with the length of time taken to complete the step. Here is one of the steps in making bread:

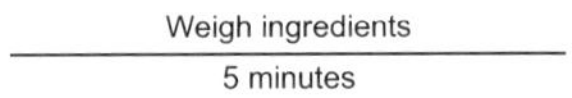

The beginning and end of a step are called **nodes** and are indicated by numbered circles:

The task begins with the lowest number and ends with highest.

Some steps cannot begin until others are completed. In the bread-making example, the ingredients must be weighed before they can be made into dough (see above diagram).

Steps that do not depend on each other can be performed at the same time. For example, a loaf tin can be greased and the oven can be warmed to the correct temperature whilst the dough is rising, as shown in the diagram below:

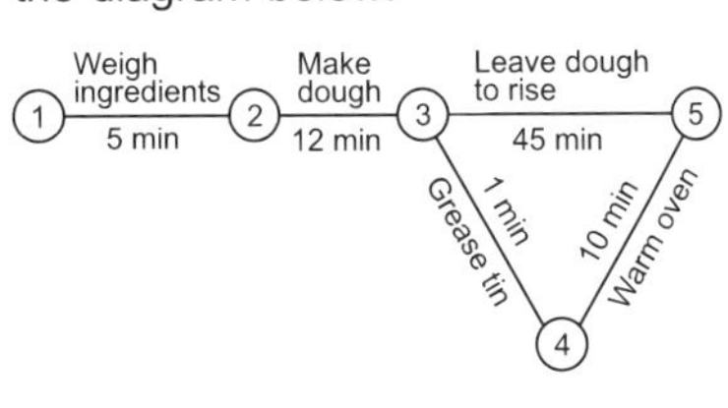

more

2. Networks (Contd)

From this diagram, you can see that greasing the loaf tin and warming the oven only take 11 minutes, compared with the 45 minutes needed for the dough to rise. So, it takes 45 minutes to progress from node 3 to node 5.

The complete network for making a loaf of bread is shown below:

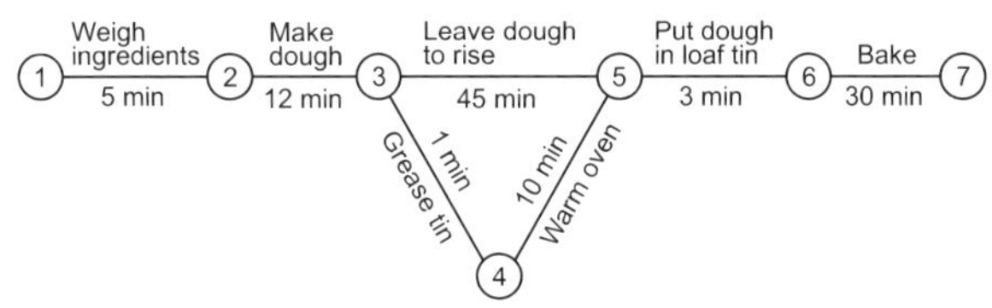

There are two **paths** from node 1 to node 7:

①–②–③–⑤–⑥–⑦

and ①–②–③–④–⑤–⑥–⑦

The total time taken for the steps in the first path is 5+12+45+3+30 = 95 minutes.

The total time taken for the steps in the second path is 5+12+1+10+3+30 = 61 minutes.

So, the shortest time in which to make a loaf of bread is 95 minutes. The path that gives this time is called the **critical path**.

3. More Complicated Example

Question *The steps needed to make an apple tart are listed opposite.*

(a) Given the following information, complete the network below.
When the pastry case has been baked, it can be filled with apples and custard.
Whilst the tart is baking, the glaze can be made.
When the tart has finished baking, it can be glazed.

Step	*Duration*
Weigh ingredients	*5 min*
Make pastry	*12 min*
Leave pastry to rest	*15 min*
Roll pastry and line dish	*3 min*
Warm oven	*10 min*
Bake pastry case	*15 min*
Prepare apples	*10 min*
Make custard	*15 min*
Fill pastry case	*10 min*
Bake tart	*20 min*
Make glaze	*5 min*
Glaze tart	*4 min*

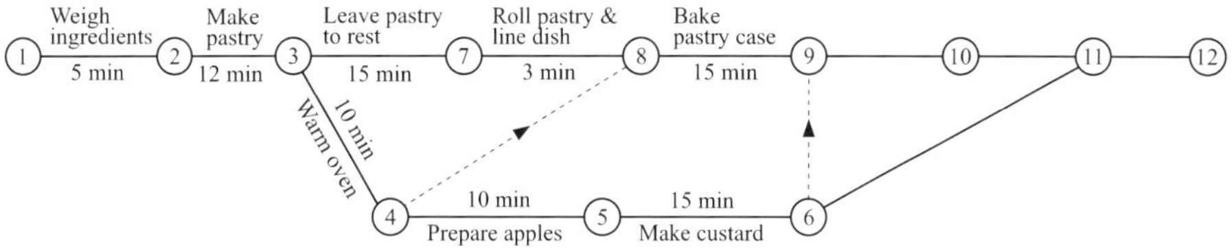

(b) A dashed line shows that one task must be completed before another can begin.
For example, the oven must be warmed before the pastry case can be baked.
Calculate the shortest time needed to make an apple tart.

The completed network is shown below

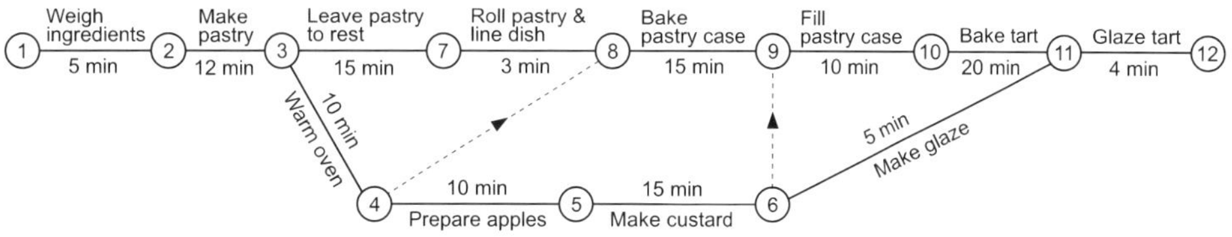

The longest path from node 1 to node 8 is: 1–2–3–7–8

The longest path from node 8 to node 9 is: 8–4–5–6–9

The longest path from node 9 to node 11 is: 9–10–11

So, the longest path from node 1 to node 12 (the critical path) is: 1–2–3–7–8–4–5–6–9–10–11–12

The duration of this path is: 5+12+15+3+10+15+10+20+4 = 94 minutes.

So, the shortest time in which to make an apple tart using this method is 94 minutes.

Examination Questions

The following examination questions have been selected from the 1998 Specimen Papers for the London GCSE. Higher level questions and parts of questions are indicated by **H**. If you have difficulty with a question, turn to the indicated page and box for help. You will find brief solutions on pages 121–123. Edexcel Foundation, London Examinations accepts no responsibility whatsoever for the accuracy or method of working in the answers given.

Number

1. $\frac{3}{5}$ of the cost of a music tape goes to the supplier. A music tape costs £6. How much goes to the supplier? (Box 9, p17)

2. $\frac{3}{8}$ of the Highlands of Scotland is covered in forest.
   (a) (i) Change $\frac{3}{8}$ to a decimal. (Box 11, p20)
   (ii) Write down your answer to part (i) correct to 2 decimal places. (Box 3, p21)
   (b) Work out the percentage of the Highlands of Scotland NOT covered in forest.
   Here is a list of fractions, decimals and percentages:
   67% $\quad \frac{1}{2} \quad$ 0.6 $\quad$ 25% $\quad$ 0.3 $\quad \frac{3}{8}$
   (c) Rewrite the list in order of size, starting with the smallest first. (Box 1, p25)

3. The distance from the Earth to the Moon is 250 000 miles.
   (a) Express this number in standard form. (Box 4, p31)
   The distance from the Earth to the Sun is 9.3×10^7 miles.
   (b) Calculate the value of the expression
   $$\frac{\text{Distance from the Earth to the Moon}}{\text{Distance from the Earth to the Sun}}$$
   giving your answer in standard form. (Box 8, p32)

4. Jake buys a television. He pays £125 deposit and 12 monthly instalments of £29.62.
   (a) Work out the total amount that Jake pays. (Box 8, p19)
   In Cooper's store, BRITE colour televisions are priced at £390 each in 1996. Lyn pays cash and is given a discount of 12%.
   (b) Calculate the amount that Lyn pays. (Box 6, p26)
   The price of a BRITE colour television has risen by 4% since 1995 to £390.
   (c) Calculate the price of the television in 1995. (Box 7, p26)

5. **H**

Diagram NOT accurately drawn

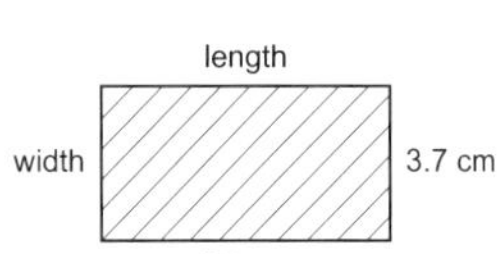

The dimensions of the rectangle are given correct to 1 decimal place.
   (a) Calculate the lower bound of the area of the rectangle. (Box 7, p22)
   Rectangles with the same dimensions as the one in part (a) are to be cut from a large piece of card of length 1 metre, correct to the nearest centimetre and width exactly 4 cm.

Diagram NOT accurately drawn

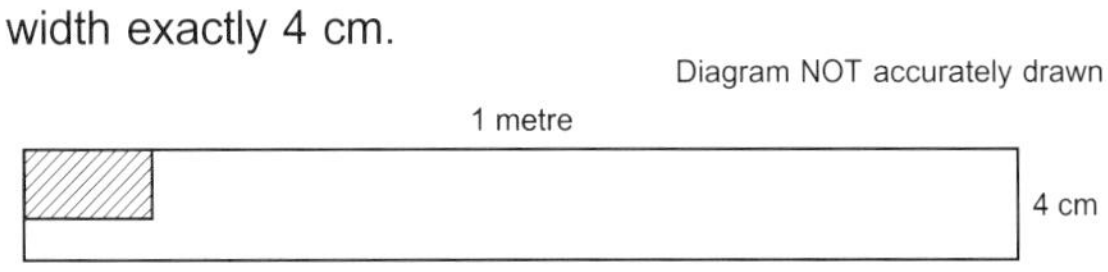

5. (b) Calculate the maximum number of rectangles that could be cut from the large piece of card. (Box 7, p22)

6. Ivan builds fences in different lengths using pieces of wood.

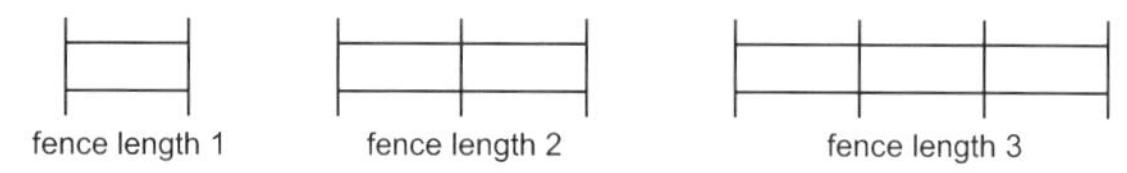

   (a) Sketch fence length 5. (Box 1, p13)
   Ivan counted how many pieces he needed to make each fence length. He then drew the table below:

Fence length	1	2	3	4	5	6
Number of pieces	4	7	10			

   (b) Complete the table to show how many pieces of wood he would use for fence lengths 4, 5 and 6. (Box 4, p14)
   (c) Explain how you would work out the number of pieces needed for fence length 25. (Box 4, p13)
   (d) Write down, in terms of n, an expression for the number of pieces of wood needed for fence length n. (Box 5, p14)

Algebra

7. Sarah uses the formula:
   $$t = \frac{2s}{u+v}$$
   She has to calculate the value of t when $s = 623.25$, $u = 11.37$ and $v = 87.22$. Sarah estimates the value of t **without using her calculator**.
   (a) (i) Write down the numbers Sarah could use in the formula to estimate the value of t. (Box 5, p21)
   (ii) Work out the estimate for the value of t that these numbers would give. (Box 5, p21)
   (b) Use your calculator to work out the actual value of t. Give your answer to an appropriate degree of accuracy. (Box 13, p24)

8. The heat setting of a gas oven is called its Gas Mark. A Gas Mark, G, may be converted to a temperature, F, in degrees Fahrenheit, using the formula:
   $$F = 25G + 250$$
   (a) Factorise completely $25G+250$. (Box 6, p41)
   A Gas Mark, G, may be converted to a temperature, C, in degrees Celsius, using the formula:
   $$C = 14G + 121$$
   (b) Make G the subject of the formula $C = 14G+121$. (Box 4, p39)

9. The air temperature, T °C, outside an aircraft at a height of h feet is given by the formula:
   $$T = 26 - \frac{h}{500}$$

The air temperature outside an aircraft is −52 °C. Calculate the height of the aircraft. (Box 3, p38)

10. The two circles have radii of R cm and r cm, where $R > r$. The shaded area, A cm^2, between the circles is given by the formula:

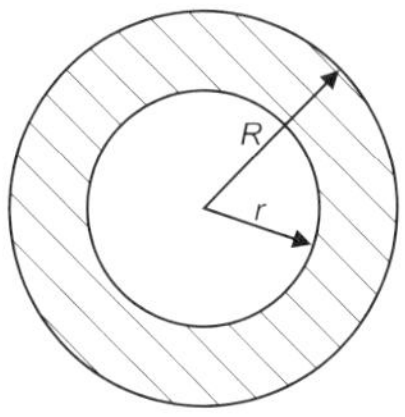

$$A = \pi(R+r)(R-r)$$

(a) Expand $(R+r)(R-r)$. Express your answer as simply as possible. (Box 3, p41)

(b) Calculate the value of r when $A = 147$ and $R = 21$. (Box 5, p39)

11. A shop sells chests of drawers in two different widths, 'standard' and 'wide'. The cost, S pounds, of a 'standard' chest of drawers with d drawers may be calculated using the formula:

$$S = 29+15d$$

A 'standard' chest of drawers costs £119.

(a) Calculate the number of drawers this chest has. (Box 7, p37)

The cost, W pounds, of a 'wide' chest of drawers may be calculated using the formula:

$$W = k+md$$

k and m are constants and d is the number of drawers. The cost of a "wide" chest of drawers with 4 drawers is £117. The cost of a "wide" chest of drawers with 6 drawers is £149.

(b) (i) Use the information to write down two equations in k and m. (Box 6, p48)

(ii) Solve the equations to find the value of m. (Box 2, p47)

12. The graph represents part of Mrs Hinton's journey from Manchester to London. Mrs Hinton stopped for a rest at a service station.

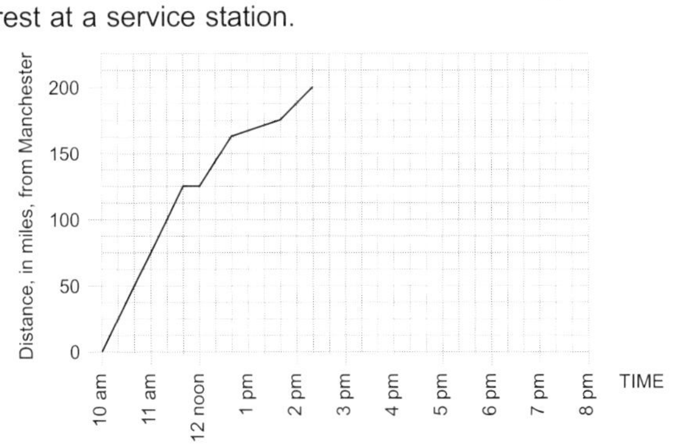

(a) (i) Write down the time at which she stopped. (Box 4, p61)

(ii) For how long did she stop? (Box 4, p61)

For part of the journey Mrs Hinton had to slow down because of a traffic queue.

(b) For how many miles did she travel at this slower speed? (Box 4, p61)

Mrs Hinton spent an hour at a meeting in London. She then returned home to Manchester, travelling at a steady speed of 50 miles per hour.

(c) Use this information to complete the graph of her journey. (Box 2, p60)

13. $y = x^3-4x-1$

(a) Complete the table of values. (Box 1, p55)

x	−2	−1	0	1	2	3
y		1				

(b) On the grid below, draw the graph of $y = x^3-4x-1$. (Box 5, p54)

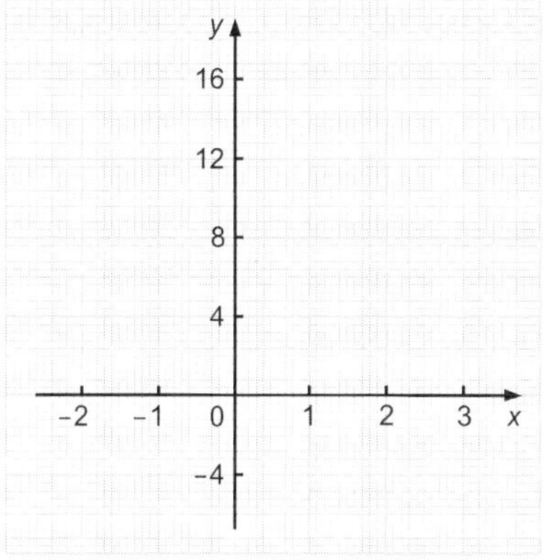

(c) By drawing a suitable straight line on the grid, solve the equation:

$$x^3-4x-1 = -2 \qquad \text{(Box 1, p63)}$$

(d) Using the method of trial and improvement, or otherwise, solve the equation $x^3-4x-1 = 30$ correct to 1 decimal place. (Box 1, p45)

14. **H** This is a sketch of the graph of $y = f(x)$, where

$$f(x) = (x+3)(x-2)(x-4)$$

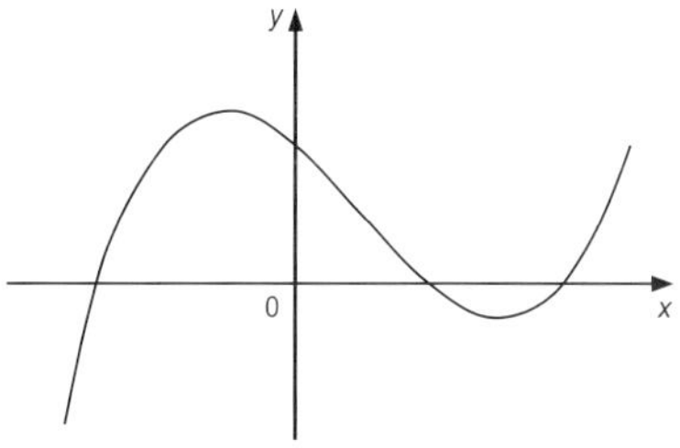

(a) Calculate the value of $f(0)$. (Box 1, p68)

(b) On the axes above sketch the graph of $y = f(-x)$. (Box 3, p68 and Box 4, p69)

(c) Describe fully the single geometric transformation which maps the graph of $y = f(x)$ onto the graph of $y = f(-x)$. (Box 4, p69)

The equation $f(x) = f(-x)$ has a solution $x = 0$. It also has a positive solution x such that

$$n < x < n+1$$

where n is a positive integer.

(d) Write down the value of n. (Box 3, p64)

Shape and Space

15. The diagram represents an L-shaped room whose corners are all right angles.

Diagram NOT accurately drawn

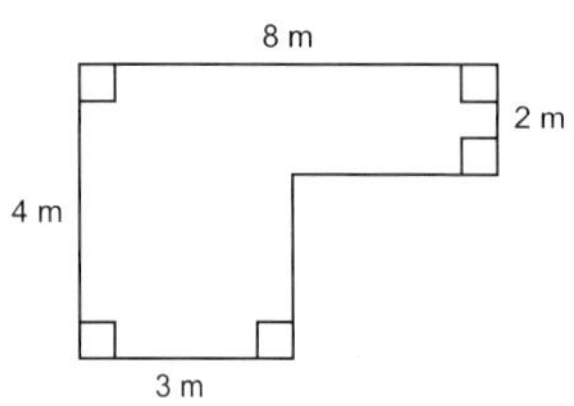

(a) Work out the area of the room. (Box 2, p70)

The height of the room is 2.7 m.

(b) Calculate the volume of the room. (Box 2, p74)

16. The diagram represents a tea packet in the shape of a cuboid.

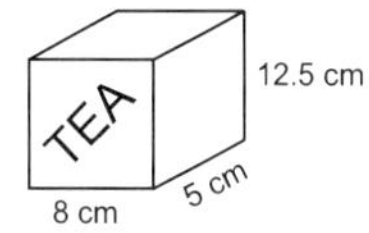

(a) Calculate the volume of the packet. (Box 2, p74)

There are 125 g in a full packet. Jason has to design a new packet that will contain 100 g of tea when full.

(b) (i) Work out the volume of the new packet. (Box 3, p27)

(ii) Express the weight of tea in the new packet as a percentage of the weight in the packet shown. (Box 3, p25)

17. The diagram represents a metal cone of height 90 cm made in two parts, labelled T and S. The top part, T, has a height of 45 cm.

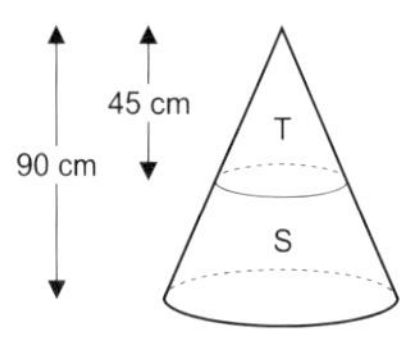

Given that the volume of S and T together is 10 500π cm^3,

(a) calculate the volume of S. (Box 7, p76)

A second cone is also made up of two parts, A and B. The radius of the top part, A, is 20 cm and the radius of the base of B is 30 cm. The height of the bottom part is 120 cm.

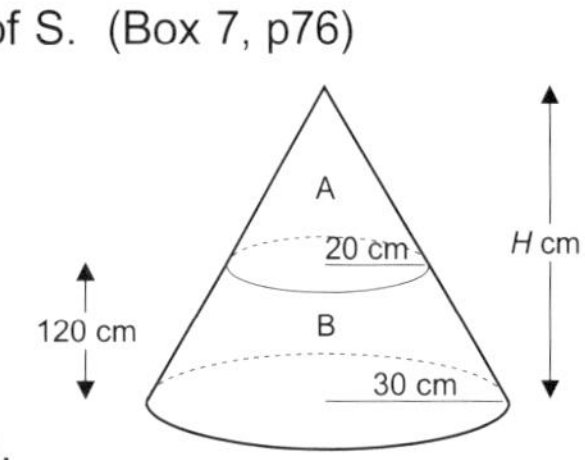

(b) Calculate the height *H* cm. (Box 6, p81)

18. The diagram shows Nelson's voyage from Great Yarmouth to position B. Nelson's boat sails due East from Great Yarmouth for 14 km to a position A. The boat then changes course and sails for 20 km due South to a position B. On a map, the distance between G and A is 56 cm.

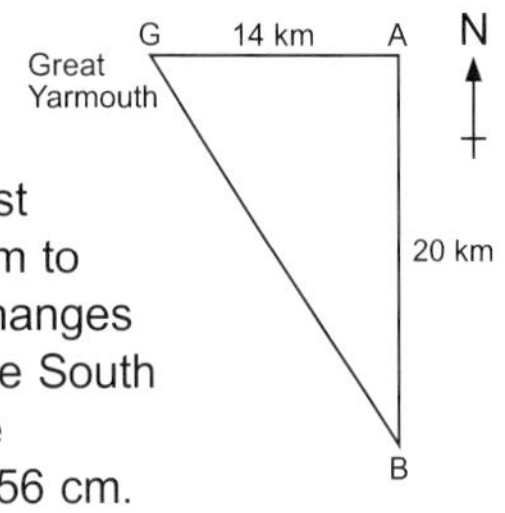

(a) Work out the scale of the map. Give your answer in the form 1:*n*, where *n* is an integer. (Box 4, p27)

(b) Calculate the distance, in km, of B from Great Yarmouth. (Box 5, p81)

(c) Calculate the bearing of Great Yarmouth from B. (Box 2, p87 and Box 8, p95)

19. ABCDE is a regular pentagon. O is the centre of the pentagon.

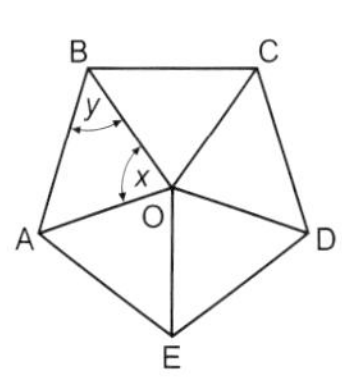

(a) Work out the value of (i) *x*, (ii) *y*. (Box 2, p83 and Box 4, p80)

(b) On the grid below, show how regular hexagons tessellate. (Box 4, p92)

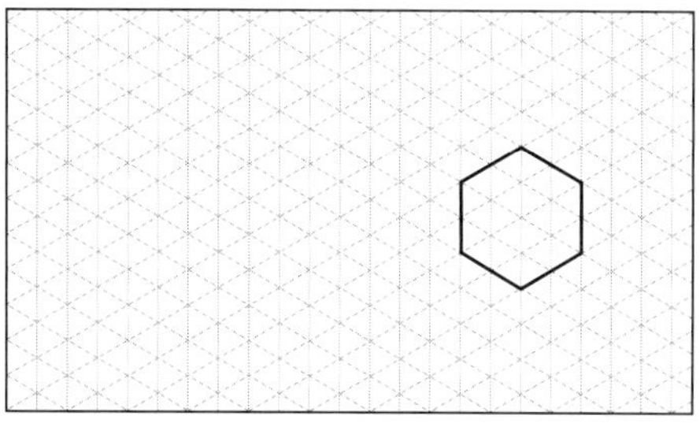

(c) Explain why regular pentagons will not tessellate. (Box 2, p92)

(d) Write down the order of rotational symmetry of a regular pentagon. (Box 2, p92)

20. (a) Reflect the triangle A in the *x*-axis. Label the reflection B. (Box 3, p89)

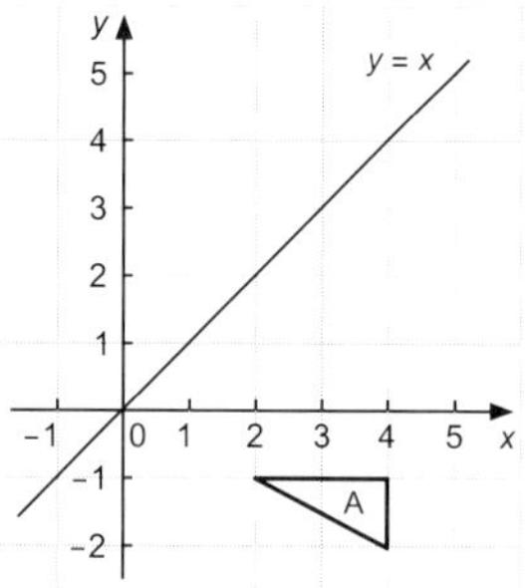

(b) Reflect the triangle B in the line $y = x$. Label the reflection C. (Box 6, p90)

(c) Describe fully the single transformation which maps triangle A onto triangle C. (Box 7, p90)

(d) Write down the equation of the line which is parallel to $y = x$ and which passes through the point (0, 8). (Box 2, p57)

This line crosses the *x*-axis at point P.

(e) Calculate the coordinates of P. (Box 5, p58)

Handling Data

21. 720 students were asked how they travelled to school. The pie chart shows the results of this survey. Work out how many of the students travelled to school by bus. (Box 4, p103)

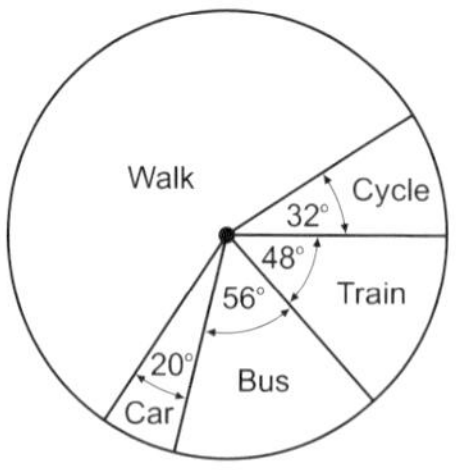

22. Ian looked at a passage from a book. He recorded the number of words in each sentence in a frequency table using class intervals 1–5, 6–10, 11–15, etc.

Class interval	Frequency *f*
1–5	16
6–10	28
11–15	26
16–20	14
21–25	10
26–30	3
31–35	1
36–40	0
41–45	2

(a) Write down

(i) the modal class interval (Box 2, p105)

(ii) the class interval in which the median lies. (Box 4, p105)

(b) Work out an estimate for the mean number of words per sentence. (Box 7, p106)

23. This table gives you the marks scored by pupils in a French test and in a German test.

French	15	35	34	23	35	27	36	34	23	24	30	40	25	35	20
German	20	37	35	25	33	30	39	36	27	20	33	35	27	32	28

(a) Work out the range of the pupils' marks in French. (Box 2, p107)

(b) Draw a scatter graph of the marks scored in the French and German tests. (Box 1, p112)

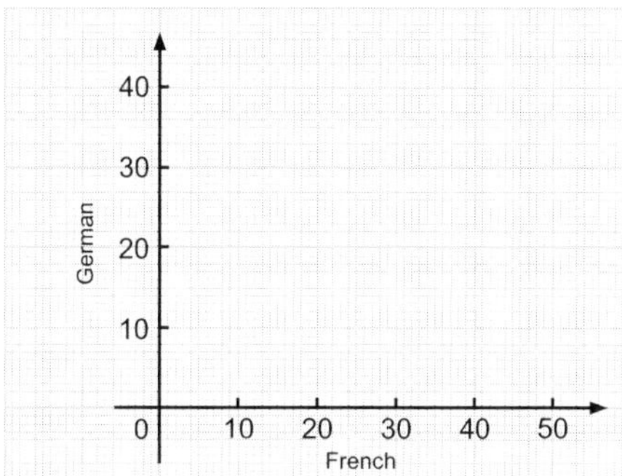

(c) Draw the line of best fit. (Box 2, p112)

(d) Use your line of best fit to estimate the mark of a pupil's test in French when their mark in German was 23. (Box 2, p112)

(e) Describe the relationship between the marks scored in the two tests. (Box 3, p112)

24. The speeds, in miles per hour (mph), of 200 cars travelling on the A320 road were measured. The results are shown in the table.

Speed (mph)	Cumulative frequency
Not exceeding 20	2
Not exceeding 25	6
Not exceeding 30	14
Not exceeding 35	28
Not exceeding 40	66
Not exceeding 45	112
Not exceeding 50	164
Not exceeding 55	196
Not exceeding 60	200

(a) On the grid below, draw a cumulative frequency graph to show these figures. (Box 2, p110)

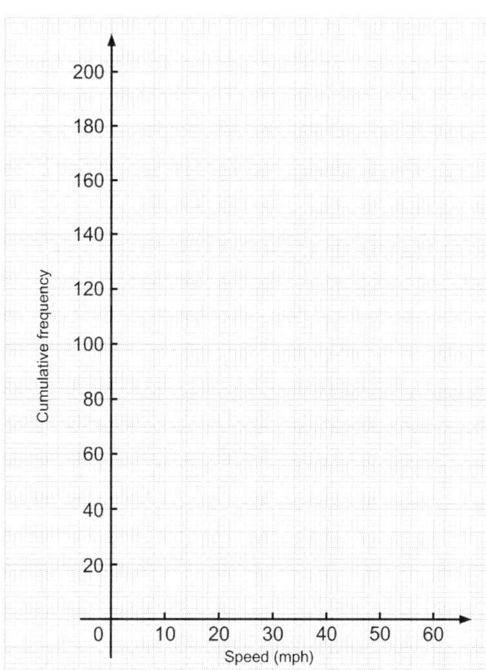

(b) Use your graph to find an estimate for
(i) the median speed (in mph) (Box 3, p110)
(ii) the interquartile range (in mph) (Box 4, p111)
(iii) the percentage of cars travelling at less than 32 miles per hour. (Box 2, p110)

25. **H** This set of marks was obtained by 10 pupils in an English GCSE examination:

73, 70, 62, 67, 69, 76, 55, 65, 61, 82

The mean of this set of marks is 68.

(a) Calculate the standard deviation of these marks. (Box 3, p107)

All ten pupils had their marks increased by 3 for good spelling.

(b) Write down for the **new** set of marks
(i) the mean (Box 6, p109)
(ii) the standard deviation. (Box 6, p109)

26.

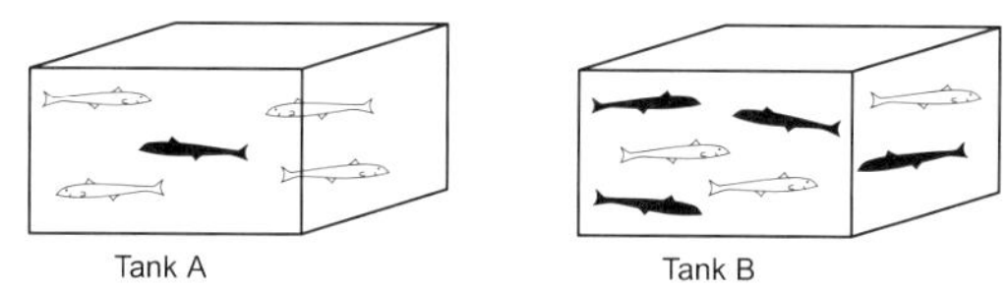

There are two fish tanks in a pet shop. In tank A there are four white fish and one black fish. In tank B there are three white fish and four black fish. One fish is to be taken out of tank A at random. One fish is to be taken out of tank B at random. Using a tree diagram, or otherwise, calculate the probability that

(a) the two fish will both be white (Box 6, p114)

(b) the two fish will both be of different colours. (Box 6, p114)

Solutions

1. $\frac{3}{5}$ of £6 = 3÷5×£6 = £3.60

2. (a) (i) $\frac{3}{8}$ = 3÷8 = 0.375 (ii) 0.375 ≈ 0.38 (2 dp)

(b) $\frac{3}{8}$ = 0.375 = 0.375×100% = 37.5% is forest.
So, 100−37.5 = 62.5% is not forest

(c) Convert percentages and fractions to decimals:

25%	0.3	$\frac{3}{8}$	$\frac{1}{2}$	0.6	67%
0.25	0.3	0.375	0.5	0.6	0.67

3. (a) 250 000 = 2.5×10^5

(b) $2.5\times10^5\div9.3\times10^7 \approx 2.69\times10^{-3}$

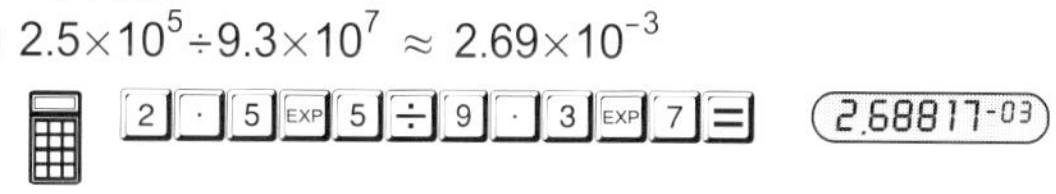

4. (a) Jake pays £125+12×£29.62 = £480.44

(b) Discount = 12% of £390 = 12÷100×£390 = £46.80
Cost = Price−Discount = £390−£46.80 = £343.20

(c) 1995 Price + Price rise = 1996 Price
100% + 4% = 104%
£? + £? = £390

104% is £390 and so1% is £390÷104 = £3.75
1995 price = 100% = 100×£3.75 = £375

5. (a) Lower bound of width = 6.25 cm
Lower bound of length = 3.65 cm
Lower bound of area = 6.25×3.65 = 22.8125 cm^2

(b) Upper bound of large piece of card = 100.5 cm
Max. no. of rectangles = 100.5÷6.25 = 16.08 ≈ 16

6. (a) & (b)

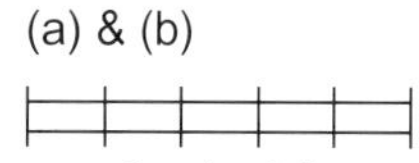

Fence length	1	2	3	4	5	6
Number of pieces	4	7	10	13	16	19

(c) No. of pieces = 3×Fence length+1 = 3×25+1 = 76

(d) No. of pieces = $3\times n+1 = 3n+1$

7. (a) (i) $s = 623.5 \approx 600$, $u = 11.37 \approx 10$, $v = 87.22 \approx 90$ (so that $u+v \approx 10+90 = 100$)

(ii) $t = \dfrac{2s}{u+v} \approx \dfrac{2\times600}{10+90} = \dfrac{1200}{100} = 12$

(b) 2 × 6 2 3 . 2 5 = ÷ (1 1 . 3 7 + 8 7 . 2 2) =   12.64327

Answer ≈ 12.64 (2 dp)

8. (a) $25G+250 = 25(G+10)$

(b) $C = 14G+121 \therefore C-121 = 14G \therefore 14G = C-121$

$\therefore G = \dfrac{C-121}{14}$

9. $T = 26 - \dfrac{h}{500} \therefore -52 = 26 - \dfrac{h}{500} \therefore -78 = -\dfrac{h}{500}$

$\therefore -78\times500 = -h \therefore -39\,000 = -h \therefore h = 39\,000$

10. (a) $(R+r)(R-r) = R(R-r)+r(R-r) = R^2-Rr+rR-r^2$
$= R^2-Rr+Rr-r^2 = R^2-r^2$

(b) $A = \pi(R+r)(R-r) = \pi(R^2-r^2)$ From part (a)

$\therefore 147 = \pi(21^2-r^2) \therefore 147\div\pi = 21^2-r^2$

$\therefore 46.79 = 441-r^2 \therefore 46.79-441 = -r^2$

$\therefore -394.21 = -r^2 \therefore r = \sqrt{394.21} \approx 19.9$ (1 dp)

11. (a) $119 = 29+15d \therefore 119-29 = 15d \therefore 90 = 15d$
$d = 90\div15 = 6$. This chest has 6 drawers

(b) (i) $117 = k+m4 \therefore k+4m = 117$ .... (A)
$149 = k+m6 \therefore k+6m = 149$ .... (B)

(ii) (B)−(A) gives: $2m = 32 \therefore m = 32\div2 = 16$

12. (a) (i) 11.40 am (ii) 20 min (b) 12.5 miles

(c) Manchester to London = 200 miles. Time taken on return = 200÷50 = 4 hours, arriving at 7.20 pm

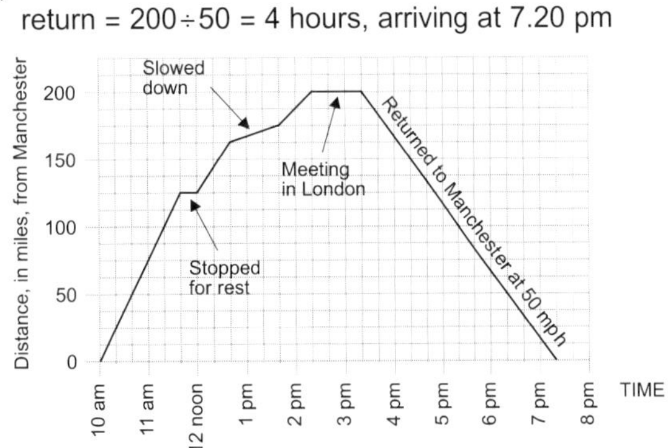

13. (a)

x	−2	−1	0	1	2	3
y	−1	1	−1	−4	−1	14

(b)

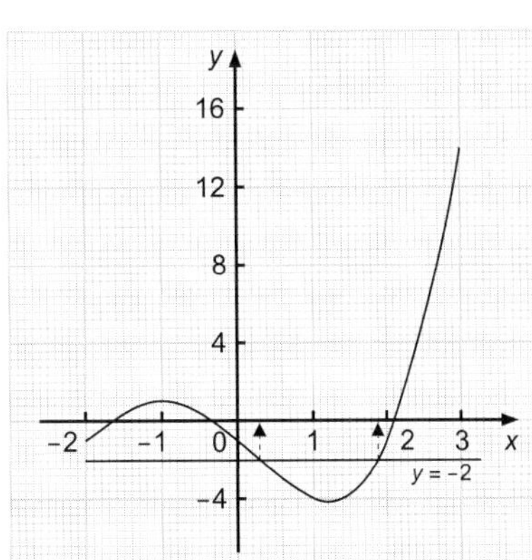

(c) Draw line $y = -2$. Gives: $x \approx 0.3, 1.9$

(d) Try $x = 4$:
$4^3-4\times4-1 = 47$ (too high)
Try $x = 3.5$, etc.
Table shows $x \approx 3.6$, correct to 1 decimal place

Guess	x^3-4x-1	Comment
4	47	Too high
3.5	27.875	Too low
3.7	34.853	Too high
3.6	31.256	Too high
3.55	29.5389	Too low
3.58	30.56271	Too high

14. (a) $f(0) = (0+3)(0-2)(0-4) = 3\times-2\times-4 = 24$

(b) & (c) Reflection in y-axis

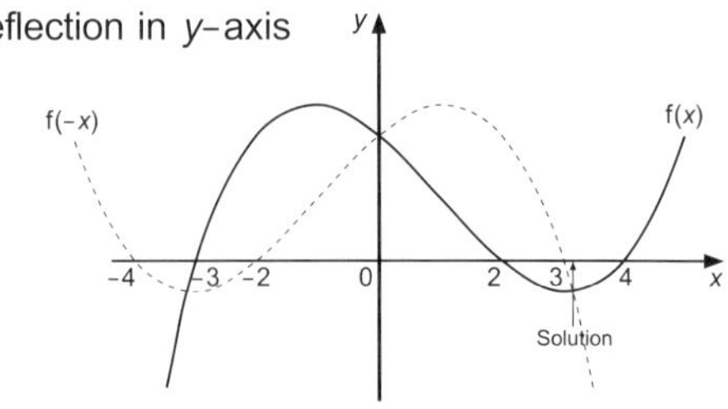

(d) $f(x)$ crosses x-axis when $f(x) = 0$, i.e. when $x = -3$, $x = 2$ and $x = 4$ (see graph). Since $f(-x)$ is a reflection in the y-axis, it crosses the x-axis at $x = -4$, $x = -2$ and $x = 3$. Graphs cross where $f(x) = f(-x)$, in between $x = 3$ and $x = 4$. So, $n = 3$

15. (a) Area $= 4\times3+5\times2 = 22$ m^2

(b) Volume of room $= 22\times2.7$
$= 59.4$ m^3

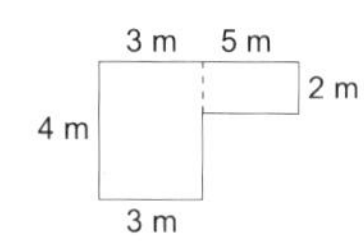

16. (a) Volume $= 8\times5\times12.5 = 500$ cm^3

(b) (i) $100:125 = 1:1.25 \therefore$ Vol. $= 500\div1.25 = 400$ cm^3
(ii) $100\div125\times100 = 80\%$

17. (a) Cones are similar.
Height of small cone : Big cone = 45 : 90 = 1 : 2
So, volume of small cone : Big cone $= 1^3:2^3 = 1:8$
$\therefore$ Volume of T $= 10\,500\pi\div8 = 4123.3$ m^3
$\therefore$ Volume of S $= 10\,500\pi-4123.3 = 28\,863.4$ m^3

(b) Large triangle is an enlargement of small triangle.
Scale factor = 30÷20 = 1.5.
$\therefore H = 1.5\times(H-120)$
$\therefore H = 1.5H-1.5\times120$
$\therefore H = 1.5H-180$
$\therefore 180 = 1.5H-H$
$\therefore 0.5H = 180 \therefore H = 180\div0.5 = 360$ cm

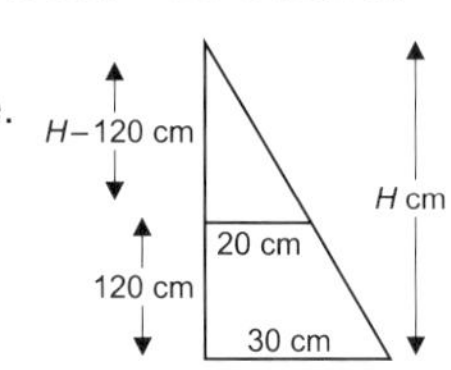

18. (a) 56 cm : 14 km = 56 : 1 400 000 = 1 : 1 400 000÷56
= 1 : 25 000

(b) Pythagoras. $BG^2 = 14^2+20^2 = 596$
$\therefore BG = \sqrt{596} = 24.4$ km

(c) $\tan B = 14\div20 = 0.7 \therefore B = \tan^{-1}0.7 \approx 35^\circ$
Bearing of G from B $\approx 360^\circ-35^\circ = 325^\circ$

19. (a) (i) $x = 360^\circ\div5 = 72^\circ$ (ii) Isosceles triangle.
$180^\circ-72^\circ = 108 \therefore y = 108^\circ\div2 = 54^\circ$

(b)

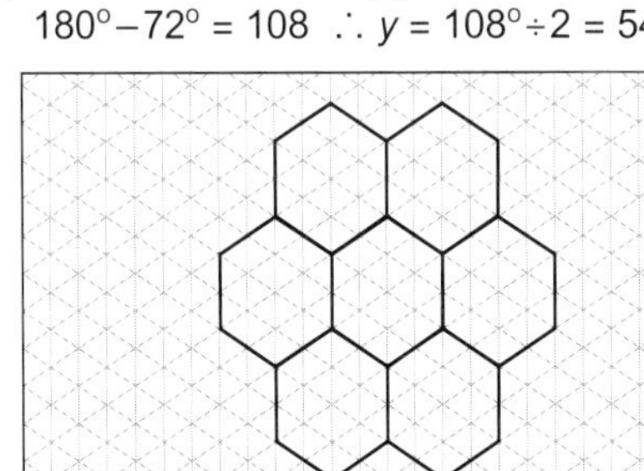

(c) Interior angle is 108°, which does not divide into 360°

(d) 5 (same as number of sides)

20. (a) & (b)

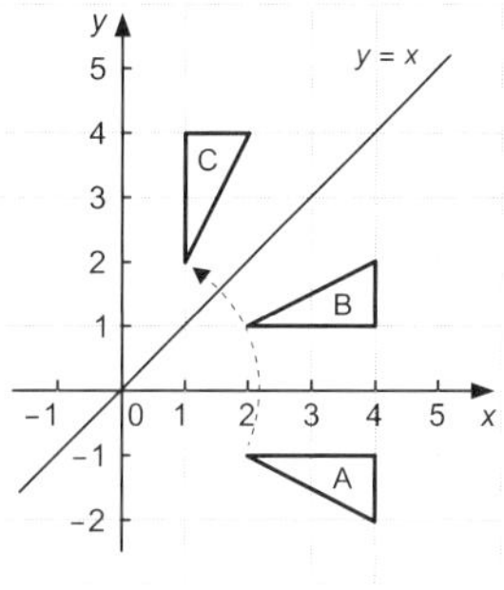

(c) 90° anticlockwise rotation about (0, 0)

(d) Gradient is 1 and y-intercept is 8, giving $y = x+8$

(e) Crosses x-axis when $y = 0$. $\therefore 0 = x+8$ $\therefore x = -8$ giving P (−8, 0)

21. 360° represents 720 students.
So, 1° represents 720÷360 = 2 students.
No. of bus students = 56° = 56×1° = 56×2 = 112

22. (a) (i) 6–10 words (highest frequency)

(ii) Total frequency = 16+28+....+2 = 100. Median lies between 50th and 51st largest values. Cumulative frequencies: 16, 16+28 = 44, 16+28+26 = 70. Median lies in class 11–15

(b) Calculate mid-class values.

$$\text{Mean} = \frac{16\times3+....+2\times43}{100}$$
$$= 12.95 \text{ words/sentence}$$

Mid-class value	Frequency f
3	16
8	28
13	26
18	14
23	10
28	3
33	1
38	0
43	2

23. (a) Range for French = 40−15 = 25

(b) & (c)

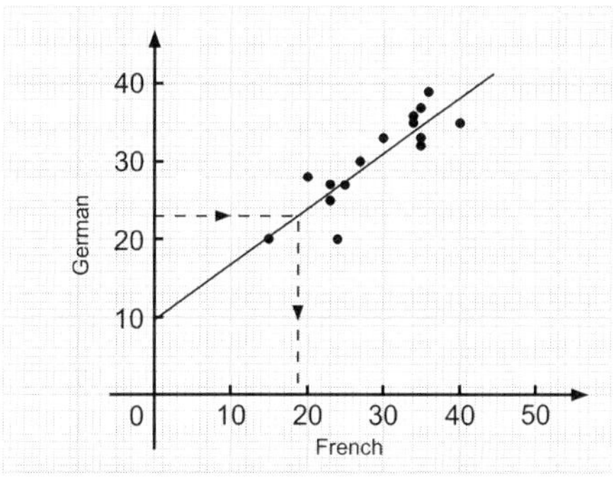

(d) From graph, French mark is approximately 19

(e) Fairly strong, positive correlation (the line of best fit has a positive gradient and the points are fairly close to this line)

24. (a)

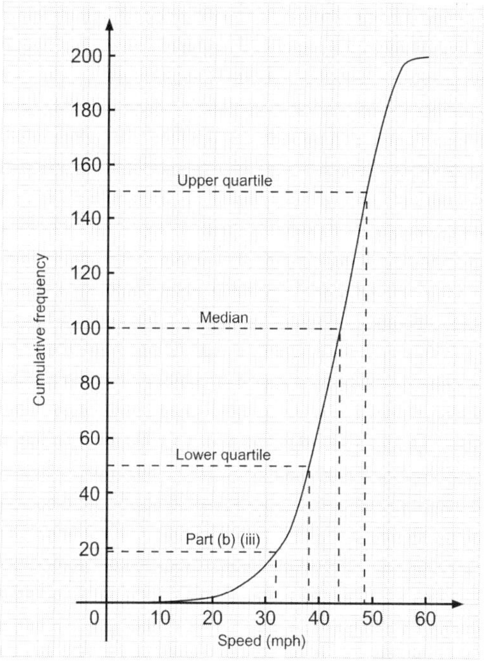

(b) (i) Median ≈ 43.5 mph

(ii) Lower quartile ≈ 38 mph
Upper quartile ≈ 48.5 mph
Interquartile range ≈ 48.5 − 38 = 10.5 mph

(iii) ≈ 19 cars out of 200 = 9.5%

25. (a) $\overline{x} = 68$ (given). Square values

Value, x	73	70	62	67	69	76	55	65	61	82
Square, x^2	5329	4900	3844	4489	4761	5776	3025	4225	3721	6724

$$\sum x^2 = 5329+4900+....+6724 = 46\,794$$

$$\text{Variance} = \frac{\sum x^2}{n} - \overline{x}^2 = \frac{46\,794}{10} - 68^2$$
$$= 4679.4 - 4624 = 55.4$$

$$\text{Standard deviation} = \sqrt{55.4} = 7.4$$

(b) (i) New mean = 68+3 = 71

(ii) Standard deviation is unchanged at 7.4

26. (a)

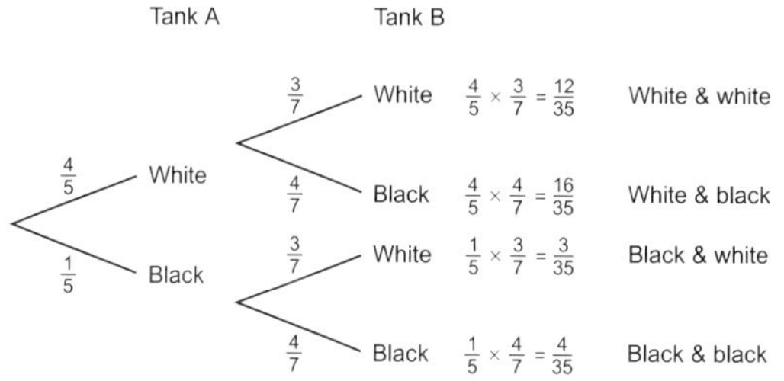

P(both white) = P(white & white)
$= \text{P(white)}\times\text{P(white)} = \frac{4}{5}\times\frac{3}{7} = \frac{12}{35}$

(b) P(different colours)
= P(white & black or black & white)
= P(white)×P(black) + P(black)×P(white)
$= \frac{4}{5}\times\frac{4}{7}+\frac{1}{5}\times\frac{3}{7}$
$= \frac{19}{35}$

INDEX

Box numbers are indicated in brackets

FORMULA SHEET

Area of a triangle $= \frac{1}{2} \times \text{Base} \times \text{Height}$

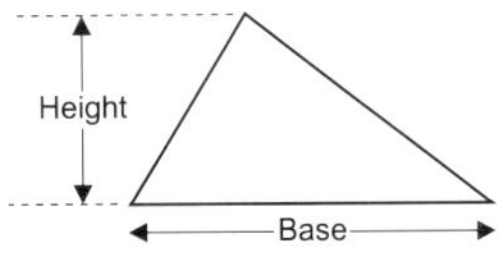

Area of a trapezium $= \frac{1}{2}(a+b)h$

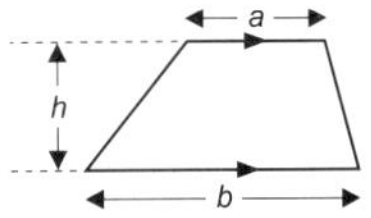

Area of a parallelogram
$= \frac{1}{2} \times \text{Base} \times \text{Height}$

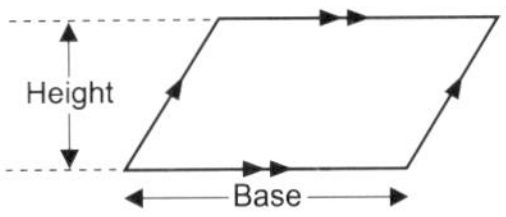

Area of a circle $= \pi \times (\text{Radius})^2 = \pi r^2$
Circumference $= 2 \times \pi \times \text{Radius} = 2\pi r$
$= \pi \times \text{Diameter} = \pi d$

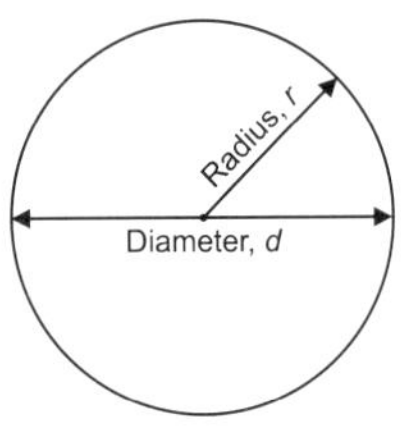

Area of a sector $= \frac{\theta}{360} \times \pi r^2$
Length of arc $= \frac{\theta}{360} \times 2\pi r$

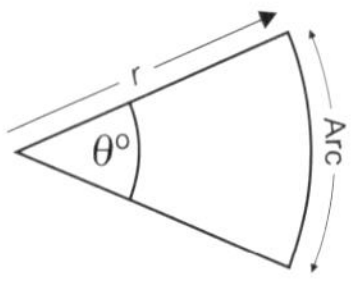

Volume of a cuboid
$= \text{Length} \times \text{Breadth} \times \text{Height}$

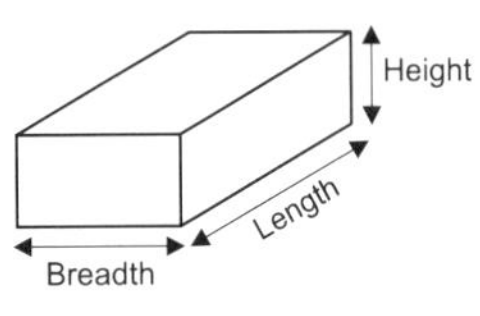

Volume of a cylinder $= \pi r^2 h$
Curved surface area $= 2\pi rh$

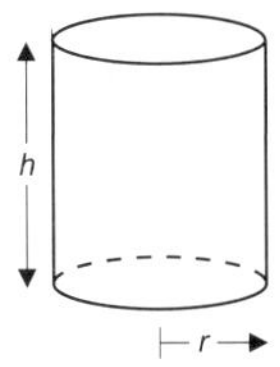

Volume of a cone $= \frac{1}{3}\pi r^2 h$
Curved surface area $= \pi rl$

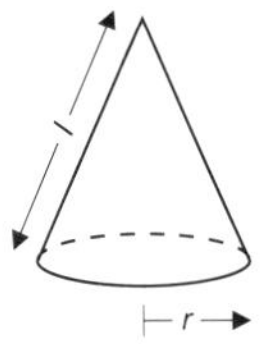

Volume of a sphere $= \frac{4}{3}\pi r^3$
Surface area $= 4\pi r^2$

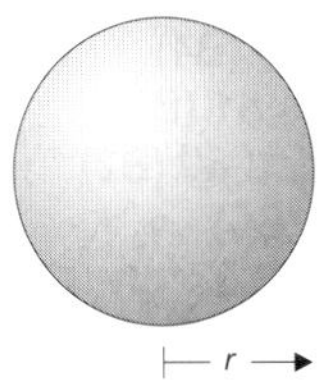

Volume of a prism
$= \text{Area of cross-section} \times \text{Length}$

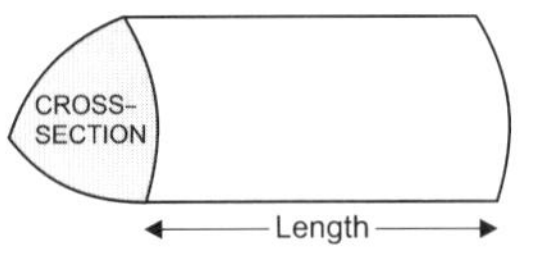

Volume of a pyramid
$= \frac{1}{3} \times \text{Area of base} \times \text{Height}$

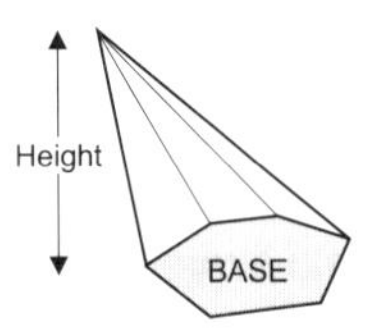

Pythagoras' theorem

$a^2 = b^2 + c^2$

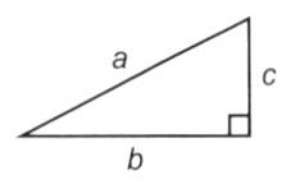

Trigonometry

$\sin\theta = \frac{\text{Opp}}{\text{Hyp}}$

$\cos\theta = \frac{\text{Adj}}{\text{Hyp}}$

$\tan\theta = \frac{\text{Opp}}{\text{Adj}}$

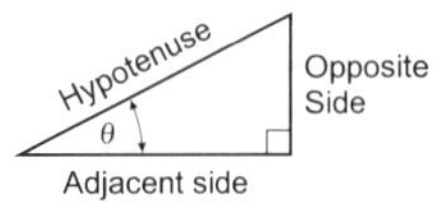

For any triangle ABC

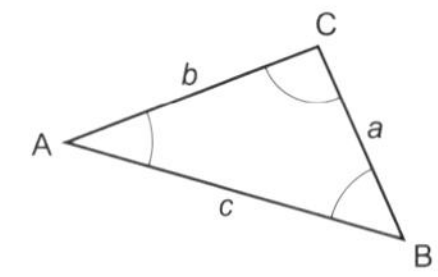

Sine rule $\frac{a}{\sin A} = \frac{b}{\sin B} = \frac{c}{\sin C}$

Cosine rule $a^2 = b^2 + c^2 - 2bc\cos A$

$\cos A = \frac{b^2+c^2-a^2}{2bc}$

Area of triangle $= \frac{1}{2}ab\sin C$

Quadratic equations
The solutions of $ax^2+bx+c=0$ where $a \neq 0$ are given by:

$$x = \frac{-b \pm \sqrt{b^2-4ac}}{2a}$$

Mean

$\bar{x} = \frac{\sum x}{n}$ or $\bar{x} = \frac{\sum fx}{n}$

Standard deviation

$s = \sqrt{\frac{\sum(x-\bar{x})^2}{n}}$ or $s = \sqrt{\frac{\sum x^2}{n} - \left(\frac{\sum x}{n}\right)^2}$

MODER SOCIETY

THIRD EDITION

Jack Nobbs

Formerly Head of Sixth Form and Director of Studies in Economics and Sociology, Hewett School, Norwich

28

Bell & Hyman

Published by BELL & HYMAN
an imprint of UNWIN HYMAN LIMITED
Denmark House, 37–39 Queen Elizabeth Street, London SE1 2QB

First published in 1976 by Allen & Unwin
(Publishers) Limited; Fourth reprint 1979
Second edition 1981; reprinted 1983
Third edition 1986; reprinted by Bell & Hyman 1987

British Library Cataloguing in Publication Data
Nobbs, Jack
Modern society: a social studies course.—3rd ed.
1. Social sciences
I. Title
301 H85
ISBN 0 7135 2808 7

Set in 9 on 11 point Melior by Columns of Reading
Printed in Great Britain by R.J. Acford Limited, Chichester, Sussex

ACKNOWLEDGEMENTS

The author and publishers would like to express their gratitude to the following for permission to reproduce photographs:

All-Sport Photographic Ltd, pp. 79, 113; Alpha Photographic Press Agency, p. 2; Antelope Films and Channel 4 Television, p. 179; Associated Press, p. 131; Bank of England, p. 152; BBC, p. 27; BBC, Alan Bleasdale, Bernard Hill and Graeme Souness, p. 87; BBC Hulton Picture Library, p. 79; Birmingham Corporation, p. 39; Martin Bond, p. 43; Kelly Cantlon, pp. 1, 12, 14, 17, 30, 37, 42, 74, 80, 94, 97, 106, 108, 109, 114, 115, 129, 143, 146, 167, cover; B. G. Chaplin, p. 93; *Daily Telegraph* Colour Library, p. 79; Department of Transport, p. 69; Elsie Timbey Collection, Society for Cultural Relations with the USSR, p. 166; European Parliament, p. 188; Format Photographs Ltd, Sheila Gray and Paul Lewis, p. 101; Kent Gavin and *The Mirror*, p. 181, cover; Sally & Richard Greenhill, p. 111; Simon Grosset, p. 106; *The Guardian*, pp. 16, 50, 72, 107; Guinness PLC, p. 137; Health Education Council, pp. 4, 53, 117; Home Office, p. 63; David Leech, p. 44; London Express News Service, p. 189; Nigel Luckhurst, p. 92; Milk Marketing Board of England and Wales, p. 127; Ministry of Defence, p. 90; Network Photographers and Mike Abrahams, pp. 57, 123, cover, Steve Benbow, pp. 44, 119, Chris Davies, pp. 37, 126; National Federation of Women's Institutes, p. 23; Tony Othen, p. 157; Photo Co-op and Janis Austin, p. 101, Gina Glover, p. 118, cover, Martin Lipson, p. 17, Vicky White, p. 99, cover, Sarah Wyld, pp. 26, 107, cover; Popperfoto, p. 5; Press Association, p. 171; Rex Features Ltd, pp. 164, 168, 176; Rolls Royce Motors, p. 90; The Salvation Army International Headquarters, p. 17; Save the Children and Mike Wells, p. 186; The Stock Exchange, p. 139; Times Newspapers Ltd, p. 67; United Nations, p. 183; Vision International and Anthea Sieveking, pp. 49, 79; World Wildlife Fund, p. 23

We are also grateful to the following for permission to reproduce the magazine covers on p. 85:

British Journal of Photography, City Limits, Do-it-Yourself, Jackie (D. C. Thomson & Co. Ltd), *Motorcycle Enthusiast* and Kerry Dunlop, *New Health, New Scientist, Practical Computing, Punch, Smash Hits, Video Today, Your Horse.*

The cartoons on pp. 3, 28, 31, 42, 52, 58, 66, 88, 103, 119, 140, 163 and 182 are reproduced by courtesy of *Punch* magazine. We are grateful to the following for permission to reproduce the remaining cartoons:

Daily Express (p. 27), Merrily Harpur of *The Guardian* (p. 33), David Austin of *The Mail on Sunday* (p. 21), C. Collins (p. 136), *Daily Mirror* (pp. 18 and 105), *The Sun* (pp. 6, 7, 39, 98 and 157). The newspaper cuttings on pp. 2, 68 and 110 are reproduced by courtesy of the *Daily Mirror*, that on p. 180 by courtesy of Michael Simmons of *The Guardian*, that on p. 116 by courtesy of *The Times/Pulse* Taylor Nelson Medical, and the remainder by courtesy of *The Times*. We are also grateful to the *Daily Mirror*, together with the *Daily Express, Daily Mail, Daily Telegraph, The Guardian, The Times, The Star* and *The Sun*, for permission to reproduce the front page of the *Mirror* on p. 82.

Written material is reproduced by courtesy of: Sphere Books (*Apple to the Core* by P. McCabe and R. D. Schonfield), University of London Press (*Children Growing Up* by John Gabriel), and the following kindly gave us permission to reproduce examination questions: Associated Examining Board, Associated Lancashire Schools Examining Board, Cambridge University Local Examinations Syndicate, East Anglian Examinations Board, East Midland Regional Examinations Board, Joint Matriculation Board, North-West Regional Examinations Board, South-East Regional Examinations Board, Welsh Joint Education Committee, West Midlands Examinations Board and Yorkshire & Humberside Regional Examinations Board.

Finally, the author wishes to express his special thanks to Virginia Davidson-Merritt for her work in editing this new third edition.

Also by Jack Nobbs

Social Economics (McGraw-Hill 1969)
Economic Problems of the 1970s (Pergamon Press 1971)
Handbook on Objective Testing – Civics (Methuen 1971)
Objective Questions in Social Economics (Methuen 1972)
Essay Techniques in Economics (McGraw-Hill 1973)
Economics for Advanced Level (McGraw-Hill 1973)
Daily Economics (with Paul Ames) (McGraw-Hill 1975)
Sociology (with Bob Hine and Margaret Flemming) (Macmillan 1975)
Introducing Social Studies (Macmillan 1979)
Sociology Workbook (with Bob Hine) (Macmillan 1979)
Sociology in Context (Macmillan 1983)

CONTENTS

1 LIVING WITH OTHERS

UNIT 1
The Individual

GCSE Aims and Objectives

Knowledge — to develop a fuller awareness of *individual* characteristics associated with personality, heredity, environment, instincts, motives, experiences, life-cycles, attitudes and community life.

Application — to analyse the reasons for the *differences* between human beings.

Evaluation — to encourage a critical awareness of the ways in which people are *changed* by modern society.

Debbie and Paul

TOPIC 1.1 WE ARE ALL DIFFERENT

If you are a girl, are you anything like the one in the picture? If you are a boy, is the youth on the right anything like you? No doubt you are better looking. One of the photographs may remind you of someone. There may be a likeness, but the person in the photograph is different from everybody else.

Now look round at the others in your group. They have different faces, yet they all have some features in common. They have eyes, ears, a mouth and a nose of sorts. No doubt they all have hair. Yet if you look at a group of old people some of the old men may be bald. Does this mean that hair is not an essential feature of human **individuals**? A war once started when a man called Jenkins had his ear cut off. It was called The War of Jenkins's Ear. An individual may lose an arm or a leg in a road accident, but we would still think of him as a human. He would still be an important individual, especially to his friends and to those who nursed him.

You rightly feel annoyed if you are not treated as an individual. We jealously guard the rights of the individual. The United Nations Universal Declaration on Human Rights affirms:

> 'All human beings are born free and equal in dignity and rights. They are endowed with reason and conscience and should act towards one another in a spirit of brotherhood'

The human race is made up of over 4,800 million people and each one has individual rights. They are all different even though they have some things in common.

Boy George became famous by being different

TOPIC 1.2 IN WHAT WAYS ARE WE DIFFERENT?

PERSONAL MEMORANDUM

Name Dick Nosh

Address Lamb House, Grub Street, S. Ham.

Height 166 cm.

Weight 57 kg.

Telephone No. S. Ham 9898

Passport No. P 643851

Driving Licence No. 307193/DI 9 XM

Size of shoes 41

Blood Group A

National Insurance No. TBP 565068

Date of birth 19·8·1968

Next of kin Mr. B.I.G. Nosh

The collection of facts shown in Dick Nosh's diary could only apply to him, although some of the details could apply to several people. It is possible that there is another boy called Dick Nosh and that several people live in Lamb House, Grub Street, South Ham. Which things in the list only apply to Dick Nosh?

The people in your class are different but they are probably dressed in a similar way. Some may be more 'way out' than others. If any one is very 'way out' you might call him a 'weirdy'. Does this suggest that because he is rather peculiar he is not acceptable to your group?

Jim, 15, misses CSE after being sent home to change

Boy in white socks barred from exam

STUDENT Jim Dorch was barred from a vital exam because he was wearing white socks.

By MIRROR REPORTER

He was ordered home to change them ten minutes before the start of the test.

But 15-year-old Jim couldn't make the three-mile round trip in time and missed taking his CSE in building. Now the unhappy youngster fears his job prospects have slumped.

Jim, of Kitchener Road, Weymouth, Dorset, said: "I spent two years working hard for the exam.

"It's crazy because I actually left the school at Easter and went back just to take the test.

"I was wearing a proper school uniform apart from the socks."

If you really study the others in your group you will find that they have not only different physical features but different characters and different personalities as well. How can a person have a strong **personality** but be a bad **character**? How does this idea apply to Adolf Hitler? Can you think of someone who has a good character but a weak personality?

We have mentioned a few ways in which people are different. Differences are not merely confined to obvious things such as height and weight, but also involve an individual's achievements, behaviour, beliefs, hopes, ideas, interests and attitudes. Which of these things applies to humans but not to other creatures?

A sea-lion can perform many tricks. His **achievements** may include walking on a moving ball or jumping through a hoop. You may say that your dog is well-behaved or badly-behaved, but are you then using human standards by which to judge his **behaviour**? If a dog could talk he might argue that from a dog's point of view it should not be considered bad behaviour to bite somebody who he suspects might be his enemy.

TOPIC 1.3 WHAT MAKES US DIFFERENT?

There are two main influences that help to make individuals different, heredity and environment.

Heredity means the things which are inherited and passed on from one generation to another. Your **environment** consists of your surroundings: your family, home, school, town, countryside, etc.

There are some important ideas or *concepts* which run through this book and which, as a student of social studies, you must try to grasp. Heredity and environment are two of these concepts. Heredity has helped to make you what you are. Do you look more like your mother or your father? If you look more like one of your parents, have you inherited any physical features from the other? Are there any birthmarks or other similarities? Have you the same colour hair or eyes? Was your father or mother tall or short at your age? Do you think intelligence is inherited? And what about temperament? If you play it cool, don't get upset and keep calm, are your parents built that way? Or are most of your family emotional people who find it difficult not to fly off the handle at the least provocation? Each day you learn new things about your environment. As their surroundings change, so people change. Your personality is influenced by people and things around you.

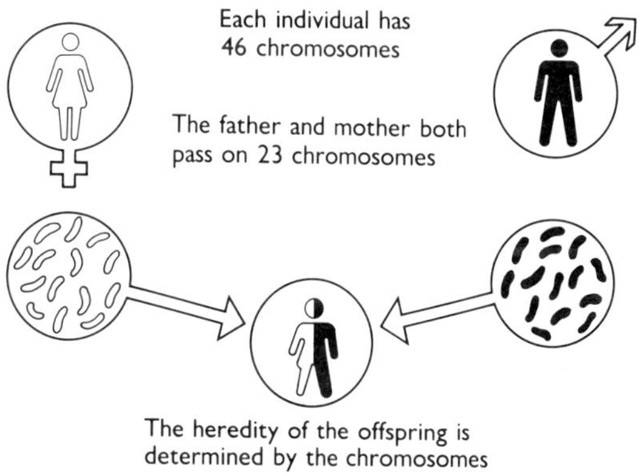

We shall continually return to the concept of environment, as we reinforce and expand our knowledge. A baby can understand some things about its environment although it does not know what the word means. At first the baby is most aware of its mother. Before very long its father and brothers and sisters become part of its life; the influence of these people depends upon how frequently they appear on the baby's scene. If the baby's father is only present for a few moments at bedtime he will be less important than a young brother or sister who is at home all day. The people and things amidst which the baby lives help to mould its individual personality. It has to learn to live with them and to live as an individual within the community. He or she has to learn to live with other people.

But there are some **influences** upon the baby's life that affect it even before it is born. A baby who remains too long in its mother's womb will suffer from shortage of oxygen, but children who are born prematurely may not be handicapped in this way, and it is known that several geniuses of the past have arrived early. In the Republic of South Africa there are children who are very intelligent because doctors supplied them with extra oxygen in the last few weeks before birth.

Recent research has shown that on average a child whose mother smokes during pregnancy will be about one centimetre shorter than normal at the age of seven and roughly four months behind others in its reading ability.

TOPIC 1.4 WHAT DETERMINES OUR ACTIONS?

All animals including humans are born with certain **instincts**. There are two main instincts from which most actions follow: preservation of self and preservation of the species.

These instincts do not have to be learnt, so we call them **innate motives**. Can you think of other forms of instinctive behaviour?

If you have ever kept gu˙ıea pigs as pets you will know that they are very timid animals. At the approach of a stranger a guinea pig will crawl under any straw it can find in its hutch and keep very quiet. There are at least three possible explanations of the guinea pig's behaviour:

1. It responds in the same way as a machine – the sight of a possible enemy sets it working.
2. It thinks things out, i.e. it says to itself, 'This chap looks like a nasty piece of work. I must make myself scarce.'
3. It acts instinctively, i.e. in the same way as we duck if somebody pretends to throw a stone at us.

Some psychologists think that animals do react in a similar way to machines, but the majority think that the behaviour of a human being is different from the working of a machine. A machine works with no purpose of its own. Your watch did not decide that it would tell the time. The purpose of its movement has been determined by the person who designed it. The guinea pig hides with a purpose in mind; it wants to be safe. It is not a machine but it is unable to think things out as humans can. It acts instinctively. Its natural instincts may be reinforced by past **experiences**. Other creatures, as well as humans, are able to conserve knowledge from the past. It has been found by experiments that if a worm is sent down a Y-shaped passage it is liable to go either way, but if it receives an electric shock when it goes to the right, after a time it will always select the left-hand route. Even worms can learn from experience.

TOPIC 1.5 WE ARE CHANGED BY OUR EXPERIENCES

One of the main reasons why humans differ is that they have had different experiences. Experiencing is vital to human existence. A famous thinker called Descartes once said something that might be interpreted as 'I experience, therefore I exist'. What did he mean? He was pointing out the essential difference between animate (or living) things and inanimate objects. Does a brick have an experience if it is thrown at the head of a soldier in Northern Ireland? Obviously not, but the soldier would have an experience if he were hit on the head by the brick. The man who threw it would have an experience of pleasure at his success or of sadness if he felt sorry for what he had done. Experiences are made up of feelings, emotions and sentiments. I do not experience a decayed tooth, but I may experience toothache. Our experiences will make us different. Look at the picture. People will vary according to the events through which they have lived.

After a 17-day ordeal, a freed American hostage experiences a mixture of emotions as he is reunited with his family

Have you ever spent some time watching from the window of a moving bus or train and yet not been able to remember seeing anything in particular? If this happens someone might say you have been daydreaming or have been in a 'brown study'. You can take things in through your **subconscious**. Jesuit priests used to say, 'Give us a child for the first seven years and we will have influenced him for the rest of his life.' About half of all we know is learnt in our first five years.

In Aldous Huxley's *Brave New World* babies are produced in test-tubes. The alpha-plus types do the brainy work while delta types are conditioned to do menial tasks without complaining. Those who control the world want deltas to work and not spend time reading books or enjoying things of natural beauty such as flowers. In his book *1984*, George Orwell described how the government kept people occupied by persuading them to hate an imaginary enemy.

In the early stages of life we are influenced a great deal by our parents. Some children are lucky enough to have parents who talk sensibly to them and see that they gain rich experiences through books, visits and holidays. Others have to struggle along without these advantages; all the more credit to them if they become all-round, well-developed individuals! In adolescence we are influenced a lot by other teenagers. At work the environment of the shop or factory and the attitude of workmates will help to mould our characters. Libraries, museums and recreation grounds help us to develop. What other things may be provided for us to help our development?

Individuals are often influenced without being aware of it. You may think your views, ideas and beliefs are quite your own and that you thought of them all yourself. Yet we have learnt in this unit that before birth, and especially in the years of childhood which are barely remembered, we were constantly being influenced, so that most of us now fit in easily with other people.

Individuals who fail to adjust to their environment are often unaware of the reasons why they are misfits. Have you ever watched a hypnotist at work? Under a state of hypnosis (or under the influence of a truth drug), an individual may reveal experiences of which he apparently knew nothing. These experiences are dug up from where they are buried in the subconscious. Nobody knows for certain just how much past experiences have modified him as an individual.

TOPIC 1.6 WHY IS HUMAN SOCIETY UNIQUE?

Man is very different from all other creatures. What makes him so special? Animals can move, see, smell, hear and touch.

Modern man is the only creature concerned about what kind of being he is. He wants to know what makes him tick.

'I wish you weren't such a boar, Henry.'

Here are some other things peculiar to man. Can you add to the list?

- Young humans *develop* more slowly than young animals
- Humans are best able to *adjust* to their environment
- Mankind has the *lowest death rate*. (A smaller proportion of the species dies annually)
- We are able to *think deeply*
- Man is able to make up a *language* which is first spoken and then written and read
- Although animals may make and use tools, only man is able to invent *machines*
- People are by nature *religious* and judge each other's behaviour and conduct
- Man organises a *society* with rules, written or unwritten, which determine the relationships between people within society

Let us compare human beings with the animal world. Can you think of an animal that can laugh? Can you think of any animals that can count? The more advanced the form of life, the longer it takes to come to maturity. It takes most human beings about a year to learn to walk, but young sheep walk within a few hours of birth. Humans cannot reproduce until their early teens. If you keep pets, how long does it take them to breed? How long is it before a young monkey is independent and able to manage without its mother? It takes longer than a rabbit, but not as long as a human.

After mating, many animals stay with one partner, as do most humans. They have families. They feed and protect their young. Animals often remain with groups of others of their kind in the same way that we have our groups and gangs. We call this being gregarious.

Animals also adapt to their environment. A chameleon changes its colour to blend with its surroundings. Many animals make crude tools.

Beavers build dams, but man is best at exploiting nature's resources. He is capable of designing a machine such as the Concorde. Why can't animals make machines? Animals are able to communicate with others of their species, but it needs a man – George Orwell in his book *Animal Farm*, for example – to give them the power to read, write and think.

TOPIC 1.7 LIFE HAS ITS CYCLES

All life has its *cycles*: we are *born*, we *reproduce*, we *die*.

Between these three main stages of the **life-cycle**, there are intermediate stages. Where would you put childhood, learning, work, courtship, marriage, parenthood and old age? Which of these things are *not* associated with the animal world?

And in between birth and death there is growth and playing, loving and fighting, working and caring. Throughout this book we shall see these aspects of **human society** recurring. We shall study the growth within the community of an individual, a group, a family, a school, a factory, and even a community of nations such as the European Economic Community (EEC). When children play they prepare for adult life; when adults play they bring back the pleasures of childhood. Our loving inspires us to do things which are *for society*: to be devoted parents, to serve the community, to protect others and keep the law, to look after the sick and the handicapped, to strive for peace. Our fighting instinct and aggression may cause us to act *against society*: to deprive others of things, to be noisy and violent, to exploit and pollute, to behave anti-socially. Organisations have sprung up to help people overcome **anti-social attitudes**. Some of these organisations are listed here.

National Society for the Prevention of Cruelty to Children
Marriage Guidance Councils
Samaritans
Alcoholics Anonymous
Narcotics Anonymous
Royal Society for the Prevention of Cruelty to Animals
Preservation of Prisoners' Rights (PROP)
Gamblers Anonymous
Amnesty International
Howard League for Penal Reform
National Association for the Care and Resettlement of Offenders

The cycles of life bring changes. Geographers study a natural cycle by which water, wind and weather change the landscape. Salmon have a fantastic life-cycle. From the river where they are born, they swim hundreds of miles out to sea. After about five years they return to where they were born and after fertilising thousands of eggs they die. It is miraculous because they are not able to observe the similar life-cycle of their predecessors. Still life goes on.

'Are you sure this is the right way?'

Do you think there is any truth in the following ideas about cycles?

1. J.B. Priestley wrote plays based upon cycles of time; people could move backwards into the past or forwards into the future.
2. Astrologers think we are influenced by regular movements of heavenly bodies. Do you believe in horoscopes?
3. Some revolutions in human history have been connected with a supposed cycle of sun-spot activity. Is this the idea of a lunatic? Even the word 'lunatic' comes from the 28-day cycle of the moon (lunar surface) around the earth!
4. The earth has cycles of extreme heat and cold; many scientists believe we are moving towards another Ice Age.
5. Stock Exchange shares tend to move in a cyclical way. Extremes of boom and slump were experienced between 1929 and 1933. This is called the trade cycle.

The effects of some of these cycles may be exaggerated but the growth of human society does indicate a cyclical pattern. It is a scientific fact that human reproduction depends upon a woman's menstrual cycle. The cycle of events which forms such a vital part of human development suggests that there is a **structure** to everything. In the same way as we study the structure of a poem or a book or a symphony we are going to examine the class structure in society, and the structure of the family, government, education, industry and other things of interest to the social scientist. We will try to discover the ideas and purposes that are vital to the life of the community. The word **community** means a group of people who share the same interests and values. It can be applied to a small group such as a village or a school, or to a large group such as a nation. In the early stages the picture may be a bit dim, but we will keep adding to our knowledge. We will go on until we get a clearer vision of society as it is and as we think it ought to be.

Explain what is meant by

individuals
personality
character
achievements
behaviour
heredity
environment
influences
instincts
innate motives
experiences
subconscious
life-cycle
human society
anti-social attitudes
structure
community

Case study

TOO OLD TO DIE

David was seven years old and his grandfather was eighty-seven. The difference in their ages seemed of little importance. David had never heard of the generation gap, and he was convinced that his grandfather was 'the greatest'.

David's father always had a lot of work to do, while his mother was often busy about the house. His grandad answered his most difficult questions with the utmost patience.

Although David was only seven he often wondered what life was all about. Why was he born? Where would he go, if anywhere, when he died? David decided to ask his grandad about the secrets of life and death. He began the conversation cautiously, pretending to be concentrating on a drawing he was doing.

'How old do people live to be, Grandad?' David asked.

Grandad thought for a minute. Why did the boy ask such profound questions? The old man had enjoyed a good life; death held no fears for him. Besides, he got so tired nowadays; he could do with a rest. Nature has her own ways of showing kindness. But how could he explain about a man's life-span to a mere seven-year-old? He resorted to an obscure reply. David would be content with that and let the matter drop.

'Well, David,' replied his grandad, 'the Good Book says that a man shall live to three score years and ten.'

'How long is that, Grandad?' asked David, without looking up from his drawing.

'It means about seventy years,' the old man answered. David continued with his drawing. There was a pause of several minutes.

The old man wondered if his grandson was really satisfied. The conversation seemed to have ended too abruptly. Surely he couldn't have got away with that one so easily? Was the boy just drawing, or was he drawing and thinking?

Suddenly David looked up from his book with a thoughtful expression.

'How old are you, Grandad?' he asked.

'Eighty-seven, my boy', the old man replied.

David studied him earnestly and looked more thoughtful than ever.

'I should think you are past dying, Grandad,' he said.

1. How would you try to explain the human life-cycle to a seven-year-old?
2. Suggest some reasons why a human life lasts, on average, approximately seventy years.
3. Why is it that the very old and the very young often get on well together?
4. In what ways have you changed during your lifetime? What are the main ways in which you could change for the better in future years?

Follow-up

1. Make a list of your main needs under the headings: physical, mental, emotional, spiritual.
2. Write a short story about an individual who behaved
   either (a) instinctively
   or (b) in accordance with what he had been taught.
3. Give the main stages in the life-cycle of *one* of the following:
   (a) a baboon (c) a frog
   (b) a butterfly (d) a wasp
4. Copy the following list of individual characteristics and then:
   (a) Put a cross against those which apply to you.
   (b) Ask a schoolfriend and one of your parents to put crosses against points which they associate with you. (Fold the paper before they make their choices, so that the other crosses are not visible.)
   (c) Compare the three assessments. What have you learnt about yourself as an individual?

Characteristic	*Self*	*School-friend*	*Parent*
conceited			
easily bored			
cheerful			
bossy			
sympathetic			
temperamental			
conscientious			
snobbish			

5. Read this extract:
   'At home the fifteen-year-old's drive for independence and his moodiness render him difficult – sensitive, touchy, with a tendency to make biting remarks. Such attacks may be expressed in loud outbursts following the imposition of a restraint or a complaint by a parent that a son or daughter is not doing his or her bit. When worsted in an argument the fifteen-year-old may rush into his bedroom, slamming the door hard as a gesture of defiance. Sometimes hostility may be expressed by a cold stare, or a verbal quip, as with the girl who said, "I see you are reading that book on the sixteen-year-old. Well, I'm going to read it too, and when I'm sixteen, I'll act just the opposite to what the stupid book says."

   'It would be incorrect to paint a wholly adverse picture of the fifteen-year-old and his parents. There are too many positives for that. The keen observer notices subtle changes taking place in parents; for example, a mother's gesture of deference to her son's greater knowledge when she asks for information on a technical matter which he has been studying. Such gestures please fifteen-year-olds immensely, and are probably an excellent way of reducing difficult behaviour of the kind described.'

   From *Children growing up* by John Gabriel

   In what ways can parents make it easier for teenagers to pass through what is often called a 'difficult period'?

Activities

1. Draw up a personal memorandum which applies to you. Add five points not included in the personal memorandum on page 12.
2. Collect advertisements from recent magazines which appeal to your
   (a) sense of *adventure*
   (b) feeling of *affection*
   (c) craving for *ambition*
   (d) dislike of *bad temper*
   (e) instinct of *curiosity*
3. Design a cartoon strip to illustrate the human life-cycle.
4. Collect photographs of relatives from both sides of your family.
   In what ways do your features resemble
   (a) your father's side?
   (b) your mother's side?
5. Cut photographs from recent newspapers illustrating examples of people who are
   (a) for society
   (b) anti-society

Examination questions

1. Describe the relationship between heredity and environment with reference to child development. Your answer should show the importance of critical stages of development.
   (Joint Matriculation Board, GCE, Social Science, 1984)

2. What problems are caused for society by the following?
   (a) A heavy smoker
   (b) A teenage runaway

   (Associated Lancashire Examining Board, CSE, Social Studies, 1984)

3. Children of today are far more 'exposed' to the influences and behaviour of the adult world through the media, particularly television and video tape.
   (a) Discuss the advantages and disadvantages of this situation and explain from the point of view of a parent and of society, how you would attempt to limit the influence of the media.
   (b) How does the position and daily life of a child in a primitive society contrast with that of your own childhood?

   (The West Midlands Examinations Board, CEE, Social Science, 1983)

4. The following passage describes the Siriono tribe of Bolivia:
   'Within this society, the formal agencies of social control are almost entirely lacking. No such thing as a police force exists. . . . The handling of one's affairs is largely an individual matter. Everyone is expected to stand up for his own rights and to fulfil his own obligations. . . . Justice is an informal and private matter. Grievances are settled between the individuals involved.'

   From A.R. Holmberg, *Nomads of the Long Bow*, 1969

   (a) (i) Among the Siriono people of Bolivia, 'the formal agencies of social control are almost entirely lacking'. What are these formal agencies in a modern industrialised society?
   (ii) Explain, using examples from one or more societies you have studied, how it is possible to maintain social control without employing these 'formal agencies'.
   (b) Why is it necessary to employ formal agencies of social control in modern industrialised societies? Why cannot justice be an informal and private matter in such societies as it is for the Siriono?

   (University of Cambridge Local Examinations Syndicate, GCE, Sociology, 1984)

5. Explain the ways in which an individual's social position is determined by his family background *and* by his education.

   (East Anglian Examinations Board, CSE, Sociology, 1983)

6. Describe fully the influence which religion can have on personal and social behaviour.

   (East Anglian Examinations Board, CSE, Social Studies, 1984)

7. What are the consequences for the individual and society of high levels of unemployment?

   (East Anglian Examinations Board, CSE, Sociology, 1984)

8. 'Deviance is created by society. I do not mean this in the way that it is ordinarily understood, in which the causes of deviance are located in the social situation of the deviant or in the "social factors" which prompt his action. I mean rather, that social groups create deviance by making the rules which when broken constitute deviance and by applying these rules to particular persons and labelling them as outsiders.'

   From H. Becker, *Outsiders*, 1966

   (a) Explain, with the help of examples, what is meant by the term 'labelling'.
   (b) (i) How have sociologists used the concept of 'labelling' to aid our understanding of crime?
   (ii) What are the limitations of the 'labelling' approach?

   (University of Cambridge, GCE, Sociology, 1984)

9. 'What do we mean by a child? Modern Western concepts of the child crumble when we begin to examine the young in other societies today. For instance, many of the normal attributes of Western childhood do not exist in the Third World. Similarly, many modern characteristics of a child seem less valid when applied to children of the past century. When in the 1850s and 1860s Henry Mayhew, the Victorian social researcher, interviewed poor children in London, he and his readers were appalled to find so few "childlike" qualities among children on London's streets. They worked, they were worldly wise, often economically independent, sometimes sexually experienced – they even *looked* older than their years. Yet they were at the age of eight or nine, children nonetheless. At this same time, many young people in the public schools were receiving a full time education at an age when many of their poorer contemporaries had been at work for a decade. Indeed, throughout this period, children at work were as commonplace as they are today in poor countries and, for much of the period, working children were thought to be economically important. Schooling and work then, do not offer a

definition of children in the past century.'

J. Walvin, *A Child's World: A Social History of English Childhood 1800-1914*

(a) Using evidence from the passage, show how the meaning of 'childhood' varied among different social classes in Victorian England.

(b) What part do (i) biological factors and (ii) social and cultural influences play in determining the definition of childhood in different societies? Illustrate your answer by reference to the passage and to other societies you have studied.

(*University of Cambridge, GCE, Sociology, 1984*)

10. How are people influenced about political matters?
What are the good and bad points of the influences that you have mentioned?

(*North West Regional Examinations Board, CSE, Government and Citizenship, 1981*)

11. 'What is considered "odd", "different" or even "deviant" behaviour varies within society. Mountaineers, parachutists and soldiers are not considered deviant by the majority of people – and yet they have chosen to behave in a very dangerous, even suicidal way. They are usually praised for their behaviour. But surely, no sane person wants to risk his/her life?

Most of us think of ourselves as "normal" human beings, yet some of us are regarded as abnormal by other people. For behaviour to be labelled "abnormal", it has to be judged as going against accepted norms. But what are these "accepted norms"? And does everyone agree with them? Many people who would judge themselves as perfectly "normal" are called "mad", "crazy", "unbalanced" by others.

It is not easy then to draw a clear line between "normal" and "abnormal" behaviour. People use different standards by which to judge behaviour.'

Adapted from Y. Beecham, J. Fiehn and J. Gates, *Deviance – themes in Sociology*, 1982

(a) According to the authors it is difficult to draw a clear line between 'normal' and 'abnormal' behaviour. Why do you think this is?

(b) Why is some behaviour which is not against the law sometimes regarded as 'deviant?'

(c) It is often the case that what is considered 'abnormal' behaviour in one society is not considered abnormal behaviour in another society. With reference to any *one* society or culture, which is different from your own, explain how this situation might arise.

(*University of Cambridge, GCE, Sociology, 1984*)

12. Most people need to conform to accepted forms of behaviour but many others do not.

(a) Explain why you think that most people do conform.

(b) Give *two* examples of non-conforming behaviour that are not against the law.

(c) Give *two* examples of non-conforming behaviour which are against the law.

(d) Name *one* organisation which works with people who engage in anti-social behaviour.

(*East Midland Regional Examinations Board for CSE, Social Studies, 1980*)

UNIT 2
The Group

GCSE Aims and Objectives

Knowledge — to promote a knowledge of how individuals are influenced by groups including peer-groups, deviant groups, delinquent groups, out-groups, in-groups, formal groups, informal groups, majority groups, nomadic groups, pastoral groups, and socio-economic groups.

Application — to develop the ability to use social science concepts such as stereotypes, dyads, roles, norms, socialisation, culture, social class and status.

Evaluation — to make critical assessments of our attitudes to group activities, why we belong to groups and how we are changed by groups customs, rules and relationships.

Paul, Gary and Kevin

TOPIC 2.1 TEENAGERS

These youngsters have a lot in common. They are about the same age. Paul and Gary are sixteen, and Kevin is fifteen. So they are teenagers. They are all boys who live on the same estate, go to the same school and share their own language or slang. Paul, Gary and Kevin enjoy each other's company. There are no people with whom they would rather be. Paul is the natural leader and initiates most of their 'rules' and customs. He likes the group to be called 'Paul's gang'. The word 'gang' is sometimes used as a term of abuse by older people who consider that a gang of teenagers must be an anti-social lot. It is true that members of a 'gang' of thieves are not the most responsible people, but a 'gang' of roadmen or building workers are a pretty useful group. So there is nothing much wrong with the word 'gang'. It depends on how it is used and the associations that go with it. The classification of a number of teenagers as an irresponsible gang is known as stereotyping. Some adults **stereotype** all teenagers as good-for-nothing vandals just because of the behaviour of a few.

In Unit 1 we considered how individuals differ. It would be very dull if an individual had to live on his own. A lively human being must have people with whom to talk, eat, play and sing. What is the use of a new joke unless you can share it with friends? You need friends to laugh with. Even our anti-social deeds are usually performed as group activities. Paul's gang has the occasional fight with another gang. The same things that stir the individual are the basis of group activity. Groups are motivated by the driving forces of loyalty, obedience, sympathy, friendship and the need to be approved of. Human motives can be considered in greater depth as we study groups. You cannot be aware of feelings such as duty, morality and respect unless you are a member of a group.

There's a lot that could be said in favour of Paul's gang. The members of the gang are more influenced by one another than by parents or school or newspapers or anything else. They are members of a **peer group**. This is a term used to describe a group of friends of similar age who meet together frequently. In their very early years the members of Paul's gang were completely dependent on their mother or father and in their primary school days they relied on their teacher. Now they look to one another for strength and support. There is nothing wrong with this. They have the same background, the same likes and dislikes. It is perfectly natural for teenagers to group together. They form just one of the many groupings in society. All groups have a structure and a purpose.

Can you identify these stereotypes?

TOPIC 2.2 ATTITUDES TOWARDS GROUPS

Like teenagers, other members of the community form groups. We may think of some of these groups as being *for society*; they are made up of people who conform to the rules and customs of present-day society. Others are *against society* because their members do not accept many of the standards and values of the society in which they live. They may be called **deviant groups**. Teenager groups who commit crimes may be called **delinquent groups**.

Ideas about deviancy depend upon the community in which we live. As individuals and communities change (see Unit 1), so the groups within a community will change, and so will our view as to whether they are deviant or not. For example, in Britain a group of republicans might be considered deviant because their ideas are at variance with the British tradition. Groups of people are considered **negative groups** or **out-groups** if they are out of step with current opinions. Today our main political groups support democratic government, so Fascist groups are deviant. In the early part of this century the Suffragettes were treated as deviants by men whose idea of democracy was that only they should have a vote. Marxist groups may be thought deviant because they want the British economy to be run on communist lines. How do we regard groups of Roman Catholics, Scientologists, Mormons, Atheists and witches? Many years ago groups of witches were put to death. Groups that are in favour because they agree with contemporary standards are called **in-groups**.

Paul and Debbie

TOPIC 2.3 WE BELONG TO MANY GROUPS

Although Paul spends much of his time with teenage friends, he belongs to many other groups in the community. Paul has joined the fan club of a famous pop group. He supports Arsenal Football Club, and he attends a karate group. Look at Paul's badges. To what other groups does he belong? Paul is also associated with lots of groups which do not have badges or emblems.

Debbie is Paul's steady girl-friend: Paul and Debbie form a group of two. Sociologists refer to a group of two as a **dyad**. This is the smallest possible group in a community. If they ever become engaged Debbie will wear a ring as a sign that their group is likely to be lasting. A marriage licence would signify a more permanent grouping. When Paul's elder sister was married recently the photographer was anxious to take a family group. Normally Paul lives with his parents, brother and younger sister. They are looked upon as a group by their neighbours. If one of the family does something wrong, Paul says the blame rubs off onto the rest of them. At the wedding the family group was extended to include not only mothers, fathers, brothers and sisters, but also uncles, aunts, cousins, nephews, nieces.

Paul's father belongs to many groups outside the family. He has joined a trade union and attends the branch meetings. He is also a member of a political party. He is one of a group of singers in a local choir and leads a church discussion group. Paul's mother helps to run a playgroup and acts as treasurer for the local branch of the WI. Paul's father and mother play different parts or **roles** depending upon which group they are attending.

Paul's mother plays many different roles

At school Paul is a member of Tutor Group 5/12. The '5' stands for fifth year; there are fourteen groups in his year and Paul is in Group 12. There is no disgrace in being in Group 12; it is a group number used for registration and Tutor Group meetings. His Tutor Group 5/12 has a member of staff who specially looks after them. Paul's school is very concerned with pastoral care, and each week his Form Tutor gives a certain time for interviews with members of his group. Tutor Group 12 has a kind of team spirit which increased when they won the fifth-year football cup last term. This was presented at the weekly assembly for fifth-year tutor groups. Even the fifth-year is a loose type of grouping with certain loyalties that are apparent when the fifth-year faces the sixth-year group or the fourth-year group. For school work, Paul is in various teaching groups where he mixes with other fifth-formers from Groups 1 to 14. The members of the teaching groups vary depending on whether the subject is Mathematics, Social Studies, English, Economics, etc. These are known as **formal groups**, because the members do not select their companions; youngsters playing in the park would make up **informal groups**.

TOPIC 2.4 GROUP CUSTOMS AND RULES

Paul's gang has no written rules but the boys have an understanding about how they will behave. They observe certain **customs**. Like all normal boys they are very interested in girls but it would be unthinkable for any of them to bring a girl-friend when their gang or group meets.

Why should anybody give up some of his individual identity to fit in with the ways of a group? Perhaps he would no longer be allowed to belong if he did not conform with the standards of the group. These standards of behaviour are known as **norms**. When trade unionists strike they expect all their fellow workers to come out. Those who remain at work are called scabs or blacklegs and they may be sent to Coventry (i.e. other workers will refuse to speak to them). This may be a childish thing to do but there is a good reason for it. Workers are less likely to win their case if they fail to stick together, and they know blacklegs will accept wage increases that the unions succeed in getting even though they have not suffered the hardships of the strikers and their families.

Striking miners on the picket line at the Prince of Wales Colliery

A group of skinheads △

National Front demonstrators. Are their banners suspect? ▷

TOPIC 2.5 OUR GROUPS CHANGE US

We are influenced by the groups to which we belong. We learn to adapt to groups both large and small, to the community as a whole, to the school, to the family. We call this process **socialisation**.

A boy who joined a group of skinheads was socialised into becoming a skinhead. He shaved his head and wore bovver boots, jeans and braces. He learnt to think and behave like a skinhead.

People who join the National Front Party mix with its supporters and accept the views of this minority group.

A member of the Salvation Army wears a uniform which identifies him with the group. He is expected to accept the customs and the values of the group. This is what socialisation is all about.

The process of socialisation carries on from the cradle to the grave. As you pass from one group to another you are continually changing. It is natural that you will join different groups. There is a life-cycle (see Unit 1) of group membership. You have no choice about your first group. You originated as a

A Salvation Army band on parade

member of a family. You were most likely born into it. It would be unusual for anyone to go to a home down the road and say. 'May I belong to your family?' You probably had little choice in selecting the childhood group with whom you played; membership of the group depended upon who lived in the neighbourhood. Would you describe your neighbourhood as working-class or middle-class? We see again the importance of environment which we studied in Unit 1.

Now you are a teenager, how much do you think you have been affected by your environment and how much by heredity? No doubt you have been influenced by a combination of both. We learn *a way of life* which is influenced by past generations. We call this **culture**. How does your school teach you the culture of your community?

You have a choice about some of the groups you belong to at school. There may be groups for drama, fencing, chess, stamp collecting, radio, etc. What activity groups exist at your school? When you leave school you may decide to work in a factory or office. You may go to college full-time or part-time. If you get married you will have to select a partner and be accepted by your partner. These changing groups will bring changing **relationships**. If you play for a football club or join a pop group your personality will be changed by others in the group. Some people are good joiners but poor stickers. They join lots of groups but don't remain with any for very long. They will be influenced by lots of people but the impression made upon them is not likely to be so lasting as that made upon an individual who is a loyal and long-standing member of a group such as a political party or a church. Whatever group you join you will be affected by relationships with other members of the group and by relationships between your group and other groups.

TOPIC 2.6 RELATIONSHIPS

Paul and Debbie usually get on well together. They share many interests. They live in the same neighbourhood, go to the same school, like the same records and pop groups. They have a good relationship. But occasionally their relationship becomes strained.

Sometimes an argument develops because Paul supports his football team on Saturday and says he doesn't want to be dragged round the shops. Debbie realises that she has to compromise to maintain a good relationship, so she has found a job on Saturday afternoons. Paul goes out with his gang less frequently because he takes Debbie to the disco; he is also studying more in the evenings because Debbie urges him to get on. They are both learning that good relationships are a matter of give and take. It's no good one of them taking everything and expecting the other to give everything. It may be fairly easy for two people to establish a good relationship, but it is more difficult as the group gets larger. Perhaps that is why some people think, 'Two's company, three's a crowd.' As groups get larger there is a greater need for compromise.

When Debbie first took Paul home to meet her parents, they were not too keen on Paul and he was a bit doubtful about them. But they are all learning to accept one another for what they are. Do you think they will change one another so that they are all more like what the others want them to be?

It cannot be expected that everything will always run smoothly. Debbie was upset when Paul went out with her best friend. In the end, she forgave him. Good relationships depend upon people not holding grudges.

Look at the strip cartoon opposite. Who do you think is mainly at fault in the quarrel? Good relationships do not mean always giving in. Have you ever found that relationships have improved after a terrific row? It seems to clear the air!

'We can't communicate?
What do you mean by that?'

TOPIC 2.7 GROUP RELATIONSHIPS

Political parties make up some of the largest groups. The following table shows the number of people who voted for the three main parties in the General Election held in 1983.

Political party	*Millions*
Conservative	13.0
Labour	8.5
Alliance	7.75

It is no good quarrelling every time you meet somebody who supports a different political party, although we must have a certain amount of disagreement in a free country. You must be ready to listen to another point of view. You might possibly be wrong! We are free to argue.

We are more likely to elect the best possible government if people air their views. There must be a lively interchange of ideas. If you travel on buses or wait in queues you may get into a discussion. It is only by having your views challenged that you are likely to appreciate the other side of the argument.

But individuals must give up some of their freedom when they join a group. If you were on an empty beach you would be free to throw stones all over the place. But if you were on a crowded beach, your stone throwing would interfere with other people's freedom to play games safely or sunbathe without getting a split head.

The groups within a community must be ready to give up some of their freedom. If we belong to **majority groups** we must safeguard minority interests. The Jew who is living in a Christian country must be allowed to worship in hs synagogue, keep his Sabbath on Saturdays and speak to his friends in Hebrew if he wants to, but what happens if, say, a Rastafarian wants to pursue a social custom which brings him into conflict with the law?

TOPIC 2.8 SOCIAL AND ECONOMIC GROUPS

The basic organisation of society is the group. Humans have always grouped together for social intercourse and for economic reasons. Early men formed **nomadic groups** which wandered over the countryside hunting, fishing and gathering food. Later, men settled down and lived in one place as **pastoral groups** or **agricultural groups**. The development of communities established in one place gave rise to a complex **group structure**.

The medieval class structure under the feudal system

After the Norman Conquest, William I divided the land of Britain among various groups. The people who came off best were the tenants-in-chief such as the abbots, barons, bishops, earls and knights. Then followed the mesne (or lesser) tenants, freemen, villeins, cottars and serfs. A social pyramid was built. Each group was of a slightly higher **social class** than the group below. They had more **status**. The poor serfs had no one below them. They were at the bottom rung of the social ladder. They were in what modern sociologists would call **socio-economic group** 5. They suffered from a birth-to-death inequality cycle. Those who are born into a poor home may suffer from a shortage of goods, an unhealthy environment and a poor educational start which may cause deprivation during their whole life-cycle. We still have a class structure today. The Registrar General of Occupations divides society into five main social groups. However, did you realise that a woman is often classified according to the occupation of her father or husband? Is this fair?

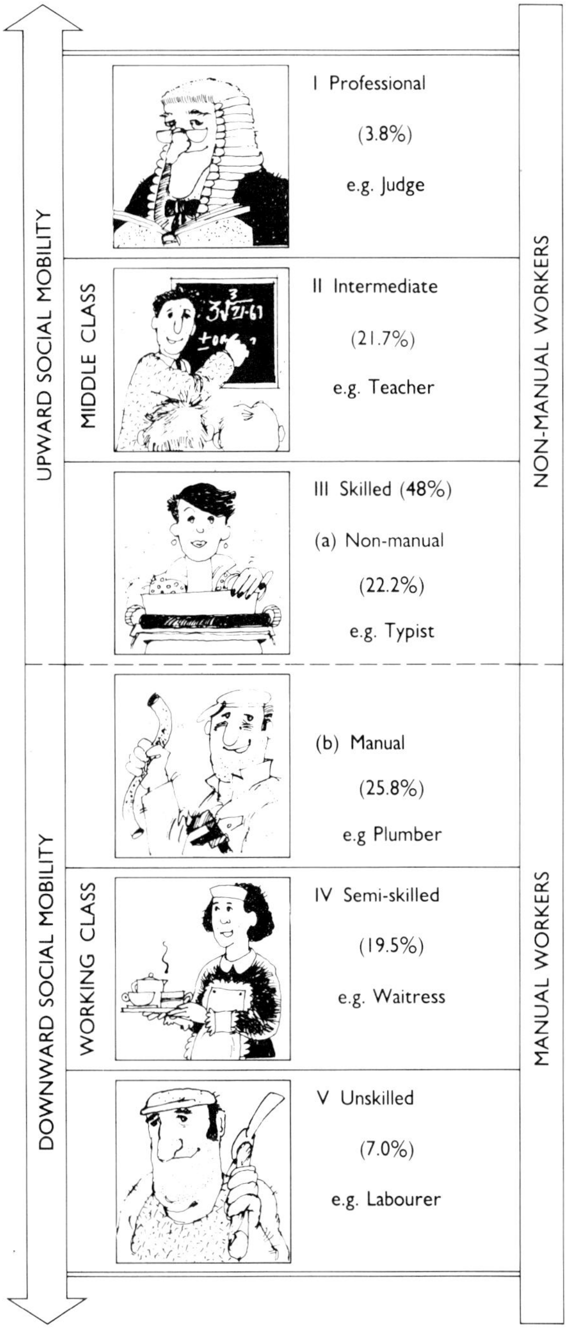

The socio-economic classes according to the Registrar General's scale of occupations

Explain what is meant by

stereotype
peer group
deviant groups
delinquent groups
negative groups
out-groups
in-groups
dyad
roles
formal groups
informal groups
customs
norms
socialisation
culture
relationships
majority groups
nomadic groups
pastoral groups
agricultural groups
group structure
social class
status
socio-economic group

Case Study

SING AS YOU GO

Trevor was walking home after a night out with the boys. They had been drinking until about 11 o'clock at The Corkscrew. It was their regular local. Then they had gone to Eric's place and had a few more pints.

Trevor was feeling pretty good – happy, but not fighting drunk. Eric and Eddy had told some good stories and he was chuckling to himself. There was a full autumn moon and Trevor kicked the leaves in the gutter. He liked to hear them rustle. Life was good. Trevor started to sing. He broke into his usual drinking song.

'There's an old mill by the stream,
Nellie Dean,
Where we used to sit . . .'

Trevor was just getting into full throttle when an upstairs window was opened noisily.

'Shut your mouth, you drunken yob; you'll wake the baby.'

'A bloke can sing in his own street, can't he?' Trevor yelled back. 'It's a free country, isn't it?'

'If it's a free country, why isn't the poor little blighter free to enjoy her shut-eye, without every hooligan in the neighbourhood yelling his top off,' shouted the woman.

1. Who do you think was more in the right, and why?
2. In what better way could the woman have handled the situation?
3. On what occasions would Trevor be allowed to sing in perfect freedom?
4. What was the most recent occasion when you restricted somebody else's freedom to do something?

Follow-up

1. Choose a group with which you are familiar. Copy the diagram (top right) and insert the following:
   (a) name of the group
   (b) badge emblems, uniforms, etc.
   (c) aims
   (d) interests
   (e) achievements, etc.

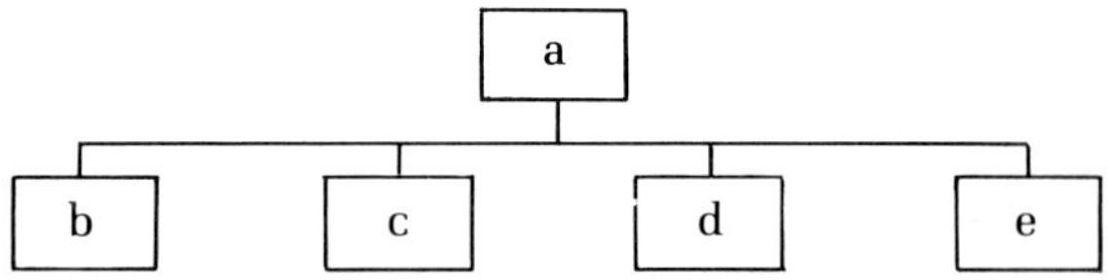

2. The example below shows how language changes when used by specific groups. Write out the passage in normal speech.
   (*A group of RAF pilots talking in 1940*)
   'Was it a good show?'
   'It was wizzo. Two jerries went for a burton.'
   'Where's groupy?'
   'Afraid he put up a black, old boy. He nearly went in the drink and then he pranged his kite near a group of sprogs who were square-bashing on jankers!'
3. Look again at the photograph of Paul's gang on page 12. Which boy do you think looks most aggressive? Which one looks most friendly? Which one looks most easily led? Select one of the boys and use your imagination to write a character study. His attitude and his facial expression should give you a guide.
4. Read this passage from Joseph Heller's *Catch 22*:
   'Colonel Cathcart stopped in his tracks. "What atheists?" he bellowed. . . . "There are no atheists in my outfit. Atheism is against the law, isn't it?"
   "I'm not sure, sir," answered the chaplain.
   "Well, I am!" the colonel declared. "I'm not going to disrupt our religious services just to accommodate a bunch of lousy atheists." '
   In what way was Cathcart being illogical?

Activities

1. Design a badge to represent one of the following:
   (a) women's lib
   (b) a snooker club
   (c) hell's angels
   (d) a youth club
2. Conduct a census of your tutor group or class, using a pre-coded questionnaire with boxes. Use the five questions below and add five of your own.
   (a) *Sex?* Male = M ☐
   Female = F
   (b) *Leisure?* Mostly at home = H ☐
   Mostly out = O

(c) *Hobbies?* Yes = Y
No = N
(d) *Pocket money?* £5 + = P
£5 − = M
(e) *Problems?* Go to parents = P
Go to teacher = T
Go to others = O

3. Interview a volunteer from your group who says he is a 'loner'. Ask him to give reasons why he dislikes joining groups. Suggest some group activities to him and try to interest him in one. If possible, tape the interview so that any weaknesses in his arguments can be examined.
4. Use local libraries, reference books, newspapers and personal contacts to find out the following about the groups whose badges/logos are illustrated below:
   (a) aims
   (b) local membership
   (c) recent activities

© WWF

Examination questions

1. The Registrar-General attempts to measure social class on the basis of occupation.
   (a) How many categories are there in the Registrar-General's scale?
   (b) What other scale exists for measuring social class by occupation?
   (c) What is meant by the term 'social mobility'?
   (d) State *four* ways in which a person can become socially mobile in Britain.
   (e) State *three* factors which influence the way people vote.
   (f) By using examples show the relationship between social class and education.

   (*North West Regional Examinations Board, CSE, Social and Community Studies, 1983*)
2. Influence is exerted over individuals and groups so that they conform to the norms of society. Various agencies are responsible to society for ensuring that order is maintained.
   (a) What is the term used to describe the influence exerted over individuals and groups?
   (b) What is meant by 'conform to the norms of society'?
   (c) Select *one* formal agency responsible for maintaining order and show how it carries out its duties.
   (d) The family, peer group and mass media encourage people to conform. Select *one* of these agencies and describe how it encourages people to conform.
   (e) What is the difference between criminal and deviant behaviour?

   (*North West Regional Examinations Board, CSE, Social and Community Studies, 1983*)
3. Social class is one way in which society may be stratified. Name and describe *two* other forms of social stratification.

   (*East Anglian Examinations Board, CSE, Sociology, 1983*)
4. Either, (a) Comment on any three social factors which may promote a person's social class mobility in Britain today.
   Or (b) State what is meant by an *ethnic group* and comment on any two social problems encountered in attempts to absorb minority ethnic groups into contemporary Britain.

   (*Welsh Joint Education Committee, GCE, Sociology, 1982*)

5. Study the following carefully. For each definition in Column A find the most appropriate term in Column B. Then record your answer by writing the letter from Column B before each number in Column A.

Column A

.... (i) A group of people characterised by intimacy and face to face association.
.... (ii) A social group based on blood relationships
.... (iii) A traditional practice in society
.... (iv) The situation where factors determining an individual's status are beyond his control
.... (v) A group of individuals of roughly the same age
.... (vi) The process of learning the ways of a society so as to be able to function within it.

Column B

a. peer group
b. socialisation
c. custom
d. secondary group
e. ascription
f. community
g. interaction
h. organization
i. primary group
j. labelling
k. social control
l. family

(*Welsh Joint Education Committee, GCE, Sociology, 1982*)

6. For *either* the local community *or* any non-British society which you have studied, describe the ways in which the young are socialised.
(*East Anglian Examinations Board, CSE, Sociology, 1983*)

7. 'A child is born into a world that already exists. From the point of view of society, the function of socialization is to transmit the culture and motivation to participate in established social relationships to new members. We may view this society from several perspectives. . . . First, there is the perspective of *norms* and *values*. A second perspective is that of *status* and *role*. Another perspective focuses on *culture and group subdivisions*. Still another perspective is that of *social change*.'
From: *The Child and Society* by F. Elkin
After reading the above passage explain, with the use of appropriate examples, what is meant by:
(a) norms
(b) values
(c) status
(d) role
(e) culture and group subdivisions
(f) social change
(*Welsh Joint Education Committee, GCE, Sociology, 1981*)

8. Explain the difference between the *two* terms in each of the following pairs.
(a) heredity and environment
(b) social role and social institution
(*Joint Matriculation Board, GCE, Social Science, 1983*)

9. It is sometimes suggested that Britain is becoming a classless society. What evidence has been put forward *to support* and *to refute* this idea?
(*Associated Examining Board, GCE, Sociology, 1982*)

10. 'Human beings need more than basic instincts in order to become full members of a human society.'
Themes in Sociology, Beecham, Fiehr and Gates
(a) In what ways are 'instincts' modified by socialisation?
(b) In what other ways does socialisation enable an individual to be accepted as a member of a particular social group?
(*Associated Examining Board, GCE, Sociology, 1982*)

11. Describe the work of *two* voluntary organisations (e.g. Oxfam) in helping to overcome such problems as poverty, hunger, and lack of education around the world.
(*East Midland Regional Examinations Board, CSE, Social Studies, 1982*)

12. 'Every society must make sure that enough children are born. Furthermore, they must acquire the personal characteristics which are necessary for them to carry out their roles in adult society. These functions – reproduction and socialisation – are the major functions which the family performs.'
Adapted from: *The Science of Society*, Stephen Cotgrove
(i) Give TWO examples of roles in adult society which are learned within the family.
(ii) What do you understand by the term *socialisation*?
(iii) List THREE other agencies in society which share with the family the function of socialising the child.
(iv) Explain why the number of children in a family has declined over the last hundred years.
(*The South-East Regional Examinations Board, CSE, Sociology, 1981*)

UNIT 3
Marriage and the Family

GCSE Aims and Objectives

Knowledge — to provide a clearer understanding of the purpose of marriage and the family, and the reasons for possible breakdowns in family relationships.

Application — to apply distinctions learned to:

(a) marriage structures (matriarchal, patriarchal, monogomy, polygamy, polygyny, polyandry, arranged marriages, endogamy);

(b) family structures (kinship, family of origin, family of procreation, nuclear family, extended family, interlocking families, child-centred families).

Evaluation — to assess critically the functions of the family, family cycles, family problems and the future of marriage as an institution.

Family group at a traditional British white wedding

TOPIC 3.1 MARRIAGE

In Britain marriage involves a group of two – a dyad. Later the group is likely to grow, because most married couples intend to have a family.

In Britain the husband traditionally goes out to work and brings home his wages to support his wife and family. However, it is increasingly common for a married woman to return to work after the birth of her children. She now has the choice whether to return to her job immediately after taking maternity leave (an employer is legally obliged to keep her job open for a certain period, if she has expressed the desire to return to work), or whether to give up her job until the children are of school age. This means that both parents often share the responsibilities of maintaining the household and bringing up the children. The greater equality of men and women is changing marriage and family structures. We use the word **matriarchal** to describe a situation where the woman is boss. There are many variations of marriage arrangements.

More than three-quarters of the people of the world live in communities where custom, law or religion determine that a man or woman has only one marriage partner. This marriage structure is called **monogamy**.

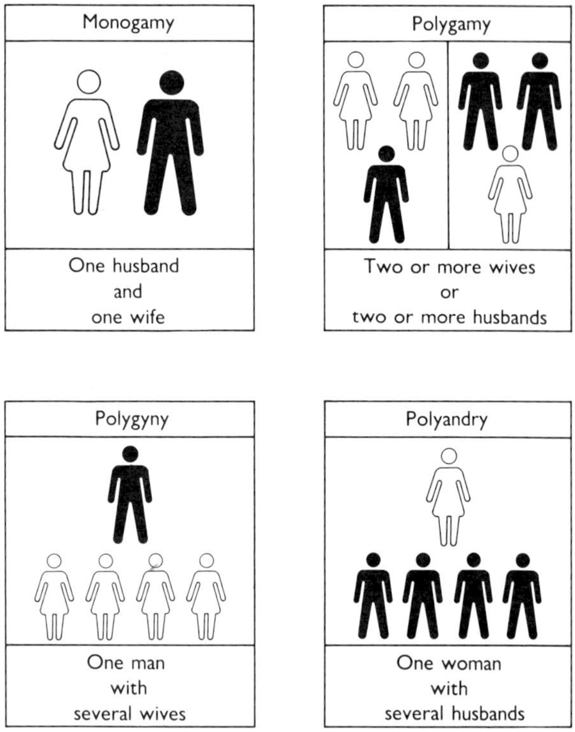

Different types of marriage structure

A Hindu wedding in London

In some communities more than two people enter into marriage. This is known as **polygamy**. When one man has several wives it is called **polygyny**. The early Mormon communities in Utah practised polygyny, and many Muslims practise it today. If there is a surplus of women some of them will be denied the natural function of having children unless men are permitted to have several wives. In this way a man may acquire not only wives, but also workers, property and status.

Some Eskimo women have more than one husband. This is known as **polyandry**. Eskimos live very dangerously and if one husband is killed, the other will look after the wife and children. The two husbands learn to share without enmity. They are friends rather than rivals. It is part of their socialisation. Good relationships are the result of compromise (see Unit 2); polyandry among Eskimos is determined by the environment (see Unit 1). Muslim wives find companionship: an older woman may be content to exercise seniority in household management while younger wives satisfy the husband's sexual desires.

Roy Potter, a Mormon from Salt Lake City, Utah, with his two wives, Vera and Mary, and his seven children, including those by his first wife Denise (now divorced)

The different structures of marriage depend upon the functions of the people concerned. The Melpa people of New Guinea are accustomed to a man having about three wives. The group forms an economic unit. The wives grow food and give the surplus to the husband who looks after the administration of the land, buildings and children. He is the chief or patriarch and the society is known as **patriarchal**.

In modern Britain you will pick your own partner, but in Victorian times the **arranged marriage** was common among the upper social classes. Even today members of the British Royal Family may be closely guided in their choice of a marriage partner and this is certainly still the case in many Asian families living in Britain. A young Pakistani girl may be married to a young man whom she has never seen, but who has been chosen for her by her parents. **Endogamy** means marriage within your own social class or group. When this is practised people can often trace back their family tree by means of a genealogical table. They are tied by **kinship**.

Who is your next of kin? When you are married, your wife or husband will be your next of kin. If you agree to ask all your family to your wedding you will be inviting all your kin. There is a structure of kinship by which blood relations are linked.

'I'm not fussy. Any rich woman who's too proud to let her husband work will do!'

TOPIC 3.2 IS THERE ANY FUTURE IN MARRIAGE?

During the 1970s, the annual number of marriages in Britain declined steadily up to 1977, but there was a slight increase in numbers between 1978 and 1980. In 1981, however, the number of marriages fell by five per cent. There has been an increase in the proportion of marriages where one or both partners have been married before. (See table.) People have been marrying later in life. In 1981 the median age at marriage for men was 25.8 years and for women it was 23.3 years compared with 24.0 and 22.0 years respectively in 1971. There has also been an increase in the number of people 'living together' before they get married. The proportion of people living in one-parent families has doubled since 1961, from only 2.5% to 5% in 1983. This trend reflects the rise in divorces, increasing numbers of births outside marriage, and fewer adoptions of illegitimate children than in the 1960s.

Can you add to this list of possible reasons for **divorce**?

Too many domestic chores	Living with parents
Falling in love with somebody else	Money worries
Few common interests	No children
Illegitimate child (before marriage)	Sex problems
	Too young
	Work (career) problems

'Calm down, Bernard! The odd dull moment is only to be expected after forty years of marriage.'

Broken marriages and unemployment blamed for increase in child abuse

Is it a good thing to try to hold together a marriage which is really on the rocks? A successful marriage must be based upon the *social emotions* which we considered in relation to the individual. It relies upon such emotions as love, kindness, understanding and sympathy. When *anti-social emotions* such as hate, envy, malice and jealousy are uppermost, the marriage is likely to be in a bad way.

You will have heard tragic stories of broken homes, abandoned and battered children or a single parent struggling to bring up children without the help of a partner. Is this worse than a patched-up marriage with continual quarrels and neglected children? It is an open-ended question. It has to be left in the air with the unfinished sentence, 'It all depends . . . .' Suggest some things it may depend on.

Many people believe that divorce is better than an unhappy marriage. A divorce rate of 93 per cent was once recorded in part of Malaya. It is simple for a Muslim male to divorce a wife. He has only to declare three times before witnesses 'I divorce you.' It's all very unjust because Muslim wives are not allowed to divorce their husbands.

The only grounds for divorce in Britain is the 'irretrievable breakdown' of the marriage. This was made the only reason for divorce by an Act of 1969. It has made divorce much easier. As long as husband and wife agree they can get a divorce without having to prove cruelty or adultery (sexual intercourse between a married man or married woman and somebody other than their marriage partner).

More people behind with mortgages as broken marriages and jobless rise

Marriages in Great Britain and the United Kingdom (Source: *Social Trends No 14*, 1984)

	Great Britain						United Kingdom
	1961	*1971*	*1976*	*1979*	*1980*	*1981*	*1981*
Marriages (thousands)							
First marriage for both partners	331	357	273	270	270	255	263
First marriage for one partner only							
Bachelor/divorced woman	11	21	30	33	33	31	32
Bachelor/widow	5	4	3	3	3	3	3
Spinster/divorced man	12	24	32	36	37	36	36
Spinster/widower	8	5	4	3	3	3	3
Second (or later) marriage for both partners							
Both divorced	5	17	34	43	45	44	44
Both widowed	10	10	9	8	8	7	7
Divorced man/widow	3	4	5	5	5	5	5
Divorced woman/widower	3	5	5	5	5	5	5
Total marriages	387	447	396	407	409	388	398
Remarriages as a percentage of all marriages	15%	20%	31%	34%	34%	34%	34%

Marriage, remarriage and divorce rates in Britain (Source: *Social Trends No. 15*, 1985)

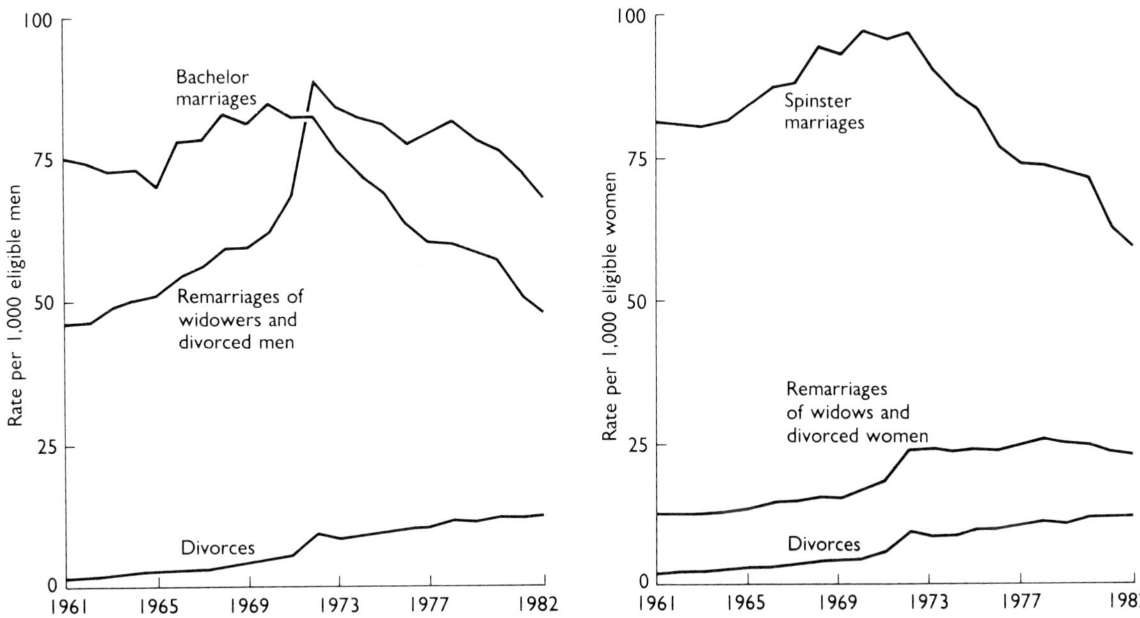

TOPIC 3.3 FAMILY CYCLES

Look at the group in the photograph. Mother, father and children are known as a **nuclear family**. This is the simplest type of family and the most common in Britain and the western world. A nuclear family is an independent unit. The members are linked by blood, live in the same home and do many things together. If one parent decided to take a job in another place they would all move and carry on their activities there. They are linked by bonds of heredity (see Unit 1) and there are close ties of loyalty between them.

As a girl grows up her loyalty to the family group will decrease. This applies to boys as well, but we will take a girl as our example. Her interests outside the family circle will include friends of her own age (peer groups) and mixed-aged groups with similar interests. Eventually she is likely to form a close relationship with a man and they will perhaps decide to get married. The family with which she started is the **family of origin**. The girl will leave home and in time will have a family of her own. This is the first round in the evolutionary family cycle.

She and her husband will start a **family of procreation**. Children born to her will regard this family as their family of origin. Notice the woman's new **role**. As a child she relied upon her parents. Now her role is that of mother and her children look to her for love and security. Her parents will enjoy their grandchildren. Baby-sitting will be a pleasure for them and of great benefit to the parents. So nuclear families become **interlocking families** as the life-cycles of several families become intertwined. Some relatives are linked by blood while others are linked by marriage.

The nuclear family was not always the most popular type in Britain. When there was a scarcity of houses, transport, and job opportunities away from home, it was customary for father, mother, children and grandparents to live together under one roof.

A nuclear family

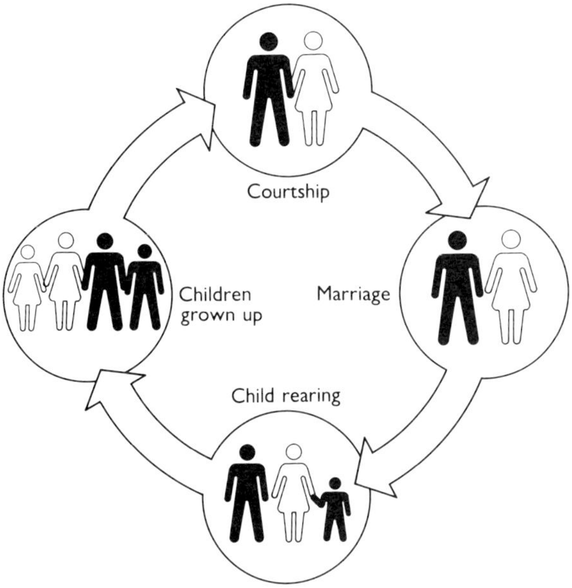

The life-cycle of a nuclear family

This large family unit is called an **extended family**. In Saxon times the family was extended to include grandparents with their children, sons' wives, daughters' husbands, grandchildren, cousins, aunts, uncles, nephews, nieces. They all lived in the long hut. It must have been quite a gathering. There are still villages where all the people are related, although in civilised societies relations usually divide up into close-knit groups with separate homes.

In working-class urban areas such as Bethnal Green in the East End of London, the extended family often continues to be more important than the nuclear family; this is bound up with the wife maintaining a close relationship with her mother after marriage.

An example of the continually changing nature of society can be seen in the different attitude towards the older generation. When the extended family was the norm, the final stage in the family cycle was for grandparents to live with their grown-up children. Nowadays there is a tendency to echo the words of an old song – 'they are far better off in a home'.

TOPIC 3.4 FUNCTIONS OF THE FAMILY

As a member of a family, what do you get out of it? Or what contribution do you make to the life of the family? The two (i.e. giving and receiving) are closely related. The more you put into a thing, the more you get out of it. The following statements indicate the importance that people attach to the family:

'She comes from a good family.'
'He's all right – I know his family.'
'She has a good family background.'
'She's let down the family.'
'She's like one of the family.'
'He's a good family man.'
'He's the black sheep of the family.'
'Blood is thicker than water.'

Which human motives (see Unit 1) do you think would make for a happy family? Family life is structured so that it fulfils certain functions. Some of the traditional functions of the family are now performed by the welfare state. The National Health Service helps in motherhood – before, during and after a birth. A pregnant woman obtains much more than material things such as vitamin pills and dried milk from an ante-natal clinic; she also gets advice,

'From the left we have number one son, number two wife, number three son and number two son with number fourteen girlfriend.'

comfort and support. The National Health Service provides maternity hospitals, doctors, midwives. After the birth a health visitor checks that everything is all right.

At one time children received nearly all their meagre education from parents, older relations or friends of the family. A son's education was mainly vocational because he was expected to follow in father's footsteps. Mother taught her daughters to cook, sew, mend and clean. This kind of education alone would be of little value today to the many boys and girls who desire to be highly skilled members of the community. They need specialist teachers and professional experts.

The table below shows some of the changes that have taken place in relation to the family.

The changing family pattern

Years ago
- Marriage partners were picked by senior relatives
- There were a large number of children, several of whom were likely to die at birth or at a very early age
- Father was head of the family and mother called him Mr
- Children feared their parents and addressed their father as Sir
- Large numbers lived together as an extended family which included a variety of relatives
- Families were independent and willing to help each other in times of poverty and other troubles

Today
- Men and women choose their own partners
- Family planning allows a small number of children at carefully chosen intervals
- Husband and wife usually treat each other as equals
- Children are put before everything else; it is a **child-centred family**
- The nuclear family lives as a separate unit apart from other relatives
- Families are self-supporting and may well look to the state (or neighbours) rather than to relatives for help

TOPIC 3.5 SETTING UP HOME

Some people make their home with a group of other families, e.g. in an Israeli kibbutz or a Chinese commune. People in the Amazon forests establish self-sufficient families so that they do not have to use money. The parents build a rough shelter and the family supply their own food and clothes.

In Western communities it is the norm for a couple to set up home together in a house or flat, and then there are essential goods to be purchased such as furniture, carpets, curtains, bedding and other linen, crockery, cutlery and a hundred and one other things which are necessary in the running of a home. To begin with a young couple probably think in terms of necessities, but they hope that later on they will be able to afford such things as central heating, a food mixer or a tumble drier.

Most couples want a home of their own. Living with relatives creates all sorts of problems. Since there is a housing shortage and your finances will be limited, it is unlikely that when you are married you will get exactly what you want. These are some of the questions you ought to consider. Will you live

1. in a rented or mortgaged house?
2. near your job?
3. in a flat, bungalow, house or what?
4. near your relatives and friends?
5. in town or country?
6. near a good shopping area?
7. where there is a good bus service?

The first question is probably the most important. At first you may be forced to rent a place because of shortage of money, but it is important to consider the advantages of both renting and buying. Nearly half the families in Britain live in rented accommodation and just over half own their home, mostly with the help of a building society mortgage (which is a loan paid back over a number of years).

Advantages of buying
- Eventually, no rent to pay
- Property increases in value
- Income tax relief
- You can sell it and buy another
- Decorate it in any way you like

Advantages of renting
- You do not need any capital
- Cheaper in the short run
- You can soon move out
- No mortgage round your neck
- The cost of repairs and maintenance is often the responsibility of the landlord

TOPIC 3.6 FAMILY PROBLEMS

Many **family problems** arise from bad relationships. Young people develop minds of their own. When very young they happily accept the rules and views of Mum and Dad. Parents and young children often show uninhibited affection. They may think the world of one other.

As children grow older they often grow away from their parents. They find friends and interests outside the family circle and are less inclined to confide in parents, although a deep affinity may remain. Why is this? It is difficult for parents to stop laying down the law when they have been doing it for years. Teenage children may think it unreasonable to be told to be in by eleven o'clock, and retaliate by giving no information about where they are going except 'out'. Family rows develop over such things as pocket money, friends and who should do domestic chores. Unless they all learn to accept their new roles, the teenager may leave home to find a place of his or her own. The estrangement may not last long and the youngster will often return to the family a much more amenable person. Parents become educated too! An individual has to be given the chance to spread his wings. Sensible parents learn not to expect too much; the problems may ease as the 'prodigal son' realises his need for the family and the family shows understanding of his adolescent problems.

Problems frequently arise through lack of money. Married couples with young families may find it a struggle to manage if they are trying to live on one salary. The **captive wife** may feel frustrated because she is prevented from furthering her career and she sorely misses her wage packet, which used to give her a sense of independence. She does not want to keep asking for extra housekeeping money, but inflation makes home management very difficult indeed. There are rates, rent or mortgage payments, electricity and gas bills, as well as food and clothes. It is even harder if 'the breadwinner' is unemployed or the family is living on social security benefits.

Relations with neighbours may not be harmonious especially if footballs land on next door's flowerbeds. Grandparents may cause friction if they spoil the children or favour particular members of the family. Difficulties can arise if a grandparent lives with the family. You should be able to add to the list below:

Things that irritate

The old
- noise (stereos)
- modern fashions
- late nights
- being ignored

The young
- nagging
- stories of aches and pains
- harping on the past
- repetitive advice

Quarrels may arise between wife and husband because of bad sexual relations. The Marriage Guidance Council has found that sex problems cause more divorces (about 30 per cent) than any other factor.

Many married people find that there has to be compromise and patience before a satisfactory sexual relationship is established. Only the minority find that it happens 'by magic'. Modern couples who are determined to find an answer to their sex problems may be willing to go together to a Family Guidance Counsellor or a Family Planning Clinic to seek advice.

Explain what is meant by

matriarchal
monogamy
polygamy
polygyny
polyandry
patriarchal
arranged marriage
endogamy
kinship
divorce
nuclear family
family of origin
family of procreation
role
interlocking families
extended family
child-centred family
family problems
captive wife

Case study

THE CAPTIVE WIFE

Liz and Pete had been to the same school, but it was only after they left that they started going out together. Liz got a job as a trainee manageress in the local supermarket and Pete was lucky to be able to work for his dad who had his own plumbing business. Both worked hard and saved hard and it wasn't long before they decided that they could afford to get married and put a deposit down on their own flat.

After a year or so Pete thought it was about time to start thinking about starting a family. Liz, too, had always thought that she would like to have children one day, but it still came as a bit of a shock when she found out that she was pregnant.

'I don't want to give up my job,' she told Pete one evening. 'If I carry on I could be a deputy manageress soon. And how on earth will we cope without my money?'

'Oh, we'll manage,' said Pete confidently (though secretly he did wonder how). 'I'll do some overtime for Dad and that'll bring in some extra cash.'

Liz and Pete had a beautiful baby girl, but after the initial novelty of being new parents wore off, they realised that they were starting to row more often. Liz missed her friends at work – and Pete. He worked such long hours these days and when he came home he was too tired to talk to Liz; he just sat slumped in front of the TV. Liz was fed up of being stuck at home all day with the baby. She felt like a prisoner in her own home.

One evening it all became too much. Liz burst into tears and ran into the bedroom. Pete suddenly realised that he had hardly said two words to Liz all evening, and ran after her. He hugged her and said, 'Come on, 'phone your Mum to see if she can babysit tomorrow night and we'll go out for a meal to talk things over. It'll do you good to get out.'

1. Suggest some ways in which couples like Liz and Pete might solve the dilemma of 'the captive wife'.
2. What are the biggest problems to be faced by a couple starting married life together?
3. Describe three possible kinds of trouble which might bring a married couple closer together.

Follow-up

1. Do you think the husband or wife is more likely to be affected by each of the points below?
   (a) expected to be faithful
   (b) nurse you when ill
   (c) dislike being tied down
   (d) financial worries
   (e) home of one's own
   (f) independence from in-laws
   (g) making important decisions
   (h) baby-sitters
   (i) choosing holidays
   (j) decoration of the house
2. Give one reason why
   (a) a woman usually changes her surname on marriage
   (b) a husband's career is often considered more important than his wife's
   (c) marriage should be 'until death us do part'
   (d) trial marriages are dangerous
   (e) only a small proportion of unhappily married couples consult a Marriage Guidance Counsellor
3. List points for and against setting up home in one of the following:
   (a) a bedsitter (c) a flat
   (b) a caravan (d) a hostel
4. Suggest one advantage and one disadvantage of living in a kibbutz for
   (a) a small child (c) a parent
   (b) a youth (d) an old person

Activities

1. Draw a cartoon strip on the subject of marriage. Here are some ideas:
   (a) courtship (f) wedding ceremony
   (b) engagement (g) confetti
   (c) best man (h) reception
   (d) wedding dress (i) wedding cake
   (e) the ring
2. Use magazine cuttings to make a collage illustrating *family problems*.
3. With photographs
   (a) make a scrapbook entitled *My Family*
   (b) illustrate your *life-cycle* so far
4. Draw your family tree going back to your great-grandparents, inserting names.

Examination questions

1. (a) Explain briefly what you understand by:
   (i) an arranged marriage
   (ii) a patriarchical family.
   (b) How adequately does the state provide assistance for the family today?
   (c) Why have divorce rates increased in Britain? Your answer should make reference to social, economic and political/legal factors.
   (The Associated Examining Board, GCE, Social Science, 1984)
2. (a) There are two main types of family structure found in our society. Name each type and define it.
   (b) What are the *four* main functions of the family in our society?
   (c) Describe the changes in the role of women in our society since the mid-nineteenth century. Explain why these changes have taken place.
   (North West Regional Examinations Board, CSE, Social and Community Studies, 1983)
3. Describe how and why children in a family may be maltreated.
   (East Anglian Examinations Board, CSE, Social Studies, 1984)
4. What are the principal health and social services and how does the provision of some of these meet the needs of our society? In what ways have these services affected the ability of the family to cope with its problems?
   (The West Midlands Examinations Board, CEE, Social Science, 1983)
5. (a) Explain carefully the differences between a nuclear and an extended family.
   (b) Why should the extended family remain an important unit in many primitive societies, but have been replaced by the nuclear family in our society?
   (The West Midlands Examinations Board, CEE, Social Science, 1983)
6. (a) Compare the merits of renting as opposed to buying accommodation.
   (b) How may renting from a housing association combine the advantages of renting and buying?
   (The Associated Examining Board, GCE, Economics (Social Economics), 1984)
7. Refer to the table below and answer the following questions.
   (i) What is meant by a 'divorce petition'?
   (ii) Why was there such a large increase in petitions in 1947?
   (iii) What trend in divorces is revealed by these figures?
   Give *four* reasons for the trend.

Divorce petitions

England and Wales	
1938	10,233
1947	48,501
1950	29,729
1954	29,036
1964	41,468
1969	61,216
1970	70,575
1972	109,822

   (iv) Under the 1969 Divorce Reform Act what reason must be given in order for a divorce to be granted?
   (v) What are some of the possible effects of marriage breakdown on family life?
   (North West Regional Examinations Board, CSE, Social and Community Studies, 1984)
8. Discuss the ways in which the family has changed in Britain over the last one hundred years.
   (North West Regional Examinations Board, CSE, Social and Community Studies, 1984)
9. In what ways has the state taken over some of the functions of the family?
   (East Anglian Examinations Board, CSE, Sociology, 1983)
10. (a) Describe the ways in which the relationships between parents and children have changed during the Twentieth Century.
    (b) Suggest reasons for the changes you have described.
    (East Anglian Examinations Board, CSE, Sociology, 1984)
11. What family problems may arise when there is a handicapped child in the family?
    (East Anglian Examinations Board, CSE, Social Studies, 1983)
12. Discuss the advantages and disadvantages of large families.
    (East Anglian Examinations Board, CSE, Social Studies, 1984)

2 LIVING IN THE COMMUNITY

UNIT 4
Urban Life

GCSE Aims and Objectives

Knowledge — to provide an understanding of what it means to live in a town with special reference to industrialisation, housing, conurbations, amenities and local government.

Application — to interpret and assess the problems of urban life (twilight zones, poverty, traffic congestion, slums, pollution and dereliction) and ways of dealing with those problems (pressure groups, pedestrian precincts, slum clearance, urban renewal and conservation).

Evaluation — to distinguish between the advantages and disadvantages of living in an urban environment.

TOPIC 4.1 INDUSTRIALISATION

Each person first sees life through the eyes of his individual self. Later he realises he is part of a group such as the family. In pre-industrial communities the basic unit of organisation was a small group of people who settled down into a rural community. Years ago it was not unusual for all the people in a village to be related or of the same kin.

In the last two centuries **urban communities** of town and city dwellers have far outnumbered those living in rural areas. The basis of the change in living habits lay in the change in the economic system or the way people work to produce things so that households can be supplied with goods and services.

The **industrial environment** of large towns has brought people both pain and pleasure. The pain has found expression in slums, large-scale unemployment, class warfare, traffic congestion, smoke and pollution. The pleasures are found in improved amenities, shopping centres, entertainment and cultural opportunities. The villager may believe that life in a big city is a rat race, but most townspeople would not change even the stress of the 'rush hour' for what they regard as life in a rural backwater.

People's regard for urban life depends a lot on their position in the class structure, i.e. their particular social class. There is often an enormous gap between rich and poor which remains in spite of the modern welfare state with all its advantages such as unemployment, sickness, death and disablement benefits, family, guardian and maternity allowances, and state pensions for the retired and the widowed.

For centuries the ownership of land was the main source of wealth, and this was especially so when people lived in rural surroundings and relied on farming to provide a living.

Landowners were the wealthy class. Even steeply progressive modern taxes, which have heavily penalised the rich, have not done much to alter the pattern of wealth distribution. In 1981, the richest 5 per cent of the adult population owned 45 per cent of all marketable wealth (*Social Trends No. 14*, 1984). This means that the vast majority (the other 95 per cent) owns only just over half the wealth, indicating a great inequality in the distribution of the country's wealth.

A few of those in the lower socio-economic classes are able to climb up the social ladder. (This is what sociologists mean by **upward social mobility**.) But the social ladder has not yet been replaced by the

A rooftop view of inner-city housing, Elephant and Castle, London

A typical new town housing estate

social highway along which all classes may pass in equal proportions and without bottlenecks and hold-ups. A very large number of people who are born in the dilapidated and derelict parts of towns known as **twilight zones** are deprived – they have poor health, poor education and poor jobs. Professor Peter Townsend has suggested that about 8 million people in Britain grow up in **poverty**.

Britain's new towns

TOPIC 4.2 URBAN HOUSING

Although there are examples of bad rural housing, some of the worst housing conditions are found in large towns and cities. Our slums go back to the early nineteenth century, when the main industrial towns increased rapidly in size leaving a trail of houses built in tight rows, many of them back-to-back, with no gardens and little sanitation. Things are often not much better today in large urban spread where towns join together to form conurbations.

Although 10.8 million dwellings have been built since World War II, millions of people are still badly housed. According to a survey made in England in 1981, 1.1 million dwellings were unfit for habitation while 900,000 houses lacked basic amenities such as flush toilets, bathrooms or running water. There are voluntary organisations, like Shelter, which act as **pressure groups** to draw attention to the plight of the inadequately housed; they exert pressure on the government to do something about it.

Slums are dwellings unfit for human habitation. What can the community do about the slums?

1. Pull down dilapidated houses and build new dwellings on the same site. This is known as **slum clearance**.
2. Modernise the existing property to bring it up to a satisfactory standard. We call this **urban renewal**.
3. Move the people
   (a) to *new towns* such as Basildon and Milton Keynes;
   (b) to *overspill towns* such as Thetford or Sudbury; these are fairly small towns which would possibly benefit from an influx of new life;
   (c) to *expanding towns* such as Ipswich or Peterborough, which are old towns with an increasing population; both are in one of the country's growth areas (East Anglia).

Life in the poorer parts of towns, especially inner cities, spoils people's health – physically, mentally and emotionally. They suffer from **deprivation**. They go without playing fields, recreation grounds – they have insufficient room to breathe. Schools in large towns are more likely to be short-staffed and housed in old buildings with overcrowded classes. In some of our larger towns and cities there are also 'ghettos' – areas in which one particular ethnic community has settled, but poor housing conditions, overcrowding and a high level of unemployment often give rise to severe social problems linked with this deprivation.

There are many towns in the world with living conditions far worse than any experienced in British conurbations. In the slums of Calcutta and Hong Kong people live in conditions similar to the worst days of our industrial revolution.

Yet we have to be careful about trying to move people in large numbers. There are thousands of people who would not change their Coronation Street for the Crossroads Motel. Those with fond associations of a crowded working-class area may not wish to leave lifelong friends. Town planning authorities may try to persuade them to move to the so-called clean air of a tenth-storey flat. Would you like to live at the top of a block of *high-rise flats*? What are the snags? A recent survey showed that about 75 per cent of young children living in high-rise flats do not have children in their own peer group (of their own age – see Unit 2) to play with.

This account of urban housing has shown that the problems are long-standing and that there is much yet to be done. Many of the worst areas are being tackled, but pockets of bad urban housing are likely to remain with us for a long time.

Since the second world war about 1.5 million dwellings have been demolished under the slum clearance programme with most households being rehoused in local authority dwellings. The Housing Act 1980 gave tenants of local authorities a right to buy their council houses if they had been tenants for more than three years. About 330,000 council houses were sold in England and Wales between 1980 and 1982. What effect, if any, do you think this is likely to have upon slum clearance in the future?

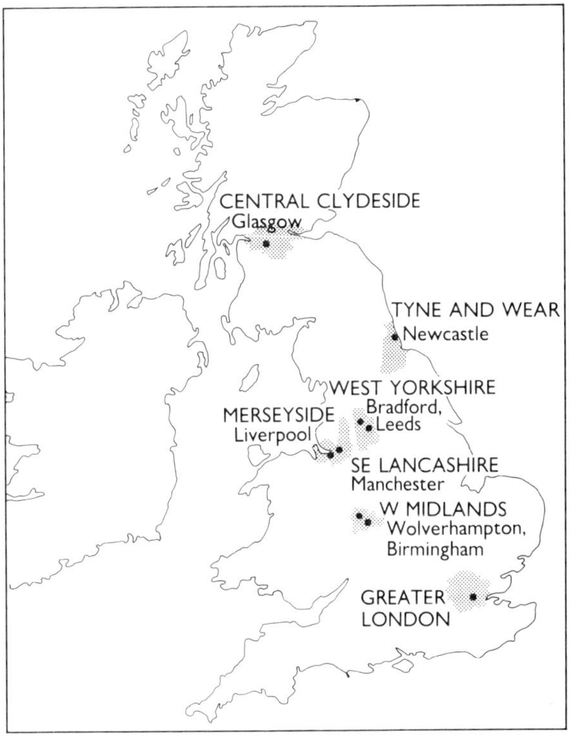

Britain's seven conurbations

TOPIC 4.3 THE CONURBATIONS

'Now 3B, let us all imagine we're in the middle of a great big jungle.'

Part of the Birmingham conurbation

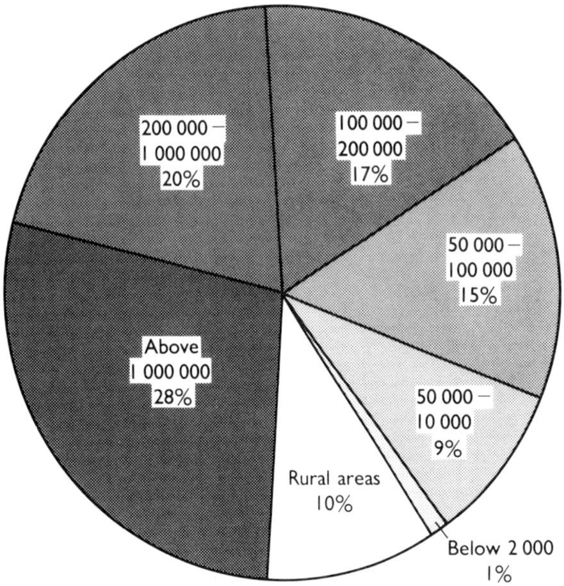

Population distribution in 1981: by type of area and size of urban area (Source: *Social Trends No 15*, 1985)

Although some of the worst housing and living conditions are found in the conurbations, some of the best facilities, amenities and services are also found in these places. The word '**conurbation**' was suggested over fifty years ago by the sociologist and town planner Patrick Geddes. The 1951 Census defined a conurbation as 'a continuously built-up area with no intervening rural land'.

About 18 million of our 56 million people live in the conurbations, so they are pretty crowded places. Some geographers have used the word **megalopolises** to describe them, and the rectangular area joining Merseyside and Thames-side has been referred to as 'the coffin'. It is wrong to think of the seven conurbations as 'much of a muchness'. London, for instance, is very different from the rest because it grew outwards from the centre. Small market towns such as Croydon and Ilford were absorbed by Greater London, but other conurbations resulted from the joining of many towns.

The life-cycles of the conurbations differ, as do the environments and attitudes of people who live in them. For example there is a distinctiveness about Birmingham. The Frenchman de Tocqueville noticed in 1835 that there was 'hardly any likeness' between Birmingham and London. He described the people of Birmingham as never having a minute to themselves. 'They work as if they must get rich the next evening and die the next day.'

Londoners tend to commute from **suburbs** which are high-density residential areas miles away from the city centre. What are the problems of suburbia? Contrast driving through Birmingham, via the M6 and Spaghetti Junction, with travelling along the underground railway tentacles of London's tube system, and you begin to appreciate the different atmosphere of urban life in two of the largest conurbations.

TOPIC 4.4 LOCAL GOVERNMENT

London, as the capital of the country and the largest city, has long had a different form of government from the rest of the country. The Greater London Council (GLC) dates from an act of 1962 which established the first type of conurbation government.

In 1974 the pattern of local government was reorganised to form six new metropolitan counties corresponding to the 5 English conurbations outside London, plus South Yorkshire. The metropolitan counties were divided into metropolitan districts. The Greater London area remained under the councils of 32 London boroughs and the City of London, and the GLC. The Conservative Government passed legislation in 1984, to abolish the GLC and the six metropolitan counties on April 1st 1986. All the metropolitan councils have been Labour controlled and their abolition is closely related to the central government's attempt to control local government expenditure. A White Paper entitled *Streamlining the Cities* contended that there was no real role for metropolitan councils. Although a previous Conservative Government had created the metropolitan councils, according to the White Paper, a later Conservative Government realised that it had been the wrong decision. Most of the services provided by the metropolitan counties have now become the responsibility of the district or borough councils. From April 1st 1986, the borough or district councils in the old metropolitan areas have taken over the functions of highways, housing, waste disposal etc. Local government functions not passed to the boroughs or districts (such as fire, police and public transport) are run by joint boards. The Conservative Government contends that the abolition of the metropolitan boroughs saves money and increases efficiency. However, there are those who believe that the reforms will end up costing more, and even some

One step at a time in rates reform

By John Carvel, Political Correspondent

Ministers agreed yesterday that rates reform will need to be undertaken more speedily in Scotland than in England and Wales. A bill for Scotland alone could be passed before the next election.

A green paper, which is unlikely to be published before the middle of January, will state the government's determination to move the burden of local taxation from property to people.

Conservatives think that the creation of the joint boards (QUALGOS – Quasi-autonomous local government organisations) will be expensive to finance. The controversy over the abolition of the metropolitan counties has caused one of the greatest political debates of the 1980s, and demonstrates how the organisation of local government is an extremely important issue which affects us all.

When local government was re-organised, in 1974, the whole pattern of local councils and services in England and Wales was drastically altered. Towns other than the large conurbations lost the right to govern themselves and were swallowed by 52 larger **county councils** within which there are 375 **district councils**. The old county borough councils (such as Bristol and Norwich) lost their control over important services such as education and housing. The 52 'shire counties' are largely Conservative controlled. Local government services are financed mainly from the **rates** and a central government grant. Councils also receive money from rents and amenities such as car parks, markets and playing fields. The central government uses about 10 per cent of its taxation revenue to help councils. In recent years the idea of introducing a local income tax (LIT), to replace rates, has been suggested. This would be contrary to the Conservative Government's attempts to cut taxes. However, all the other political parties support LIT as do local authority associations including the Association of Metropolitan Authorities, the Association of County Councils and the Association of District Councils.

Central government has accused local councils of over-spending and this particularly applies to urban councils. In 1983-4, central government accused 16 local authorities of over-spending, but the authorities themselves argued that over-spending is a subjective term because they had been elected by local people to provide adequate services. In 1984, the Conservative Government issued a white paper which outlined a **rate capping** scheme. Overspenders will be hit by a penalty system which will result in a loss of central government revenue according to how much is overspent. The White Paper, *Rates*, contends that 'we live in a unitary and not a federal state. Although local authorities are responsible to the electorate, they derive their power from parliament.' However, there are many people who oppose unitary government because they believe that local democracy should be strong and that as many people as possible should be involved in our local democracy.

TOPIC 4.5 URBAN AMENITIES

If you live in a town or city you will have many advantages not enjoyed by country folk. Whether you appreciate them is a different matter; town people are so accustomed to all the benefits that they tend to take them for granted. Let us consider some of the **amenities** of urban life.

No doubt your house is supplied with running water and the dustbins are emptied regularly. How far is the nearest street light from your home? If you want to go into the centre of the town you travel along well-surfaced, frequently swept streets, possibly bordered by trees. It is unlikely that you will walk for more than five minutes before finding a bus stop. You will be safeguarded by patrolling police, either on foot or in a panda car. Leaving the residential area you will reach a shopping centre where most everyday requirements can be bought.

It is being realised more and more that towns are for the people and not for the motor car. Modern **pedestrian precincts** allow people to move around

Milton Keynes shopping centre

the shopping area without worrying about the traffic. Car parks allow people to leave their vehicles and shop at leisure. They are urged to 'park and ride' to the shops by means of public transport. A large shopping precinct will attract many competitive stores.

The offices of the council and local newspaper, cinemas and theatres, hotels and restaurants, the Citizens' Advice Bureau, banks, insurance companies, estate agents and building societies are likely to be found near the town centre. So are the main libraries, art galleries and museums. All these places provide both services and employment. There are many opportunities for jobs in a town, e.g. at hospitals and airports and on industrial estates. Parks and golf courses allow town people to spend their leisure in the open air.

Most towns have problems of noise, congestion and pollution, but they are very convenient places in which to live and work.

TOPIC 4.6 URBAN PROBLEMS

Town planners have to grapple with the problems of urban living:

1. **Traffic congestion** means traffic jams, wasted time and frayed tempers.
2. **Town boundaries** are usually limited by historical considerations. Most modern towns have tightly drawn boundaries and the counties and rural areas naturally resist urban encroachment. Yet there are often thousands of people who benefit from the nearest town. They may work there, or use its services and entertainments – but officially they live outside its boundaries and contribute little or nothing to the services they use.
3. **Suburbia** is often thought of as an insulting term. Life in the suburbs may mean few shops or amenities and the lack of a social nucleus. It is difficult to foster a sense of community when so many people are away from the area all day.
4. **Pollution**. Prophets of doom say that man may not be able to see the sun in the twenty-first century and that ultimately life on earth will be impossible. The prophets will be right if we continue to live selfishly, just for today. We must think of future generations and conserve our resources.

Those who live in urban areas are worst affected. Towns were often sited in a certain place because of the need for water, good communications and industry. These are pollution hazards.

(a) **Water pollution** has increased with industrialisation and overcrowded towns. We have purification and filtration plants where water is treated chemically; but it is difficult to cope with industrial effluents from our factories – ammonia, benzene, detergents, lead and sulphuric acid. If you were to fall into certain rivers or canals there would be a rush to hospital to save your life.

(b) **Air pollution** such as smoke and fumes from transport and industry pollute our air. Cars pollute urban motorways with carbon monoxide exhaust fumes.

(c) **Industrial pollution** is caused by our aim to produce goods at a low cost. It is cheaper to dump potassium cyanide, used in steel production, than to recycle it for future use. Britain is notorious for its industrial pollution; we are regarded by our North European neighbours as 'the dirty old man of Europe' since it has now been proved that it is mainly the emissions from our factory chimneys which causes 'acid rain' and the resultant destruction of forests and pollution of lakes in Scandinavia and Germany. Japan and the United States have strict laws for noxious exhaust fumes but they have not yet been introduced in Europe. Non-returnable bottles are dumped in their millions because it is cheaper than collecting them for sterilisation. Some firms wrap goods in plastic which it is difficult to burn or destroy. Another modern pollution problem is the transport and disposal of nuclear waste as it poses many potential dangers to the community as well as to the environment. Highly radioactive *spent nuclear fuel* is transported from nuclear power stations by road or rail in 'nuclear waste flasks' to Windscale (or Sellafield) where it is reprocessed. The resulting *nuclear waste* then has to be disposed of either by dumping in the sea or burial underground at nuclear waste dumps such as Drigg in Cumbria. We are building up trouble for the future in this way. Have we a duty to future generations? It may be easier for industrial organisations to leave slag-heaps, buildings, canals, and railways and construct new ones, rather than clean up the **dereliction** of the past. But dereliction presents dangers. Children may wander into dangerous buildings or fall into canals.

Spent nuclear fuel flask being transported through a built-up area

The answer to pollution lies in **conservation**. Water must be kept clean and must not be wasted. Chemical fertilisers, sewage and factory waste should not be discharged into water supplies. The international oil crisis of the 1970s made everyone acutely aware of the shortage of our natural resources, and since then there have been many campaigns to encourage people to conserve energy – electricity, gas and coal. Each one of us must be vigilant. Even if we cannot do much on a large scale, at least we can pick up our empty Coke cans and make use of the bottle and can banks provided.

Recently, concern has also been shown about the importance of conserving our rural environment. Specially-designated areas of outstanding natural beauty, National Parks and the work of organisations such as the National Trust and other environmental pressure groups now provide some protection of this most precious 'resource' – the countryside.

Land can be reinstated after the laying of a pipeline

Cleaning oil from a polluted beach after a tanker oil spill

Explain what is meant by

urban communities
industrial environment
upward social mobility
twilight zones
poverty
pressure groups
slums
slum clearance
urban renewal
deprivation
conurbation
megalopolises
suburbs
county councils
district councils
rates
rate-capping
amenities
pedestrian precincts
traffic congestion
town boundaries
suburbia
pollution
water pollution
air pollution
industrial pollution
dereliction
conservation

Case study

THE URBAN POVERTY TRAP

'Where's Sharon Williams again today?' asked Miss Baker, the fourth form tutor.

'Has anyone seen her?'

'No, miss, I ain't seen her for days,' said Jenny, the girl who usually sat next to Sharon in class. 'Maybe her Mum's ill again.'

'Well, I haven't had a note from her parents, I'd better go round and see them.'

As Miss Baker made her way towards the block of council flats on Brownhill Road, she couldn't help feeling slightly nervous. Most of the flats were boarded up and covered with graffiti, and there were piles of rubbish and broken glass everywhere. There was an overpowering stench of urine in the hallway and all the way up the stairs to the ninth floor (the lift was out of order). Sharon herself answered the door, holding a rather grubby-looking toddler in her arms, and very nearly slammed the door in Miss Baker's face, but Miss Baker deftly wedged her foot in the way.

'Hello Sharon, I've come to see why you haven't been in school lately. Is your mother or father at home?' asked Miss Baker.

'I ain't got a dad,' answered Sharon sullenly, 'and me Mum's in bed. It's her who's ill.'

'Oh, I see,' said Miss Baker, 'would it be all right if I had a word with her?'

'I s'pose so. Mum, it's me teacher come to see you,' Sharon called.

Miss Baker then followed Sharon, who was still holding the toddler, into a damp-smelling, scantily-furnished room. As she walked in, a lump came into her throat. There, huddled in blankets on the sofa, lay a pale-faced woman, probably only in her early thirties but looking a lot older, holding a very young baby in her arms. A television was blaring in one corner of the room, and two more children were intently watching some cartoon film.

'I'm sorry our Sharon hasn't been in school,' the woman said weakly, 'but I needed her here to look after the little ones while I'm laid up.'

'But she should be in school, Mrs Williams. It isn't her job to look after the children.' But as Miss Baker spoke, she knew in her heart that anything she said wouldn't help the situation. She'd seen it all before. The eldest child of a large, one-parent family, living in poor housing conditions, playing truant from school, Sharon's life would almost inevitably follow the same pattern as her mother's. She, too, would probably get pregnant at sixteen in an attempt to leave home and start her own life, but more than likely she would fall into the same 'urban poverty trap' (as the sociologists called it) – a trap from which there was very little chance of escape.

What special extra help do you think could be given to the following?

(a) Very large families caught in the 'poverty trap'
(b) Men who have been unemployed for a long time, and who have families to keep
(c) Illegitimate children who feel an unjustifiable sense of shame

Follow-up

1. What is meant by the following?
   (a) ribbon development
   (b) a community spirit
   (c) squatting
   (d) rent rebates
2. Here are some possible reasons for people being evicted from rented property. Choose one and write an imaginary case study.
   (a) keeping pets
   (b) arrears of rent
   (c) having children
3. Pollution is often caused because things are deliberately made not to last ('planned obsolescence'). Give examples of ten such things and suggest remedies.
4. Conservation involves attacking those things that are polluting the earth. Suggest some action which could be taken against
   (a) exhaust fumes
   (b) noisy aircraft
   (c) plastic bottles
5. Write an essay entitled 'Living in the Year 2001'.

Activities

1. Carry out some research at your local reference library to find out how your town has grown since the 1801 Census. Illustrate this growth by rough maps of the area in 1801 and today.
2. Prepare plans for a group of sheltered homes for old people.
3. Compile a questionnaire and use it in a nearby street which includes a variety of houses. Start with these questions and add six of your own.
   (a) age?
   (b) detached, semi-detached or terrace?
   (c) owned or rented?
   (d) central heating?
4. Visit your local town or city hall and find out about future
   (a) slum clearance schemes
   (b) council house programmes
   (c) traffic plans
   (d) car park arrangements
5. Take a good look at your town centre and check the number of
   (a) department stores
   (b) supermarkets
   (c) multi-storey car parks
   (d) large office blocks
   Write 'I' against those increasing in number and 'D' against those that are decreasing.

Examination questions

1. The Town and Country Planning Act makes it compulsory for the public to be consulted before major development takes place. Using examples, indicate the advantages and disadvantages of such an Act, with reference to transport, housing, pollution and other possible problems.
   (*East Anglian Examinations Board, CSE, Social Studies, 1984*)
2. What in recent years has caused air pollution? What is being done to improve the situation?
   (*East Anglian Examinations Board, CSE, Social Studies, 1983*)
3. The New Towns Act of 1946 paved the way for the creation of 23 new towns. Why were these new towns built, and why, in recent years, has there been a move to redevelop older towns?
   (*East Anglian Examinations Board, CSE, Social Studies, 1983*)
4. Discuss the social problems of inner-city areas.
   (*East Anglian Examinations Board, CSE, Sociology, 1984*)
5. (a) In what ways is the modern city congested?
   (b) What problems does this congestion cause for city dwellers?
   (c) Outline some of the attempts by government to solve the problems of congestion. Comment on the effectiveness of these methods.
   (*East Anglian Examinations Board, CSE, Sociology, 1983*)
6. Read the statements which follow, then use them to help you write an essay on the changing pattern of pollution in the United Kingdom, giving reasons for the changes.
   (a) The burning of various fuels and the discharge of industrial waste into rivers are two main sources of pollution.
   (b) Domestic coal-burning still contributes greatly to the output of smoke in the United Kingdom, but has been greatly reduced over the past 20 years.

(c) There has been an increase in the output of sulphur dioxide into the atmosphere, from electricity power stations.
(d) There has been an increase in carbon monoxide in the atmosphere from vehicles using petrol and diesel oil.
(e) In 1970, 54% of non-tidal rivers were classified as unpolluted, 36% as partially polluted, and 10% as grossly polluted.
(f) In 1980, 76% of non-tidal rivers were classified as unpolluted, 22% as partially polluted, and 2% as grossly polluted.

(*East Midland Regional Examinations Board, CSE, Social Studies, 1984*)

7. (a) What is the 'Party System' in Local Government?
   (b) What are the advantages of this system?
   (c) What are the disadvantages of this system?

   (*Associated Lancashire Examining Board, CSE, Social Studies, 1984*)

8. Look carefully at the following groups, each of which can lead to the pollution of the environment:

Group 1	*Group 2*	*Group 3*
factories housing	cars buses trains aeroplanes	restaurants chip shops takeaways

Group 4	*Group 5*
market shops	brick walls iron bridges

   (a) Take each group in turn and name the main types of pollution you would expect.
   (b) Should pollution be reduced in society? Give suitable reasons.

   (*Associated Lancashire Schools Examining Board, CSE, Social Studies, 1983*)

9. (a) What is Local Government?
   (b) In what ways is it different from Central Government?
   (c) In what ways does the general public appear to show a lack of interest and enthusiasm for Local Government?

   (*Associated Lancashire Schools Examining Board, CSE, Social Studies, 1983*)

10. Describe the main housing problems in Britain today and explain how they can be reduced. Your answer should include reference to: finance, slum clearance, amenities, new developments etc.

    (*North West Regional Examinations Board, CSE, Social and Community Studies, 1984*)

11. Describe the problems facing urban communities today. What can be done to solve these problems both by the 'State' and individuals?

    (*The West Midlands Examination Board, CSE, Social Studies, 1984*)

12. Either, (a) What is meant by *urbanisation*? Discuss the respective roles of the New Towns and Green Belts in coping with the problems of urbanisation.
    or, (b) Describe and comment upon the forms of local social stratification found in both a small community and in a large city.

    (*Welsh Joint Education Committee, GCE, Sociology, 1981*)

UNIT 5
Population of Britain

> **GCSE Aims and Objectives**
>
> **Knowledge** — to promote knowledge of demographic problems with special reference to the problems of Britain.
>
> **Application** — to interpret data in relation to life expectancy, infant mortality, birth rate, death rate, dependency ratio, working population and migration.
>
> **Evaluation** — to make sensible judgements regarding population growth, population distribution (age and sex), race relations and population projections.

TOPIC 5.1 POPULATION GROWTH

Since the recording of population statistics began, there has been a steady increase in the population of the United Kingdom; the birth rate has nearly always exceeded the death rate. Early figures were based on contemporary estimates, but from the beginning of the nineteenth century more accurate information has been available. Most of it is based upon two main sources:

(1) the census taken regularly every ten years since 1901 (apart from 1941 when there was no census because of the war) – the most recent census was in April 1981.
(2) the registration of births, deaths and marriages since 1836.

In the period 1975-78, for the first time since records were made, the population fell slightly, reflecting a temporary decline in the birth rate. The upward trend was resumed in 1979 and it is likely that population growth will take place at a much slower rate than was expected a few years ago. Britain's population is expected to be 56.4 million in 1986, 58.4 million in 2001 and 59.6 million in 2016. (Source: *Britain 1984*, HMSO).

The main reason for the steady growth in population has been increased **life expectancy**. Men can expect to live on average until they are about 70 while women are likely to make 76. At least six revolutionary movements in the last 200 years have helped to prolong human life.

These revolutions were not separate events which took place over a certain length of time. They contributed to one another and are still going on.

Revolutions that have improved life expectancy

Revolution	*Improvements*
Agricultural	More food and greater variety
Industrial	Cotton and synthetic clothes, chemicals, soap, power, etc.
Medical	Curative and preventive medicine (inoculation, vaccination, anaesthetics, penicillin, antibiotics, etc.)
Scientific	Applied science gives a pleasant and easier life
Transport	Railways, motor cars, aircraft and hovercraft speedily move goods and people
Trade	Exporting and importing make for higher living standards

Demography is an important feature of Social Studies. This is the study by statistical methods of population growth, birth rates, death rates, fertility and mortality. There has been a marked decrease in **infant mortality**, i.e. the number of children per 1,000 born alive but who die before reaching the age of one.

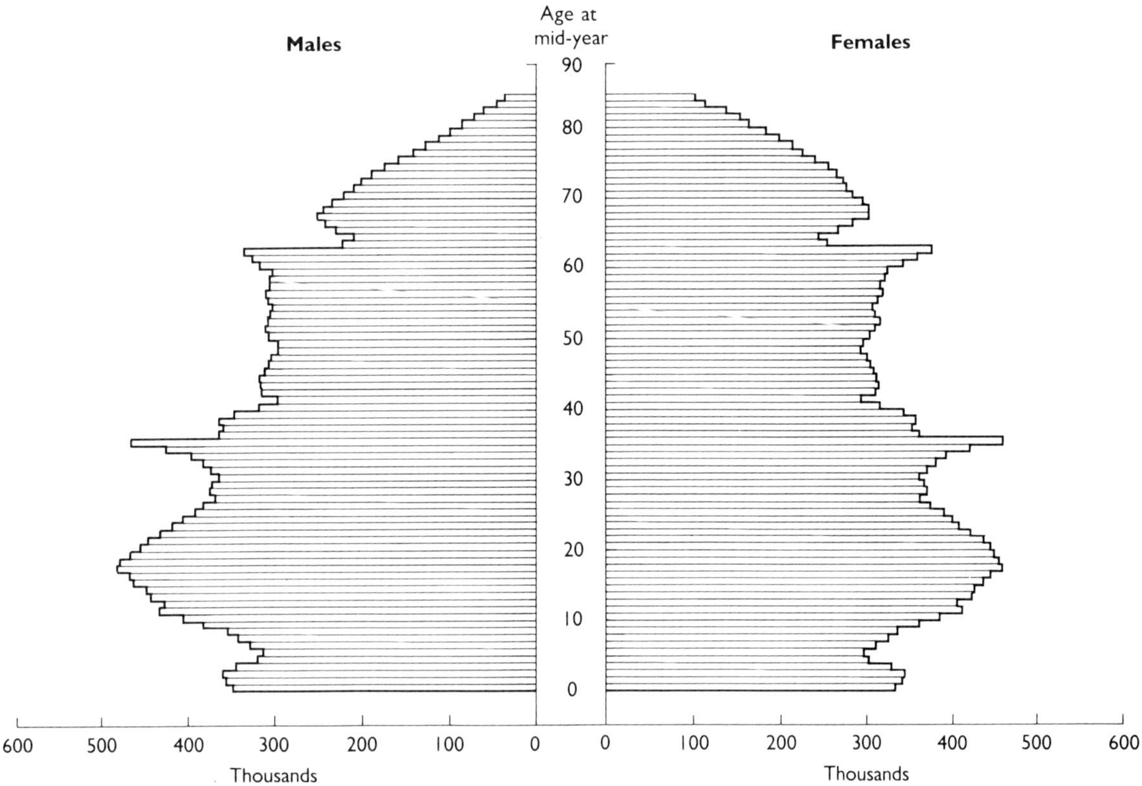

Population by sex and age, Britain 1982 (Source: *Social Trends No 14*, 1984)

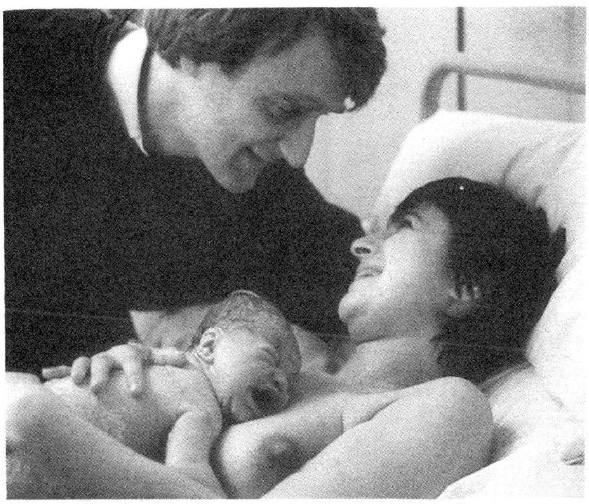

Most babies in Britain are born in hospital. Why?

A century ago the British infant mortality rate was about 150 per 1,000 and it did not really begin to improve much until the twentieth century.

In 1981 it was about 11 per 1,000, which is negligible compared with India's 94 per 1,000. Only a few countries have a better record than Britain. Improved infant mortality rates signify economic and social progress. There are more deaths of young children among the working classes because they are less fortunate than the middle classes in the following ways:

(a) income
(b) standards of hygiene
(c) midwifery and hospital services
(d) housing
(e) general living conditions.

Recalling our basic concepts of heredity and environment, note the marked differences in infant mortality for different races in the Republic of South Africa:

Asians	65
Blacks and Coloureds	107
Whites	28

Rates of infant mortality are closely linked with environment. Countries that have experienced a steady rate of economic growth for a number of years usually enjoy a high standard of living and good social conditions. There is a far greater chance of babies living beyond the age of one in developed countries such as the United States, the Soviet Union and most West European countries, than in South East Asia, Latin America and the Caribbean. An increase in population is usually brought about if

1. the birth rate exceeds the death rate; or
2. the number of people entering a country is greater than the number leaving.

In the 1980s Britain's birth rate is not much greater than the death rate, and there is a *net outflow* of migrants, i.e. more are leaving than arriving.

TOPIC 5.2 THE BIRTH RATE

The **crude birth rate** is the annual number of live births per 1,000 people. It is 'crude' because it does not give an accurate statistic of **fertility**, which is best measured in relation to women of child-bearing age, not to the population as a whole.

The present birth rate in Britain is the lowest since births were first registered. About a century ago the crude birth rate was over 34 per 1,000 but in 1982 it had fallen to 12.6 per 1,000.

The diagram on page 51 shows that the birth rate decreased until about 1940 when it was not much different from the present rate. It was so low at this time that a Royal Commission on population (1945 to 1949) investigated the position and recommended steps to prevent the total population from declining. Some of the measures adopted included:

1. Payment of family allowances, now called child benefit allowances.
2. Parents allowed increased tax allowances for their children.
3. Improved services for mothers such as ante-natal clinics, home helps and nurseries.

After World War II the birth rate rose. Husbands and wives, who had been unable to establish permanent homes because of forced separation, now settled down to rear families. A bulge was produced in the birth rate curve; a smaller bulge took place when those born in the immediate post-war period had families of their own. The birth rate, which had dropped again to 16 per 1,000 in 1951, increased to 18 between 1960 and 1965. We are experiencing very low population growth, but it is impossible to predict accurately what will happen in the future. For instance, important issues are now raised by the

Mr Ian Craft, director of gynaecology at Cromwell Hospital, at a party with mothers and some of the 58 test tube babies born under his 'in vitro' fertilisation programme

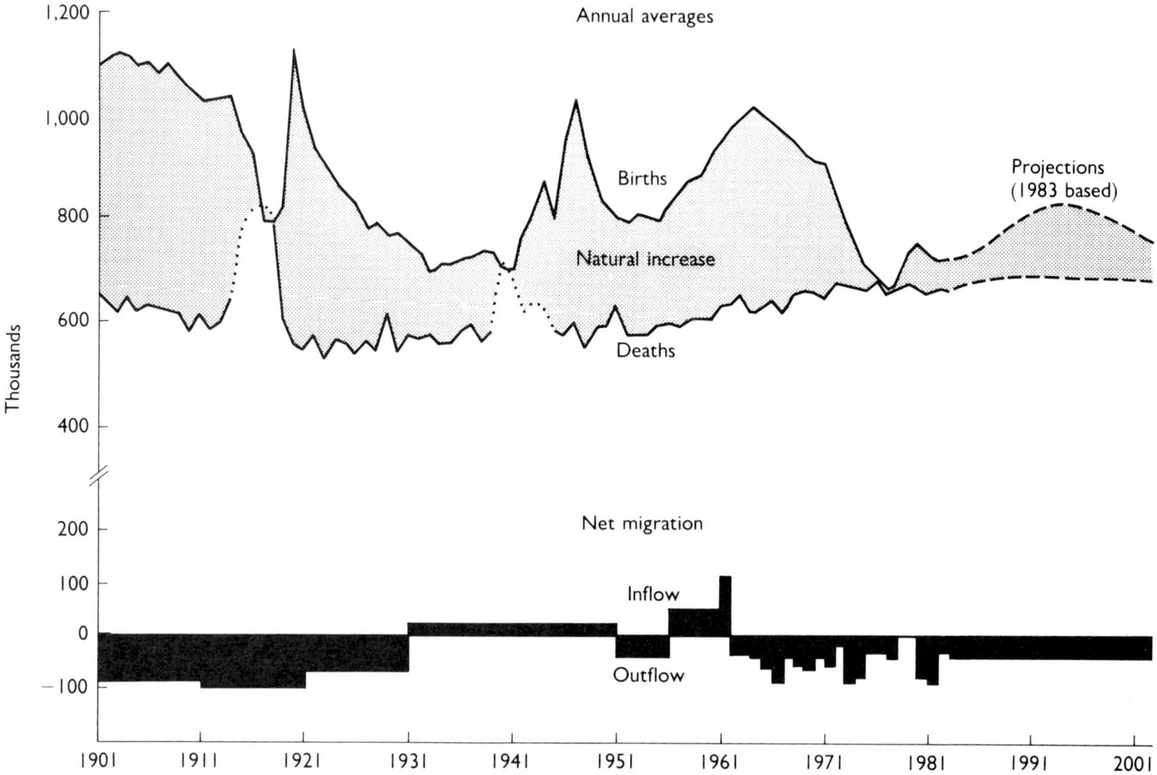

Population changes and projections in the United Kingdom (Source: *Social Trends No 15*, 1985)

pioneering success of birth methods such as artificial insemination by donor (AID) and 'test-tube babies'. What are they?

Possible reasons for a fall in the birth rate

- Improved methods of contraception
- Advice from family planning centres
- Emancipation of women
- Effect of propaganda and organisations such as Population Concern
- More women out at work
- Television
- A drop in fertility (and virility?)
- Legal abortion
- Gloomy economic prospects
- Easier divorce

Changes in the birth rate influence **family sizes**. When the birth rate declines there is likely to be a decrease in the average size of families. Family sizes will depend also on how long people live and especially on infant mortality.

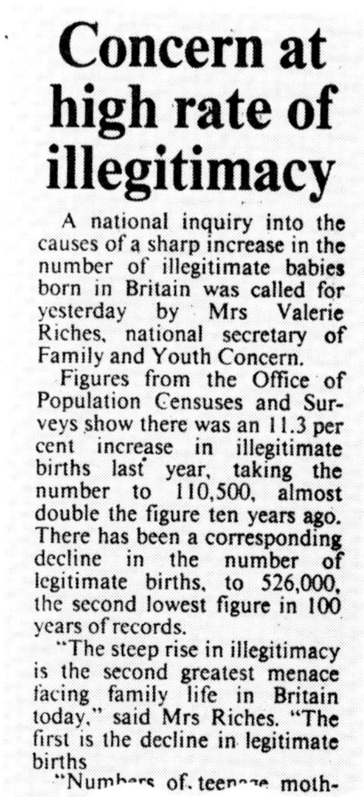

Concern at high rate of illegitimacy

A national inquiry into the causes of a sharp increase in the number of illegitimate babies born in Britain was called for yesterday by Mrs Valerie Riches, national secretary of Family and Youth Concern.

Figures from the Office of Population Censuses and Surveys show there was an 11.3 per cent increase in illegitimate births last year, taking the number to 110,500, almost double the figure ten years ago. There has been a corresponding decline in the number of legitimate births, to 526,000, the second lowest figure in 100 years of records.

"The steep rise in illegitimacy is the second greatest menace facing family life in Britain today," said Mrs Riches. "The first is the decline in legitimate births

"Numbers of teenage moth-

TOPIC 5.3 THE DEATH RATE

The **crude death rate** is the annual number of deaths per 1,000 of the population. Although the death rate has been lower than the birth rate for well over a century at least, the change in the death rate has not been so marked as the change in the birth rate. In 1901 the death rate was 16 per 1,000 while today it is about 12. The death rate is higher for males than for females.

There are two reasons in particular why the death rate has been almost static for so many years:

1. many of the main killer diseases remain active: heart disease, cancer, pneumonia, bronchitis.
2. although some diseases have been considerably reduced, e.g. tuberculosis, compensatory causes of death such as motor accidents contribute to the maintenance of the death rate at a constant level.

If the death rate were to remain at 12 per 1,000 and the birth rate were to drop below this figure, Britain would be faced with the prospect of a declining population.

Possible advantages of a decreasing population

- More space per head
- Decreased danger from pollution
- Less traffic congestion
- More land for agriculture
- Fewer people to feed
- Reduction in imports
- Spare buildings, e.g. houses and factories
- Less strain on health services
- Smaller classes at school
- Improved quality of life

See if you can list ten disadvantages.

'We're in luck, Joe – there's a vacancy!'

There have been four stages in the life-cycle of population growth in Britain and they have been linked with changes in the birth rate and the death rate:

Stage	*Period*	*Birth rate*	*Death rate*
1	Up to 1760	High	High
2	1760 to 1880	Stable	Falling
3	1880 to 1920	Falling	Falling
4	1920 to today	Stable	Stable

TOPIC 5.4 THE SEXES

There are statistical differences between the life-cycles of males and females. In Britain 106 boys are born for every 100 girls, but the *mortality rate* for males is higher at all ages, so among the whole population there are only 94 males for every 100 females.

Why do women live longer than men? Some reasons are suggested below, but you ought to challenge them for yourself. Some are included merely to start an argument.

Do you think there is a modern trend towards similarity between **male and female life-cycles**? The proportion of women dying from lung cancer and in road accidents is increasing. Now that both sexes follow more closely the traditional male life-cycle, men and women are likely to suffer from similar accidents and diseases. Male and female life expectancy may come more into line. At present far more men than women die of bronchitis, lung cancer and heart disease.

Suggested reasons why women live longer than men

- Women's physical make-up allows greater longevity
- Men have more economic and financial pressures
- Men make too much of 'retirement' and often die soon after
- Women do not have such dangerous jobs, e.g. mining or deep-sea fishing
- Men participate more in risky sports
- It is safer for women in house or office than for men in heavy industry
- World War II took a greater toll of men
- Men pursue their pleasures more earnestly and persistently, e.g. sport

PACESETTERS
DON'T
SMOKE
HEALTH EDUCATION COUNCIL

TOPIC 5.5 YOUNG, OLD AND THOSE IN BETWEEN

It is convenient to divide the population into three **age groups**:

Classification	*Age*	*Percentage* (1981)
Young (at school)	Under 16	22
Available for work	16 to 64	63
Old (retired)	65 and over	15

Many classifications used by social scientists are arbitrary. Dividing lines between groups are subjective or largely a matter of opinion; in which of the five social classes would you place *yourself*? (See Unit 2.)

If you are fifteen years old you may not regard yourself as all that 'young', especially compared with children at primary school. Sixty-five-year-olds do not regard themselves as old by modern standards of life expectancy. We tend to think of middle age as being ten years older than we are at the moment!

The majority of those classified as young and old are dependants, and the **dependency ratio** is of great importance in the structure of society. Economists refer to the under-sixteens and those undertaking further full-time education as being not **gainfully employed**. Although some people go on working after the age of sixty-five most of this age group are not gainfully employed.

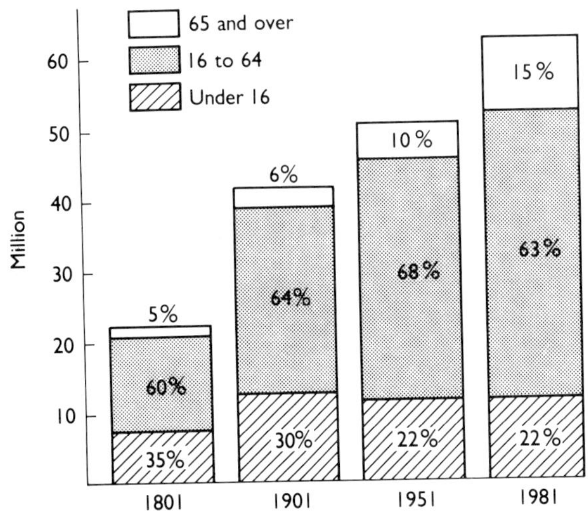

Distribution of the population in the UK by age groups

They have to be kept by the **working population** (i.e. those employed or registered for work). In the United Kingdom the working population makes up 26 of the 56 million people. The dependent population (i.e. children under school-leaving age, women aged 60 or over and men aged 65 or over) reached a peak of 72 per cent of the population of working age in the UK in 1974, shortly after the school leaving age had been raised to 16 (See Figure: *The dependent population*). The dependent population is projected to continue to decline as a proportion of the population of working age until about 1990 and then rise slightly, but it is not expected to rise above 67 per cent again this century.

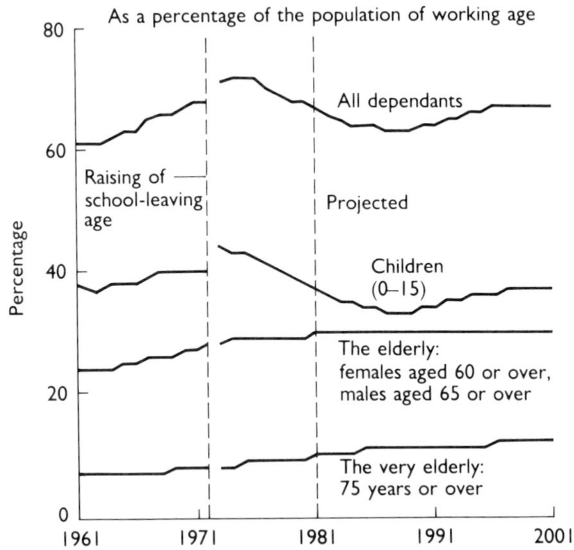

The dependent population in the UK (Source: *Social Trends No 14*, 1984)

As more than half of the population are dependent it is necessary to make the best possible use of those available for work. The table suggests some ways. You should be able to suggest other useful ideas.

Keeping up productivity

- Keep number unemployed to a minimum
- Improved industrial relations to prevent strikes and lockouts
- Modern equipment and productive methods
- Good working conditions
- Encouragement of job satisfaction
- Contented work force – reasonable pay, hours, holidays, amenities, etc.

Do you think people are too old to work at 65? Find out how old President Reagan was when he was re-elected in 1984 and also how old Winston Churchill was when he first became prime minister in 1940 and when he became prime minister again in 1951.

Most advanced industrial countries, such as Britain, have experienced an ageing population. However, the percentage of those over sixty-five will not change significantly over the next thirty-five years. The following table shows some possible social consequences of an ageing population:

- Increased state expenditure on retirement pensions
- More welfare and health services necessary
- Special homes for the elderly
- The gainfully employed must work harder to keep a larger proportion of dependants
- Extension of voluntary services, e.g. meals on wheels, looking after the lonely, etc.

TOPIC 5.6 MIGRATION: EMIGRANTS AND IMMIGRANTS

Migration includes **emigrants** (people leaving the country to live abroad) and **immigrants** (people arriving to live in this country). For centuries we have had more people emigrating than immigrating, although there were two short periods when the flow was in the opposite direction:

1. The 1930s, following the Great Slump and political persecution in Europe.
2. The 1950s, when a large number of coloured immigrants arrived from new commonwealth countries.

In 1948, the first large batch of 400 West Indians arrived to an official welcome. After that, there was such a large net inward migration of coloured people (402,700 Indians, Pakistanis and West Indians from 1955 to June 1962) that the Commonwealth Immigrants Act was passed in 1962. Before this Act thousands of unskilled coloured workers entered Britain. The 1962 Act introduced a strict system of immigration vouchers, and since then the majority of commonwealth immigrants have been skilled.

The main social problem concerning the migration of peoples in recent years has been that of **race relations** following the admission of so many coloured immigrants since 1955. Overall numbers of immigrants from the New Commonwealth and Pakistan (NCWP) have declined from a peak of 55,000 in 1976 to 30,000 in 1982 (Source: *Social Trends No. 14*, 1984). Attempts at estimating the

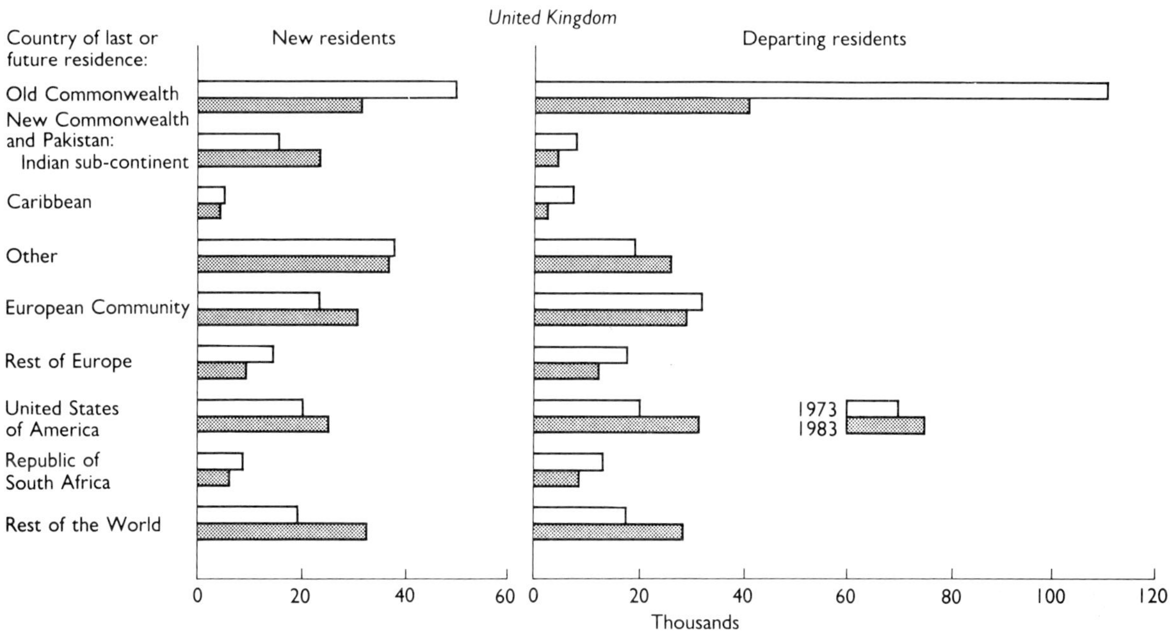

Migrants: by country of last or intended future residence, 1973 and 1983 (Source: *Social Trends No 15*, 1985)

future size of the coloured population have caused much controversy. In 1964 the Economist Intelligence Unit estimated that there would be 3.2 million coloured people in Britain by AD 2002.

Whether there are 2, 3 or 4 million by the end of the century, the important thing is that they must be assimilated and we must learn to live together in a **multi-racial society**. At all costs we must avoid racism and racist attitudes cannot be tolerated.

Ways of helping coloured immigrants to settle

- Adequate housing
- Additional teachers in immigrant areas
- Extra classes in English language
- Highly trained welfare workers, e.g. health visitors
- Education of white population to overcome prejudice
- Legislation to prevent racial discrimination
- Ensure immigrants have a job to come to
- Give consideration to minority tastes and interests
- Encourage community homes and international clubs
- Government grants to aid areas where there is a large proportion of coloured immigrants

Do you think the effect of environment is more important than heredity? We know that working-class youngsters get to the top in many walks of life, and there is little reason to think that coloured immigrants cannot be successfully absorbed into every part of British life. They are already making an important contribution to the community. The health and transport services would almost break down without them.

The Race Relations Act of 1965 established a Race Relations Board and made it 'unlawful to discriminate on grounds of colour, race, or ethnic or national origin in places of public resort'. The Race Relations Act 1968 was much wider in its scope and made it 'unlawful to discriminate in the provision of goods, facilities and services, in employment, housing and advertising'. It also set up a Community Relations Commission (CRC) to help promote racial harmony.

The Race Relations Act of 1976 replaced the Race Relations Board and the CRC by a new body, the Commission for Racial Equality. Instead of concentrating on individual complaints it tackles wider issues of racial discrimination.

The Notting Hill Carnival, held annually in a part of London where there is a large proportion of coloured people, hit the headlines in 1981 when the celebrations suddenly transformed into an ugly street riot.

Racial violence in London

Victims feel helpless in face of daily harassment

By Pat Healy, Race Relations Correspondent

The spate of arson attacks on the homes and businesses of Asian families in London has highlighted dramatically the growth of racial incidents in the capital, and the helplessness that many victims feel.

Last year, the Metropolitan Police recorded 1,289 racial incidents, which is considered by monitoring groups in the London boroughs most affected to underestimate seriously the number of attacks.

While the force was registerring 319 incidents in Tower Hamlets, Community Alliance for Police Accountability (CAPA), the local group operating with limited resources, recorded more than 400.

The Newham Monitoring Project recorded 144 incidents last year, while the police registered 146 for the whole of K Division, which comprises Newham, Havering and Barking.

The police and monitoring groups have different views on what should be registered as a "racial" incident. But there is little doubt that the most serious attacks, which have resulted in families being burnt to death in their homes and a number of near-fatal stabbings, are the extreme end of a pattern of violence and harassment.

In some areas, it has led to Asian families being reluctant to leave their homes for fear of verbal abuse and physical violence from white youths.

About 30 Bengali familes on the Ocean estate in Tower Hamlets have requested transfers to inferior accommodation in the Mile End Road or Whitechapel where they would feel safer. They include "Mr B", one of many victims refusing to be identified, who is now so scared of the abuse he and his family experience that he returns home by taxi if he is out later than 7pm.

"Mr B" came to Britain from Bangladesh in 1958. He says he did not experience the same trouble when he lived in Manchester and Birmingham. Whenever he goes out, white boys and girls in their mid-teens yell "Paki, Paki, Paki" at him, spit and throw stones, he says.

His wife was set upon by youths a few weeks ago in an attack with sexual overtones, which he hesitates to specify because there are young children in the room. He pours out a catalogue of incidents affecting his children, neighbours and friends.

Few of the incidents result in serious injury; it is the cumulative effect of daily harassment that makes life so unbearable.

At a neighbouring flat, a Bengali girl, aged 11 says: "They throw stones at us all the time, even at the little ones. They swear at us."

The girl's mother took her baby, aged six months, shopping a few weeks ago and was attacked by a gang of six youths.

The Home Office recognized the pattern in its survey on racial attacks, which found in 1981 that black people were much more likely to suffer assaults, with the rate for Asians 50 times as high as for whites, and for Afro-Caribbeans more than 36 times as high.

"The frequency of such attacks, often of a particularly insidious nature, and the depth of feeling and concern they generate in the ethnic minority communities, are a matter of fact, and not opinion," the survey said.

The Home Office said yesterday that the Home Secretary and Metropolitan Police Commissioner were very concerned about the level of racial attacks, and dealing with the problem had been made a high priority this year. New recording methods and more effective action was being introduced, particularly in Newham and Tower Hamlets.

The effect on the ground of these new methods is disputed by the victims and the monitoring groups. Three months after the force had assured CAPA that it would take more seriously common assaults where there was a racial overtone, a head teacher complaining about an attack on primary school pupils by older children was told to take out a private prosecution.

The Greenwich Action Committee Against Racist Attacks claimed yesterday that in two recent incidents involving serious injury and damage, the victims had been threatened with arrest, while the attackers were allowed to go free.

● The Commission for Racial Equality (CRE) has called for urgent action regarding racial attacks. In a joint statement last night, Sir Kenneth Newman, the Metropolitan Police Commissioner, and Mr Peter Newsam, chairman of CRE, expressed "particular concern about the recent spate of arson attacks in London".

Dancing in the streets at the Notting Hill carnival

Since then, the organisers and the police have made special efforts to ensure that the event goes ahead smoothly, and now blacks and whites join in the exuberant carnival atmosphere, bearing witness to the fact that good race relations now exist in that area. However, the sudden outbreaks of street violence in Handsworth and Tottenham in 1985 are evidence that work still needs to be done to help integrate young blacks, especially those living in socially deprived areas.

TOPIC 5.7 POPULATION TRENDS

The most accurate trends are to be found in the **population projections** of the Office of Population and Censuses. Since 1920 long-term estimates of population changes have been used as a basis for government planning.

A projection is not the same as a mere forecast or prediction. It is based upon the *latest* figures of birth rate, death rate, migration, etc. These statistics may well change during the period for which the projection has been made; the projection may sow the seeds of its own undoing, because it may influence events. Individuals, the community and government policies may all be changed by a knowledge of the projections. A projection which showed the trend to be towards a large net inflow of immigrants might bring about an increase in emigration. A projection cannot take account of unforeseen circumstances.

Population Projection No. 4 prepared by the Office of Population and Censuses covered the period AD 1973–2013. Using 1973 as the base year it projected the population of Britain to the year AD 2011 as about 60 million.

The projection indicated the following trends:

(a) *Expectation of life of babies born in 1973.*

	Assuming present conditions (in years)	*Assuming 2012–13 conditions (in years)*
Males	68.7	71.2
Females	75.1	77.6

Assuming that present trends continue, people will live a little longer. However, there could be new cures for cancer and heart disease that would improve conditions beyond present expectations. There is no real evidence that this is likely, so the population projection shows mortality trends according to what is expected and not according to what one hopes will happen. 'Guesstimates' may possibly turn out to be more accurate than projections, but if we are to be scientific we must accept the evidence as we find it and not as we hope to find it.

(b) *Migration.* Britain has returned to the traditional migration pattern of a net outflow of people. There was a large inflow of coloured immigrants from 1955 to 1962, but this trend has been arrested by the 1962 Act. Over the decade 1972 to 1981 there was a net loss of 0.4 million people from the UK through migration – 1.9 million immigrants and 2.3 million emigrants.

(c) *Fertility.* Today families are smaller than they were a century ago.

Number of children per family

Year of marriage	*Number of children*
1861–9	6.2
1871	5.9
1890–9	4.1
1900–9	3.3
1941	2.0
1951	2.2
1981	1.75

The total period of fertility rate (an indication of the average size of family) has, in general, been declining and in 1982 was 1.75, which is below the level of 2.1 required for the replacement of the population. Several factors have contributed to the relatively low birth rate. Contraception has become more widespread and effective, making it easier to plan family sizes, and voluntary sterilisation of men and women has also become more common. A larger number of pregnancies (about 40 per cent of those conceived outside marriage and 8 per cent of those conceived within marriage) are ended by legal abortion in Britain.

The postponement of conception within marriage has led to an increase, to 25, in the average age at which women have their first child in marriage. There is a preference, in modern society, for smaller families, especially two children families and this has led to a significant decline in the proportion of families with four or more children.

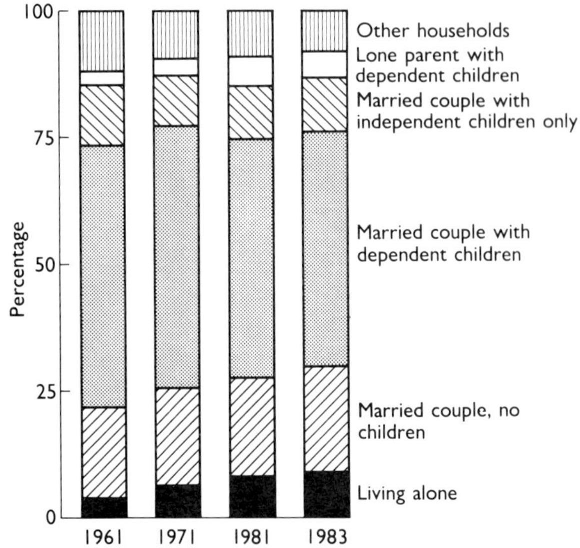

People in Britain: by type of household and family in which they live (Source: *Social Trends No 15*, 1985)

'And this time, lets not leave it so late deciding on names.'

Explain what is meant by

life expectancy	dependency ratio
demography	gainfully employed
infant mortality	working population
crude birth rate	migration
fertility	emigrants
family sizes	immigrants
crude death rate	multi-racial society
male and female life-cycles	race relations
age groups	population projections

Case study

ONLY HALF DEAD

The Medical Officer of Health was giving his annual report to the Happy Valley Council. He was a conscientious man and his report was delivered in precise terms. The MOH had no time to waste on flowery language. He was a believer in statistical data. In his opinion this gave a clearer indication of public health in Happy Valley than accounts of the work done in ante-natal clinics and old people's homes.

'The improvement in the health of the inhabitants of Happy Valley,' he announced, 'is indicated by the fall in the crude death rate from 15.5 per 1,000 last year to 12.5 per 1,000 this year.'

Councillor Bill Huggins looked puzzled. Why couldn't local government officers speak in the language ordinary blokes could understand? He didn't intend to sit there being puzzled by figures. He got to his feet.

'Councillor Huggins,' said the Mayor.

'Mr Chairman,' said Bill, 'would the Medical Officer of Health be good enough to explain what he means by these figures? Could we just know how many people died?'

The Mayor looked sorry for Bill. Surely everybody was capable of interpreting birth-rate and death-rate statistics.

'Well, Councillor Huggins,' he said, 'the MOH means that out of every 1,000 people, 12 died and 5 nearly died.'

1. What did the MOH really mean? Why is it more useful to know the crude death rate than the number who died?
2. What is meant by the following?
   (a) life expectancy
   (b) dependency ratio
   (c) the rate of natural increase
   (d) the rate of population growth
   (e) general fertility rate
3. Find out the main duties of Medical Officers of Health in accordance with the Public Health Act of 1875.
4. Make a rough graph of recent changes in birth rates and death rates in Britain. Indicate the point of zero population growth.

Follow-up

1. Suggest some tips which would give a young person a better chance of living to a 'good old age'.
2. How would you make a young overseas visitor feel at home in this country?
3. Family planning involves using methods of birth control (contraception). In this connection select two of the ideas below and suggest some problems that might be involved.
   (a) expense
   (b) religious beliefs
   (c) health
   (d) agreement between husband and wife
4. A coloured man and a white woman living in your neighbourhood decide to get married. Describe the possible attitudes of
   (a) neighbours
   (b) friends
   (c) relations
   (d) workmates
   (e) enemies

Activities

1. Use the *Monthly Digest of Statistics* (Central Statistical Office), or any other relevant publication available at your local reference library, to find the latest statistics for the United Kingdom relating to the number of
   (a) males between 10 and 19 years
   (b) females between 10 and 19 years
   (c) live births
   (d) marriages
   (e) deaths
2. Conduct a survey among members of your group to find the number of children in
   (a) their grandfather's family
   (b) their father's family
   (c) their own family
3. Use one of the annual issues of *Social Trends* to find the number of immigrants to Britain in recent years from
   (a) Africa
   (b) India
   (c) Jamaica
   (d) the rest of the West Indies
   (e) Pakistan

   Compare the totals for the two most recent periods. Use the statistics to make a pie diagram.

Examination questions

1. (a) What is racial prejudice?
   (b) What attempts have been made to reduce racial prejudice in this country?
   (c) Why does racial prejudice continue to exist in our society?
   (*Associated Lancashire Examining Board, CSE, Social Studies, 1983*)
2. (a) Explain the term *immigrant*.
   (b) For what reasons do immigrants come to the United Kingdom?
   (c) Why have governments attempted to restrict immigration?
   (d) In what ways have governments restricted immigration since 1962?
   (*East Anglian Examinations Board, CSE, Sociology, 1983*)
3. For *either* the local community *or* any non-British society which you have studied, describe and explain the ways in which any *one* NAMED minority group is treated.
   (*East Anglian Examinations Board, CSE, Sociology, 1983*)
4. Study the table below and then answer the questions which follow it.

United Kingdom: death and infant mortality rates to nearest whole number

	1881	1931	1981
Death rate	20	12	12
Infant mortality rate	137	67	11

   (a) Explain the term *death rate*.
   (b) Explain the term *infant mortality rate*.
   (c) Suggest reasons for the changes since 1881 in
      (i) the death rate.
      (ii) the infant mortality rate.
   (*East Anglian Examinations Board, CSE, Sociology, 1984*)
5. Explain, giving reasons, why the average West European family has decreased in size.
   (*East Anglian Examinations Board, CSE, Social Studies, 1983*)
6. Over the last 100 years, there have been significant changes in the age structure and the occupational structure of the population of modern Britain. Taking *either* age structure *or* occupational structure briefly describe these changes. Show how and why they have taken place and examine the likely consequences.
   (*Associated Examining Board, GCE, Sociology, 1982*)
7. (a) What is an ageing population?
   (b) What are its probable causes?
   (c) What are its probable economic consequences?
   (*Associated Examining Board, GCE, Social Economics, 1982*)
8. Describe the reasons that cause people to emigrate to another country. What problems do immigrants to a foreign country face?
   (*East Midland Regional Examinations Board, CSE, Social Studies, 1982*)
9. World population is increasing. What is the cause of this increase? Describe the advantages and disadvantages of a growing population.
   (*East Midland Regional Examinations Board, CSE, Social Studies, 1982*)
10. Give an account of how the birth rate, the death rate, and the rate of migration affect the size of population of Great Britain and of one *named* developing nation.
   (*East Midland Regional Examinations Board, CSE, Social Studies, 1981*)
11. (a) Give reasons why prejudice and discrimination exist in our present-day multi-racial society.
   (b) Explain what measures you think are necessary to overcome these forms of prejudice and discrimination.
   (*Yorkshire Regional Examinations Board, CSE, Humanities, 1981*)
12.

Census Charts the Ageing British People

'During the 1970s, nearly a million more people became pensioners, and by 1981 nearly a quarter of all households consisted solely of elderly people.

'This meant that there had been a 10% increase in Britain's pensioners, with roughly one in six people qualified for retirement and a state pension. Twice as many women as men lived beyond their 75th birthday . . . .

'Despite the fact that 22.3% of the total population was under 16 years old when the Census was taken last April there was nevertheless a 12% drop compared with 10 years earlier . . . .

'One of the largest sociological changes was the 30% increase in the number of households occupied by a person living alone . . . . Nearly a quarter of the population lived alone by the end of the decade.'

From *The Guardian*, 30 June 1982

(a) Using the information in the passage, describe how the shape of the UK 'population pyramid' changed between 1971 and 1981. How does the UK pyramid differ from the pyramid for a country like India?

(b) (i) Explain how the changes in the UK pattern affect the 'dependency ratio'.

(ii) How might these changes influence social planning?

(*University of Cambridge, GCE 'O' Level Sociology, 1984*)

UNIT 6

Law and Order

GCSE Aims and Objectives

Knowledge — to promote a fuller comprehension of the law, the working of the courts and the role of the police in modern society.

Application — to analyse the effects of the British legal system and distinguish between types of law (common, statute and case; criminal and civil), offences (indictable and non-indictable), courts (magistrates, juvenile, crown, coroners, county and high courts).

Evaluation — to evaluate critically the force of the law, the reasons for breaches of the law and methods of punishing offenders.

TOPIC 6.1 IS LAW AND ORDER BREAKING DOWN?

Look at these newspaper cuttings. Do they give you the idea of a lawless society? Many people have the impression that law and order is breaking down because those who break the law get so much publicity. Some newspapers thrive on sensationalism. Their readers are not interested in the millions of people who keep the law. People who keep the law do not make *news*. Stories of hijacking, kidnapping, violence, hooliganism and mugging are given a great deal of publicity. Although there is evidence of increases in more *serious* crimes, the mass of the people are law-abiding citizens.

Some reasons why people break the law

- For 'kicks', e.g. drugs, muggings, etc.
- Unsettled home background, especially in an urban environment
- Greed: 'have-nots' against the 'haves'
- Revolt of individualists against restrictions
- Easily led: mob behaviour by weak characters
- Bad example from parents
- Decline of religion and moral standards
- Abnormal instincts, e.g. perverts
- Think they will 'get away with it'
- Inadequate, physically or mentally

No doubt you have sometimes been told, 'You can't do that. It's against the law.' The speaker means the action you are contemplating is against the body of rules which we must obey if we are to stay out of trouble. If we were in a different country we would have to obey a different set of rules. We live in a democracy and the majority of our laws have been made *by* the people *for* the benefit and protection of the people. Laws are not made simply for the sake of restricting people.

Our laws are considered sensible by the British people or they would not be obeyed. Nearly everybody agrees that it is wrong to kill a fellow human, so we accept that murder must be made illegal.

In a totalitarian state, where a dictator lays down laws, the majority of people may not agree but they obey just the same. The Germans obeyed Hitler because he had the Gestapo and the SS to make sure his orders were carried out. Do you think the German people believed it morally right to exterminate millions of Jews in gas chambers? Is it true to say that people get the government and the law they deserve?

The law banning alcohol at football matches was made for the protection of the people

Mr Bumble the Beadle in *Oliver Twist* said, 'The law is an ass.' If people think that laws are stupid or unjust, they may disobey them. Young people commonly break the law which forbids them to buy alcoholic drinks in public houses until they are eighteen. In 1984–5, the National Union of Mineworkers, and other trade unions, defied the law as amended by the Employment Acts of 1980 and 1982 which limited picketing to the picket's own place of work and restricted the scope for secondary action such as blacking and sympathetic strikes. The Tolpuddle Martyrs challenged the anti-trade union laws in 1834 and were sentenced to seven years' transportation, while the suffragettes willingly went to prison in their fight to secure votes for women.

If we do not like certain laws we must work to get them amended. We can do this by working for the election of a government which will change the law.

In Britain we have three main types of law:

1. **Common law** has been passed down through the ages. For centuries there have been rules commonly recognised to be law. These were rules or laws common to the whole country.
2. **Statute law** consists of laws passed by Parliament and *written* in statute or law books. Every Act of Parliament adds to statute law.

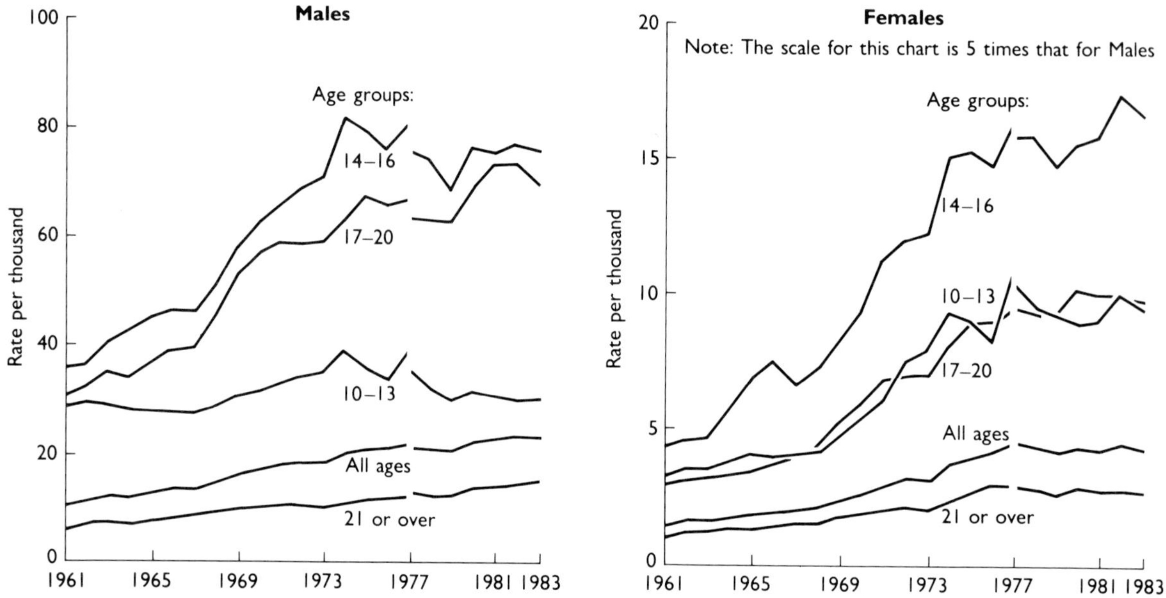

Offenders found guilty of, or cautioned for indictable offences: by sex and age in England and Wales, 1961-1983 (Source: *Social Trends No 15*, 1985)

3. **Case law** is so called because a court case has taken place and a judge has given a ruling on a point of law. The judge bases his decision on the law of the land and the interpretation previously given to it by other judges. The judge's decision forms the basis of future law.

Our laws are also divided in another way: into **criminal law** and **civil law**.

TOPIC 6.2 CRIMINAL LAW

Criminal law concerns offences committed by individuals against the community. These range from serious crimes such as murder to larceny or petty thefts. A person can only be convicted of theft if it can be proved

(a) that he intended to deprive another person of his property permanently;
(b) that he took the property without the owner's consent.

If an owner willingly gives up his property, but is persuaded to do so by deceitful means, it is called fraud.

If a person causes the death of another but it can be proved that he did not intentionally intend to kill, the crime is known as manslaughter.

Actions which break the law are known as offences and are of two types: indictable and non-indictable. **Indictable offences** were once tried before a jury but may now be tried by magistrates if the accused agrees. **Non-indictable offences** are heard in a magistrates' court with no jury. The most numerous indictable offences involve theft, while traffic offences comprise a very large proportion of non-indictable offences.

There are five main types of offence:

1. *Offences against persons*, e.g. assault, wounding (where the skin is broken), grievous bodily harm (GBH) where severe injuries are inflicted, manslaughter and murder.
2. *Offences against property*, e.g. theft, fraud, forgery, malicious damage, and robbery.
3. *Traffic offences*, e.g. failing to obey traffic signs, riding a bicycle dangerously, having faulty brakes and riding on the footpath.
4. *Sexual offences*, e.g. bigamy, rape and incest.
5. *Offences against the state*, e.g. perjury (telling a lie on oath), treason or unlawful assembly.

Some crimes cut across these divisions, e.g. robbery with violence. The robbery is a criminal offence against *property* while the violence is a criminal offence against the *person*. Some of the Great Train Robbers were given thirty years' imprisonment for this offence.

Notifiable offences recorded by the police for 1982. (England and Wales)

	Thousands
Violence against the person	108·7
Sexual offences	19.7
Burglary	810.6
Robbery	22.8
Theft and handling stolen goods	1,755.9
Fraud and forgery	123.1
Criminal damage	417.8
Other offences	3.8
TOTAL	3,262.4

(Source: Home Office, *Annual Abstract of Statistics*, 1984)

TOPIC 6.3 CIVIL LAW

The Civil law deals with the rights, duties and responsibilities of individual members of the community to one another. A civil wrong or injury inflicted by one person on another is known as a *tort*. When two people are in dispute over something they do not usually bring the law into it. If you quarrel with neighbours the matter is likely to be settled without involving any legal authority. If a quarrel gets out of hand an individual may be forced to seek *legal advice*. If your father finds that a next-door neighbour has moved the fence, so taking some of your garden, a solicitor's letter is likely to get him to move the fence back. If a builder provides a £100 estimate for repairs and sends in a bill for £150, the matter can usually be settled *out of court* through solicitors. If the difference between two individuals is very great, it may have to be settled in court.

Throughout this book we have been confronted with the problems experienced by a community of individuals trying to live in harmony. When individuals are in dispute about a **contract** or agreement, then it is the function of civil law to decide on the rights and wrongs of the case. Compensation or damages may be awarded under civil law to any individual who has suffered in any way through the actions of another. Contracts are involved in

1. family law – marriage and divorce
2. company law – formation and dissolving of companies
3. law of trusts – a trustee deals with the property of others
4. law of succession – inheritance

Civil law is concerned with documents ensuring that individuals secure benefits, services or conveniences to which they are legally entitled. For instance, civil law lays down requirements for making *wills*.

Recorded and unrecorded crime in England and Wales 1981 (Source: *Social Trends No 15*, 1985)

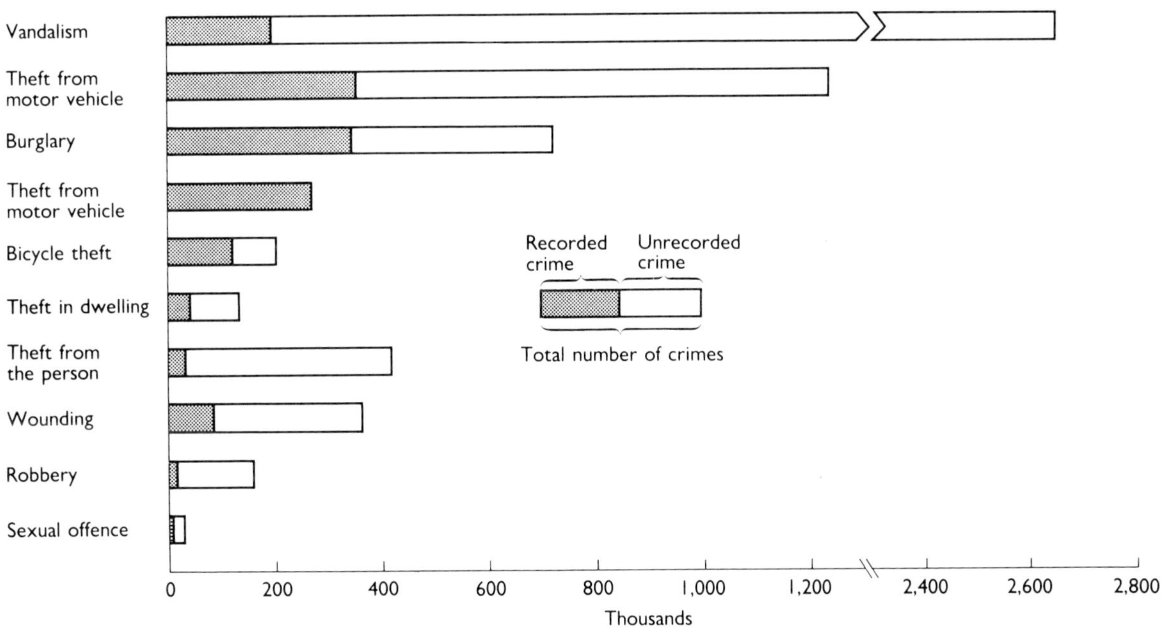

When a person dies, everything he leaves comes under his *estate* and is shared out according to wishes expressed in his will. A person's will is a legal document.

Other examples of civil law include:

1. *Nuisance.* No individual should interfere with other people's freedom to enjoy reasonable rights.
2. *Negligence.* It may be a civil offence if a person is injured by the action or lack of action of others.
3. *Trespass.* It is not a criminal offence to trespass but a trespasser can be sued in the civil courts.
4. *Slander and libel.* Slander means that one person damages another's reputation by making a false statement; libel is a similar offence but the statement is in writing.

Some cases involve both civil and criminal law. If a spectator at a football match beats up a rival supporter so that he is unable to work for several weeks, then the person who caused the injury could be sued by

1. the opposing supporter under *civil law* for damages; or
2. the Crown under *criminal law* for assault and battery.

The civil case would be heard in a civil court while the criminal case would be tried in a criminal court.

TOPIC 6.4 CRIMINAL COURTS

If you have ever been in a courtroom, it is likely that you have been to a **magistrates' court**. There are 23,000 magistrates and they deal with 98 per cent of criminal cases. Magistrates have limited powers of punishment, e.g. they cannot impose more than twelve months' imprisonment, but they investigate more serious offences before passing them to a crown court.

There is no jury in a magistrates' court. Cases are tried usually by three magistrates, although between two and seven may sit. The magistrates base their decision upon a simple majority. If they are in doubt about the accused's innocence or guilt they retire to make their decision. As they are not legal experts, magistrates are advised by the *Clerk to the Justices*. Magistrates, or **justices of the peace** (JPs) as they are also called, are not paid. The appointment of magistrates is sometimes criticised as being shrouded in secrecy, but any individual or group can recommend a person to be a magistrate.

Let us imagine that you are charged with cycling the wrong way up a one-way street. The charge is read out by the Clerk of the Court and you are asked to plead 'guilty' or 'not guilty'. If you plead 'guilty' the police officer prosecuting gives a brief account of the incident. You are asked if there is anything you wish to say in defence (perhaps you did not know it was a one-way street!). The magistrates then decide on the penalty, e.g. a £50 fine, and ask you if you can pay. If you cannot pay immediately the magistrates are likely to allow you a certain period (e.g. twenty-eight days) or they may listen to an account of your means and ask you to suggest weekly amounts. If you offer £5 a week the magistrates may accept or they may insist upon £10 a week and make an order to that effect.

If an accused person pleads 'not guilty' a **trial** is necessary. The prosecuting solicitor describes details of the offence, reads out any statement made by the accused and calls **witnesses** who give evidence on oath in support of the allegations. These witnesses are questioned by the solicitors for the prosecution and the defence. If the accused has no solicitor he may question the witnesses himself. The defending solicitor may be able to call witnesses to support the defence's plea of 'not guilty'. The accused will probably go into the *witness box*, because evidence given by him on **oath** carries more weight with the magistrates than casual statements. The accused can then be cross-examined by the prosecuting solicitor. The prosecution close their case and the defence solicitor puts forward a plea on behalf of the accused. If the magistrates decide there is no case to answer he will

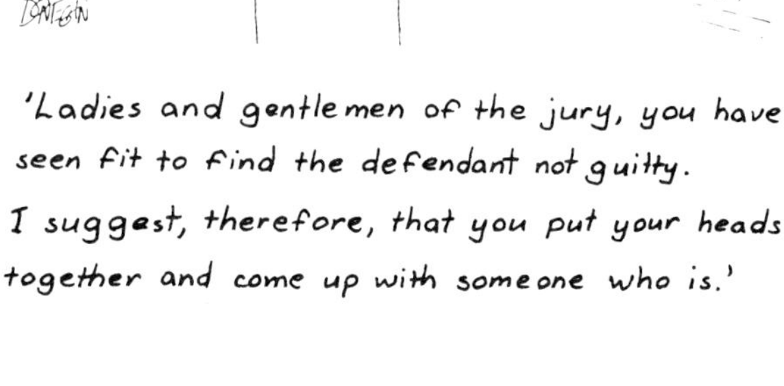

'Ladies and gentlemen of the jury, you have seen fit to find the defendant not guilty. I suggest, therefore, that you put your heads together and come up with someone who is.'

be acquitted. If they decide to convict, consideration is given to the accused's history and his means before **sentencing** takes place.

For and against the magisterial system

For
- Cheap – JPs are not paid
- Without voluntary work much less would be done for the community
- Magistrates are drawn from people experienced in the ordinary business of living
- People uncluttered with legal trappings can best decide on rights and wrongs of a case
- A magistrates' court is better than a 'police court'. JPs are impartial
- Usually men and women sit together in judgement
- Since 1966 they are trained for the job

Against
- Three people try a case which could be decided easily by one paid stipendiary magistrate
- Difficult to find people of the right quality, intelligence and compassion
- Too much secrecy about their selection
- Represent the middle class and have little appreciation of the struggles of the poor
- Too soft with their sentencing (say some of the police)
- Too easily influenced by the police
- Don't move with the times; average age too old

There has been an attempt in recent years to get more people with a working-class background upon the bench of magistrates, and since 1966 there has been compulsory training for new justices. JPs are not expected to be legal experts.

If a person accused of a crime cannot be brought before a magistrates' court within one day, and the offence is not serious, the police must release him on **bail**. He is then discharged temporarily but must appear at a police station or court at an appointed time. Magistrates also have the discretion to grant bail in accordance with the provisions of the Bail Act of 1976. They will not grant bail to people who are unlikely to turn up at a future court hearing or who may interfere with police inquiries if released.

Ajeet Harbans Singh became the first female Indian JP in Britain in 1985

If John Smith is arrested by the police and charged with murdering his wife, although the crime is so serious that it warrants a sentence more serious than magistrates can impose, proceedings would still be heard before magistrates in the first place. These are known as committal proceedings.

If the magistrates find there is a case to answer they will pass it on to a crown court. These courts were established by the 1971 Courts Act. There are crown courts at ninety centrally situated towns. Crown courts are divided into three tiers.

High court judges deal with the most serious cases while circuit judges deal with more serious cases than recorders.

Tier	*No. of centres*	*Officials*	*Cases*
First	24	High court or circuit judges	Criminal and civil
Second	19	High court or circuit judges	Criminal only
Third	46	Circuit judges and recorder	Criminal only

If the accused pleads 'guilty', crown court procedure is very similar to that of the magistrates' court, but the big difference occurs if there is a plea of 'not guilty' and the trial takes place before a jury. Twelve men and women aged between twenty-one and sixty are chosen from a group summoned to appear as jurors. They listen carefully to the evidence and return a verdict on questions of *fact*. As long as ten of the twelve agree, a verdict can be reached. During his summing-up, the judge will give directions to the jury on points of law. If they find the accused guilty, the penalty is a matter for the judge.

A person convicted by a magistrates' court may **appeal** to the Crown Court against the sentence. Appeals from the Crown Court, either against conviction or against sentence, are usually made to the Court of Appeal (Criminal Division). The ultimate appeal is to the law lords in the House of Lords.

TOPIC 6.5 OTHER COURTS

Juvenile courts deal with young persons under the age of seventeen, although no child under ten years of age can be held to be criminally responsible in England and Wales. Young offenders are dealt with in a humane way. The magistrates pay great regard to the welfare of the offender.

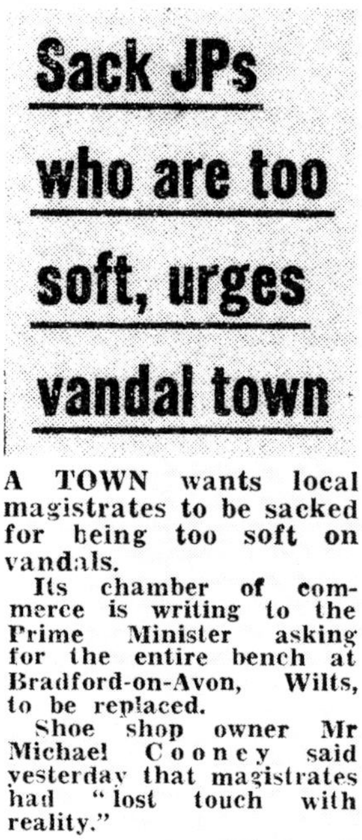

Sack JPs who are too soft, urges vandal town

A TOWN wants local magistrates to be sacked for being too soft on vandals.

Its chamber of commerce is writing to the Prime Minister asking for the entire bench at Bradford-on-Avon, Wilts, to be replaced.

Shoe shop owner Mr Michael Cooney said yesterday that magistrates had "lost touch with reality."

Special points about juvenile courts

- No more than three magistrates
- There must be male and female magistrates
- Held separately from magistrates' court and usually in a different room
- Informal procedure
- Journalists cannot reveal names, addresses, school of the accused, or anything which leads to identification
- Public not present

Under the Children and Young Persons Act of 1969 there are two types of procedure in juvenile courts:

1. *Care proceedings*, if it is thought the young person is beyond control.
2. *Criminal proceedings*, if the young person has committed criminal offences.

Procedure	*Possible decisions*
Care proceedings	Put in the care of local authority Parents bound over Payment (damages or compensation) Guardianship order
Criminal proceedings	Any of the decisions above Absolute discharge Conditional discharge Fine (up to £50) Attendance centre Detention centre Borstal training

A severe penalty for a young person is *Borstal training* which can last as long as two years. The first Borstal was opened at a place of that name near Rochester in Kent. The sentence is indeterminate, which means that if the young person responds to the training he may be freed before his sentence is completed. All those leaving Borstal are released on licence; if they break the law they may be recalled to serve for six months, or the rest of their sentence, whichever is the longer. The strict discipline of **detention centres** provides what is often referred to as 'the short sharp shock treatment'. Life in such centres involves adherence to a strict routine, which comprises a full working week including rigorous physical exercise, so that the young offenders are never left idle. However, although they often emerge fit and healthy, it has not yet been proved that this treatment is an effective deterrent; a high proportion of the youngers re-offend as soon as they are released. *Attendance centres* are sometimes used as a punishment for soccer hooligans. They may have to report at 3 p.m. on Saturdays to some place designated by the court. There they will have to do physical training or some social work. *Remand homes* house young people waiting for a report to be made upon them. Children who have been committed to the care of a local authority, but who have not committed any criminal act, may spend some time in a remand home.

Coroner's courts. About 350 **coroner's courts** hold *inquests* on unexplained deaths. A coroner should be a doctor or a member of the legal profession. A coroner's court does not try a case, but merely attempts to establish the facts. A post-mortem may be ordered on the occasion of a sudden death; an inquest will then be held if death was not from natural causes. If it is found at the inquest that Mrs Smith's death was due to murder by her husband, then Mr Smith will go for trial in the criminal courts.

Civil courts. Minor civil cases are dealt with by **county courts**, while more serious cases go to the **high courts**.

There are 375 county courts in England and Wales with over 100 county court judges. Each judge has a *circuit* covering as many as twelve towns. In busy towns there are daily sittings, but other county courts are held weekly or monthly. County courts deal with cases where the disputed sum is less than £750, hire purchase disputes, adoption and divorce cases, etc. A person wishing to bring a civil action has to decide whether to take it to a county court or a high court.

Since the 1970 Administration of Justice Act, the branches of the High Court are:

1. Family Division, dealing with adoption, wardship of minors, guardianship and matrimonial disputes.
2. Chancery Division, dealing with wills and estates.
3. Queen's Bench Division, dealing with the greatest number of cases, including all those not allocated to other divisions of the high court.

TOPIC 6.6 PUNISHING OFFENDERS

Why do we punish offenders? It is not just because their behaviour is different from everybody else's, i.e. the norm. We know that people's behaviour may be

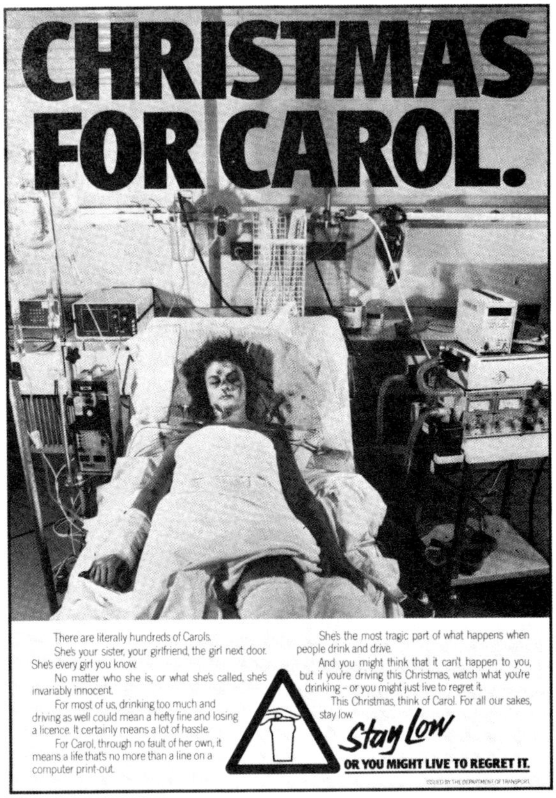

Campaigns make people realise the harm that drinking and driving causes to innocent members of the community

social or anti-social, i.e. *for the community* or *against the community*. Society does not object to a Sikh wearing a turban. But it did object when Sikhs wore turbans on motor cycles instead of the crash helmets that were required by law. Is it anti-social not to wear a crash-helmet? A Sikh might argue that he was not doing anybody else any harm. Yet if he had a serious accident an ambulance might have to attend. If treated in hospital he would take up medical services which could be used for other members of the community. Similarly, driving while under the influence of alcohol is illegal because a drunken driver can easily cause death and injury to ordinary members of the public, as well as to himself and his passengers. We are all interrelated and we are faced again with the problem of having to restrict the freedom of one person for the sake of the community.

We punish people for many different reasons:

1. To deter offenders, e.g. more road deaths occur with fast driving, so if a person is fined for speeding he may think twice before he speeds again.
2. To protect the rest of the community, e.g. a man may be bound over to keep the peace.
3. To improve an individual, e.g. Borstal training includes learning a useful craft as well as learning to behave.
4. To give another chance, e.g. a suspended sentence.
5. To help, e.g. probation officers give guidance.
6. To make good any damage done, e.g. compensation to victims.
7. To ensure safety of the public, e.g. a driver may lose his licence because he drives dangerously.
8. To allow society to take its revenge, e.g. capital punishment (hanging).

I am not too happy about the last reason. Are you? Although Parliament suspended the death penalty for murder in 1970, following the Brighton Hotel bomb outrage of 1984, opinion polls indicated that a majority of the general public were in favour of hanging IRA bombers. Do you believe in hanging terrorists who place bombs near innocent people? Would hanging stop the bombing or bring more terror? No one has been hanged in Britain for several years, but legally it is still a possible sentence for piracy and for setting fire to ships in naval dockyards.

Fine default by jobless often means jail term

By Our Home Affairs Correspondent

Many unemployed fine defaulters have little or no prospect of paying them, the National Association for the care and Resettlement of Offenders said yesterday.

Studies showed that 60 per cent of those fined for indictable offences are unemployed, the association said. There are 200,000 fines a year in England and Wales for indictable offences.

Unemployed offenders were less likely to be fined than others, but were more likely to be imprisoned for default accoring to one Home Office study. That was in 1978, before the present levels of joblessness. The study was based on questionnaires to 368 people fined for offences.

It is estimated that in 1983 400 men and 120 women were imprisoned for failing to pay fines imposed for using a television set without a licence, a Nacro briefing paper said yesterday.

The paper says that a quarter of all sentenced offenders entering prison and 45 per cent of those given sentences of six months or less are imprisoned for fine defualt.

A total of 23,241 people were received into prison as fine defaulters, and there were on average 900 fine defaulters in prison on any one day.

The fine is the most common sentence used by the courts. In 1983, 1,699,900 people, 84 per cent of all persons convicted, were fined, the briefing paper says.

Of people convicted of indictable offences 43 per cent were fined, more than a million of them for motoring offences.

The number of fines for indictable offences has fallen from 218,100 in 1977 to 199,000 in 1983. As a percentage of all sentences that is a drop form 51.2 per cent to 43.1 per cent.

Fine Default: Nacro, 169 Clapham Road, London SW9 OPU (free).

The hanging argument

For

- The murderer cannot kill again
- It deters others
- An imprisoned murderer might escape
- Cheaper than keeping murderers in prison
- Retribution – a life for a life
- Majority of people think murderers should hang
- Protects police and prison officers
- Gives consolation to relatives of the victim

Against

- Wrong person may be hanged
- Statistics indicate it is not a deterrent
- If it is wrong to kill, then the state should not do it
- Terrible effect on those who carry out sentence
- Against Christian ethics – i.e. belief in love (not an eye for an eye)
- Purpose of punishment should be to rehabilitate
- Brings more violence, e.g. if IRA terrorists were hanged
- Murderer is often mentally ill

Sentences in Britain range from conditional discharge to imprisonment for life:

1. *Discharge.* An absolute discharge carries no punishment, but it is more usual to have a conditional discharge, which is conditional on there being no further offence committed for a period not exceeding three years.
2. *Fines.* These are the commonest punishments. If the maximum fine fixed by law is £100, for a first offence, the actual fine could be about £20, depending upon circumstances and previous record.
3. *Probation.* The offender agrees to be the subject of a **probation order** for a certain period, e.g. two years. He promises to live an industrious life, to report to his probation officer regularly and accept his advice.

4. *Suspended sentence.* If an offender is found guilty of another offence during a period of suspended sentence, he is liable to be made to serve the original sentence. He is put on trial for a period which can be up to three years.
5. *Custodial sentence.* Offenders are not usually sent to prison for a first offence unless it is very serious. Those serving more than one month (apart from 'lifers') have one-third of their sentence remitted unless remission is reduced for a breach of prison discipline. *Parole* was introduced by the Criminal Justice Act of 1967; this means that a prisoner who has served one-third of his sentence or twelve months (whichever is the longer) may be released on licence if he satisfies the Parole Board that he has made progress in prison and his release will not endanger the community. The custodial sentences for young people include detention centres for males from three weeks to four months and youth custody sentences for both sexes.

Victims of crime to get new deal

By Peter Evans
Home Affairs Correspondent

A new deal for victims under which reparation can be agreed with the offender who has wronged them was launched yesterday by Mr David Mellor, Under Secretary of State at the Home Office. If mediation leads to agreed reparation, the court will take it into account in passing sentence.

The experiment, one of four the Government is funding, is at Wolverhampton. Mr Mellor hoped that in some cases at least the court would feel able to pass a lesser sentence than it might otherwise have done to mark the offender's efforts.

"If not, however, the victim will still benefit. And so, too, I hope, will the offender, through the realization of what he has done."

Referring generally to reparation, Mr Mellor said that in some cases the victim may be satisfied with an apology, once he is reassured that the offender is genuinely sorry for what he has done.

The Home Office divides prisons into four categories ranging from *maximum security prisons* in Category A to *open prisons* in Category D. Prisons such as Wandsworth and Parkhurst were made maximum security prisons after the Mountbatten Report. Dangerous criminals and those serving long sentences are held in a carefully guarded 'prison within a prison'. Prisoners who are unlikely to try to escape, such as those who have made one slip in an otherwise unblemished life, may be sent to an open prison such as Ashwell prison which has boundaries but no prison walls. Although a few new prisons with modern facilities have been built recently to cope with the increase in numbers of criminals receiving custodial sentences, the majority were built over a hundred years ago, and conditions are primitive. This situation is worsened by the problem of overcrowding. Sometimes up to six prisoners are confined for 23 hours a day in cells which were intended to take just two inmates.

What do you think of the ideas set out below?

Some possible changes in treatment of offenders

- More community service instead of fines or imprisonment (this was started under the Criminal Justice Act of 1972)
- Indeterminate sentences, i.e. until authorities are satisfied that prisoner has been rehabilitated
- Prisoners to do more useful work and be better paid so that they have a greater chance when released; this is being tried at Coldingley
- Family courts where magistrates and parents discuss a case together
- Increased opportunities for educational training and psychological treatment (Grendon Underwood is a small psychiatric prison hospital)
- Spend more on probation and after care (in 1985–6, Home Office budget provision was £2.8 billion for the police, £590 million for the prison service and £175 million for probation and after care)
- Less use of the dock in courts; it has been described as a giant cage
- Extension of legal aid
- Youth Treatment Centres for children severely disturbed and disruptive; three centres are in use for offenders aged 12–19
- Improved prisons to prevent overcrowding
- Offenders to meet their victims; direct confrontation induces a sense of shame.

TOPIC 6.7 THE POLICE

Have you any rules at your school or youth club? It is no good making rules which cannot be enforced. We saw in Unit 2 how even teenage gangs have rules or unwritten laws. The main job of the police is to make sure that the law is kept and that people behave in an orderly way. There can be no worthwhile economic or social life without order. The continuation of socially useful work in hospitals, schools, youth clubs, community centres, etc., necessitates the maintenance of law and order.

A policeman is like a football referee. When you play without a referee arguments develop:

'It wasn't in!'

'That was a dirty foul.'

'I'm not going to play, then!'

At professional football matches the referee comes in for a lot of 'aggro', but his role is to see that the game continues in an orderly way. The same is true of the police in modern society.

Probation officers criticize police

From Peter Evans, Home Affairs Correspondent, Bridlington

Probation officers from trouble-torn Handsworth and Tottenham made allegations of police misconduct to their annual conference at the weekend.

Mr Basil Hylton, who is responsible for the Handsworth area, said the police there could take pride in having pioneered community policing.

But they could also demonstrate themselves as agents of provocation and even violence, yet be assured of "business as usual" with agencies such as the probation service.

"We are not anti-policing, but we wish police to be servants of the community as a whole – and to be upholders of, and adherents to, the law," Mr Hylton told the National Association of Probation Officers (Napo) meeting at Bridlington.

Police in violent clashes with pickets during the miners' strike of 1984

The police force is made up of different branches – uniformed police, the CID, mounted police and river police – but they are all concerned with the job of seeing that the law is kept. It is a thankless job, especially as they are often understaffed. In England and Wales at the end of 1984 there were 120,573 police officers, giving a ratio of 1:411 of police to population.

The police now have to cope with endless different situations, anything from sieges and street riots to violence on union picket lines, as well as their normal duties. An added problem is the increase in criminals' use of firearms (see figure below). This means that the police, not usually armed, are increasingly forced to use firearms to protect themselves and the public. However, there have been cases of mistaken identity where an innocent person has been shot by the police, and such cases have naturally caused public outrage.

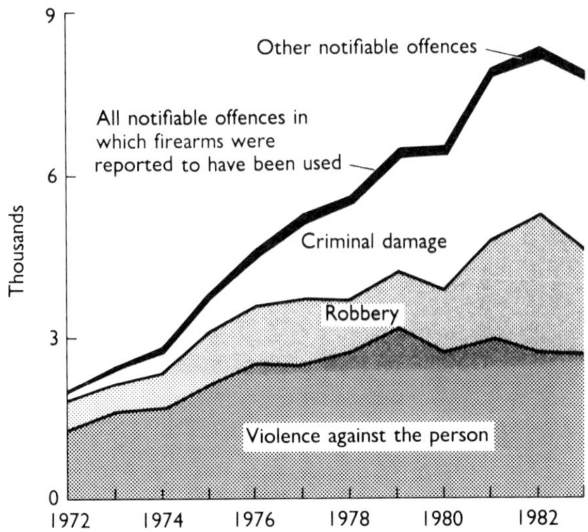

Criminals' use of firearms: by type of offence (Source: *Social Trends No 15*, 1985)

The police are there to see that rules are kept, but we often act instinctively against rules. If a notice in a bus says 'Do not spit' we may feel like spitting just to show them!

If we are to obey the rules we have to appreciate the purpose of them. This is why we have studied the law before turning to those who have the difficult job of seeing that people keep it. If we don't want to be murdered, we will support the police because they try to stop murders. The police are our paid representatives in preserving law and order, but in the end it is the job of each of us to see the law is kept.

The duties of the police are so numerous that we can only look at a few:

1. *Crime prevention.* During their training the police learn that **crime prevention** is their primary duty. If the police can prevent crimes they do not have to catch criminals.
2. *Public relations.* A peaceful atmosphere in the community makes the job of the police easier. They are there to help *most* people and hinder only a *few* (i.e. the few who are *anti-society*). A policeman might help you to find your way when lost. They try to win the understanding of young people.
3. *Traffic.* We live in an urban environment with about 20 million motor vehicles, in addition to millions of cyclists and pedestrians. If the police did not regulate traffic there would be chaos. One of the problems of the police in recent years has been antagonism between them and motorists, caused by the increasing number of traffic offences. This is hardly the fault of the police. They have to keep traffic moving. About 5,500 people die in road accidents every year and over 300,000 are injured, 72,000 of them seriously, so there must be laws to prevent road-users behaving foolishly. The police have to prosecute motorists who break speed limits. Those who drink and drive are *anti-society*. Many recent laws such as the breathalyser, minimum tread on tyres, compulsory seat belts and MOT certificates have burdened the police with additional responsibilities. But the laws are passed by Parliament which is elected by the people. The police do not make the laws; they have the unenviable job of enforcing them.

   It is often not appreciated that the police deal with far more traffic offences by verbal cautions than by arrests or summonses.
4. *General help for the public.* By 1985 there were about 12,000 policewomen and they are especially important in times of distress.
5. *Police patrols.* The policeman on the beat is trained to deal with a thousand and one things. He has to check that business premises are locked; he may have to arrest a murderer. Although panda cars and walkie-talkie devices improve communications, the commonplace job of the uniformed policeman on the beat is still vital. The village policeman keeps a friendly eye on rural life and often knows all the people in the village.

A friendly policeman on the beat is an important member of the community

In England and Wales, the police are assisted in their work by 16,000 *special constables* and 5,000 *traffic wardens*. There are also 39,000 civilian staff who are not members of the force and about 500 police cadets.

A chief constable is responsible for all decisions relating to his force. The local **Police Authority** (two-thirds local councillors and one-third JPs) looks after equipment and accommodation but has no control over the functions of the police.

Chief constables are responsible to the **Home Secretary** for the smooth running of the force. The Home Secretary has direct control over the *Metropolitan Police* but in practice the Commissioner of Police is the officer responsible for Greater London. Who is the current Commissioner of Police?

Explain what is meant by

common law
statute law
case law
criminal law
civil law
indictable offences
non-indictable offences
contract
magistrates' court
justices of the peace
trial
witnesses
oath
sentencing
bail
appeal
juvenile courts
detention centres
coroner's courts
county courts
high courts
probation order
crime prevention
chief constable
Policy Authority
Home Secretary

Case study

CRIME DOESN'T PAY

Andy Summers was what was known as a problem child. Now aged sixteen, he looks like a typical boy of his generation with his spiky, dyed hair and trendy clothes, but a few years ago nobody quite knew what to do with him. Coming from a broken home, he was in and out of care as a young child, and in and out of trouble as a teenager.

It all started when he realised how easy it was to steal sweets and cigarettes from the local paper shop. All his friends did it too, but it was Andy who got caught. He got off with a warning the first time, but when he tried to steal a car he was caught redhanded by a policeman on the beat; Andy ended up in a detention centre. He hated it there; he had to have his hair cut short and he was on the go from 6 a.m until lights out at 9 p.m. He stayed there for three months and as he left, he swore he would never go back.

But Andy wasn't known for his luck. After only two months of going straight, he was up before a juvenile court, this time for breaking and entering. Another four-month spell in detention centre convinced Andy that he had to go straight this time, as he was put on probation. On his first weekly visit, the probation officer asked,

'Assuming that you don't want to spend the rest of your life in prison, Andy, what are you going to do with yourself?'

'You tell me, mate!' retorted Andy. 'How will I ever get a job with my record?'

'Well, I do have one suggestion. There's a new coffee bar that's just opened in town, to keep kids like you off the streets, in fact. It's funded by the social services department. We need some more help with clearing the tables and washing up. It's dirty work – do you think you'll stick it?'

Andy made a face, but he thought anything was worth a try. In fact, after his first day, he decided that he quite liked it. After his experience of detention centres, he was used to the dirty work, and here he could chat to his mates who came in for coffee and listen to his favourite records on the juke box all day!

At the end of the week, he got his first wage-packet. It wasn't much, but for Andy it was his first honestly-earned money, and it gave him a real thrill to count out the fivers. For Andy, it was definitely a case of 'crime doesn't pay, but honesty does!'

1. What can society do to help the 'problem children' like Andy?
2. What reasons can you put forward for the probation service keeping in contact with people released from jail?
3. Give three reasons why it is difficult for an ex-prisoner to hold down a job. Name some organisations which help ex-prisoners.

Follow-up

1. Give five possible reasons why more people seem to be breaking the law.
2. List laws which you consider
   (a) are out-of-date
   (b) ought to be passed
3. Why is mugging such a 'bad' crime?
4. Write notes to be used in a debate on the motion 'Criminals are made, not born'.
5. Suggest reasons why women commit fewer crimes than men.

Activities

1. Write an account of a visit to a court.
2. Make a map of a policeman's beat in your area.
3. Use dictionaries and/or reference books to find the meaning of:
   (a) legal separation
   (b) paternity
   (c) stipendiary magistrates
   (d) summary punishment
4. Carry out a survey of your group to find their ideas on attitudes towards the police.

Examination questions

1. (a) Name and briefly describe the four main functions of punishment.
   (b) What have been the main changes in policing methods which have occurred in this country during the past 25 years?

   (*Associated Lancashire Schools Examining Board, CSE, Social Studies, 1983*)

2. According to official statistics most crime is committed by young working-class males. Why is this so?

   (*East Anglian Examinations Board, CSE, Sociology, 1984*)

3. (a) Describe the difference between deviance and crime.
   (b) How do the following factors contribute to delinquent behaviour:
      (i) family;
      (ii) peer group;
      (iii) environment?
   (c) Suggest why it is that reported crimes of violence have increased since 1960.

   (*The Associated Examining Board, GCE, Social Science, 1983*)

4. What methods of punishment and reform are available for sentencing those who commit crime? Suggest with explanations, possible ways in which these methods could be improved.

   (*East Anglian Examinations Board, CSE, Social Studies, 1984*)

5. Football supporters and political extremists are often the cause of crowd violence and vandalism. Explain with reasons why you think this is so, and describe how such behaviour should be prevented or punished.

   (*East Anglian Examinations Board, CSE, Social Studies, 1984*)

6. Fully explain *three* different types of criminal behaviour and the normal punishment which follows. Discuss whether the punishments fit the crimes.

   (*East Anglian Examinations Board, CSE, Social Studies, 1983*)

7. (a) Write what you know about *three* of the following:
      Borstal
      Open prison
      Suspended sentence
      Crown Court
      Probation Officer
      Community Service Order
   (b) A fifteen year old steals some money. Briefly describe what must happen before that young person is convicted.
   (c) Do you think that the present Government's intention to make punishment more severe for some young offenders is likely to reduce crime? Give reasons for your answer.

   (*The South-East Regional Examinations Board, CSE, Civics – Citizenship and Social Affairs, 1981*)

8. Kevin Smith is an unemployed seventeen year old who likes to drink beer in local pubs with his friends. Unfortunately for him, beer is expensive and his money from Social Security is not very much, so he sometimes steals money and other things from people's houses. One night, as he is sitting with a pint of beer in the Red Lion, the police come in and tell him that he must come with them to the police station in order to answer some questions.
   (a) For what offence, other than theft, could the police charge Kevin?
   (b) If the police decided to charge Kevin with theft, at which Court would he first appear?
   (c) Decribe what might happen in Court if the offence of theft is:
      (i) only a small one
      (ii) very serious.
   (d) In what ways might Kevin be at a disadvantage in trying to defend himself against the police accusations? What can he do in order to help defend himself?
   (e) Do you think that the police should be allowed to break the law if it results in more convictions? Give reasons for your answer.

   (*The South-East Regional Examinations Board, CSE, Civics – Citizenship and Social Affairs, 1983*)

9. Crime is now generally agreed to be a major problem. The number of crimes known to the police has increased greatly in recent years. There has also been a large increase in the number of court cases and in the number of persons in prison.
   (a) What *three* reasons are given for regarding crime as a problem?
   (b) Give *three* reasons why crime rates may have increased.
   (c) Name any *one* type of law-breaking activity and describe how the police try to prevent or control it.

(d) Describe *two* alternative sentences which can be given by a Court in order to keep offenders out of prison.
(e) Why do some people believe that it does not help to send certain types of offender to prison? Your answer should consider the reasons why we send people to prison.

(*The South-East Regional Examinations Board, CSE, Civics – Citizenship and Social Affairs, 1982*)

10. (a) Describe the work of *three* specialist departments of the police.
(b) Criticism of the police has grown in recent years. Briefly describe *two* events which have encouraged these criticisms.
(c) What would you say to a critic in defence of the police? In what ways might the work of the police be more difficult than in the past?

(*The South-East Regional Examinations Board, CSE, Civics – Citizenship and Social Affairs, 1981*)

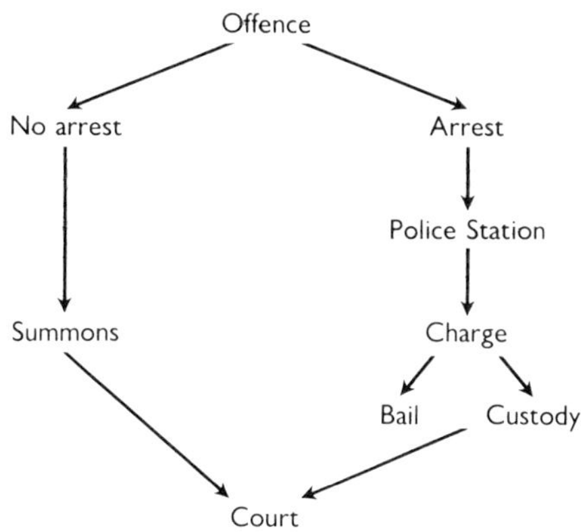

11. (a) Using the above chart, explain what might happen to an adult who is found in suspicous circumstances near the scene of a **serious** crime. At each stage you should say what are the rights of the accused person.
(b) Describe **four** possible sentences which a Court could impose if the accused is found guilty of the serious crime.
(c) If **you** were the Judge, what things would you take into account when deciding on the most suitable sentence?

(*The South-East Regional Examinations Board, CSE, Civics – Citizenship and Social Affairs, 1982*)

12. Sid Jones is accused of arson and has to appear at the Crown Court. Describe the legal and financial help he could receive in answering the charge. Explain how he could appeal, if he were to be convicted.

(*East Midland Regional Examinations Board, CSE, Social Studies, 1981*)

3 LIVING AND LEARNING

UNIT 7

Communications and the Mass Media

GCSE Aims and Objectives

Knowledge — to comprehend the significance of methods of communication and the influence of the mass media in modern life.

Application — to inculcate an ability to identify the influence on society of newspapers, magazines, television, radio, cinema and advertising.

Evaluation — to make judgements about the ways in which our opinions are influenced by conventions, editorials, leader writers, press campaigns, broadcasting, documentaries and the hidden persuaders.

TOPIC 7.1 COMMUNICATION WITH OTHERS

In the early stages of life, **communication** is a simple process. Mother and baby communicate by noises and signs. The cries of the baby tell its mother that it is hungry. A smile is a sign that the food tastes all right while a grimace may mean, 'It's nasty, take it away.'

It is important that the method of communication is understood by the *sender* and the *receiver*. A picture may put over a message better than words.

The method by which people communicate a message is known as the **medium**. The plural of 'medium' is **media**. This is a word we use to describe the many ways by which people communicate. You should be able to add to this list of ways of communicating:

signs	sounds
words	drawings
photographs	music

Can you communicate by actions? If you marched out of the room and banged the door, both the noise and the action would communicate your displeasure. If somebody hit you or kissed you, would you get the message?

Two people can have personal communication by speaking face-to-face or telephoning or writing a letter. These are media for passing messages. When very many people are influenced by methods of communication we call them **mass media**. A huge industry has grown up by which messages are communicated to masses of people by newspapers, magazines, television, radio, films, advertising, etc.

We will return to some of the concepts which have threaded their way through this book from the first unit. If you are to develop as an individual, with your own distinctive personality and character, you must be on your guard against the power of the mass media. People may behave in an abnormal way when they act *en masse*. They are influenced by the crowd. Have you ever been part of a crowd which shouted 'Off, off', at a footballer or hurled abuse at the referee? You were acting instinctively, motivated by group behaviour. In the environment of the crowd you experienced a loss of individual characteristics. In

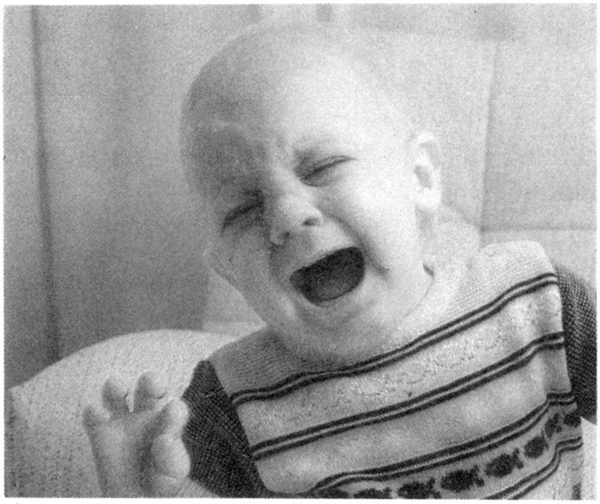
a)

b)

c)

d)

e)

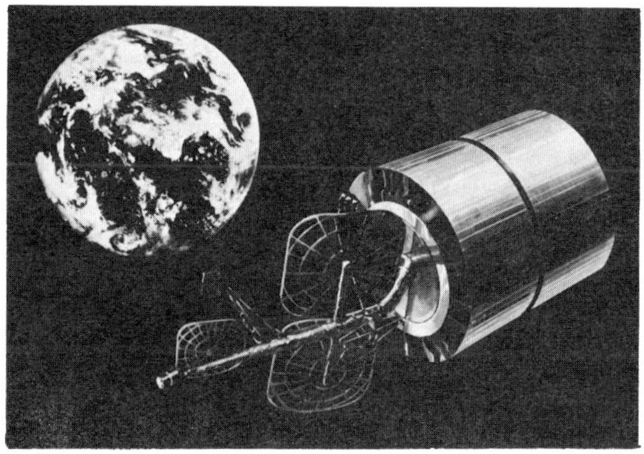
f)

Some methods of communication. Can you name them?

your heart you may not have really agreed with the crowd. It is easy to become caught up in mass hysteria.

Let us consider a few people who have stood aside from the mass. In the sixteenth century Copernicus wrote a book to communicate the fact that the planets, including the earth, revolved round the sun. His book was not published until after his death because he thought the authorities might put him to death for disagreeing with the **conventional opinion** of the day. People were unwilling to change their ideas; but changes in ideas and people are inevitable, as we have seen.

After the death of Copernicus, his belief that the earth revolved round the sun was proved right by Galileo who was imprisoned for his pains. Galileo was compelled to sign a confession that he was wrong. Even while he was signing, the earth continued to travel round the sun!

In the seventeenth century William Harvey demonstrated the circulation of the blood. The mass of doctors disagreed and he was struck off the Medical Register. Even while they were striking out his name, the heart continued to pump blood round the body as it always has done and always will.

These three men proved that the individual may be right while the masses are wrong. Keep this thought in mind as you study the mass media.

TOPIC 7.2 NEWSPAPERS

What newspapers do you take in your home? Jot down all the newspapers you can. Have you listed ten? If so you have done quite well, but it may surprise you to know that there are over 1,300 newspapers published in Britain. You could hardly be expected to know them all, especially as many of them are **specialised newspapers** whose circulation is limited to a certain region of the country or to a special interest such as a particular sport, hobby or business activity.

There are more papers bought by people in Britain than by citizens of any other nation of the world. Our papers are not the biggest in the world – some American papers have been published with over 1,000 pages – but our *News of the World* has a circulation of about 4.75 million copies throughout Britain. As most of these copies go into households where they are available for three or four people, probably about 15 million people see the *News of the World*. This paper contains quite a lot of stories about sex and violence. Does this tell you anything about the British people? Are they sex-mad and violent? Do they love reading about these things? Is it a way of getting these things out of their system?

A newspaper vendor is a familiar sight in the streets of our towns and cities

We can divide up our newspapers in various ways. All of these are useful classifications, although you will see that many of the newspapers fall into more than one group.

Classification of newspapers

1. *By taste*

So-called popular papers	*So-called quality papers*
Daily Express	*Financial Times*
Daily Mail	*Daily Telegraph*
Daily Mirror	*Guardian*
Daily Star	*Observer*
The Mail on Sunday	*Sunday Telegraph*
Morning Star	*The Sunday Times*
News of the World	*The Times*
Sunday Express	
Sunday Mirror	
Sun	
Sunday People	

2. *By layout or format*

Tabloids (short and fat)	*Broadsheets (long and thin)*
Daily Express	*Financial Times*
Daily Mail	*Morning Star*
Daily Mirror	*Daily Telegraph*
Daily Star	*Sunday Express*
The Mail on Sunday	*Guardian*
Sunday Mirror	*Observer*
Sun	*Sunday Telegraph*
Sunday People	*The Sunday Times*

3. *By region*

Northern England	*Scotland*
Liverpool Daily Post	*Daily Record* (Glasgow)
Morning Telegraph (Sheffield)	*Press and Journal* (Aberdeen)
Yorkshire Post (Leeds)	*The Glasgow Herald*
Manchester Evening News	*The Scotsman* (Edinburgh)
Northern Ireland	***Wales***
Irish News	*Western Mail* (Cardiff)
Belfast News-letter	

If your region is mentioned, see if you can add some more names. Include evening and Sunday papers. If your region is not mentioned, make a list of newspapers circulating in your area.

4. *By political opinion.* Ask your parents and other adults if they disagree with the table which classifies papers according to **political opinions**. This is very controversial. Have a look at a copy of *The Times* and see if you think it is really independent and unbiased.

Are there any good reasons for believing that we have had more Conservative than Labour governments in the last seventy years because we have more Conservative papers? Why do fewer papers support Labour? In what ways can a paper do more harm than good to the political party it supports? The table below gives circulation figures for our main national newspapers.

National Newspapers	Circulation average Oct. 1984–March 1985
NATIONAL DAILIES	
Daily Express	1,943,271
Daily Mail	1,844,617
Daily Mirror	3,389,728
Daily Star	1,555,406
The Daily Telegraph	1,226,359
Financial Times	224,931
The Guardian	485,391
Morning Star	30,345
The Sun	4,060,441
The Times	467,763
NATIONAL SUNDAYS	
The Mail on Sunday	1,615,828
News of the World	4,750,346
The Observer	754,986
Sunday Express	2,479,190
The Sunday People	3,147,224
Sunday Mirror	3,331,526
Sunday Telegraph	695,056
The Sunday Times	1,271,393

TOPIC 7.3 INFLUENCE OF THE PRESS

Throughout this book we have learnt about things which influence us – affect our experiences and mould our characters. Newspapers that are read by millions must exert a very great influence. We shall consider ten ways in which the press (newspapers, journals, magazines, etc.) influences the masses. Don't forget that the masses include you and me.

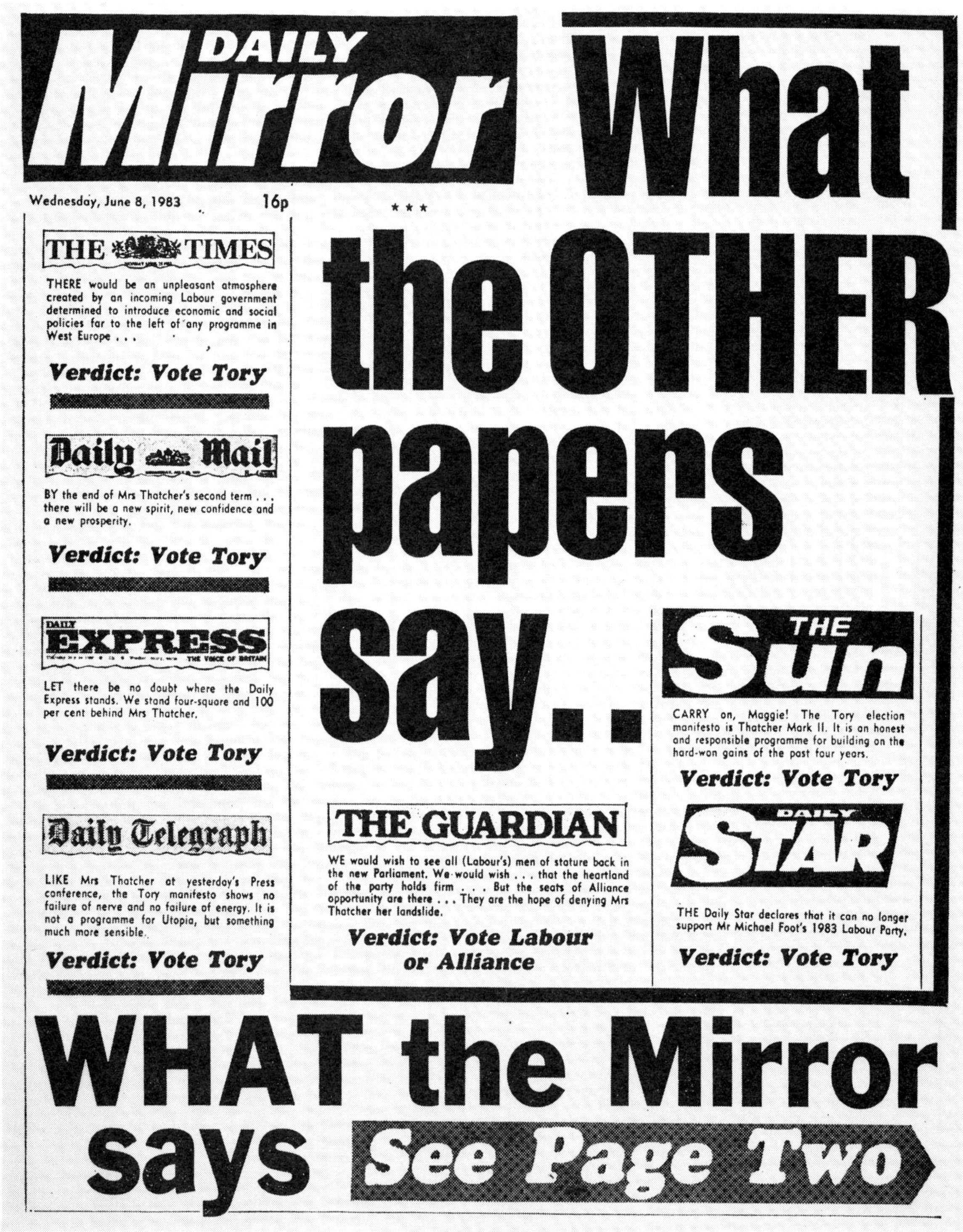

DAILY Mirror

Wednesday, June 8, 1983 16p

What the OTHER papers say...

THE TIMES

THERE would be an unpleasant atmosphere created by an incoming Labour government determined to introduce economic and social policies far to the left of any programme in West Europe . . .

Verdict: Vote Tory

Daily Mail

BY the end of Mrs Thatcher's second term . . . there will be a new spirit, new confidence and a new prosperity.

Verdict: Vote Tory

DAILY EXPRESS

LET there be no doubt where the Daily Express stands. We stand four-square and 100 per cent behind Mrs Thatcher.

Verdict: Vote Tory

Daily Telegraph

LIKE Mrs Thatcher at yesterday's Press conference, the Tory manifesto shows no failure of nerve and no failure of energy. It is not a programme for Utopia, but something much more sensible.

Verdict: Vote Tory

THE GUARDIAN

WE would wish to see all (Labour's) men of stature back in the new Parliament. We would wish . . . that the heartland of the party holds firm . . . But the seats of Alliance opportunity are there . . . They are the hope of denying Mrs Thatcher her landslide.

Verdict: Vote Labour or Alliance

THE Sun

CARRY on, Maggie! The Tory election manifesto is Thatcher Mark II. It is an honest and responsible programme for building on the hard-won gains of the past four years.

Verdict: Vote Tory

DAILY STAR

THE Daily Star declares that it can no longer support Mr Michael Foot's 1983 Labour Party.

Verdict: Vote Tory

WHAT the Mirror says See Page Two

The front page of the *Daily Mirror* on the day before the 1983 General Election

Classification of newspapers by political opinion

Conservative	*Labour*	*Liberal/SDP Alliance*	*Communist*	*Independent*
Daily Express *Daily Mail* *Daily Star* *Financial Times* *The Mail on Sunday* *News of the World* *Sun* *Sunday Express* *Daily Telegraph* *Sunday Telegraph*	*Daily Mirror* *Sunday Mirror* *Sunday People*	*Guardian*	*Morning Star*	*Observer* *The Times*

1. *Headlines.* Look at the headlines in a selection of daily papers. Which one do you think is the most sensational? Do any of them show bias or prejudice? Are there any which are ambiguous, i.e. have a double meaning?

2. *Layout.* Some popular papers have very large headlines but very few facts. The squat tabloid layout of the *Daily Mirror*, the *Daily Star* and the *Sun* does not leave much space for words; they have huge headlines and photographs. These newspapers exert their influence by setting things out sensationally in bold print. How far is it true to say there are newspapers which cater for those

who can read and think;
who can read but not *think;*
who don't want to *read or think?*

3. *Illustrations.* We saw that communication is possible without words at all. Newspapers can influence their readers by vivid photographs, humorous cartoons and other illustrations.

4. *Selection of material.* Newspapers have far more material submitted to them than they can possibly print. Three British press agencies collect news and sell it to newspapers, Reuters Ltd, the Exchange Telegraph Co. Ltd, and the Press Association Ltd. About 2 million words of general news, sports and other reports are received in London every day over a world network of cable and radio circuits. A newspaper also has its own **correspondents** and journalists plus freelance agencies which sell news. Sub-editors sift news for national dailies. We can only read what is selected for us, and we never see items which papers disregard and do not print.

5. *Readers' letters.* It is very easy for the editor to select letters supporting a view which is in line with the policy of the paper. A similar purpose can be achieved by printing a letter which expresses stupid arguments for the other side. Readers may write long letters which have to be abridged. If a vital paragraph is omitted in editing, the sense of the message is altered. The editor's word is final. The message becomes his rather than that of the letter writer. The editor has control over the message, because he has control over the medium.

6. *Editorials.* Most papers have an **editorial**, or column called a leader because it is written by a **leader writer**. It provides an opportunity to state a point of view in accord with the policy of the paper and the **newspaper proprietor** or owners. We pride ourselves on having a **free press** in Britain. It is free from government control, unlike the Soviet papers, *Pravda* and *Isvestia*. Any Briton is free to start a newspaper and convey any message so long as it is legal. But it costs a lot of money to run a paper; the privilege of a free press may well be the privilege of rich men to influence mass opinion. The history of newspapers is full of accounts of powerful men – Northcliffe, Beaverbrook, Kemsley, King, Thomson, Murdoch, Matthews and Maxwell.

7. *Comments.*

> 'News is sacred: comment is free'
> (C.P. Scott, ex-editor of *The Guardian*)

Newspapers mix **comment** with news. A straightforward account could be written about Mrs Thatcher winning the General Election in May 1983, but it would be easy to intermingle these simple facts with

(a) favourable comments about her businesslike approach and her tenacity of purpose
or

(b) adverse comments referring to an apparently pompous attitude or her 'posh' voice.

In a well-written piece it is almost impossible for all except the most discerning reader to pick out biased comments from straightforward facts.

8. *Exaggerations.* The popular papers include a great many **human interest stories**. There are spicy stories about unusual happenings. It is easy to find a story about 'the headmaster's son who went wrong'. Many other fathers have sons who kick over the traces, but sub-editors select stories for **sensationalism** and a touch of scandal. Have you ever listened to two or three people giving an account of a road accident? Could Mr Arthur Scargill (the President of the National Union of Mineworkers) have expected to secure a fair trial when he appeared before a stipendiary magistrate in December 1984, after all the bad publicity to which he had been subjected in the British press?

9. *Campaigns.* A newspaper can act as a pressure group. *The Sunday Times* conducted a long and successful **campaign** on behalf of thalidomide children.

10. *Personal attacks.* Newspapers may make personal attacks upon a prominent figure in an attempt to rouse public opinion against him. Tony Benn has been subjected to intensive press criticism in recent years.

The **Press Council** provides the greatest safeguard against abuse by newspapers. Since 1953 it has supervised professional standards and investigated many complaints against newspapers.

TOPIC 7.4 MAGAZINES

Ten specialised periodicals

Angling	*Anglers' Mail*
Birds	*Birds*
Coins	*Coin Monthly*
Music	*Music and Musicians*
Motoring	*Motorist*
Computers	*Practical Computing*
Soccer	*Football Monthly*
Cricket	*The Cricketer International*
Stamps	*Stamp Monthly*
Trains	*Railway Magazine*

Do you take any regular magazines? The girls have a wide choice from such popular magazines as *Jackie*, *Just Seventeen*, *Mizz* and *My Guy*. If you are a pop fan, you might read *Melody Maker*, *New Musical Express*, *Smash Hits* or *No 1*. Perhaps you have one specific interest and take a periodical like *Bike*. There are specialised magazines to cater for almost every interest.

What would you expect to find in a good teenage magazine? Arrange these features in the order of your choice. Your order of preference will give you an idea of the items which are likely to influence you most. We ought to be aware of things which are likely to play a part in moulding our personalities.

Contents of a teenage magazine

stories	mysteries
serials	career guidance
fashion	'help' letters
features	joke page
competitions	real-life stories
quizzes	picture serials
horoscope	pop pin-ups
poems	gossip columns
advertisements	music
humorous cartoons	romances
word games	adventure stories
puzzles	plays
sport	recipes
health hints	chat columns
beauty hints	pen pals
pets	reviews

The weekly periodicals with the highest sales are *Radio Times* and *TV Times Magazine* which have circulations of over 3 million, and *Woman's Weekly*, *Woman's Own*, *Woman* and *My Weekly* with circulations in the 800,000 to 1.4 million range. The leading journals of opinion are *The Economist*, *New Statesman*, *Spectator*, *Tribune*, *New Society* and *New Scientist*. *Punch* is the leading humorous magazine while *Private Eye* is a satirical fortnightly magazine.

TOPIC 7.5 TELEVISION AND RADIO

The two major bodies in charge of **broadcasting** in Britain are the British Broadcasting Corporation (BBC), and the Independent Broadcasting Authority (IBA).

THE BEST SELLING DIY MAGAZINE
FITTING A WASTE UNIT follow our picture guide
Do it yourself
KITCHEN PLANNING super design ideas
SHOWERS a real turn on

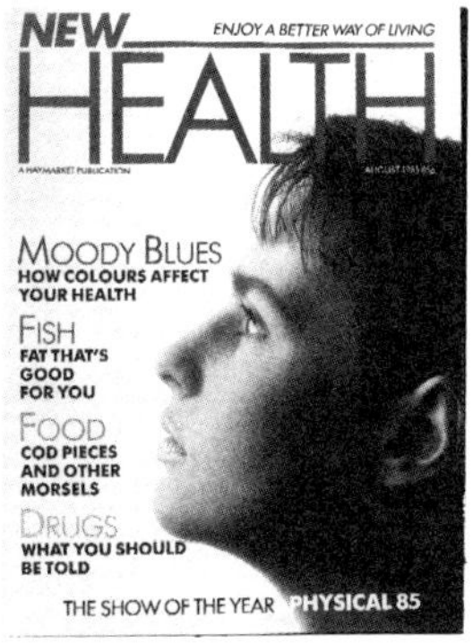
NEW
ENJOY A BETTER WAY OF LIVING
HEALTH
MOODY BLUES
HOW COLOURS AFFECT YOUR HEALTH
FISH
FAT THAT'S GOOD FOR YOU
FOOD
COD PIECES AND OTHER MORSELS
DRUGS
WHAT YOU SHOULD BE TOLD
THE SHOW OF THE YEAR PHYSICAL 85

CITY LIMITS
FAME – the truth!
WAYNE EAGLING hobbles on

video TODAY
POLICE ACADEMY
Summer Holidays 1985
STEVE MARTIN
VIDEODROME
– The full story

BJP

newscientist
HOW TO PICK THE PERSONALITY FOR THE JOB
The world as a telescope
Can medical research survive the cuts?

ENTHUSIAST

Jackie
22p
FREE
INSIDE
BOOK!
Win A Weekend In PARIS
In Our Great COVER GIRL COMPETITION!
36 pages of looking good

Smash HITS
"MEAT IS MURDER!"
PLUS
GARY KEMP
BRUCE SPRINGSTEEN
KING

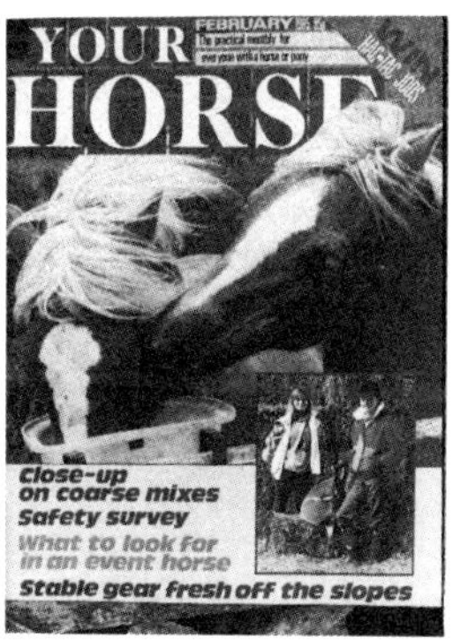
YOUR HORSE
Close-up on coarse mixes
Safety survey
What to look for in an event horse
Stable gear fresh off the slopes

PRACTICAL COMPUTING
FOR BUSINESS AND PROFESSIONAL MICRO USERS
SPECIAL: BUSINESS GRAPHICS

Punch
WEEKLY 80p

The BBC operates two national television services, four national radio services, three national regional radio stations (BBC Radio Wales, Ulster and Scotland) and over thirty local radio stations. It also broadcasts to many foreign countries. The BBC started a regular television service in 1936 and by 1967 provided the British people with the first European colour service through the medium of BBC 1. BBC 2 began in 1964 and went over to colour in 1967.

BBC television and radio programmes cater for all tastes.

Medium	*Type of programme*
Television	**Majority interests:**
BBC 1	light entertainment, sport, general interest, children's programmes, outside broadcasts
BBC 2	**Minority interests:** documentaries, serious drama, music, Open University programmes
Radio	
Radio 1	pop music
Radio 2	light music and sport
Radio 3	classical music, intellectual, artistic material
Radio 4	news, drama, talks, panel games

In the 1960s the activity of 'pirate' radio stations, ships such as *Radio Caroline* and *Radio London*, indicated a big demand, especially from young people, for continuous pop music. The result was the birth of Radio 1 in 1967. The 'pirates' proved that many firms were willing to spend large sums on advertising through the medium of commercial radio. This resulted in the Sound Broadcasting Act of 1972 which legalised independent radio, and there are now 47 different stations. Which are your local BBC and independent radio stations?

Television has not killed radio, but nowadays people tend to listen to specialised programmes. They settle down for an evening's television but the radio is switched on for 'The Archers' or 'Woman's Hour' or 'Any Questions'. Do you like listening to pop music? Why do some young people leave the transistor on to provide a background of non-stop pop?

Is it the responsibility of the BBC to raise people's standards or just to provide them with what they want? Should broadcasting authorities try to introduce people to ballet, classical music, modern art and other minority tastes? Lord Reith, the first Director-General of the BBC, believed it was his task to raise people's standards and introduce them to new cultural activities so that their range of pleasures was widened. Perhaps the best thing is a compromise. If

Television viewing in the UK: by sex and age (Source: *Social Trends No 15*, 1985)

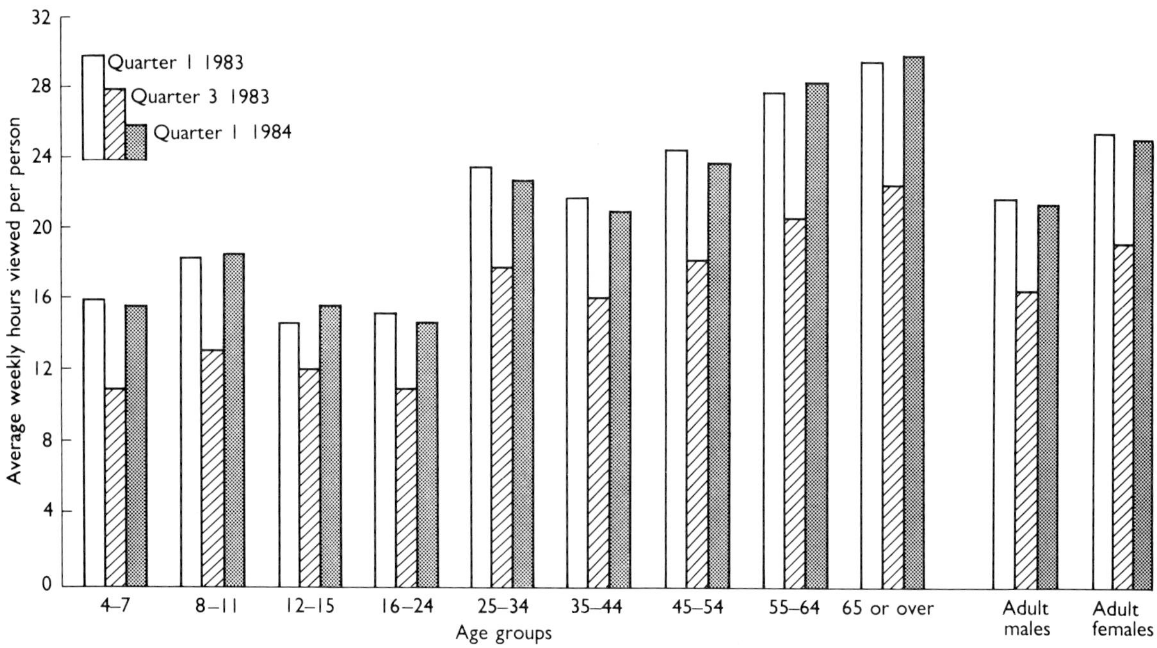

Alan Bleasdale's *Boys From the Black Stuff* TV series starring Bernard Hill as Yosser Hughes (pictured here with Graeme Souness) brought home the tragic problems, but also the resilience and humour of unemployed people in Liverpool

radio and television pandered to mass opinion all the time there would be no coverage of minority interests. The BBC's revenue comes from those who pay licence fees, and that includes all sorts of people.

The Independent Broadcasting Authority (IBA) was known as the Independent Television Authority until 1972 when its responsibilities were extended to include sound broadcasting. The main function of the IBA is 'to provide public television services of information, education and entertainment'.

The IBA owns the transmitting stations, and is responsible for what is shown on ITV, but the programmes are produced by fifteen regional ITV programme companies plus ITN, TVAM and Channel 4 who make annual rental payments to the IBA. The companies charge firms who advertise their products on television. Those who buy goods and services advertised in this way pay indirectly for both advertisements and the programmes; perhaps it is less painful than paying licence fees. Do you think advertisements distract from programmes or do you like advertising jingles? No doubt it depends on the programme being watched. It's all a question of taste.

The weekly programme output of ITV amounted to over 100 hours a week in 1985.

The make-up of **audience profiles** is influenced largely by (a) sex, (b) social class and (c) age.

How many hours a week do you spend watching television? The most popular programmes are series such as *Coronation Street*, *East Enders*, *Dallas* and *Dynasty* which are sometimes called **soap operas**. What would you say to someone who said that watching such programmes is a time-wasting pursuit?

Many television programmes serve a social function. Television has drawn attention to social evils by screening compassionate **documentaries**. The plight of the starving people of Ethiopia was first brought to the attention of the public by a BBC TV special news report in 1984. What social documentaries and visual case studies have you seen on television in recent months?

By 1985 about 20 million households had TV sets, including about 16 million with colour sets. Television viewing is by far the most popular leisure pastime in Britain, average viewing time per person being about 20 hours a week. Virtually everyone has access to TV. Over twenty per cent of households have two or more receivers. The growth in ownership of video cassette recorders and equipment (including home computers) for playing television games has increased the choice of entertainment available in the home.

'Tape it and I'll watch it tonight.'

TOPIC 7.6 THE CINEMA

Although attendances have declined rapidly since the heyday of the between-the-wars Hollywood movie, young people are still greatly attracted to the cinema. Television is a godsend to old people and to young marrieds, who find it difficult to get baby-sitters and cannot afford regular cinema visits on top of a heavy mortgage, but the cinema provides a meeting-place for young people. In most towns it is one of the main places for teenage entertainment.

Film making has become increasingly costly, and cinema seats consequently more expensive, so that in recent years the number of admissions has more than halved. Admissions to cinemas in Great Britain numbered only 60 million in 1982, about 65 per cent fewer than in 1971. The number of cinemas fell by just under 45 per cent over the same period. However, the number of screens has not shown the same fall because of the modern use of multi-screen cinemas following the conversion of large single screen cinemas where capacity was too great for the numbers attending. In 1982 twin and triple screen complexes accounted for 60 per cent of all screens and a similar proportion of cinema admissions.

Towns that had between twelve and twenty cinemas in 1950 often had about 3 by 1985. Many of the cinemas have become bingo halls or warehouses.

Cinemas have fought back by

1. providing 'multi-cinemas', with three or four films being shown in small cinemas
2. concentrating upon expensive spectaculars
3. making use of larger screens.

As it dates back to the late nineteenth century, the film is one of the oldest forms of mass media, and thousands of films are available for hire from film libraries. Many documentaries have great educational influences but some of the films which appeal most to the masses are those about

(a) gangsters
(b) horror stories
(c) romance
(d) the wild west
(e) war.

Many of these films present unreal situations. Westerns portray familiar themes of rustling and gambling culminating in the crunch when the goodie is faster on the draw than the baddie. Familiarity breeds contempt and an interesting breakaway from typical sequences has been the **real-life films** where the director has turned his back on the artificiality of American war heroes and the tinselled lives of romantic heroines to concentrate upon living drama. Many of these films have centred on a working-class background. Perhaps you have seen some of them on television.

Sociologists are naturally interested in whether action portrayed on the screen affects real-life behaviour. We saw in Unit 1 that it is often impossible to say why we behave as we do. If you watch gangster films, are you likely to be a more violent person? There is little real evidence that many people commit evil actions merely because they saw something similar on the screen (although recently there has been concern about the effect of 'video nasties' on very young children). Indeed a film may have a **cathartic effect**: you may get some hate out of your system by watching others perform violent acts on the screen. We don't all turn into vampires because we watch Count Dracula!

TOPIC 7.7 ADVERTISING

The purpose of advertising is to influence the behaviour of the masses to serve the ends of the advertisers. Modern **persuaders** want us to support their causes, or buy their products, or they want to combat the advertising of competitors. All the mass media can be used: national and provincial newspapers; magazines, journals and periodicals; cinema, radio and television. To these we can add hoardings, posters, sandwich men, sky-writing, handbills, pamphlets, leaflets, vouchers, etc. Can you think of others?

The persuaders will appeal to those basic emotions and instincts which motivate each one of us. Throughout our human life-span these things drive us on with greater or lesser effect, depending on age, background, education, health, intellect, sex and social class.

Advertisers use the mass media to great effect for good or for ill. There are both advantages and disadvantages in advertising.

How £4,055 million was spent by 'the persuaders' in 1984

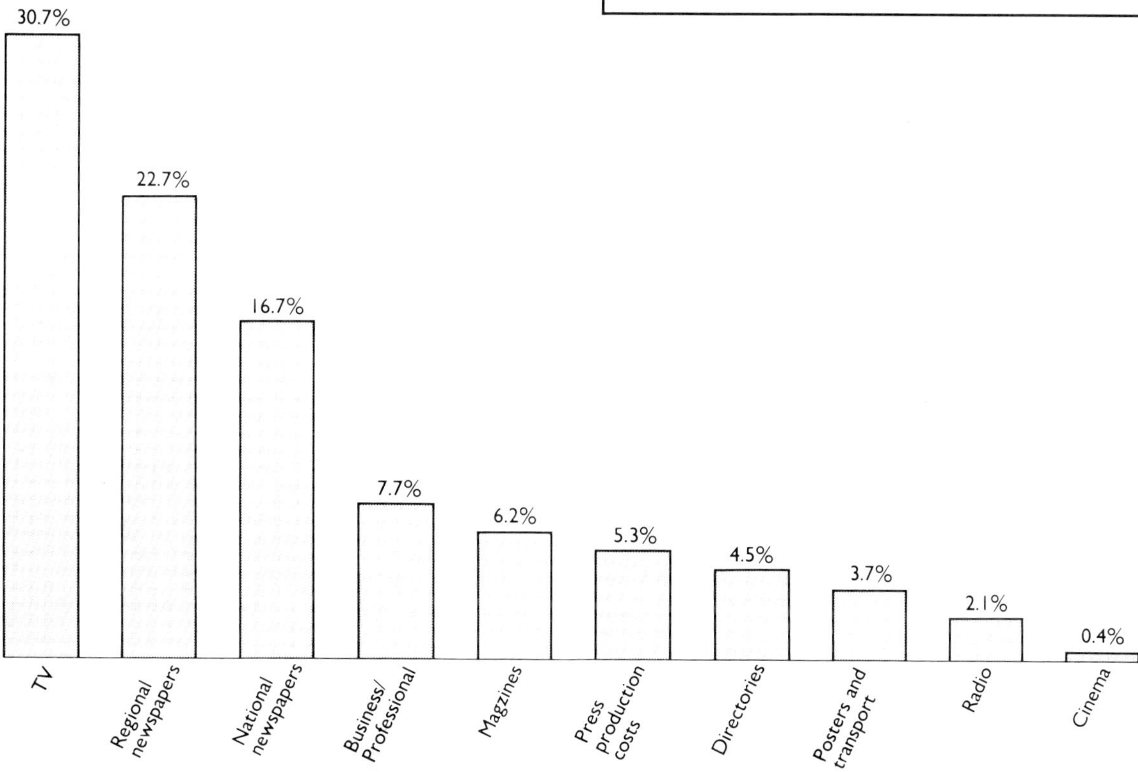

How advertising helps

- Informs us about products and prices
- Ensures keen competition
- Makes possible large sales and lower production costs
- Pays 50 per cent of the cost of newspapers
- Pays for commercial radio and television
- Enables more people to be employed
- Raises standards of living
- Attracts people to support worthy causes (not all advertising is commercial)

How advertising hinders

- Misleads consumers with wild claims
- Persuades people to buy things they don't need and can't afford
- Increases prices (over £4,000 million a year is spent on advertising)
- Endangers health, e.g. adverts for cigarettes and alcoholic drinks
- Spoils ITV and radio programmes by interrupting them
- Wastes paper, time and money
- Damages our language and our environment
- Gets on people's nerves

Illustrated above are two very different types of advertisement. To which human instincts do you think they appeal, and for what type of person are the advertisements intended? It is interesting to note the various methods used by the persuaders in order to sell or promote their product. Do you think that these particular examples are successful in achieving their objective? Perhaps you can suggest some appropriate newspapers or magazines in which you might expect to find these advertisements.

Human attributes to which the persuaders appeal

Sense of adventure	*Materialistic motives*	*Sex instinct*
ambition	greed	femininity
danger	comfort	glamour
power	snobbery	manliness
speed	status	maternity

Some of the methods of the persuaders

Before-and-after approach
Colourful appeal
Distinctive brand name
Doubt if you can afford it!
Everybody's buying it
Humorous cartoons or verses
Ostentation – snob appeal
Packaging and wrapping
Recommendation by a famous person
Repetition, e.g. jingles
Upper-middle-class way of life
Pretending to 'knock the product'
You can't manage without it!

Explain what is meant by

communication
medium
media
mass media
conventional opinion
specialised newspapers
political opinions
correspondents
editorial
leader-writer
newspaper proprietor
free press
comment
human interest stories
sensationalism
campaign
Press Council
broadcasting
majority interests
minority interests
audience profiles
soap operas
documentaries
real-life films
cathartic effect
persuaders

Case study

IN A WORLD OF HIS OWN

Danny Jones had always been crazy about computers – ever since he'd first seen a TV advert for a micro when he was ten years old. He begged his parents to buy him one for Christmas. Although Mr and Mrs Jones thought it was rather expensive, they reasoned that it was a good educational 'toy', and a far better investment than some of those 'faddy' toys which the children seemed to get bored with after a few weeks. So Danny's wish was granted.

As soon as he unwrapped his Unicorn QL on Christmas morning, Danny was in his element. He tapped away on the keyboard all day and he quickly mastered the basic 'language' to do all sorts of clever things on it. Danny's parents were delighted, too; the computer kept him quiet for hours on end!

However, after a few weeks, Mrs Jones began to get a bit worried. Danny had spent the whole of the holidays playing with his computer from the moment he got up in the morning until he fell into bed exhausted each night. He didn't seem bothered about playing with his friends any more, and he'd become very pale and withdrawn. It was almost as if he was so used to 'talking' to his computer, that he'd forgotten how to communicate in human language. Mrs Jones decided to have a word with his teacher.

When Mrs Jones explained the situation, Danny's teacher looked thoughtful. 'Ah, that explains why Danny has been so quiet this term. He's been in a world of his own – the times I've caught him daydreaming!' she smiled. 'But I think you must be aware of the danger of Danny spending too much time on his computer. You'll have to ration him from now on, otherwise he'll turn into a computer himself. Human beings need to learn how to communicate with each other as well as machines!'

1. List three advantages of communicating with words rather than signs.
2. Find out all you can about Esperanto.
3. Give two advantages and two disadvantages of an international language.

Follow-up

1. Give reasons supporting and/or opposing the idea that violence on cinema and television screens leads to violence in real life.
2. How can people be misled by advertisements?
3. Write down ten popular advertising slogans or jingles.
4. Suggest some rules that IBA should lay down in relation to
   (a) advertisements
   (b) colour prejudice
   (c) crime
   (d) sex
   (e) violence.
5. Write about your favourite
   (a) cartoon character
   (b) comic actor
   (c) TV cop
   (d) pop group
   (e) science-fiction programme.

Activities

1. The prefix 'tele-' comes from Greek and means 'far'. What is meant by the following:
   (a) telecommunications
   (b) telemessage
   (c) telegraph
   (d) telepathy
2. A table giving the circulation figures of newspapers can easily be found in *Britain 1985*. This is an annual HMSO publication. Use the latest edition in your reference library to find the current circulation figures.
3. Find the *dates* of these important developments in communications and mass media:
   (a) William Caxton's printing press
   (b) *Daily Courant* – the first English paper
   (c) Edison's gramophone
   (d) Marconi's first wireless message
   (e) John Logie Baird demonstrating television
   (f) the first BBC broadcast
   (g) the beginning of the BBC's television service
   (h) the advent of colour TV
   (i) the start of ITV transmissions
4. Conduct a survey of your group to discover
   (a) the newspapers they take
   (b) the TV programmes they watch
   (c) the radio programmes they listen to.

Examination questions

1. The British Code of Advertising Practice stated that *all advertising should be legal, clean, honest and truthful*.
   (a) Why is such a code of practice necessary?
   (b) In what ways is advertising controlled in Britain?
   (*East Anglian Examinations Board, CSE, Social Studies, 1984*)
2. (a) Explain the meaning of the word propaganda.
   (b) The Argentine and British versions of the battle for the Falklands differ greatly. What reasons do we have to believe that our own is the true version and not propaganda?
   (*East Anglian Examinations Board, CSE, Social Studies, 1983*)
3. (a) List *four* agencies of the mass media.
   (b) To what extent are producers of goods and services able to influence patterns of consumption?
   (c) To what extent do the mass media influence political attitudes?
   (*The Associated Examining Board, GCE, Social Science, 1984*)
4. Write an essay describing how the various forms of mass media can be controlled and censored.
   (*East Midland Regional Examinations Board, CSE, Social Studies, 1984*)
5. (a) What are the mass media?
   (b) State which features of the mass media have the greatest influence on your life and explain why.
   (*East Anglian Examinations Board, CSE, Social Studies, 1984*)
6. (a) Explain the term 'mass media', giving at least *three* examples.
   (b) Discuss the influences of any *one* of the mass media.
   (*East Anglian Examinations Board, CSE, Sociology, 1983*)
7. (a) Give TWO ways in which 'Video' has influenced people's television viewing habits.
   (b) What are the main disadvantages of the 'action replay' which is used frequently on television?
   (*Associated Lancashire Schools Examining Board, CSE, Social Studies, 1984*)
8. (a) Name THREE examples of New Technology which have improved communications during the past few years.
   (b) Briefly explain how any ONE of these examples of New Technology has improved our everyday life.
   (*Associated Lancashire Schools Examining Board, CSE, Social Studies, 1984*)
9. (a) Give THREE examples of the types of book which are referred to as 'popular literature'.
   (b) Give THREE reasons why radio has continued to exist as a powerful force in our society.
   (c) What is bias? Why is it a problem in society?
   (*Associated Lancashire Schools Examining Board, CSE, Social Studies, 1983*)
10. (a) How useful to society is the advertising industry?
    (b) What harm to society might be caused by the advertising industry?
    (*Associated Lancashire Schools Examining Board, CSE, Social Studies, 1984*)
11. (a) In what ways has the cinema industry attempted to survive in modern society?
    (b) Do you think that the cinema industry will still exist by the end of this century? Give suitable reasons for your answer.
    (*Associated Lancashire Schools Examining Board, CSE, Social Studies, 1984*)
12. (a) What is communication?
    (b) Briefly name and explain the major means of communication in our society.
    (c) In what ways might our society be different without the availability of many forms of mass communication?
    (*Associated Lancashire Schools Examining Board, CSE, Social Studies, 1983*)

UNIT 8

Education

GCSE Aims and Objectives

Knowledge — to clarify understanding of the educational process, the functions of education, the educational system and education after leaving school.

Application — to study different types of education in practice including formal and informal education, nursery education, primary education, secondary education, further education and higher education.

Evaluation — to distinguish between the merits and demerits of different educational methods and concepts.

Student participation is an important part of education

Using a computer is now one of the basic skills taught in schools

TOPIC 8.1 WHO ARE THE EDUCATORS?

The process of education goes on throughout the life-cycle. We never stop learning; consequently education does not stop either. All those who train and instruct others are **educators**.

Your first educators were your parents, to be followed by the rest of your family, relations and friends. This was **informal education**: there were no organised lessons. Gradually and often unknowingly you learnt from others by copying, practising, failing sometimes but eventually succeeding. Pleased with your achievement you did it again and again.

Informally, we gain knowledge, acquire skills and learn the values of the community. This is socialisation.

At about the age of five the majority of children in Britain go to school and start their **formal education**. This takes place in schools and similar institutions. Formal education has been available for everybody in Britain for only just over a century. Until Forster's Education Act of 1870 there was no **compulsory education** and even the school boards were unable to get all children to school. An Act of 1876 compelled parents to see that their children received **elementary education** in basic subjects such as 'the 3 Rs' (reading, writing and arithmetic).

Some methods of informal education

Informal educator	*Material taught*
Parents	Language, counting, reading, conduct, socialisation
Brothers and sisters	Living with others, sharing, compromising
Games	A boy is a man in the making: our games fit us for a man's (or woman's) role
Close friends	Loyalty, companionship, forgiveness
Baby-minding	Looking after children, responsibility, patience
Mass media	Information, discrimination
Youth activities	Acceptance of leadership, co-operation, initiative
Community service	Caring for others, practical work, kindness

In pre-industrial societies, boys were educated in many informal ways:

1. Their fathers taught them their trades or how to hunt.
2. The priests taught them about religion and how to conduct themselves.
3. The warriors taught them how to fight in battle.

Girls had a more specialised role than that fulfilled by females today. The word 'unisex' had not been invented. A girl's mother taught her to perform domestic duties and to look after younger children.

There are many ways in which we are informally educated today.

TOPIC 8.2 WHY BE EDUCATED?

Apart from the obvious benefits of education in promoting the fullest development of each individual, the many functions of education help people to adjust to community life.

1. **Cultural function.** Sociologists apply the term culture to standards of conduct and values accepted by a particular society. As a multi-racial society Britain has now assimilated many different subcultures; each ethnic community retains its distinctive character – its ethical codes, customs, festivals and outlook – and at the same time accepts the standards of the British way of life.

2. **Economic function.** A modern industrial society must have educated people who can read, write, calculate, interpret diagrams, and transmit and understand instructions. Schools teach basic skills. Do you think that too little attention is paid to the 3 Rs today? Job skills are taught at the place of work or at institutions of further education.

3. **Continuity function.** People are educated according to existing patterns of life. Although changes must take place if society is not to decay and die, education has a stabilising effect. Parents teach children what their parents taught them. Has your mother ever said, 'Well, my mother taught me . . .'? Or has your father ever said, 'My father used to say . . .'? Education tends to be traditional, and this applies especially to schools. They are freer places than in the past, but the organisation of assemblies, classes, teaching groups and examinations follows similar patterns to past practices. Most of us rebel at times against formal education, but we would learn very little unless there was organisation.

4. **Law and order function.** Even a family cannot exist without law and order. A baby has to learn that he is not expected to throw his food over the rest of the family. The laws are unwritten but still valid. Schools continue the process by impressing on pupils that laws and orders must be observed. We saw in Unit 2 that an individual's freedom is limited by his environment. The chief authority at school is the head teacher, and the rest of the staff support him in maintaining sufficient law and order for education to take place.

5. **Political function.** If Britain is to remain a democracy schools must teach pupils to play an active part in politics. Have you ever had a mock election at your school, or do you elect form representatives or a School Council? We want people to stand for Parliament or for the local council. People should vote at election time and play an active part between elections.

TOPIC 8.3 THE EDUCATIONAL SYSTEM IN BRITAIN

The **educational system** is illustrated on page 96. You are at the half-way house, having worked through the pre-school and primary stages to reach the secondary stage. Your informal education will go on as long as you go on! In 1982, about £14,000 million was spent upon the provision of education. The likely number of school pupils, at each stage, is shown in the table.

Projected number of pupils and students (*in thousands*)

Type of Schools	*1986*	*1991*
Public sector		
Primary schools	3,930	4,401
Secondary schools	4,063	3,283
Assisted and independent schools	485	464
Special schools	145	147
All schools – total pupils	8,623	8,295

The provision of far more **nursery schools** was one of the main policies of a ten-year programme (*Education: A Framework for Expansion*) announced by the Government in 1972. The plan promised nursery education by 1983 to all 2–5-year-olds whose parents wanted it for them, but this plan has not been fulfilled. Some areas are very badly served.

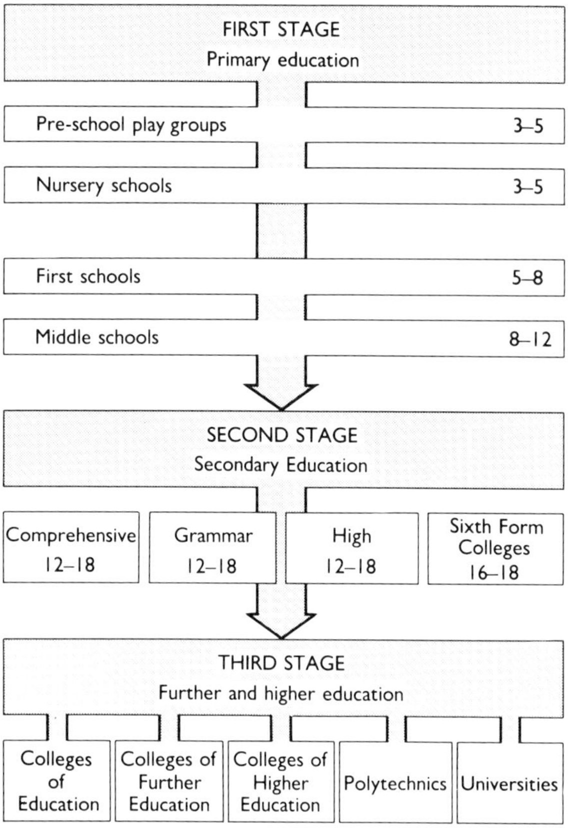

The structure of education in England and Wales

TOPIC 8.4 NURSERY EDUCATION

Nursery education is supplied by

1. pre-school **play groups**
2. free nursery schools run by local education authorities (LEAs)
3. nursery schools where parents pay.

Do you think very young children are better off being formally educated in organised nurseries or at home with their mother (or father)? Recall our investigation into environment in Unit 1 and social classes in Unit 2. An under-five-year-old who is lucky enough to have attentive, sensible, well-educated parents and instructive toys might be better off at home, epecially if there were other youngsters of his own age for him to play with. Research has shown that attendance at nursery schools has no significant effect on primary school attainment. By the end of the first term at primary school those who had not attended nurseries had caught up with ex-nursery pupils and adjusted to formal education. The short-term gains from nursery education were wiped out.

There is a continuing debate between those people who believe that the value of nursery education can be overstated and others who argue strongly that the provision of nursery education for all young children helps to counterbalance any social inequalities in their home environment. What do you think?

Benefits claimed for nursery education

- Makes up for deprivation in children from a poor environment
- Gives earlier contact with others from same peer group
- Picks out those who require early treatment because they are handicapped physically, mentally or psychologically
- Makes young children realise the need for more co-operation and less selfishness
- Provides socialisation opportunities not found in the home

Have you ever been to a play group? Are the methods used more informal or formal? Mothers help in the running of play groups but the National Children's Bureau has pointed out that the needs of deprived children are inadequately met if mothers are not suitable to help.

Do you know any children who are left with a child-minder because there is no provision for nursery education? As more mothers go out to work, provision has to be made for their children to be looked after. Do you think these children are better off with a child-minder, in a play group or in a nursery school?

TOPIC 8.5 PRIMARY EDUCATION

Primary education may be divided into

1. *First schools* from 5 to 8 years
2. *Middle schools* from 8 to 12 years.

The extension of primary education to the age of 12, or even 13 in some areas, has several advantages:

1. Interesting, individualistic and self-expression work can be continued before being replaced by more formalised secondary education.

2. There may be a carry-over of the enthusiasm usually found in primary education.
3. It allows an extra year at the primary stage without detracting from the secondary stage. This is possible because ROSLA (raising of the school leaving age) has extended the secondary stage.
4. Children may benefit from **neighbourhood schools** which serve a particular catchment area.
5. Full benefit can be taken of the abolition of the 11+ examination with inhibited junior schools previously.

There have been complaints in recent years about the standards of reading and writing. In 1975 the Bullock Report stated that research evidence pointed to a decline in educational standards.

The Report contained 333 recommendations but unfortunately they have not all been put into practice. In recent years the education budget has been substantially reduced, which has placed a great strain on resources, including teachers! Have you felt the effect of any cutbacks? Have you any criticisms about the educational standards in your school?

Some recommendations of the Bullock Report

- Close checks on reading standards
- Improved methods of testing and recording
- National centre for language in education
- Detailed reading policy in every school
- Screening to pick out backward readers
- Extra help for slow readers
- Local authority clinics to assist schools to teach reading
- Parents' evenings to gain their co-operation in English teaching
- Well-stocked libraries
- Improved teacher-training to ensure that all teachers know how to teach the English language

Is there any truth in the following statements? Children at first and middle schools

1. spend too much time playing
2. have insufficient discipline
3. ought to have homework

Children 'learning through play' at a nursery school

4. spend too long on subsidiary topics or projects
5. are not sufficiently subjected to parental pressure
6. watch too many visual programmes of entertainment rather than educational value
7. are tested irregularly and not compelled to learn
8. are no longer expected to reach any standard (e.g. the 11+ examination)
9. fall victims to untried modern methods and experiments
10. are often taught by teachers without degrees.

You have probably rejected the last point. There are more important qualities necessary for teaching young children than high academic qualifications. What about patience, enthusiasm, sympathy and an appreciation of the difficulties of younger children?

It is right that middle schools should be places of keen experimentation. Mankind progresses through *change*. If one experiment fails, another should be tried.

An example of a brave experiment was the teaching of a second language to primary school children. It seemed a good idea. Many English people are poor linguists. However, after a ten-year experiment the National Foundation for Educational Research reported that it seemed a waste of time to teach French to 8–11-year-olds. The NFER studied 18,000 children who had been taught French; it was found that when they arrived at secondary school half of them had had enough and were hostile towards learning any more of the language. Can you suggest any reasons why this might have been the fault of

(a) the teachers
(b) the methods
(c) the pupils?

'Well, how do you like being in a comprehensive school?'

One of the greatest changes has been the abolition of the 11+ examination. Many educationalists believed that if schools were freed from this examination the 'C' stream would no longer be sent to dig the school garden while the 'A' stream was processed to gain 11+ passes by which the school was judged! Comprehensive schooling has meant the end of the 11+, but there is a danger that standards in schools may fall. What ways can you suggest of maintaining high standards without an examination?

TOPIC 8.6 SECONDARY EDUCATION

You are now in the secondary sector of education. The chances are that you are at a comprehensive school.

Percentage of secondary pupils in comprehensives (England)

1950	0.3
1960	4.7
1971	34.4
1976	68.8
1982	82.9

(*Source: Social Trends No 14*, 1984)

The 1944 Education Act outlined three main types of secondary school:

1. modern
2. technical
3. grammar.

In the last ten years these have largely been replaced by **comprehensive schools**. Councils who had not adopted the comprehensive system by 1974 were instructed to do so. Many advocates of comprehensives think that the Labour Government ought to have made comprehensives compulsory by law, since a small number of local authorities have refused to adopt the system.

It is difficult to put personal feelings on one side, especially if you are at a comprehensive, modern or grammar school, but we must try to look at the matter impartially.

You may think that some of these are self-cancelling arguments. If so, it proves a point. There are no simple 'yes' or 'no' answers. Comprehensives provide many advantages but it would be wrong to

The great comprehensive debate

For	*Against*
• More resources, e.g. language laboratories, careers rooms, computer terminals	• Large educational factories lacking the personal touch
• Children from all types of background learn to mix socially	• Extinction of private grammar schools with great traditions
• Greater opportunities for late developers, e.g. in the 13–15 age group	• Lower standards; the less able may hold back those with high ability
• Pastoral care and counselling for pupils who are deprived, handicapped or unstable	• Lack of discipline with subsequent hooliganism, vandalism, bullying and truancy
• Removal of stigma: 80 per cent used to be classified as 11+ failures	• Decline in manners, and in dress; school uniform is fast disappearing
• New-style sixth forms offer pupils chance to take GCSE subjects before going on to higher studies	• Less likely to provide intensive education for country's future leaders

think the perfect solution has been reached in the structure of secondary education. The wheel often turns full circle. Perhaps the next generation will reject large schools and return to a belief in the idea that 'small is beautiful'.

The modern trend in secondary education is away from **streaming** into classes where those of similar ability are taught together. The 1944 Act replaced the 3 Rs by the 3 As, i.e. education according to

(a) age
(b) ability
(c) aptitude.

An increasing number of schools are organising **mixed ability groups** where the dullest and the brightest learn together. A mixed ability class might include pupils with **intelligence quotients** (IQs) ranging from 80 to 140. An IQ over 100 denotes above average intelligence; below 100 means less than average.

$$IQ = \frac{\text{Mental age}}{\text{Actual age}} \times 100$$

Therefore if your mental age is 18 years (i.e. you are as clever as an average 18-year-old), but your real age is 15, your IQ will be 120:

$$\frac{18}{15} \times 100 = 120$$

Mixed ability groups require new teaching methods. A teacher cannot stand in front of a class and explain the same thing to less able and very able pupils. The teacher talks to them in small groups and uses work-cards to keep pupils active. Materials can be shared and **team teaching** practised. Mixed ability groups may improve discipline because problem children are scattered and no longer mass together.

One of the main arguments for abolishing streaming is to provide **equal opportunity**. Does streaming in comprehensives constitute a form of selection as bad as segregating pupils into separate schools?

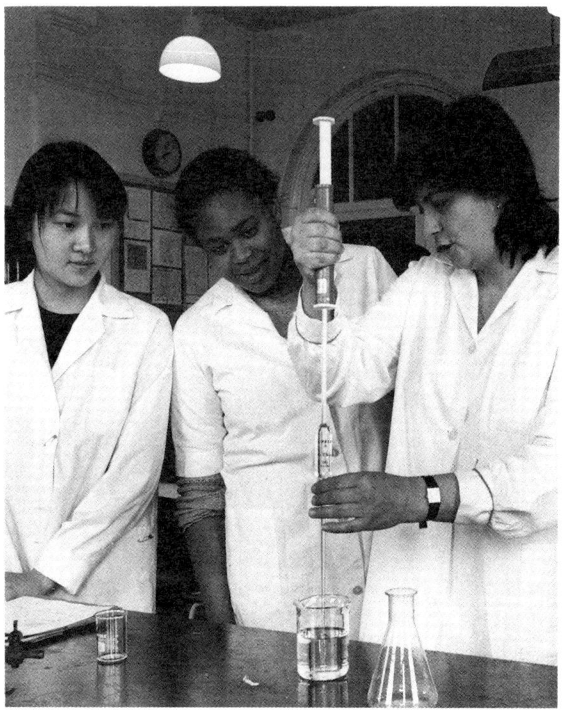

Teaching science by experiment

In 1972–3 the **school leaving age** was raised to sixteen. There was opposition to ROSLA from

1. pupils who thought the extra year was a waste of time
2. employers who lost low-cost young workers
3. parents who had little interest in education and wanted more household income
4. teachers who found 15–16-year-olds difficult.

Does it seem logical that less time should be spent upon those who find learning difficult?

Education for what? This is the vexed question that surrounds the raising of the school leaving age. There is a need to spend time on basic skills, but pupils must be motivated by something more than just reading, writing and arithmetic. Vocational incentives should be tapped as much as possible. The increase in unemployment, especially among young people, has meant that many pupils now opt to stay on in the Open Sixth. It is important that the extra year at school should not be wasted.

Some suggestions for the extra year (Open Sixth)

- New subjects related to life, e.g. motor mechanics, design for living, child care, home management, office practice and typing, social studies, etc.
- Work experience with local firms
- Help with transition from school to work, e.g. social economics classes on employment, taxation, social security, etc.
- Community studies of the neighbourhood, e.g. housing, shopping, transport, local government, etc.
- Worthwhile extended visits
- Good outside speakers
- Improve the leavers' qualifications – most pupils can take the GCSE in a few subjects
- Encourage discrimination in the use of leisure time

TOPIC 8.7 EDUCATION AFTER SCHOOL

One of the most important ways in which young people can secure education after school is through **Youth Training Schemes**. Organised by the Manpower Services Commission (MSC), YTS is available for employed and unemployed 16 year olds and some unemployed 17 year olds. The initial scheme started in September 1983 and aimed at finding places for

Youth Training Scheme

460,000 young people. YTS replaced the Youth Opportunities Programme (YOP) and differed from it by offering education and training to all young people leaving school, whether they had jobs or not. The original YTS scheme lasted for one year and aimed to train youngsters in skills related to a range of jobs, including periods of planned work experience and a minimum of 13 weeks off-the-job training. YTS schemes have been criticised for providing cheap labour (a trainee received £26.25 a week compared with £17.30 a week on supplementary benefit in 1985) and for not ensuring regular work at the end of the scheme. However, current figures show that about two-thirds of all YTS trainees who have completed schemes are now in full-time regular employment.

The Manpower Services Commission has now introduced a new two year Youth Training Scheme. The number of training places was increased by about 200,000 places and as from April 1987 only approved training organisations (that is organisations with approved sets of standards and trainers) may take part in the scheme. There is a training standards advisory service (similar to HM Inspectorate in schools) to monitor training. A minimum of 20 weeks off-the-job training is provided over the two years. Weekly allowances have been increased to £27.30 for first year trainees and £35 for those in their second year. Trainees will receive a certificate as proof of their skills and training.

The number of apprenticeships has been reduced in recent years, but there are still young people who are bound to a master who acts as their teacher and employer. They are bound by an agreement called an indenture. **Apprenticeship** usually begins at 16 and lasts for a fixed period of 3 or 5 years. Local technical colleges offer day release courses for young workers i.e. a day's full time study at the firm's expense. It may be a condition of apprenticeship that attendance is made at day release classes and probably at evening classes as well. Such tuition is essential in order to pass a City and Guilds or Ordinary National Certificate examination.

Since October 1983 the Business and Technician Education Council (BTEC) has been responsible for business and technician courses. BTEC courses are available at two levels and may be undertaken at technical colleges, colleges of further education and polytechnics.

1. *National Certificates or Diplomas*
These are unit-based examinations. A certain prescribed number of units are necessary for the award of a certificate, while an increased number of units secures a diploma. The normal entry for the BTEC National is 3 to 4 O levels, sometimes in specified subjects. Courses normally last two years and are arranged on a full-time basis for diploma candidates.

2. *Higher National Certificates or Diplomas*
These are also unit-based and involve two or three years full time study, sometimes on a sandwich course basis. Entry to Higher National Certificate courses is by means of BTEC national or one or two A levels. The BTEC Higher National approaches degree level.

Those who wish to go on to a full degree course must pass five GCE subjects, including two at advanced level. The institutions of higher education will usually demand high A level grades.

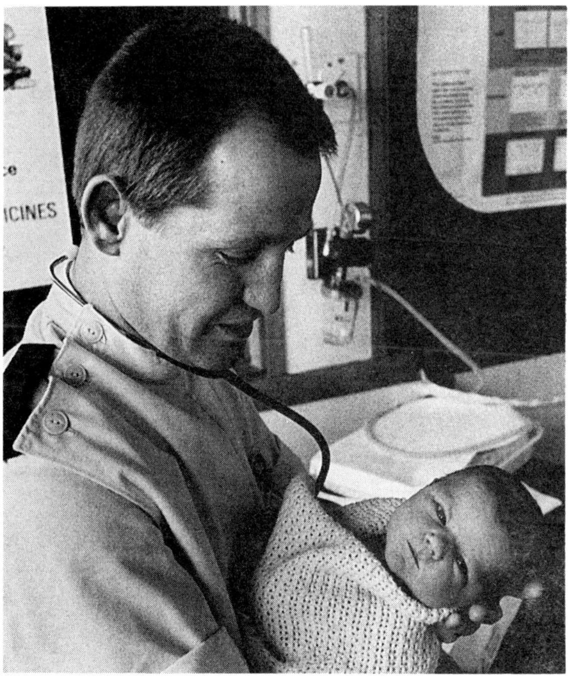

Not many men would choose midwifery as a career . . .

You don't see many female bricklayers either. Why?

Higher Educational Institutions and Courses

Institutions	*Courses*
44 universities	Degree and post graduate
30 polytechnics	Degrees, post graduate, BTEC etc.
Colleges of higher education	Certificate of Education, Dip.H.E. and B.Ed., other degrees.

Adult education

Local authority courses
Workers' Educational Association (WEA)
Residential colleges
Extra-mural departments of universities
Voluntary bodies, e.g. WI, YMCA
Youth groups
Industrial organisations, e.g. CBI, TUC
Village colleges
Community centres
Open university

Explain what is meant by

educators
informal education
formal education
compulsory education
elementary education
cultural function
economic function
continuity function
law and order function
political function
educational system
nursery schools
nursery education
play groups
primary education
neighbourhood schools
secondary education
comprehensive schools
streaming
mixed ability groups
intelligence quotients
team teaching
equal opportunity
school leaving age
Youth Training Schemes
apprenticeship

Case study

LIFE BEGINS AT SIXTEEN

'You've got to make up your mind soon, Karen,' said Jack Brown to his daughter one evening. 'You'll be sixteen next month and you've got to decide one way or the other. Either you stay on at school and get some more qualifications under your belt or you make an effort to get a job. No daughter of mine is going on the dole.'

'But Dad, it's impossible to get a job. It's not worth even applying because there's always hundreds of people like me all going after one job – and most likely they've all got more 'O' levels than me. I don't stand a chance,' answered Karen despondently.

Poor Karen didn't know what to do. Her father just didn't understand what it was like. When he left school it was really easy to get an apprenticeship, and nobody expected you to have work experience for your first job, as they seemed to want these days.

Nevertheless, Karen wrote off to several banks, near to where she lived, asking for a job. They all wrote back very politely saying they hadn't got any places, but they were enclosing a YTS leaflet for her information. Karen thought that Youth Training Schemes were considered to be the last resort; she'd heard bad rumours about them being cheap labour and so on. But she had nothing to lose by having a go, and she could always drop out if she didn't like it. Karen sent off her application and she felt quite pleased when she heard she'd been accepted for a scheme at a local bank. Her dad was pleased too. Of course, he thought that £27 was a jolly good wage, when you were being trained on the job. 'It's like being paid to go to school,' was what he kept saying.

Karen was the only person on her particular YTS at the bank, and she was a bit lonely at first. But the staff were all very friendly and they soon accepted her. After only a couple of months, the manager asked if she would like to stay on as a proper employee. When Karen told her parents that evening, they were naturally delighted.

'You've obviously worked hard and proved you're a reliable person, Karen. I'm proud of you, love,' said Jack Brown fondly to his daughter. 'And now you've got a real start on the career ladder, I bet.

1. Some people are very sceptical about YTSs, saying that they provide false employment for youngsters (and therefore false unemployment statistics). What do you think?
2. Suggest some reasons why work training schemes should be made compulsory for all school-leavers.
3. Suggest three possible reasons for a lack of communication between parents and their children.
4. In what ways can the 'generation gap' be bridged?

Follow-up

1. Explain what you understand by five of the following:
   (a) audio-visual aids
   (b) curriculum
   (c) intelligence quotient
   (d) integrated studies
   (e) interdisciplinary studies
   (f) project work
   (g) social education
   (h) syllabus
   (i) team teaching
   (j) timetables
2. What are the advantages and disadvantages of continuous assessment compared with traditional examinations?
3. Suggest five reasons why some schoolgirls are becoming more aggressive than the average schoolboy.
4. List ways in which
   (a) co-operation
   (b) competition
   can be used to help educational development.
5. (a) What difficulties are likely to be met during the transition from school to work?
   (b) What job would you like to do and to what extent is your school taking this into consideration?

'Shall we put down "some form of community service," Bowker?'

Activities

1. Either make a study of a young brother or sister, or visit a friend's home where there are young children, and study some 'educational' toys. Write an account of the skills which these toys are designed to develop.
2. Over a period of one month collect newspaper and magazine pictures on the subject of education. Arrange them in order to show the various stages of formal education.
3. Find out about
   (a) the work of your own school's Parent–Teacher Association
   (b) methods of teaching reading practised in local schools.
4. List some important points connected with your own educational future after talking to an older person who is following a course which fits in with your ambitions.

Examination questions

1. Explain fully, possible reasons for the existence of private schools which command high fees, in a country where the state provides a free comprehensive education.
   (*East Anglian Examinations Board, CSE, Social Studies, 1983*)
2. Explain what is meant by a progressive school and describe how it differs from one which is traditional.
   (*East Anglian Examinations Board, CSE, Social Studies, 1984*)
3. (a) You have been asked to find out information about local facilities for pre-school children in your area. Outline how you would carry out a survey to find out the following:
      (i) What exists already for pre-school children in your area?
      (ii) What is thought about existing facilities in terms of their location and availability?
      (iii) What further facilities ought to be provided?
   (b) What might be the major difficulties you would come across in successfully carrying out this survey?
   (*Associated Lancashire Schools Examining Board, CSE, Social Studies, 1984*)
4. The provision of education is one of the most expensive and important services for which Local Government is responsible.
   Explain how education from a child's earliest age is organised, provided, and administered in an

Authority. In your answer, you may wish to take into account some of the following:
education committee
different age sectors of educational provision
different intellectual levels of educational provision
voluntary and maintained sectors of educational provision.

(*Associated Lancashire Schools Examining Board, CSE, Social Studies, 1983*)

5. (a) Name *four* functions of education in a modern society.
(b) Explain the main reasons why the Conservative Party has generally favoured the retention of public and grammar schools, while the Labour Party has tended to support moves to comprehensive education.
(c) Why are middle class children more likely than working class children to enter higher education?

(*The Associated Examining Board, GCE, Social Science, 1983*)

6. If you asked most people what schools are supposed to do in our society, they would probably say, 'teach children subjects like Maths, English and Geography.' This is partly true but schools also do a number of other things which help society in the socialisation of a new generation of future adult workers and citizens.
(a) What is meant by the term 'socialisation'?
(b) How does the school assist in the socialisation of children?
(c) Which stage of education caters for each of the following age groups?
(i) 3 – 5 year olds.
(ii) 5 – 11 year olds.
(iii) 11 – 16 year olds.
(d) Which types of school form the system of tripartite secondary education?
(e) (i) The 1980 Education Act requires schools to provide parents and the community with certain items of information in writing. State *three* of these items of information.
(ii) Suggest *four* other ways by which members of the community are informed of what children do at school.

(*North West Regional Examinations Board, CSE, Social and Community Studies, 1983*)

7. *Either*, (a) Comment on any *three* aspects of the school as a social organisation
Or, (b) Explain what is meant by *equal educational opportunity* and comment on any *two* sources of inequality and suggest how these might be alleviated or overcome.

(*Welsh Joint Education Committee, GCE, Sociology, 1981*)

8. (a) What are the problems associated with the tripartite system?
(b) To what extent has the comprehensive system solved these problems?

(*Associated Examining Board, GCE, Sociology, 1982*)

9. Describe the different ways secondary school pupils receive advice and help in finding out about future careers.

(*East Midland Regional Examinations Board, CSE, Social Studies, 1982*)

10. Teenagers receive their education both in a formal and informal manner. Explain what these terms mean and describe how these two types of education affect you.

(*East Midland Regional Examinations Board, CSE, Social Studies, 1982*)

11. How are children of school age encouraged to develop an appreciation of the arts? What facilities for doing this are there, both in and outside the school?

(*East Midland Regional Examinations Board, CSE, Social Studies, 1982*)

12. Describe the educational facilities offered by the State for children aged between five and sixteen in your area. What educational responsibilities do parents have and how do they contribute to the cost of State education?

(*East Midland Regional Examinations Board, CSE, Social Studies, 1981*)

UNIT 9
Youth

GCSE Aims and Objectives

Knowledge — to comprehend the significance of youth culture, pop music, youth organisations, the use of leisure, adolescent problems and juvenile delinquency.

Application — to interpret the importance of the 'youth explosion' through a study of youth sub-cultures (punks, hell's angels, skinheads, etc.), youth organisations (boy scouts, girl guides, youth clubs) and general difficulties experienced in 'growing up'.

Evaluation — to evaluate the part played in community life by young people.

TOPIC 9.1 YOUTH CULTURE

In what ways are young people different from the rest of the community? Have they a different culture? To sociologists the word culture means a code of ideas, beliefs and manners, a distinctive way of life. In the same way that a nation has its own special way of life separating it from other nations – in dress, language, feeling of togetherness and aspirations – so youth has its own life-style dividing it from the rest of society. It is a culture within a culture: a **sub-culture**.

Youth culture is a sub-culture because young people remain part of the community even though they separate themselves from it in many ways. It may help you to understand this concept if you think how the British people jealously guard their way of life although Britain is a member of the European Economic Community; one of the strengths of the EEC is the variety of cultures within it. Young people may sow their wild oats, but after a few years they return to the community and most of them become fully integrated. The length of time during which young people separate themselves from the mainstream of community life depends upon their background and physical make-up.

The generation gap may be narrow or wide. Some young people get on well with the older generation including their parents. Some parents lack understanding and fail to communicate with their offspring. The extent of the **communication gap** depends upon both sides. If there is an atmosphere of give-and-take then the gap may be bridged easily. But it is important that young people should 'find' themselves. They must establish a separate **identity** of their own.

There is no other period in life when a group separates itself from the rest of the community quite so much as during youth. This period usually begins in the early teens and lasts into the early twenties.

Adolescence is a time of change in the human body, and it is a time when youth wants to change the world. We all want to put the world right when we are young. It is a good thing that the community should have to face **youth protests**. It is more worrying if old-fogeyism starts at sixteen! The important thing is to

'I must be getting old – I'm turning the volume down without even being told.'

Students demonstrating against educational cuts

find the right things for which it is worth organising **demos**, and not to give up a struggle for a cause which you believe is worthy and right.

It is sad to see people of middle age, who once had high ideals, but who have forgotten the wrongs existing in society and no longer protest about such things as the plight of lonely old people or the millions of refugees in the world. There is no reason why pupils at school should not protest, as long as they do it sensibly. A serious, thoughtful protest is far more likely to get results than a rowdy, violent one which brings authority down on the heads of the protesters and probably causes people to lose what privileges they had. The great Indian statesman Mahatma Gandhi won independence for India, not by war but by passive civil disobedience.

People are likely to respect us if we protest about great issues rather than about selfish things affecting us alone. Some things are worth getting steamed up about.

TOPIC 9.2 THE YOUTH EXPLOSION

Young people have always sought a life of their own. Approaching adult status, they cling less to their families and seek companions of their own age group. They separate themselves from those younger or older than their peer group. Later they return to the community and settle down as adult members.

Since World War II the distinctive identity of youth has exploded. One reason for this is that the mass media have exploited youth for commercial ends. Young people have far more to spend than ever before and there are plenty of people ready to relieve them of their money. The word *teenager* has become associated with certain trends in fashion, music, language, magazines and entertainment. There are many **youth sub-cultures**; groups such as punks, Hell's Angels, skinheads, mods, 'headbangers' and casuals are all instantly recognisable by their distinctive style of dress and their tastes in pop music. Can you name and describe any other youth sub-cultures? There is also a trend for young people to try and be as individualistic as possible, in the way they dress, their behaviour and their musical tastes. Do you like to be different?

A skinhead couple

Punks – instantly recognisable by their hair styles and earrings!

Mods at a scooter rally on the Isle of Wight

Breakdancing is one of the latest youth crazes to arrive from America

TOPIC 9.3 POP

You will know that pop music is a vital part of the life of young people today. It was not always so. Not many years ago, boys in particular spent much of their spare time taking part in sport and in youth organisations such as the Boy Scouts and Boys' Brigade. Transistor radios and stereos have changed much of that. If you have struggled to be served around a record counter in a large store you will know all about it.

Do your parents realise the importance of music in the lives of young people? Do you see them reading *Melody Maker*, *New Musical Express*, *Smash Hits* or *No 1*? How much do they spend on records each week?

It is impossible to give an exact date to the **pop revolution**. Your grandparents may regard George Formby as the original Beatle! You wouldn't regard Bing Crosby and Frank Sinatra as pop stars, would you? What about Elvis Presley or Bill Haley and the Comets? Before you were born they were the pop idols of the fifties. Sometimes we get reminders of the activities of the youth of yesterday by revivals of old crazes such as rock-and-roll.

The key date in the commanding influence of pop music upon youth was 1964 and the key influence was the Beatles. When Brian Epstein arrived in the United States with the Beatles in 1964, the seventeen-year-olds were the largest single age group in that country for the first time.

> *'By virtue of sheer numbers and spending power, American teenagers now were more easily able to diffuse their causes, ideas and adolescent enthusiasms among other age groups. The timing of the Beatles' arrival in New York could not have been better.'*
>
> (Apple to the Core *by Peter McCabe and Robert D. Schonfeld)*

Since those days of Beatlemania in the early 'sixties, a wide variety of pop cultures have developed on both sides of the Atlantic. Bob Dylan and Joan Baez became the idols of the hippie era with their protest music of the late 'sixties. The Who, The Rolling Stones, Elton John and David Bowie – to name a few – became the superstars of the 'seventies, while the 1980s have heralded a new wave in music with the fantastic successes of groups like Duran Duran, Spandau Ballet, Wham and Culture Club

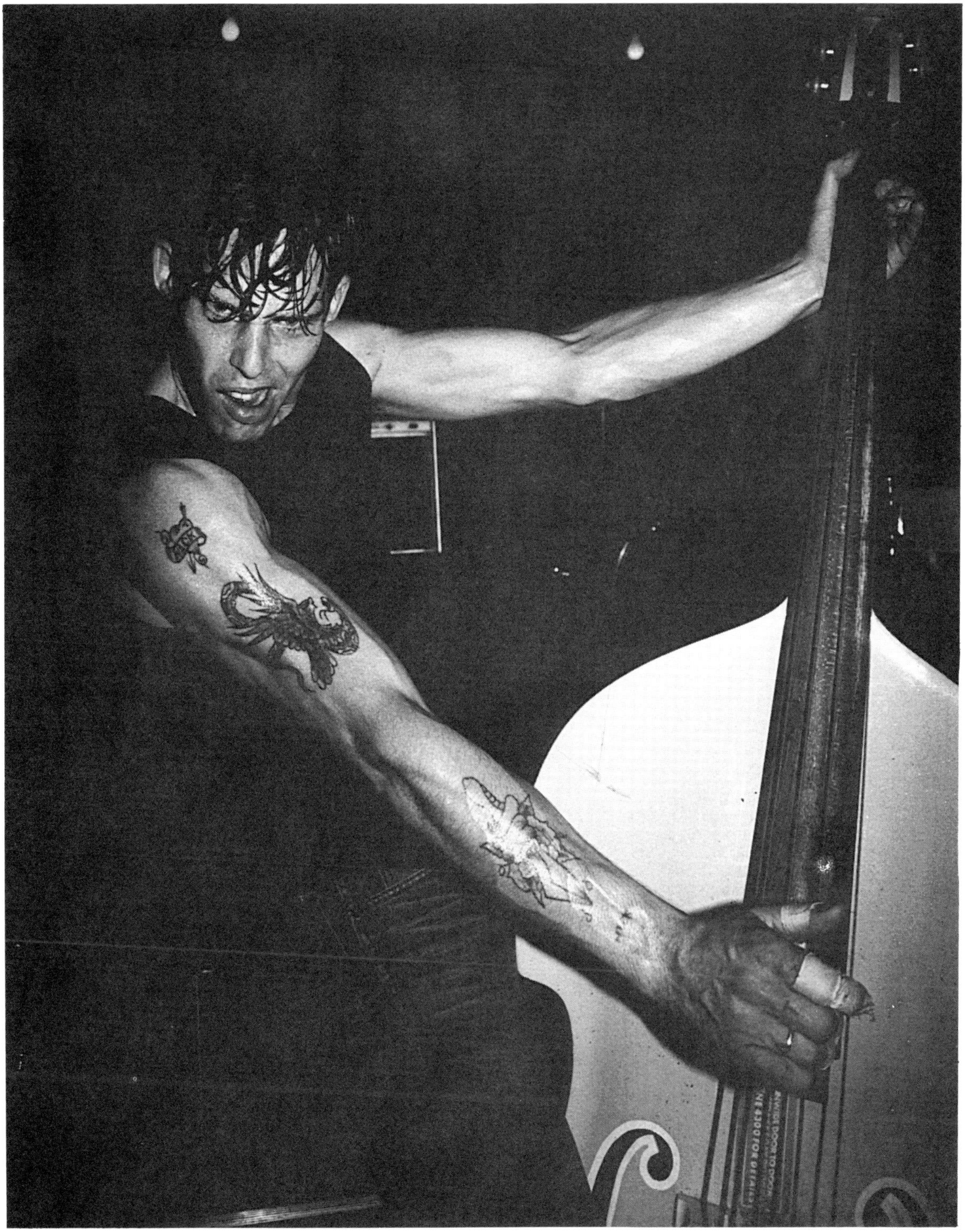

A 'rockabilly' musician at a live gig

(with Boy George). Black pop music has always retained its own distinctive sounds of Blues, Soul, Reggae and Tamla Motown, with Michael Jackson being one of the success phenomenons of the disco scene.

A huge industry has grown up around the business of making and advertising records. The music industry has given employment to groups, disc jockeys, promoters of concerts and television programmes such as 'Top of the Pops'. Have you been to a concert or a pop festival? Which records are in the Top Ten this week and who is top of the charts?

The extent of the pop revolution is best illustrated by the fact that a group of British pop stars, organised by Bob Geldof under the name of 'Band Aid', recorded a song to raise money for the starving millions of Ethiopia, and it became the bestselling record of all time. A subsequent 'Live Aid' concert held simultaneously in Wembley, London, and Philadelphia, USA, raised more than £40 million and was seen by more than 1.5 billion people in 140 countries, including China.

TOPIC 9.4 YOUTH ORGANISATIONS

It would be wrong to think that the youth revolution with its pop culture has replaced traditional **youth organisations**. Voluntary organisations provide the largest share of all youth activity and have the loyal support of 5 million young people.

Why be a member of a youth club?

- Activities
  (a) social – dances, discos and socials
  (b) physical – games, sports and recreation
  (c) educational – talks, visits and discussions
- Meeting other young people and forming new friendships
- A valuable change from staying at home and watching television
- Participation in running something
  (a) serve on committees
  (b) organise tournaments
  (c) serve refreshments
  (d) arrange the club programme, etc.
- More satisfaction than just roaming the streets
- Club leaders are people who care about youth and are worth following
- Scope for contribution to life of the community; a good club will look outwards as well as inwards

The Scout who wears a skirt

THE lads in the Rover Scouts will welcome an unusual new recruit tonight—a girl.

Beverley Bamford, 19, is joining the Scouts with her fiancee, Gavin Martin, 21, after spending 18 months helping to run the St. Ambrose Cubs in Bristol.

Beverley, thought to be Britain's first girl Rover Scout, is looking forward to mucking in on camping trips.

"But I will be wearing a skirt," she said.

The *Scout* movement, started by Lord Baden-Powell, is the most famous world-wide youth organisation. With its ideals and its motto of 'Be prepared' it has been copied even by communist countries.

Fifty national youth organisations belong to the National Council for Voluntary Youth Services. Some of the biggest voluntary youth organisations are shown in the table.

Large youth organisations

Organisation	*Membership (Thousands)*
Cub Scouts	302
Brownie Guides	419
Scouts	231
Girl Guides	336
Boys' Brigade	145
Girls' Brigade	104
Methodist Association of Youth Clubs	125
National Association of Boys' Clubs	181
National Association of Youth Clubs	730

(*Source*: *Social Trends No. 14*, 1984)

Are you a member of a youth organisation? A survey on the **Youth Service** showed that two-thirds of young people between the ages of 14 and 20 were members of some club or association. Organised youth activities are most popular with the 14–16 age group. It is hard to pass through school without being involved in some organised activity. It is likely that you belong to a club and join with members who are interested in radio or chess or stamps or model aircraft or some other hobby. List the clubs or activities at your school. To which do you belong?

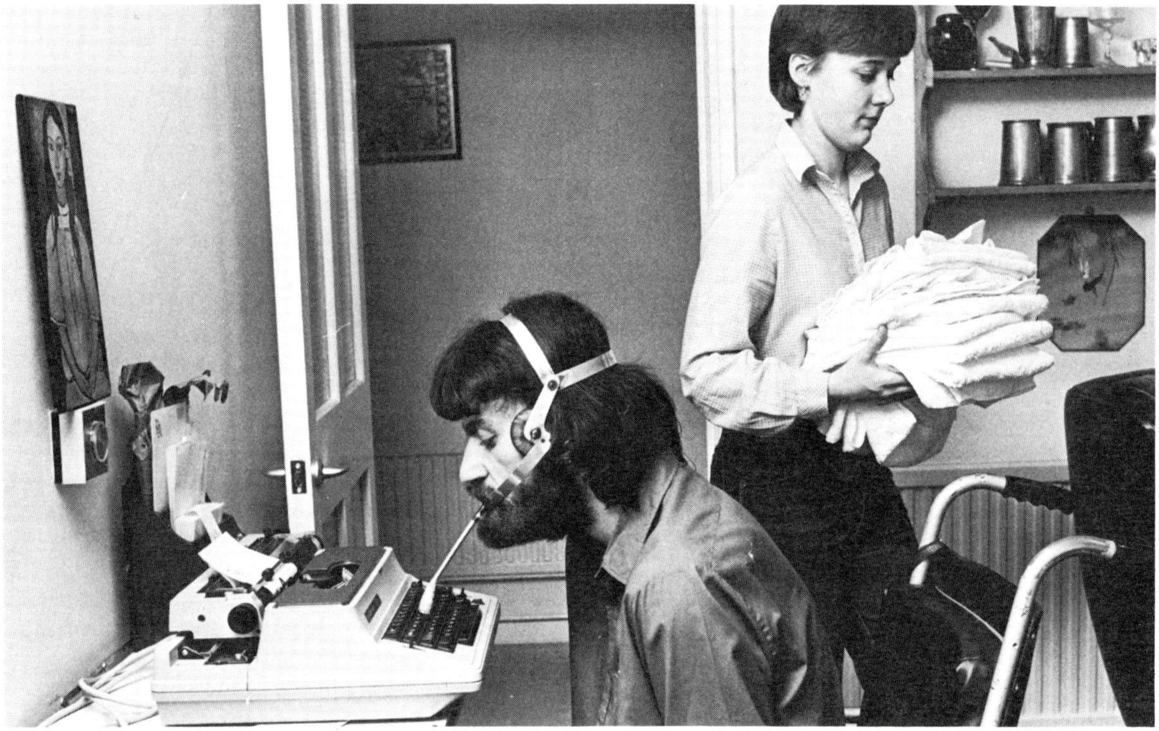

Helping the disabled is an important form of community service

A successful club will have a clear purpose. If it is a fishing club members will like to hear talks about fishing, watch demonstrations of equipment and take part in competitions. People who just drift into a club in an aimless way will not stick. If nothing is provided except a record player then the club is likely to flop. Members of a good youth organisation will be actively involved.

Voluntary youth groups in the community are helped by local education authorities who lend premises or equipment and make money grants. Local authorities have a **youth organiser** and a **youth committee** upon which voluntary groups are represented. In Britain there are 600 full-time youth officers, 1,100 youth workers, 16,500 part-time workers and 400,000 unpaid volunteers.

The best youth clubs introduce a measure of democracy into the running of the club. Although the youth leader must *lead*, he will be assisted by a committee elected by the members. A good club will supply its own leaders from the ranks, and it will be the ambition of those who have the welfare of the club at heart to grow up to be assistant leaders. The club is likely to be more alive and in touch with youth if it is mainly run for youth by youth.

However, the youth services draw upon the ideas and initiative of full- and part-time, paid and voluntary workers while catering for an enormous variety of tastes and types of young people within many different structures, the 'club' being just one example.

The diversity of Britain's 'Youth Service'

- Traditional, single-sexed movements with uniforms and leaders, e.g. Scouts and Girl Guides
- Social Clubs catering for girls and boys
- Voluntary sports associations, e.g. local football leagues
- Arts organisations, e.g. drama clubs
- Community service groups
- Activity and adventure schemes such as the Duke of Edinburgh Award and Operation Raleigh

Are there any other provisions for young people in your area? Don't forget that there are also local authority services such as swimming pools and sports centres which all cater for young people's leisure activities. How do you spend your leisure?

TOPIC 9.5 LEISURE

The *Oxford Dictionary* gives the meaning of leisure as 'free time, time at one's own disposal'. It is the time you have for yourself. The youth of today have far more **leisure** time than young people had years ago.

There is such a wide variety of leisure pursuits in the modern world that you have the opportunity to be very selective. Technology has changed our lives, and it has changed the way we spend our leisure.

Inventions have changed leisure patterns

- Improved methods of travel, e.g. bicycle, motorbike, car, railways, aircraft, hovercraft, etc.
- Broadcasting – television and radio
- Cinema and photography in general
- Records – from the gramophone to hi-fi stereo
- Mechnical aids – juke boxes, pin machines and video games
- Building constructions – swimming pools, sports centres, discos

The way people spend their leisure depends upon environment and social background. This is easy to see if you look at the extremes of class divisions. Most working-class youths spend their leisure in urban pursuits. They go 'down town' with their friends or meet them there. Their leisure time is likely to be spent in a crowd. There are plenty of people of the same age to form teams for competitive games, even though the number of recreation grounds is limited.

The professional football team, representing the town, attracts thousands of supporters, many of whom live for the matches. Life is occupied by pre-match anticipation and after-match activities. In the evenings the dance halls and the discos are crowded, while cinemas provide another place of meeting.

Urban working-class youths spend much of their leisure in spectator pursuits. They let off steam by shouting for their team. If they want to get into the countryside they have to wait for weekends or holidays so they can enjoy the open air and go fishing, cycling, camping, canoeing, rock climbing, orienteering, etc. These are all **active pursuits** and provide a welcome change from the **passive pursuits** of watching football or films. Leisure activities which provide 'kicks' give relief from repetitive factory work.

Upper-middle-class youths have parents who can provide leisure pursuits needing more space. They may have opportunities for sailing, golfing or gliding. Having a lot of money does not necessarily bring happiness. A person who has so much money that he has done almost everything may finish up like the young girl who said:

> *'I'm so bored with talking about holidays. It always causes an argument. Mother wants to go on a cruise. Father wants to motor to Rome, but we would rather sunbathe on the beach. We can't make up our minds so I don't think we will go anywhere.'*

Yet youths from all social classes can enjoy participating in sports activities. Sports such as squash, running and wind-surfing have gained huge popularity in recent years, but the traditional sports such as football, cricket, tennis and swimming still attract large numbers of enthusiasts too. Music – collecting records and going to watch live performances at concerts and festivals – is another common leisure pursuit of young people, since most have access to a stereo and spend a large amount of their pocket money or wages on records and cassettes.

Families spend £5bn on home electronics in leisure time boom

By Bill Johnstone, Technology Correspondent

The British spend two thirds of their leisure time, about 33 hours a week, using home electronic gadgets for entertainment and spend more than £5 billion each year for the privilege, more than is spent on holidays both at home and abroad.

The details are disclosed in a report by the Economist Intelligence Unit published yesterday. The study shows that the home is increasingly becoming the venue for family entertainment with television, radio, video, computers and disc players playing a significant role. The substantial sales of home computers has encouraged the purchase of second television sets as has the price reductions on most television models.

The researchers conclude: "As a form of entertainment, home electronics is most akin to other passive activities, such as visiting the cinema or the theatre, or watching sport. But home electronics is a much larger market than these and it is also growing while they are declining in real terms and, to a great extent, home entertainment is replacing passive activities outside the home."

The report compares the boom in the home entertainment market to the declining cinema audiences and the heavy concentration of the theatre in London's West End, which is increasingly becoming a tourist attraction. Also while many sports crowds are dwindling the interest in television-related or patronized sport is on the increase. This is another significant factor in the growth of the home entertainment sector.

The report says: "Sports attendances are static or in decline for the traditional sectors but there are buoyant areas in this wide classification linked to fashion and television coverage: golf, horse-jumping, snooker."

As a consequence the British spend 33 hours on average using their home electronic equipment – five times more than they spend on gardening, the nearest spare time rival for spare time, the report says.

The entire profile of what type of electronics is in the home and the products being purchased has changed in the past four years. The study highlights the growth in home computers, the remarkable expansion of the British video recorder market and the decline in the television rental market. The latter is under assault from the manufacturers of inexpensive and reliable sets. Renting in the long term appears to make little sense to many British viewers.

The change in profile has meant that the British consumer spends two thirds more on home electronics than four years ago. The sales of second television sets, now in 42 per cent of households, and home computefs which are in a quarter of homes, account for a substantial proportion of that growth. Other factors are the popularity of audio-video recorders and personal stereo equipment.

The television however is still the most important feature in the home of the 1980s and is being used to display computer programs, watch video films and live television broadcasts.

Running is a popular sport for all classes

The differences between the sexes does not cause such a dividing line between leisure pursuits as it once did. Boys are normally not interested in cookery or dressmaking, but in these days of unisex there is a greater conformity of leisure activities. Boys are better at rugby while more girls prefer horse-riding, but both sexes share in most amusements. Choirs and drama groups are more successful when both sexes participate.

Leisure is no longer confined to taking it easy or unhurried pastimes. Hobbies such as collecting coins and stamps are very popular, but many leisure attractions are connected with speed. An increasing number of people are interested (either as participants or spectators) in motor racing, speedway, stock cars and go-carts. Is this because we live in a hurried age?

It may surprise you to find that leisure and work are closely linked. The best type of professional footballer enjoys the game, whether he plays it for money or love. A qualified music teacher may spend his spare time conducting an orchestra or playing an instrument.

In the next unit we shall be looking at work. It is important to find some employment which provides job satisfaction. If you are interested in animals, then veterinary work, or looking after animals in any way, will give you a great deal of pleasure. You may find yourself happier at work than at other times. When your work is leisure, and your leisure is work, you have found the secret of a happy life.

TOPIC 9.6 HEALTH PROBLEMS

The business of growing up is bound to produce health problems. Some of these are not worth worrying about as they soon pass. What can adolescents do if they are called pimply teenagers? It helps a lot to wash the face frequently, with medicated soap if the skin is greasy, to avoid starchy food and squeezing or scratching the pimples!

Young people, especially if they are dating, want to guard against BO (body odour). Human beings have 2 million sweat glands producing two pints of fluid a day. Frequent washing and clean underclothes help a lot. If bad breath is caused by decayed teeth it can only be corrected by a visit to the dentist. But there are far more important health problems than facial spots or BO. One of these is the possibility of deafness which has increased enormously in recent years. It is difficult to convince young people who enjoy discos that the noise may damage their ears for the rest of their lives. Three other health hazards which people face are:

(1) drugs (including solvent abuse)
(2) tobacco
(3) alcohol.

These are not problems of young people alone. Alcohol is really a drug. So is nicotine which is found in tobacco. It is arguable whether alcohol and tobacco are worse evils than **drugs** such as amphetamines and cannabis. Cigarettes and alcohol are **socially accepted drugs**. This does not make them any better for the human body, but it does make them legal.

Drugs. A great problem with all these things is the question of dependence. If young people become **escapists** from everyday problems by taking drugs, there is always the chance that they will become dependent on them. It is a vicious circle. People who take drugs often start with problems. After they have taken drugs they have far more problems.

A drug is a chemical substance affecting the brain and nervous system. **Drug addicts** are people who become hooked. Addicts (or 'junkies') may have to seek supplies from a 'pusher' and the criminal underworld. So their addiction will bring them into trouble with the law. In the case of heroin and other hard drugs it may eventually mean death.

'Drug' is a very misused word. Sick people are prescribed drugs by qualified doctors. Drugs such as antibiotics, anaesthetics, antihistamines and vaccines are used to prevent or treat diseases. Habit-forming drugs which adversely affect the nervous system are dangerous. The best advice to anybody thinking of taking such drugs is 'Don't start.'

'Chasing the dragon' – inhaling heroin

Drugs coma boy aged 14 is dead

By Robin Young

Jason Fitzsimmons, the b aged 14 who had been in a co in a Liverpool hospital si Monday after taking an c dose of heroin, cannabis methadone, died yesterday

His parents, who li Norris Green, Liverpo kept vigil at his bedsid he was in intensive c life-support machine. uncle, Mr Tony Mu that the parents unable to face the switch the machin though the docto them he is really d

The boy had weeks to becor heroin. Methad prescribed to a wean them fr

Big rise in number of children sniffing glue

By Robin Young

The incidence of glue-sniffing among schoolchildren may have more than doubled in a year, according to figures produced yesterday by the National Campaign Against Solvent Abuse.

The campaign, formed by social workers and parents of glue-sniffers four years ago, revealed the results of a 1984-85 student survey in which nearly 17 per cent of respondents – one-eighth of the girls and one-fifth of the boys – admitted they had tried solvents.

That was more than double the figure (7.6 per cent) for the same fourth-year secondary school age group in 1983-84.

It is right that you should want to know about such things. Ignorance is the main enemy. Plenty of drug addicts have said that if they had known the whole story from the beginning, they would not have started. The vast majority of those addicted to hard drugs started with soft drugs such as marijuana. Drugs have fascinated mankind for generations. They can be used for good or evil.

Would you open a medicine chest and experiment by taking some of the contents? You are not likely to be as stupid as that. It is dangerous to dabble in things which even medical experts do not fully understand.

Drug addiction is becoming a more serious problem. The Home Office recorded 1,549 known addicts in 1971 and 4,371 in 1982, giving a 35% increase, but the number of press accounts indicates that there are more cases than officially recorded.

Great suffering is caused to an addict who cannot get supplies of his drug. He may suffer from such **withdrawal symptoms** as breathing difficulties, variable heart beats, vomiting, unease, depression, delusions, and may even become violent. British law allows addicts to be registered and have drugs prescribed under the National Health Service. This measure has been criticised as encouraging drug-taking, but it seems better than the American method of refusing supplies; addicts who cannot get a 'fix' may resort to suicide, violence or a life of crime. The addict becomes penniless as he is at the mercy of ruthless 'pushers', and if the addict is a parent, the children may go without food. He becomes careless about his habits and personal cleanliness. Drug clinics do a fine job, but the best cure is prevention – never to start. In 1985 there was a considerable increase in public concern about the high levels of addiction spread across all social classes and regions of Great Britain. This was reflected in an extensive, government-sponsored advertising campaign warning of the dangers of drug abuse, as well as TV programmes such as 'Drugwatch', which gave addicts advice about where they could go for help.

Another form of abuse practised by an alarmingly high proportion of very young people is glue-sniffing or solvent abuse, as it is technically known. Many tragic cases of deaths caused by continual glue-sniffing have been publicised by the media, and there is now increasing pressure on shops and stores to restrict the sales of petroleum-based glues to prevent young people from being able to obtain these highly dangerous solvents so easily.

A young glue sniffer. It may seem harmless, but sniffing glue can cause serious health problems and even death

Abused drugs

Drugs	*Effect*	*Common names*
Stimulants (e.g. amphetamines)	Temporary feeling of wellbeing, but later depression	Speed (e.g. Black bombers, Dex or Blues)
Depressant or sleeping tablets (e.g. Mandrax and barbiturates)	Remove anxiety, but usually lead to being dependent	Mandries, Nems or Downers
Analgesics or opiates (e.g. heroin)	Sense of power followed by great dependence	'H' (or Horse) or Smack
Hallucinogens (e.g. Lysergy)	Provide a trip with hallucinations and delusions. May be frightening and cause brain damage	LSD (lysergic acid diethylamide)
Soporific (e.g. cannabis)	Feeling of well-being	Marijuana, Hash or Pot

Tobacco is 'third on danger list'

Doctors rate tobacco as the third most dangerous drug behind heroin and cocaine, ahead of alcohol, glue or cannabis, according to a survey published in the doctors' newspaper, *Pulse*, today.

The survey of 100 family doctors found the numbers of those giving up smoking has increased considerably. Half had smoked at some time but given up.

The overwhelming evidence of smoking's link with lung cancer, heart disease and other conditions were the key reasons.

Tobacco. It is all too easy for adults to complain about young people taking drugs, while they themselves have been captured by the smoking habit. Smoking and drug-taking can both be addictive and dangerous. The leaves of the tobacco plant contain nicotine which is very poisonous. Fortunately about 90 per cent of nicotine is either burned or puffed out with the smoke, but that still leaves 10 per cent to be absorbed by the body.

Cigarettes provide the greatest danger for smokers. About 90 per cent of those who die from lung cancer are heavy smokers of cigarettes (over forty a day). Lung cancer is the form of cancer that is increasing most rapidly, and although more men than women die from this disease, women are catching up fast as they adopt men's habits.

You may hear people argue in favour of drugs such as tobacco and alcohol on the grounds that these drugs have a social purpose. They may say it is easier to make friends if you offer somebody a cigarette or a drink and that these things are social devices for removing **inhibitions** and making people feel more at ease. There is some truth in all this, so a good motto might be 'all things in moderation'. Would you like leprosy or lung cancer in moderation?

Alcohol. Those addicted to alcohol may become alcoholics or die from alcoholic poisoning, apart from the dangers of losing self-control so that they become fighting drunk or commit some sexual act which they spend a lifetime regretting. The community faces great dangers from drunken drivers.

There are probably over three-quarters of a million alcoholics in Britain; in 1979 over 13,000 people suffering from alcoholism were admitted to mental hospitals. The Road Research Laboratory found that 48 per cent of road deaths during a Christmas holiday period were caused by people who had been drinking alcohol.

Alcoholics Anonymous is a group of people who try to help one another to stop drinking. The advice of this society is that those struggling to give up drink *must not drink alcohol at all.*

Alcohol has harmful effects medically

- Impairs the nervous system and causes slow reactions, poor vision, blurred speech and loss of judgement
- Abnormal behaviour such as giggling, crying, aggression and rowdyism
- Lack of muscle co-ordination so that the person stumbles, falls, or drives a car carelessly
- The skin blood vessels carry an abnormal amount of blood (notice the red noses and flushed faces of heavy drinkers), and the subsequent decrease in blood supply to the liver and kidneys may be dangerous. The kidneys also suffer from lack of water because more water is sent to the skin (note the clammy flesh of the drunk)
- If alcoholism develops, the victim craves for alcoholic drink even in the early mornings. One small drink may now cause loss of control and the person becomes pathetic and depraved
- Alcoholism can develop into 'DTs' or *delirium tremens* because the brain cells are damaged; the victim has hallucinations of frightening things

Conclusion. You are bound to face problems during that stage of the human life-cycle called **puberty**. That is the stage when the sex organs mature. Puberty comes to girls usually two years before it affects boys. Girls have the problem of menstruation at the age of about eleven. Boys are usually about thirteen when sperms become active as the testes mature; the larynx enlarges and they may be embarrassed as their voices 'break'. Both girls and boys find hair growing over the pubis and under the armpits. Boys and girls become more self-conscious and more conscious of each other as these physical changes take place. All this is part of growing up and is not to be feared.

The best way to face health hazards is to live a full life so that you do not feel the need for artificial stimulants or depressants.

Eight pints of beer and four large whiskies a day aren't doing her any good.
It's your children who suffer if you drink too much.
You may not physically harm them, though an alarming number of heavy drinkers do. But you almost certainly neglect them.
The more time you spend with a glass in your hand, the less time you can devote to your family.
That's not all. Children of excessive drinkers are more likely to become alcoholics themselves later.
Drink too much and it's your family that pays for the last round.
If you're worried about the amount you drink, see your doctor or contact the North East Council on Alcoholism. Mea House, Ellison Place, Newcastle-upon-Tyne. NE1 8XS.
Tel: Newcastle-upon-Tyne 20797.
The Health Education Council
78 New Oxford Street, London WC1A 1AH.

Most of us have aggressive instincts

TOPIC 9.7 JUVENILE DELINQUENCY

The majority of young people go straight. It is only a small minority who go wrong.

Maladjusted children find it difficult to fit into the community. If we look at the problem of **juvenile delinquency** with our basic concepts of heredity and environment in mind, we find that the main cause of juvenile crime is a failure to adjust to the environment, i.e. to rules laid down by society. Where a large number of children from one family finish up before the court, the underlying reasons are nearly always found in their surroundings.

Most of us have aggressive instincts. We learnt in our study of the individual that human action is guided by motives such as aggression which can be *anti-society*. During a war the aggressive instinct is useful to the state and the violent young man may become a hero for king and country.

When there is no war on, it is difficult to harness the energy of these young people in a useful way. In medieval times tournaments were arranged to allow individual human combat. It has been a custom of communities the world over to have **tests of manhood** by which young men proved themselves. Do you think it is one of the problems of our society that there is no such test?

A depressingly familiar sight in urban areas and large housing estates today is the work of vandals. Graffiti, vandalised bus shelters, telephone boxes and public conveniences are all results of young people venting their aggressive instincts. Young boys especially seem to think it makes them more 'macho' or masculine if they show off to their mates how easy it is to break something. Or maybe it's because they are bored and have nothing useful to do? What do you think are the causes of vandalism and the sort of hooliganism that we see at football matches? Can society do anything to prevent it?

Uncurbed aggressive impulses lie dormant in many boys today, and also in girls, though to a lesser extent. But in these days of sexual equality some girls may display the same *anti-community* attitudes as boys.

By the time the child gets to the court much of the damage has been done. But attempts are made at a juvenile court to look with a caring nature upon those brought before it. The parents are brought in to consult with the magistrates, and the newspapers are not allowed to print names of offenders because that would make more difficult the job of reclaiming them for society. The main object of the juvenile court is not to punish but to help the young person to readjust to society.

Juvenile delinquency often occurs in derelict areas

But the seeds of juvenile delinquency are sown at an early age. The lack of a warm parent-infant relationship may prevent the child from maturing into responsible adulthood. A youngster who is deprived of a normal upbringing is likely to find life lonely and meaningless. On the other hand, some children may go wrong because of soft, doting parents. Children can have things too easy.

'He doesn't seem to turn to Teddy for comfort any more.'

Some causes of juvenile delinquency

- Instincts of aggression, acquisition, etc.
- Lack of parental affection
- Poverty in the home
- Illegitimacy
- Brought up by single parent
- Examples from brothers and sisters
- Bad urban area
- Feeling of insecurity and injustice
- Emotional disturbance
- Over-indulgence by well-to-do parents
- Inadequacy, mental and physical
- Neglect in upbringing
- Loneliness – few friends
- Weak parental discipline
- Misunderstood at school
- Injustice of a society in which wealth and income are unfairly distributed
- Boredom

Explain what is meant by

sub-culture
communication gap
identity
youth protests
demos
youth sub-cultures
pop revolution
youth organisations
The Youth Service
youth organiser
youth committee
leisure
active pursuits
passive pursuits
drugs
socially accepted drugs
escapists
drug addicts
withdrawal symptoms
inhibitions
puberty
juvenile delinquency
tests of manhood

Case study

ONE MAN IN A BOOT

Kevin Smith was seventeen. It had been his lifetime ambition to be a professional footballer and play for his local team, the United. He had already signed terms as an apprentice professional and was well known at the ground.

The manager was a hard man and much of the club's success in recent years had been achieved through his devotion to the club. He had no time for what he called young undisciplined softies. 'What I want,' he would declare, 'are young men of character.'

Kevin found he was expected to do many odd jobs at the ground. On Mondays he rarely saw a football as he swept the stands, picked up litter, carried crates of empty bottles and collected the players' kit for laundering. Tuesday was the best day of the week because he trained with the first team. Kevin had been playing in the Reserves for two months and had the makings of a star. He had scored in every game and his name was beginning to be mentioned in connection with a first-team place.

The manager was shrewd in his treatment of young players. His idea of a good player was one who played for the *team* and who was not selfish with the ball to gain glory for himself. He had ticked Kevin off several times for hogging the ball. Kevin began to think the manager did not like him. He was wrong. The manager liked Kevin enough to be determined to make a good footballer out of him – for the United's sake, not Kevin's! He had to find some way of making Kevin realise that he still had a lot to learn. He had to take the youngster down a peg.

'You're still a one-footed player,' he snapped after Tuesday's training session, when Kevin had thought he had done particularly well. 'You'll never be any good to United until you can kick with your left boot. I'll give you some special training next Tuesday.'

On the following Monday the manager immersed a football in a pail of water and left it there until the next day. As Kevin ran out for his training session with the first team he was feeling very pleased with himself. He had scored a hat trick for the Reserves on Saturday, he had a pair of new boots and the first team skipper had just congratulated him. Kevin had forgotten the manager's threats.

'Smith!' said the manager sharply, before a ball had been kicked, 'go and take off your right boot, sock and shinpad.'

'But I . . . ,' began Kevin.

'Don't argue with me, lad, or you'll get my boot behind you.'

Kevin trooped off dejectedly and returned with a naked right foot.

'Now you can play in the practice game,' said the manager, 'and don't chicken out of anything.' He threw Kevin the ball. It felt like a lump of lead.

Kevin kicked off gently with his left foot, received a return pass and went racing down the field. He veered over to the right of the goal and shot with his trusty right. Then he yelled. He thought his right foot had come off; he sank to the ground.

'Get up, you lazy blighter,' shouted the manager. 'You're not on a picnic.'

By the end of the practice game Kevin was kicking quite well with his left. He had even scored with a left-foot drive. The manager put his arm round him on the way back to the dressing room.

'That was pretty good, Kevin lad,' he said softly, 'We'll make a man of you yet. By the way, I'm giving you a chance in the first team on Saturday.'

1. Make a list of things you do not do well, and which you try to avoid practising (e.g. mathematics, French, science?).
2. Why do so many youth organisations such as Scouts and Guides stress discipline, uniforms and a code of conduct?
3. Relate in your own words an example of how something 'made a man (or woman) of you'.
4. Try to find examples of various tests or initiation ceremonies practised in other societies to mark a young man's 'coming to manhood'.

Follow-up

1. Suggest five ways in which your culture differs from that of your teacher because unlike you he or she is an adult.
2. Write a case study describing the activities of a young acquaintance who has been in trouble with the police. Use fictitious names.
3. 'The teenager is an international phenomenon: from San Francisco to London, from Leningrad to South Africa, the youth revolution is continuous.' (Kenneth Leech in *Youthquake*, 1973.) Describe youth activities that you have heard about in other countries.

Activities

1. Find out all you can about
   (a) the NSPCC
   (b) the National Campaign against Solvent Abuse
2. Visit a disco or youth club in the role of an observer, not a participant. Make notes about any distinctive
   (a) activities
   (b) attitudes
   (c) clothes
   (d) language (not just swearing).
3. Collect cuttings of some youth protests (e.g. about students' grants). Divide them into those that are
   (a) intended to get something *for the people concerned*
   (b) intended to get something *for other people*.
   Assess your findings.
4. Make a survey of TV programmes and list ways in which they attempt to influence young people.
5. Use your public reference library and list five books about youth. Write a book review of *one* of them.

Examination questions

1. What aspects of the modern pop scene may be considered harmful to young people?
   (East Anglian Examinations Board, CSE, Social Studies, 1983)
2. Why has drug addiction become much more commonplace in recent years?
   (East Anglian Examination Board, CSE, Social Studies, 1983)
3. Either, (a) Describe what is meant by *youth cultures* and comment on any two possible reasons for such cultures.
   Or, (b) Describe what is meant by, and comment on any two aspects of the family's function of socialising the young child.
   (Welsh Joint Education Committee, GCE, Sociology, 1982)
4. Discuss *three* factors which affect the way in which an individual uses his leisure.
   (East Anglian Examinations Board, CSE, Sociology, 1984)
5. Discuss the problems young people face between the ages of eleven and sixteen. Consider such factors of leisure time, school and work.
   (East Midland Regional Examinations Board, CSE, Social Studies, 1981)
6. (a) State *two* difficulties or problems which face teenagers.
   (b) Write a paragraph of advice on coping with these two problems.
   (c) Give an account of the qualities you think are needed to be an ideal parent of two adolescent children.
   (Welsh Joint Education Committee, CSE, Social Studies, 1981)
7. (a) Outline some of the areas in which there has been an expansion in the provision of leisure activities over the last four years.
   (b) Suggest areas in which further expansion is desirable and outline how you would organise this.
   (Yorkshire Regional Examinations Board, CSE, Humanities, 1981)
8. Give examples of and comment on positive contributions that can be made by young people today
   (a) in the family
   (b) in youth clubs
   (c) at school
   (d) at work.
   (Yorkshire Regional Examinations Board, CSE, Humanities, 1980)
9. 'Definite links found between cigarette smoking and lung cancer, and heart disease.'
   'Smoking cigarettes can seriously damage your health.'
   The above statements are two of many which officially claim that smoking cigarettes is dangerous to health.
   (a) Why do people smoke cigarettes?

(b) Why do smokers continue to disregard official health warnings?
(c) How would you go about convincing a class of sixteen year-olds in a school that either they should give up smoking or not start to smoke at all?

(Associated Lancashire Schools Examining Board, CSE, Social Studies, 1982)

10. Every year the media is full of stories about football hooliganism. This behaviour seems to occur in almost every part of Britain, and affects the 'supporters' of almost every major football club.
   (a) Name the main types of football hooliganism which occur.
   (b) Other sports attract as many followers as football, yet hooliganism does not appear to occur on anything like the same scale. Give four reasons why you think that hooliganism seems to occur more widely at football matches.
   (c) How do you think football hooliganism might be reduced? Give suitable reasons for your views.

   (Associated Lancashire Schools Examining Board, CSE, Social Studies, 1982)

11. Drug abuse is on the increase in the United Kingdom, although it is itself part of a world-wide problem.
   (a) What is drug abuse?
   (b) Why is drug taking so widespread in our society?
   (c) How do you think the problem of drug taking might be reduced in our society?

   (Associated Lancashire Schools Examining Board, CSE, Social Studies, 1982)

12. (a) Distinguish between work and leisure.
   (b) What is meant by job satisfaction?
   (c) Select *two* different occupations and compare the job satisfaction of each from *four* points of view.

   (Associated Examining Board, GCE 'O' level Economics/Social Economics, 1982)

UNIT 10
Industry

GCSE Aims and Objectives

Knowledge — to provide an understanding of the world of work, the location of industry, the production of goods, modern technology and the protection of consumers.

Application — to apply knowledge gained of industrial activities to such spheres as manufacturing, commerce, services, development areas, specialisation, automation etc.

Evaluation — to assess the value of changes in modern industrial development.

Built by robots – automated assembly lines have revolutionised the car industry

TOPIC 10.1 WHAT IS AN INDUSTRY?

If your teacher says you are an industrious person he means that you are good at working. You may be good at writing or woodwork or social studies. Similarly, when the economist uses the word **industry**, he means all types of work. He includes workers who earn their living in factories, offices or shops, plus nurses, teachers, policemen and all those who work with people.

Here is a useful threefold division of workers:

1. **Industrial workers** are concerned with the production of goods. They include people who produce goods from which other things are made: miners who get the coal and farmers who grow wheat which will eventually be made into bread. The 6 million men and women who earn their living in our factories helping to manufacture goods are also classed as industrial workers.
2. **Commercial workers** help to pass goods to consumers. Under this heading we include people who work in warehouses and shops, or who transport the goods. Workers in banks are important because there would be no goods produced without money.
3. **Service workers** do not produce goods but they provide valuable services such as repairing cars or serving meals in restaurants.

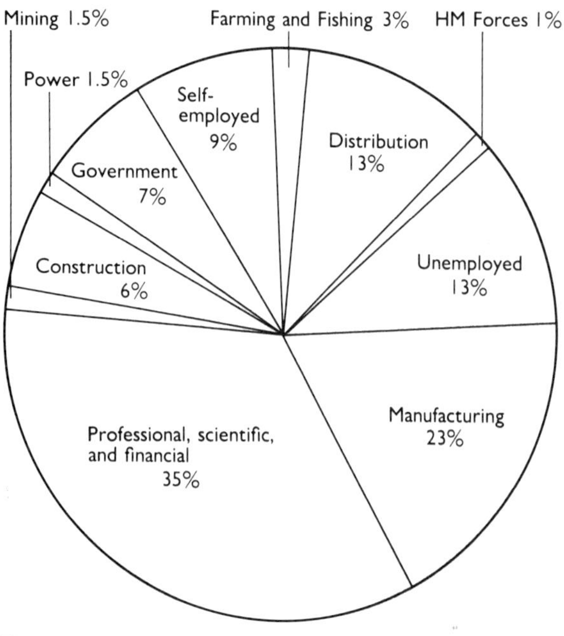

The working population, 1984

Of the 26.5 million people in the **labour force**, over 23 million are in work. This is not very different from the total in work 30 years ago, but the similarity in numbers masks big changes in the nature and location of jobs. In 1955, manufacturing accounted for 40 per cent of employees in employment; in 1984 it was 23 per cent. Yet the service sector has grown from 45 to 65 per cent. This dramatic change has been matched by a shift in the types of jobs; even in the manufacturing industry a much higher proportion of people are now in non-manual occupations, particularly scientific, technological and professional ones. The number of self-employed has grown markedly too – from 1.7 million in the 1950s to 2.4 million in 1984. You will see the reasons for these changes in the rest of this unit.

Steel mill 'to close with loss of 800 jobs'

With the shift in the occupational balance, there has also been a shift in the geographical balance, hitting hard those regions most dependent on the older industries. But other regions meanwhile have profited from the growth of new services and high-technology manufacturing.

Work for another 1,000 at microchip plant

Some industries are situated in a particular area, but others (such as agriculture and banking) are spread all over the country. There are farmers in every county in Britain and there are banks in every town.

If an industry is located in a certain area there are good reasons for it. What industries are located in your area and why are they there rather than somewhere else?

TOPIC 10.2 LOCATION OF INDUSTRY

The siting or **location of industry** is rarely a simple business. An industry is found in a certain district for a number of reasons. Although Britain has many

The shift towards a service economy; change in employment in England and Wales, 1971-1981 (Source: *Social Trends No 15*, 1985)

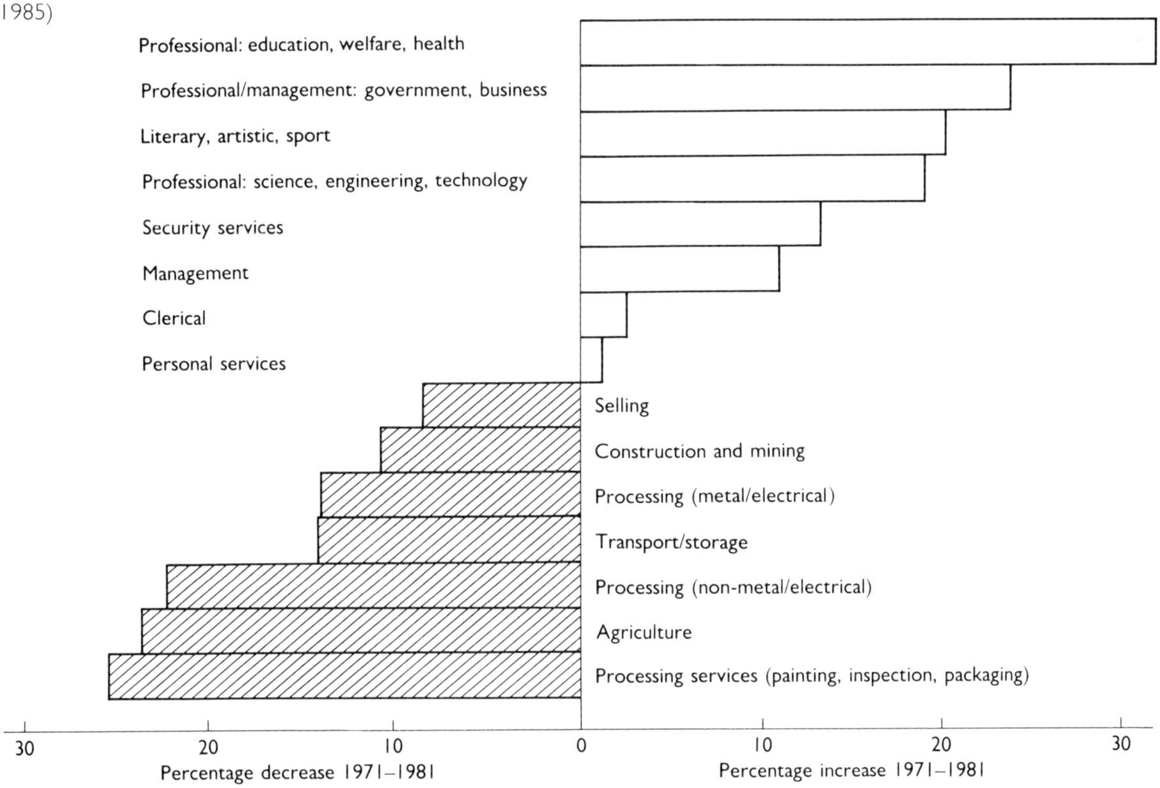

industries we have no large supplies of **raw materials** such as wool or iron ore with which goods are made. Why, then, have we a woollen manufacturing industry and an iron and steel industry?

Years ago we had large supplies of home-produced wool; in the eighteenth century there were more sheep than people in Britain! During the Industrial Revolution, iron ore and coal were mined together. Today we import from overseas most of the wool used in our woollen mills and most of the ores used in our steel rolling mills. So an industry may continue in an area even though many of the original advantages have disappeared – we call this **industrial inertia**. The industry carries on because the factories are still there and so are the people trained to do the job.

Raw materials must be transported, so industries have grown up where there are good **means of communication**. Rivers and harbours are natural geographical advantages. Try to think of at least five towns situated on a river (use an atlas if necessary). Many towns are sited on a river estuary and take their name from the river, e.g. Yarmouth at the mouth of the Yare, and Exmouth at the mouth of the Exe. Rivers are useful for moving raw materials to factories or for moving finished goods to consumers.

In the eighteenth and nineteenth centuries a large number of canals were constructed; many of these linked rivers and were very useful for the movement of goods. Josiah Wedgwood, the famous potter, arranged for an engineer called James Brindley to take a branch of the Grand Trunk Canal, joining the rivers Trent and Mersey, into one of his factories at Etruria in the potteries.

Raw cotton was transported by sea to Liverpoool at the mouth of the river Mersey. Lancashire did not produce any raw cotton, so the industry had to rely on good transport. An examination of the cotton industry shows how a number of factors combined to bring about the siting of the industry.

The increasing population provided a cheap labour force. From the Pennine streams came abundant water supplies for use in bleaching, dyeing and other processes. Water-power was used to drive factory machines. When steam-power replaced water-power essential coal supplies were near at hand. Even the climate played a part. The Manchester area is renowned for its heavy rainfall, and the damp atmosphere helped to stop the thread breaking while the cotton was being spun.

In the modern world different factors influence industrial location. The south-east of England is an area of great expansion because of the **pull of the market**. It is useful to have people near at hand to buy the goods. One-sixth of the population of England lives in the London conurbation and there is also the nearness to Europe and the markets of the EEC. Electric power has meant that industries do not have to be sited near coalfields because electricity can be transmitted hundreds of miles by the national grid.

This means that the traditional, industrial regions have suffered a tremendous decline, leaving derelict factories and high unemployment. For many years the government has helped these regions known as **development areas** by means of financial grants and loans, and since 1981, particularly hard-hit areas such as Clydebank, Tyneside, Salford and the Lower Swansea Valley have been designated as **enterprise zones**. In 1984 there were 25 of these zones, and for ten years all new and existing industrial and commercial businesses within these areas are allowed to benefit from certain exemptions designed to promote industrial development and investment.

Benefits which apply in Enterprise Zones

- Exemption from rates
- 100 per cent tax allowances on capital expenditure
- Employers exempt from industrial training schemes
- Simplified and speedier planning permission procedures
- Applications from firms in EZs for certain customs facilities are given priority
- Government requests for statistical information are reduced

With measures such as these, the government influences old and new industrial locations and it is the biggest single force in the siting of industry in modern Britain.

TOPIC 10.3 PRODUCTION OF GOODS

Most goods are produced in factories where a large number of people work together. If one man is working on his own he spends a lot of time putting down one tool to pick up another, and moving between different jobs. A more productive idea is to

Workers on an assembly line packing fish-fingers

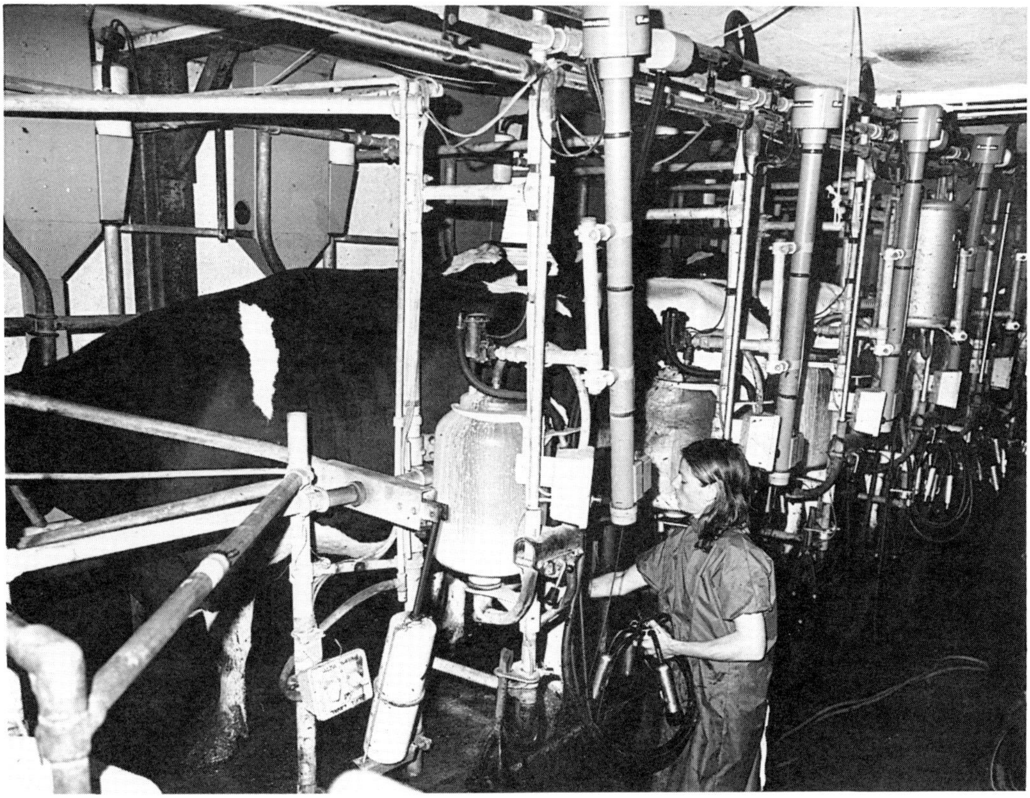

Mechanisation has transformed the traditional milking parlour

organise industry so that each worker concentrates on one task. This is called **specialisation** or the **division of labour**. A worker's task may only take a few seconds, but the article could not be produced without him. Henry Ford was one of the first successful factory owners to use the **conveyor belt system** where the work was conveyed past each employee.

Disadvantages of division of labour

- Workers become bored by repetitive tasks
- If a bottleneck develops (e.g. through a strike of a few men or a shortage of raw materials), the whole factory will stop production
- Craftsmanship has largely disappeared
- Dull uniformity of things produced, e.g. standardised models of cars
- There is a big gap between the bosses and the workers, who may never see the people who are really responsible for the factory
- A worker feels little job-satisfaction because he does not produce anything on his own

Advantages of division of labour

- More goods are produced
- It is not necessary to spend a long time training workers
- Many workers are pleased not to have to do complicated jobs
- Workers of limited ability can be employed
- The large numbers employed have amenities such as works canteens, playing fields and medical treatment
- Workers co-operate in large-scale joint production
- Piecework rates, bonuses and other incentives keep people working hard
- Employment can be found for more workers
- The increased number of workers brings an increase in the demand for goods

Division of labour involves **mechanisation** or the use of machinery. Sometimes, however, instead of work being divided up amongst a lot of workers using different machines, the complete article is produced by one enormous machine which performs all the production processes. This is known as **automation**.

TOPIC 10.4 AUTOMATION

The Industrial Revolution brought **large scale production** with workers using machines to help with the work. This process has been continued by automation but the big difference is that instead of a large number of workers being employed, the work is done by an automatic machine.

Automation refers to production by electronically controlled robot machinery. Raw materials can be fed in at one end and finished articles taken out at the other end. Even the feeding in of the raw materials and the removal of the finished goods can be done automatically. We call this **integration** and it allows production to be carried on completely by automation.

Automation has been going on for over thirty years. In the 1950s the Soviet Union had a fully automatic plant making aluminium pistons for lorry engines. By 1960 eight men produced all the pistons for Russia's motor car industry.

Computers are an important part of automation. In a few seconds they make calculations which could take a brilliant mathematician many months. Once a computer is correctly programmed it is far more accurate than a human being.

Disadvantages of automation

- Extra leisure time may cause problems such as unrest and crime
- Fewer people will be required for production so there will be redundancies and mass unemployment, especially among people too old to be retrained for new jobs
- Old skilled jobs will disappear
- It will be difficult to provide very simple jobs similar to those associated with division of labour
- Automated processes are very costly in the first place
- Only large monopolies can afford expensive automation equipment so small businesses may be eliminated
- Trade unions will find it difficult to protect their members when thousands of workers are no longer required

Computers are specially suitable for office work. Large computerised offices are working successfully in the Driver and Vehicle Licensing Section of the Department of the Environment at Swansea, and in the National Giro Centre at Bootle. In fact, most modern offices now have computers, so it is important to learn how to use them. Do you learn computer studies in your school?

Automation means that there will be less demand for labour (i.e. workers). It may be possible to prevent wholesale unemployment by offering more educational opportunities to young people (so that they stay at schools or colleges longer), or by allowing people to retire earlier (say, at fifty years of age). It would not be very surprising if the working week was cut to twenty hours.

Advantages of automation

- An increase in the production of goods and therefore goods can be priced more competitively
- An improvement in the standard of living
- Much of the drudgery taken from routine everyday work
- Extension of the power of humans:
  (a) computers 'extend' our brains
  (b) automated telecommunications 'extend' our ears and tongues
- More resources freed for improvements in education, medicine, etc.
- More leisure available

The last point in the table is of great interest to social scientists.

Automation will bring *more* leisure, but what *sort* of leisure? We saw in Unit 9, that there are many ways of using leisure time. If people work only twenty hours a week, they will have a great deal of spare time. It may happen that only the highly qualified or **specific workers** will find work, and that the people with most extra leisure time will be those least equipped to use it properly. But what do we mean by *properly*?

Who will decide how people should use their spare time? After all, it is the time they have for themselves. Is it better to listen to classical music and go to the opera, or to play bingo and watch 'Coronation Street'? We have learnt from the beginning of this book that each person is different and deserves to be accepted as an individual human being in his or her own right.

Space Invaders lure you into wasting time – and money!

Most of us would agree that it is better to be constructive than destructive. It is better to *do* something, than to *destroy* something. We would probably also agree that it is better to be active than passive. Those who have done both usually agree that there is more enjoyment in spending leisure time playing games than in watching others play them.

Automation could bring problems if extra leisure periods are not used wisely and well. There will be more time available at school to teach young people how to use their leisure well. With pocket calculators readily available, it may become more important to teach pupils to appreciate music or books and art, or to perform in plays and sports, than to spend tedious hours in pursuing traditional school subjects.

If automation allows people to retire earlier they will have more time to relax and enjoy their old age.

A frightening thing about automation and leisure is the number of **synthetic amusements** people can enjoy – games devoid of skill and individualism. No doubt we have all wasted money on one-armed bandits, pinball machines or Space Invaders. Perhaps we have spent hours inserting coins, flicking a handle hoping that luck, not skill, will record three cherries and reward us with a very small prize. No doubt we know in our hearts that if we go on long enough the automatic machine is bound to win.

It is difficult for individuals to display any originality in their workplaces in an age of mechanisation and automation. So we are witnessing a craze for **do it yourself (DIY)**, as people try to make use of their creative impulses in their leisure activities. Increased leisure time could be one of the great advantages brought by automation, so long as it is used fruitfully.

TOPIC 10.5 PROTECTING CONSUMERS

The aim of production is to make things that can be used by consumers. Years ago these things were made by one man working on his own. A tailor would specialise in making a suit and a carpenter might specialise in making furniture. If the person who bought an article found anything wrong with it, he could complain to the man who made it. If the worker valued the buyer's custom, and his own reputation as a skilled craftsman, he would put the article right or make another one.

Today production is more complicated and thousands of people can be involved in manufacturing one article. One modern car has over 40,000 parts. When there are so many things that can go wrong, and it is impossible for the purchaser to appeal to any one person, **consumer protection** is necessary. From the beginning of this book we have examined the behaviour of individuals. Sometimes behaviour becomes *group behaviour* when people join with others who feel the same way about taking a particular action; sometimes behaviour is controlled by law. There are both organisations and laws which protect the consumer.

One of the first organisations to help consumers was formed in 1901. The British Standards Institution was established by Royal Charter as a voluntary non-profit-making body aimed at maintaining good standards of quality, design and performance of products. Its 4,500 committees are responsible for testing products and checking sizes of articles. For instance it is important that electric plugs should fit the sockets. The BSI has played an important part in the change-over from the old imperial system of measurements to the **metric** system.

If production has been carried out according to the standards laid down by the BSI, the product displays a 'kitemark'. See if you can discover, within the next week, at least five articles which display this mark. While you are searching you will probably find different marks indicating that products have been approved by other organisations, e.g. an approval mark can be seen on electrical equipment considered safe by the Electricity Council. If you look carefully at a gas cooker you should find a mark indicating that it accords with the specifications of the Gas Corporation. A seal of approval is awarded by the Design Council to high-quality, well-designed work.

The Consumers' Association is a private organisation supported by members' subscriptions. Its monthly magazine, *Which?* gives detailed results of thorough tests on various products, concluding with a recommended 'best buy'.

'Watchdog' programmes on radio and television also play an important role in protecting the consumer. BBC TV's 'That's Life' programme is immensely popular as viewers like to see crooked practices exposed and things being put to right.

Some Acts of Parliament which protect consumers

Act of Parliament	*Date*	*Provisions*
Sale of Goods Act	1893	Protects consumers if goods do not come up to scratch
Hire Purchase Act	1965	Protects people who buy on credit
Trade Descriptions Act	1968	Prevents goods being wrongly described
Unsolicited Goods and Services Act	1971	Prevents goods being sent to people who do not want them
Fair Trading Act	1973	Director General of Fair Trading reviews unfair practices and investigates complaints
Supply of Goods (Implied Terms) Act	1973	Consumers prevented from signing away rights
Consumer Credit Act	1974	Credit organisations must reveal the true rate of interest
Unit Pricing Act	1975	Price must be shown according to recognised units
Consumer Safety Act	1978	Directed against unsafe goods, e.g. children's nightdresses in case of fire
Food and Drugs Act	1982	Safeguards consumers with regard to purity of food and description of drugs

There are three important groups of local government officers who help to protect consumers:

1. *Public health inspectors*, who examine places where food is prepared or sold and ensure that adequate hygienic standards are maintained and that 'sell by' dates on fresh food are strictly enforced.
2. *Inspectors of weights and measures*, who check the accuracy of scales, petrol pumps and other means of measuring goods. Since the 1963 Weights and Measures Act the quantity of all prepacked food must be stated clearly on the package, and the 1975 Unit Pricing Act states that the packet must be clearly marked according to price per unit weight, e.g. pounds or kilograms.
3. *The public analyst*, who visits food premises and checks the exact content of tins, jars and packets.

Explain what is meant by

industry
industrial workers
commercial workers
service workers
labour force
location of industry
raw materials
industrial inertia
means of communication
pull of the market
development areas
enterprise zones
specialisation
division of labour
conveyor belt system
mechanisation
automation
large-scale production
integration
specific workers
synthetic amusements
do-it-yourself (DIY)
consumer protection

Quality control on wine. Some imported wines have been found to contain additives usually found in anti-freeze

Case study

FEET OF CLAY

Once upon a time there lived a very wise old man. He was called a sage. This man was so clever that he made a machine which could make real live men from clay.

These little people worked for the sage and he ordered them to produce a large quantity of goods. They constructed dams to supply power and transported goods all over the country so that the people shared in the new wealth.

The country became rich and the king commanded the sage to make more of these men from clay. Soon nobody had to work. But still the people grumbled and clamoured for the sage to make more slaves from clay.

These man-made workers laboured very quickly, but they also grew very quickly, and the sage found it difficult to keep them under control. When they reached the height of average men the sage wrote *dead* on their foreheads and they died.

When the people of the country kept demanding more slaves, it became impossible for the sage to get round to all of his creations and write *dead* on their foreheads. One day he discovered that some of the slaves had grown so fast, and had become so tall, that he could not reach their foreheads to write *dead*.

The workers made from clay refused to work. They were out of control.

Eventually, one of the slaves grew so strong that he rebelled and killed the sage. The rest of the people in the country were very frightened but had no idea what to do. In the end the man-made workers killed all the greedy people and took over the country for themselves.

1. In what ways might automation take over from human beings?
2. Why do some experts argue that computers can only do what they are programmed to do?
3. Is it possible for a computer to beat an expert at a game of chess?
4. What are the dangers to mankind of using computers to control ballistic missiles with nuclear warheads?

Follow-up

1. Copy out the following table and add to it.

Industrial workers	*Commercial workers*	*Service workers*
fitter	insurance agent	laundrette assistant

2. Give one advantage and one disadvantage of the following:
   (a) piece-work
   (b) overtime
   (c) night work
   (d) assembly-line work
3. Prepare a script for an interview between an employer and a prospective employee.
4. The following are approval marks of consumer protection organisations or trade associations. What protection do you think they give?

CERTIFIED TO BRITISH STANDARD

BS ••••
Lic. No ••••

BEAB
Approved

BEAB
Approved
via CCA

5. What protection is offered by the following?
   (a) Weights and Measures departments
   (b) Consumers' Consultative Councils
   (c) Public Health departments
   (d) Race Relations Board
   (e) Housing aid centres

Activities

1. Visit a factory in your area and give reasons why it is sited there.
2. Draw a cartoon strip illustrating how the division of labour might be used to make one of the following:
   (a) a pencil
   (b) a match
   (c) a cigarette
3. Many interesting stories are told of computers that go wrong, e.g. the teacher who was paid £600,000 pension one month instead of £60. Collect some examples for yourself. Ask parents, relations and friends. How can the use of computers be justified?
4. Design a job advertisement including the following:
   (a) type of work
   (b) name and address of firm
   (c) pay (including overtime)
   (d) hours and conditions of work
   (e) prospects
5. Collect press cuttings of recent examples of automation and sort them into these groups:
   (a) primary work
   (b) manufacturing work
   (c) commercial work
   (d) service trades

Examination questions

1. Either, (a) 'The pattern of industry has changed in the past hundred years or so.' Comment on any of three ways in which this change has been reflected in the occupational structure.
   Or, (b) Comment on any *three* social groups whose occupational choices will be severely curtailed during a period of high unemployment.
   (*Welsh Joint Education Committee, GCE, Sociology, 1982*)
2. (a) What is meant by the chain of distribution?
   (b) How has this been affected by the growth of hypermarkets and mail order selling?
   (*University of Cambridge Local Examinations Syndicate, GCE, Economics, 1984*)
3. (a) Define 'the working population' and 'the geographical distribution of the population'.
   (b) Describe and explain the changes that have taken place in these recent years.
   (*University of Cambridge Local Examinations Syndicate, GCE, Economics, 1984*)
4. (a) What are the differences between Enterprise Zones and Special Development Areas?
   (b) What are the main aims of regional policy in the United Kingdom?
   (*The Associated Examining Board, GCE, Economics (Social Economics), 1984*)
5. The following is an extract from an interview with a production-line worker:
   'The point about this place is that the work destroys you. It destroys you physically and mentally. The biggest problem is getting people to accept it, to accept being here day in and day out. So you've got low morale from every point of view. You're inflicted with a place of work and you've got to adapt to it because it won't adapt to you. So morale is terribly low. But you can build on this you see. You can't get much lower than hell and so you can increase your morale through the union. Pull together sort of thing rather than dog eats dog. . . . We're all in it together like. That's where the union comes in.'
   From H. Benyon, *Working for Ford*, 1973.
   (a) (i) What aspects of work on a production-line could be said to 'destroy you physically and mentally'?
   (ii) 'You've got to adapt to it because it won't adapt to you.' Do you agree? What changes could raise the morale of production-line workers?

(b) How would you expect an office worker's view of his/her union to differ from this production-line worker's view of his union?

(University of Cambridge Local Examinations Syndicate, GCE, Sociology, 1984)

6. (a) Describe the part played by local authorities in the protection of consumers' interests.
   (b) Explain how, whilst shopping at the January sales, the public's interest can be protected by
      (i) government legislation,
      (ii) voluntary consumer organisations.

(Associated Examining Board, GCE, Social Economics, 1983)

7. Name an industry which is declining and explain the causes and effects of its decline. What can be done to help an area with declining industries?

(East Midland Regional Examinations Board, CSE, Social Studies, 1982)

8. Compare the advantages and disadvantages of road and rail transport for industry and the public.

(East Midland Regional Examinations Board, CSE, Social Studies, 1982)

9. Explain what is meant by mass production, looking at its benefits and drawbacks.

(East Midland Regional Examinations Board, CSE, Social Studies, 1982)

10. Give an account of the effects, both good and bad, that automation may have on our lives. You would be wise to consider such aspects as education, family and social life, employment and the standard of living.

(East Midland Regional Examinations Board, CSE, Social Studies, 1981)

11. Write an account of a firm you have studied, including such factors as raw materials, amount of trade, markets, types of skill required. What changes do you foresee for the firm in the future?

(East Midland Regional Examinations Board, CSE, Social Studies, 1981)

12. (a) Explain *three* major disadvantages of sole proprietorship as a form of business organisation.
    (b) Account for the fact that sole proprietorships are the most common form of business organisation.

(Associated Examining Board, GCE 'O' level Economics/Social Economics, 1982)

UNIT 11
The British Economy

GCSE Aims and Objectives

Knowledge — to comprehend how the British economy works with special relation to the public and private sectors, employment and unemployment.

Application — to analyse the detailed working of the economy and the part played by sole proprietors, partnerships, companies, the stock exchange, nationalised industries, privatised concerns and the trade unions.

Evaluation — to make appropriate judgements about the state and future of the British economy.

TOPIC 11.1 WHAT IS THE ECONOMY?

The word **economy** was originally used to mean the administration and organisation of households. Your household is probably made up of your mother, father, brothers or sisters and yourself. You all need certain things, e.g. beds and chairs. In addition, each member of the household has individual wants. A young man may want a motor bike, while a young woman may want cosmetics. A need is not quite the same as a want. We all *need* food and drink. We *want* lots of other things such as records and clothes in the latest fashion, although we can get by without them.

There are many different types of organisation that supply things to households. Just look around your area. There are small firms run by one person and there are larger organisations owned by a group of people. These large firms may be private companies like the General Electric Company or nationalised corporations like the Post Office.

There is a tendency towards monopoly organisations to develop in a modern economy. Strictly speaking, in economic terms, a **monopoly** exists where there is a single producer and where no adequate substitute is available. However, in practice, perfect monopoly does not exist; although the Gas Corporation has a virtual monopoly in the supply of gas there are substitutes available, e.g. one can choose electricity for cooking, central heating, refrigeration, lighting etc. Although monopoly conditions provide the advantages of large-scale productive methods, there are disadvantages of monopoly such as:

(a) output may be restricted so that higher prices may be charged
(b) prices may be raised by persuasive advertising
(c) consumers' choice is limited
(d) monopolies have less incentive to adopt new economical methods of production than do firms in a competitive market.

The first step in the successful control of monopolies is to ensure that activities that work 'against the public interest' are investigated and recorded, so that any necessary action may be taken. The Monopolies Commission set up under the Monopolies Act of 1948, has examined many monopolistic practices in such goods and services as soapless detergents, newspapers and films. Since 1973, investigations by the Monopolies Commission have been the responsibility of the Director General of Fair Trading.

The Director General of Fair Trading is now in charge of a vast organisation which supervises:

(a) The Monopolies and Mergers Commission with power to investigate practices against the public interest undertaken by private firms and nationalised industries.

(b) The Consumer Protective Advisory Council which recommends to the Secretary of State for Industry that any unfair practice against consumers should cease
(c) The Registry of Restrictive Practices which lists all restrictive trading agreements including those practised by service industries such as banking and insurance companies.

The monopoly power of nationalised industries is controlled by:

(a) the House of Commons Select Committee on Nationalised Industries which examines the annual reports and accounts of all public monopolies.
(b) Consumer Councils which enable the public to express their complaints.

A modern use of the term economy applies to the administration and organisation of the community. Communal needs arise from the demands of all households. We are going to examine two types of economic organisation supplying the needs of the community:

1. private businesses
2. nationalised industries.

Within these categories there are:

(a) big and small concerns
(b) organisations representing employers and employees
(c) producers for the home market or overseas markets
(d) firms specialising in production or distribution.

TOPIC 11.2 PRIVATE BUSINESS

The **private sector** of the economy includes both the shop at the street corner and the largest firm in the country. About 75 per cent of the British economy is still privately owned. Who are the owners?

1. The **sole proprietor** owns his business. As the owner he does the main organising, takes the risks and profits, but has to bear any losses. The one-man business is found mostly at the distribution end of the production chain. He is a most important link because the whole purpose of production is to supply consumers. The small shopkeeper is vital to the economy, for without him goods would not reach consumers. There are

'They've been bullying him again, Mrs Dawson...'

still hundreds of thousands of successful one-man businesses, especially in the retail and service trades.
2. **Partnerships** are formed when several people combine to run a business. They have two big advantages over sole proprietors:
   (a) more capital (money)
   (b) greater expertise – partners specialise in different things.

   Next time you walk through the streets of your town have a look at the nameplates of business partnerships. This will help you to remember the types of business which are organised as partnerships.

   A partnership is a convenient arrangement for doctors and dentists because it is necessary to have one partner available in times of emergency. It is also very convenient for a firm of solicitors to have partners who are experienced in conveyancing (handling the deeds connected with property deals), while other partners are experts at representing clients in courts of law.
3. **Companies** are the largest form of private business organisation and some are very large indeed. Companies which own businesses operating all over the world are known as **multi-nationals.** Large **public companies** are owned by many **investors** or **shareholders** who have bought shares in the companies.

Largest companies (1982)

Ranked by turnover	*£ million*
British Petroleum Company	34,583
Shell Transport and Trading	21,910
BAT Industries	11,318
Imperial Chemical Industries	7,358
Shell U.K.	6,665
Esso Petroleum Company	6,109
Unilever PLC	5,447

(*Source*: *Britain 1984*, HMSO)

Large-scale production requires far more capital than could possibly be supplied by one man or a partnership. British Petroleum is a huge oil company while ICI makes hundreds of different products based upon chemicals.

These public companies concentrate mainly on production but they are also engaged in storing and selling their products. You will have seen petrol pumps bearing the brand signs of BP and Shell. Even Marks & Spencer does not just own department stores. The goods it sells under the brand name of St Michael are made in factories supervised and checked by the company.

How is a company organised? There is an annual meeting of shareholders which elects a **board of directors**. Shareholders vote according to the value of the shares they own. Those with most shares have the biggest say in running the company. Only a few shareholders attend the annual meeting, so directors are appointed by those with the greatest interest. These are likely to be the largest shareholders. If you were a company director you would be expected to attend meetings of the board. A *company chairman* is appointed by the board, and he is the most important man in the company. Being a company director may not be a full-time job and many businessmen are directors of several companies. A full-time *managing director* or general manager is appointed to look after the day-to-day running of the business. He will co-ordinate the various activities of the company.

A *chief accountant* will be appointed to supervise the financial side. He has the responsible job of seeing that the company's money is spent carefully, books are properly kept and the accounts are bal-

Directors oust Beecham chairman

By Cliff Feltham

Sir Ronald Halstead, chairman of the Beecham Group, was ousted in a boardroom coup yesterday as the company announced disappointing profits. A stunned City immediately wiped £280 million off the value of its shares, which tumbled 33p to 293p.

Sir Ronald, who was also chief executive, will probably receive nearly £600,000 in compensation for the loss of the two posts which he has held only since July last year.

Highly persuasive advertising campaigns are now a common feature of major take-over bids anced. The *personnel officer* is in charge of the appointment of staff, their general welfare, working conditions and problems that arise at work. He may have to deal with trade union officials representing the workers. The *sales manager* is in charge of selling the product while the *marketing manager* is responsible for co-ordinating the advertising campaigns and market research. A public company invites the general public to invest in its business by selling shares on the *stock exchange*. Anybody can buy shares in a public company if he has the money. Many investors buy and sell shares through their bank, although the bank will have to contact a *stockbroker*.

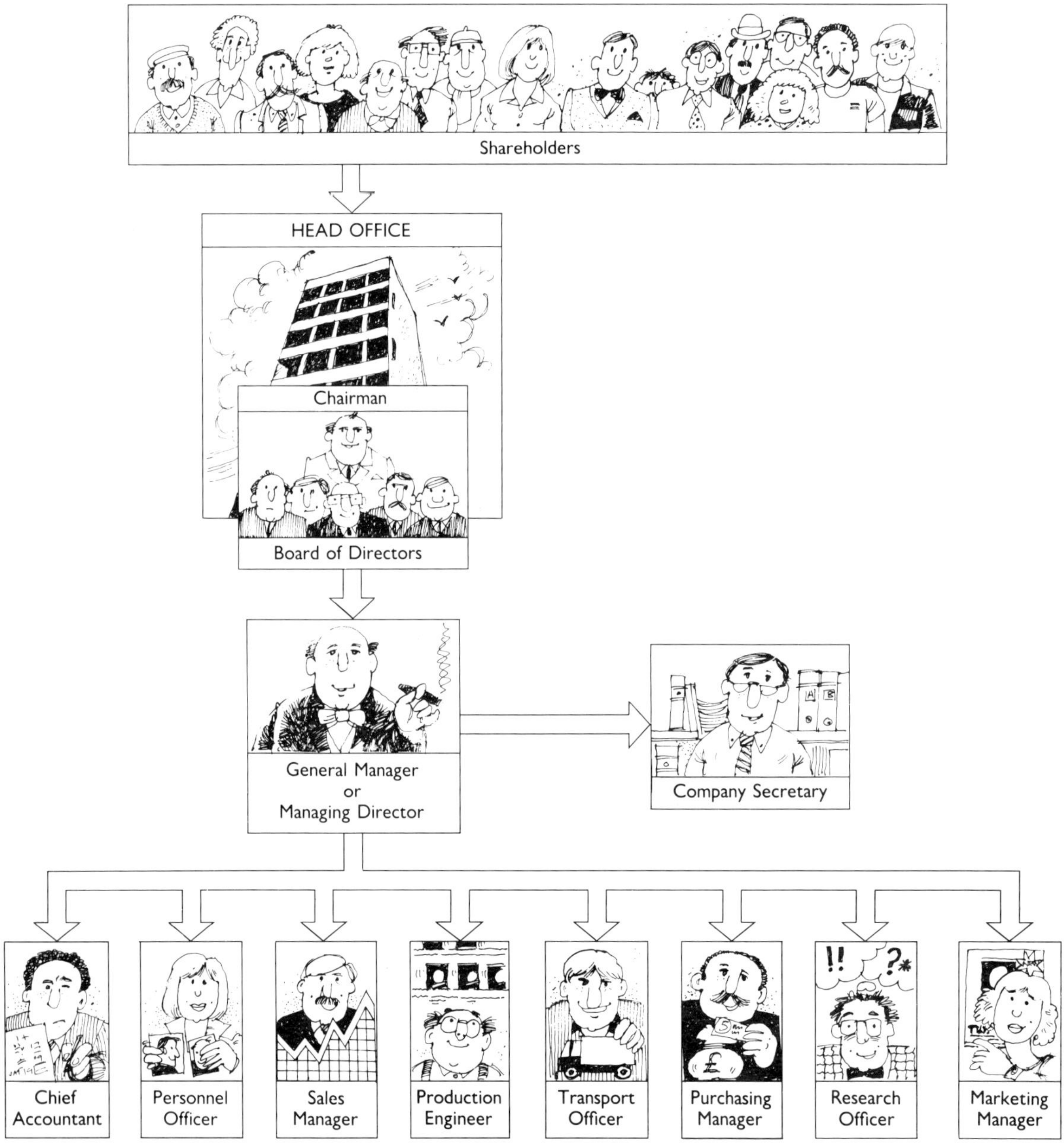

Company organisation

The Stock Exchange

Imagine you owned shares in a public company and wanted to sell them. You could go direct to a firm of stockbrokers or contact your bank manager who would pass on your instructions to a broker. Look at the photograph of the trading floor of the Stock Exchange. It seems a muddle, but the broker knows his way about. There are sections specialising in shipping, breweries, oil shares, etc.

Let us assume that you have 100 shares in Jaguar, the car company. Perhaps you bought them in September 1985 when the price was 286 pence a share. It is the broker's task to secure the best possible deal for you as his client. The broker contacts a *jobber* who buys and sells shares. The jobber quotes two prices – a buying price and a selling price. The latest Jaguar price might be 300 pence for buying and 320 pence for selling. The difference in the two prices is known as the jobber's turn. It is his profit. If this is the most favourable quotation the stockbroker can obtain, he will make a deal and the jobber will buy your shares at 300 pence each or £3,000 total. Even after paying the broker's commission you will have made a profit. But investing on the Stock Market does not always pay. If share prices fall investors may make a loss. Even if they make a profit there will be capital gains tax to pay.

There is far more money invested in public companies in Britain than in any other form of economic organisation.

Functions of the Stock Exchange

1. *Primary market*: Enables companies and governments to raise money (to expand businesses etc.). As a result
   - (a) employment is created for workers
   - (b) institutions such as insurance companies and trade unions are encouraged to invest their spare cash in industry
2. *Secondary market*: Offers a market-place for people who want to buy or sell shares

'Now, stockholders, if you have all drained your glasses, I'm afraid we have some sobering news for you!'

At the end of 1982 there were about 900,000 companies on the companies' register of which about 6,000 were public limited companies. Private companies are usually small family concerns. The maximum number of members or shareholders is fifty. The shares of private companies are not sold on the Stock Exchange. Have you ever seen the abbreviation 'Ltd' or 'plc' after the name of a firm? They mean 'limited liability company' and 'public limited company' respectively. Both private and public companies enjoy **limited liability**. This means that if you invest £50 in a company your liability is limited to the sum of £50. If the firm goes bankrupt you only lose £50. That may seem a lot, but before the Limited Liability Acts of 1855–6, shareholders in a company that went into liquidation could be made to sell their property to pay the debts of the company. In those days it was not unknown for unlucky investors to have to sell their furniture to help pay off the company's debts.

TOPIC 11.3 NATIONALISED INDUSTRIES

A **nationalised industry** is owned by the state. About a quarter of the British economy is now under state control and included in what is called the **public sector** of the economy. The word 'nationalisation' was originally used to indicate the transfer to state ownership of a privately owned industry. In recent years an industry such as the Post Office, which has always been organised by the state, has been said to be nationalised because it has been placed under a public corporation and is no longer run by civil servants. Postal services were nationalised by an Act of 1969 and a separate board was appointed to run the organisation as a state industry. Previously the Post Office had been directly controlled by a cabinet minister known as the Postmaster-General. Since 1969 the Post Office Board has been responsible for the day-to-day running of the business. There is an annual parliamentary debate about the Post Office, just as there are debates about the accounts of all the nationalised industries, but MPs are not entitled to ask questions about the detailed running of any nationalised industry. Anybody who has complaints about a nationalised industry must contact the *consumers' consultative council* for that industry.

In any mixed economy, where some industries are privately owned and some are publicly owned, it is important that the state should be responsible for the **basic industries** on which other industries depend. The government aims to exercise control over the economy by planning these essential industries. It tries to ensure that there is sufficient money to provide modern equipment. Since 1945 most *key industries* in the British economy have been nationalised. The citizens of the country own these industries collectively. The owners of industries that are nationalised are compensated according to the value of the shares owned.

The responsibility for nationalised industries lies with Parliament but everyday organisation is undertaken by a board. The appropriate minister appoints the chairman and members of the board.

Industries nationalised since 1945

Industry	*Authority*
Coal	National Coal Board
Electricity	Electricity Council
Gas	Gas Corporation
Railways	British Railways Board
Civil Aviation	British Airways
Steel	British Steel Corporation
Post Office	Post Office Board
Central Bank	Bank of England Court
Oil	British National Oil Corporation
Aircraft	British Aerospace
Shipbuilding	British Shipbuilders

During the 1980s, the Conservative Government has embarked upon a policy of **privatisation** whereby certain industries have been sold to private shareholders and non-government institutions. Privatisation is sometimes referred to as denationalisation, but it is really more than that because it refers not only to the return of public sector industries and services to the private sector, but also to the transfer of public services which were never in private hands. The most notable assets which the government has sold off include parts of Britoil, British Aerospace and Cable and Wireless. After the successful sale of 51 per cent of British Telecom (BT) to the private sector in 1984, it was suggested that privatisation might take place next in British Airways, the Gas Corporation and the Trustees Savings Bank. It is argued that privatisation could benefit the consumers (who may receive a more efficient service) and the taxpayer (who will benefit from the proceeds of the sale). However, if the privatised industry had been making a profit and contributing to government funds, then the taxpayer would lose.

Nationalisation arguments

In favour
- State ownership of basic industries allows proper planning of the economy
- Government money is necessary to modernise industrial equipment
- Nationalised industries can co-operate easily in international schemes
- Some industries are better organised on a national basis, e.g. railways, Post Office, gas and electric super-grids

Against
- Nationalisation restricts the working of competition throughout the economy
- Nationalised organisations are very large and bureaucratic with too much red tape
- Some nationalised industries have to be subsidised by the taxpayer
- Local feelings may be overridden in the interests of national schemes

One of the problems of a mixed economy is the proportion of the mixture. Should the state control 20 per cent, 25 per cent or 50 per cent? If we had a communist state, like the People's Republic of China or the Soviet Union, the government would control almost all the economy. In these countries some property is still owned by individual members of the community. The co-operative principle is used on the collective farms and communes, but a communist would still wish to have his own toothbrush. Capitalism and communism are forms of economic organisation.

It is very wrong to think that nationalised industries always run at a loss. The Electricity Council and the Gas Corporation run very successful nationalised industries and make large profits. The two nationalised industries in recent years that have been faced with most difficulties have been British Steel (at one time losing nearly £1million a day) and the coal mining industry which has been faced with the problems of uneconomic pits and the resultant miners' strike of 1984/5.

Largest nationalised industries (1983)

Ranked by turnover	*£ million*
Electricity Council and Boards	8,417
British Telecom (privatised 1984)	5,708
British National Oil Corporation	5,671
British Gas Corporation	5,199
National Coal Board	4,727
British Steel Corporation	3,443

Source: *Britain 1984*, HMSO)

£6bn sale of state assets

By ROLAND GRIBBEN, Business Correspondent

AN acceleration in the sale of state assets with British Gas as the centrepiece was signalled by the Government in the Queen's Speech yesterday.

The Government hopes to raise between £4½ billion to £6 billion a year over the next two financial years from the ambitious programme to help finance tax cuts and return more assets to the private sector.

TOPIC 11.4 TRADE UNIONS

So far in this unit we have looked at economic organisations without considering the workers. Employees have their organisations too, the **trade unions**. Trade unions are associations of workpeople who join together to look after their common interests. Nowadays the majority of people work for large organisations and they feel the need to combine in strength to see that they get a square deal.

There are different kinds of industries within the economy and there are also different types of trade union. There are large unions such as the Transport and General Workers' Union with 1½ million members, and tiny unions such as the Jewish Bakers with about a dozen members. Nearly half the workers in the country belong to a trade union. There are ninety-one unions representing nearly 10 million workers which are affiliated to the TUC.

Trade union aims

- Good wages so that workers have a high standard of living
- Reasonable hours of work with sufficient holidays
- More pay for overtime and piecework
- Benefits for members who are sick, retired or on strike
- Educational schemes, especially for apprentices
- Control of the supply of labour to maintain employment
- Political action in support of causes in which the union believes
- To give working people a voice on national bodies such as the National Economic and Development Council (NEDC)
- Combination with other trade unions
- Participation in talks on the economy with the government and employers

The **Trades Union Congress** is the annual meeting of the trade union movement. All the main unions are associate members of the TUC and send a number of delegates to Congress according to the size of their membership. Congress takes place in September each year, usually at some large seaside resort where there are hotels vacated by holidaymakers. It lasts for a week and discusses important topical questions affecting the British economy. A **General Council of the TUC** is elected to carry on the work throughout the year.

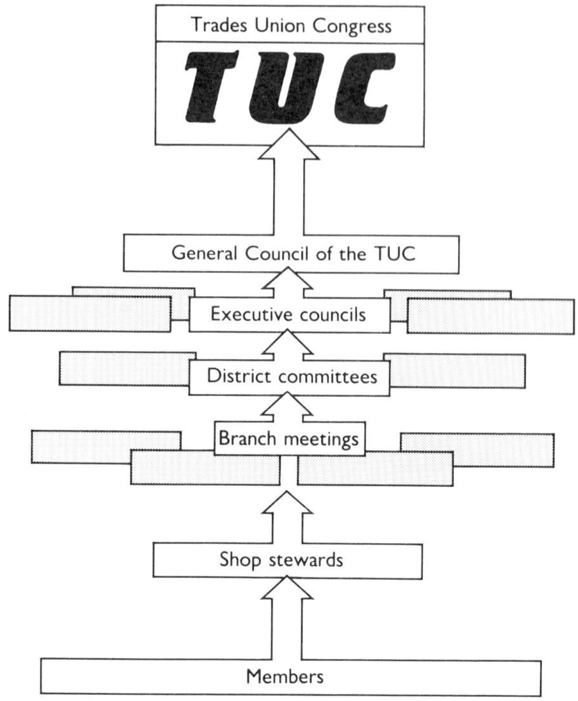

Types of union

Craft unions	Small unions of skilled men, e.g. woodworkers, plumbers, musicians and locomotive drivers
Industrial unions	Represent all the workers in one industry, irrespective of their job, e.g. National Union of Railwaymen
General unions	Largest unions speaking for a variety of workers both skilled and unskilled, e.g. the General and Municipal Workers' Union
White-collar unions	Workers on the staff side are represented by such unions as the Association of Scientific, Technical and Managerial Staffs (ASTMS)

Although the TUC discusses all sorts of social, political and economic questions such as the Common Market and inflation, the real purpose of a

The TUC headquarters building in London is often featured on television during pay negotiations or industrial action

trade union is to secure better wages and working conditions. To do this the unions carry out **collective bargaining** and negotiate with employers. It is not surprising that opposing interests result in clashes. The union wants to secure as high a wage as possible for its members and the employers want to hire workers as cheaply as possible. This means that at the beginning of wage talks there is likely to be a gap between the two sides. Employers' and employees' interests have to be brought together because both sides will prosper only if the economy runs smoothly.

A trade union will work out its wage claim bearing in mind such things as inflation (rising prices), increased productivity, the economic situation and whether or not employers can afford to pay. If employers continue to reject the unions' claims there are weapons which the unions can use to bring pressure to bear.

Trade union pressures

Method	*Actions*
Work to rule	Work is carried out in accordance with every minor rule
Go slow	Workers will take their time
Closed shop	Workers will only work with fellow trade unionists
Demarcation dispute	It is insisted that only a trained worker will do a job that could be done by another
Strike	If wage negotiations break down completely, trade unions will withdraw their labour
Picketing	Strikers' representatives dissuade employees who do not wish to strike

New miners' union brings threat of split TUC

Liverpool unions call for indefinite strike in figh[t] [a]gains[t] ...

...ll cost of pit strike ...ar £6bn, says NCB

Acas talks to unions in teachers' dispute

There is a great need to improve **industrial relations** in Britain. When trade unions take industrial action they are accused of holding the community to ransom over a matter which could easily have been settled sensibly. Both sides of industry are to blame in stubbornly refusing to use the instruments of conciliation and **arbitration** whereby an independent third party is asked to give a judgement. In the end disputes are settled by talking rather than by strikes of workers or lockouts by employers.

The 1971 Industrial Relations Act was a failure. The policy of fining or imprisoning trade unionists who went on strike resulted in a worsening of industrial relations.

The Employment Protection Act 1975 placed the Advisory, Conciliation and Arbitration Service (ACAS) on a statutory basis and established an Employment Appeal Tribunal to hear appeals from industrial tribunals relating to unfair dismissals, equal pay, sex discrimination and redundancy payments. The Employment Act 1980 made funds available to encourage the wider use of secret union ballots and limited lawful picketing to the picket's own place of work, thus restricting secondary picketing and sympathetic strikes. The Employment Act 1982 removed legal protection from all closed shops and ended the position which gave trade unions almost complete immunity from civil actions.

Do you think trade unions are too powerful? Perhaps trade unions do have too much say in a modern community, but remember that before trade unions became so powerful working people were very badly off indeed. Ask your parents or grandparents about the conditions that prevailed in the years between the world wars, or do some individual research in your library on the General Strike of 1926.

Working days lost

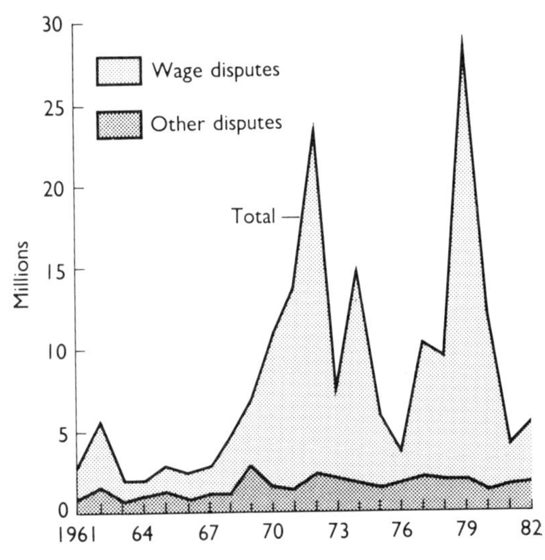

Number of stoppages

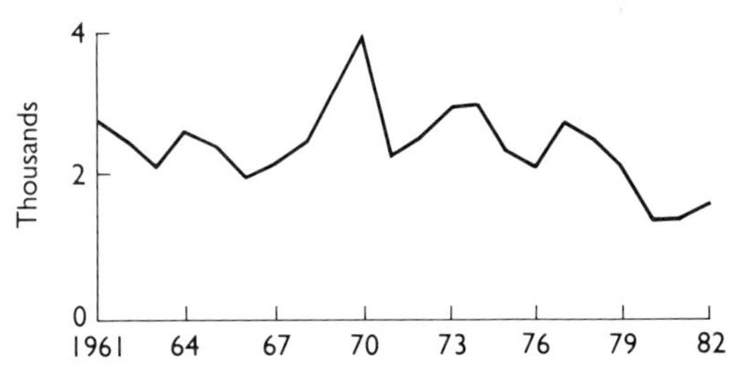

Industrial disputes in the U.K. (Source: *Social Trends No 14*, 1984)

TOPIC 11.5 UNEMPLOYMENT

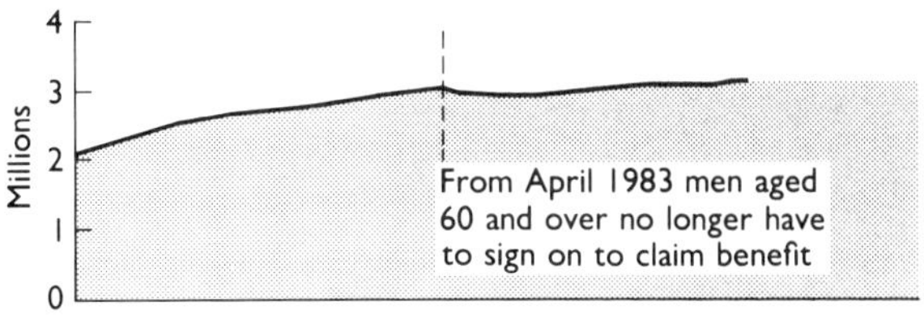

U.K. registered unemployed (excluding school-leavers)

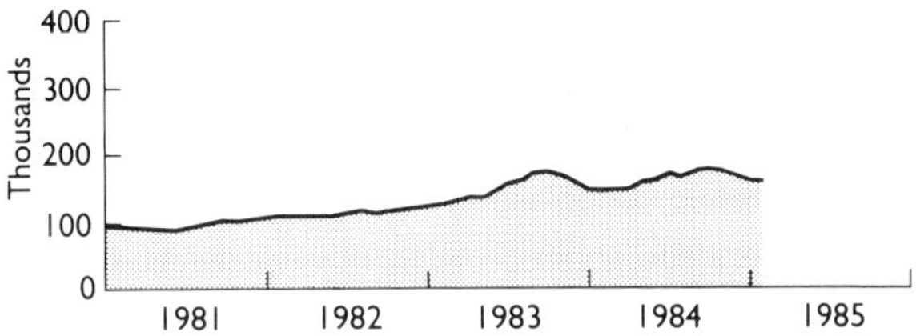

Stock of unfilled vacancies (notified to job centres)
(Source: Economic Progress Report, March/April, 1985)

Unemployment is the greatest problem of the British economy in the 1980s. Between 1978 and 1985, over two million workers have lost their jobs, bringing the total number of unemployed to more than 3.3 million.

A quarter of 18-year-olds unemployed

More than a quarter of 18 to 19-year-olds are now out of work, according to the latest annual report from the Manpower Services Commission published yesterday.

The document

There have been periods in the past when mass unemployment has occurred. In the 1930s, Britain suffered from having about three million people on the dole and the unemployment rate was usually above 10 per cent. This was often thought to be caused by the trade cycle, but mass unemployment seemed to have been conquered after World War II when we moved towards a planned economy and full employment. It was largely to prevent mass unemployment that an economist called Lord Keynes advocated 'public works' such as putting the unemployed to work constructing roads and making public parks.

Of course it would not be possible to have no unemployment. There will always be people who:

(1) are casually or seasonally unemployed;
(2) suffer from changes in demand for the products or services which they supply;
(3) work in declining industries;
(4) are unable to work because of some physical or mental illness;
(5) do not hold their jobs because they are incompetent or workshy.

Lord Beveridge in *Full Employment in a Free Society* (Allen & Unwin) accepted that if a country could keep 97 per cent of its working population employed, then that constituted a state of **full employment**. From 1945 to 1970, Britain enjoyed a state of full employment with fewer than three per cent unemployed. Compare the state of society then, when only one in thirty-three workers were unemployed, with society today when one in seven of those registered for work are unemployed. There are politicians who have declared that the economy will work better if it is slimmed down and workers fear unemployment. However, most people realise that unemployment is bad for a man, his family and the nation. Apart from the cost of unemployment benefit, a man may go 'rusty' and lose specific skills; he may even eventually lose the inclination to work.

So what has gone wrong in modern society to bring such a large increase in unemployment? There are those who would argue that unemployment is one of the hallmarks of capitalism and a free-for-all market economy; other capitalist countries are suffering from unemployment and recession with peak unemployment percentages in the 1980s, e.g. Belgium (15.4%), Holland (14.8%), Canada (12.7%), Italy (11.6%), U.S.A. (10.7%) and West Germany (8.1%). There are other people who blame monetarist economic policies whereby governments try to control the money supply too rigidly and are unwilling to reflate the economy and increase government expenditure as Lord Keynes suggested. Others consider that the main cause of unemployment in modern society is technological change (such as automation, robots, computers and the micro-chip revolution) which made so many workers **redundant**. Some jobs have undoubtedly been lost because of the advance of modern technology and this particularly applies to British manufacturing industry where most of the two million jobs were lost during 1978–85. Over 50,000 jobs were lost in local and national government services and this reflects reductions in government spending.

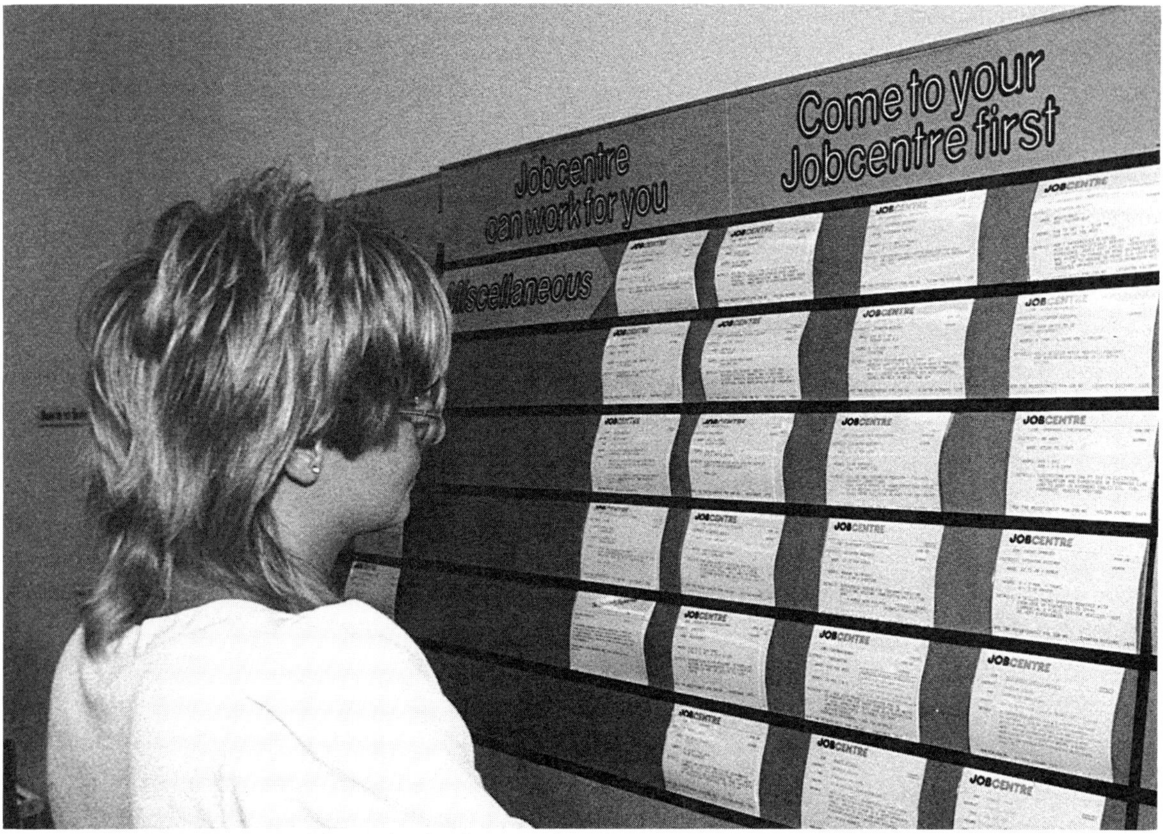

Visits to the Jobcentre are part of the routine for the unemployed

What can be done to cut the number of unemployed? The continued decline of old staple industries means that we must look for new employment opportunities in the service industries. In a modern society, the majority of workers are likely to be employed in supplying the public with services rather than with goods. In recent years, there has been a considerable growth in the number of people self-employed and these workers mostly supply services. Other measures which can be taken to decrease unemployment include a shorter working week, longer holidays, earlier retirement and work-sharing. The EEC Commission is supporting shorter working hours; in France the reduction of the working week to 39 hours in 1982 was estimated to have created 70,000 jobs and President Mitterand expressed the intention of cutting the working week to 35 hours in 1985. What do you think of the idea that retirement pensions should be paid to men at 60 in the same way they are paid to women? If we do not pursue adventurous policies to cut unemployment we may be forced to redefine full employment in terms of 6 or 8 per cent and even this level may be difficult to maintain.

Explain what is meant by

economy
monopoly
private sector
sole proprietor
partnerships
companies
multi-nationals
public companies
investors
shareholders
board of directors
limited liability
nationalised industry
public sector
basic industries
privatisation
trade unions
Trades Union Congress
General Council of the TUC
collective bargaining
industrial relations
arbitration
full employment
redundant

Unemployment in Britain 1978 to 1984

	1978	*1979*	*1980*	*1981*	*1982*	*1983*	*1984*
No. unemployed (million)	1.2	1.3	2.1	2.7	3.3	3.1	3.2
Percentage of workforce	5.4	6.0	7.7	11.1	12.8	12.9	13.4

Case study

THE STRIKE THAT LIT THE FUSE OF CONFLICT

The miners' strike of 1984–5 was one of the longest and bitterest industrial disputes in British history. Not only did it cost lives, thousands of injuries, nearly 10,000 arrests and more than £7 billion of taxpayers' money, but its implications for Britain's economic, political and social structure are far-reaching.

Throughout the 358-day strike it was common for the mass media to regard the dispute as a head-on clash between the left-wing revolutionary leader of the National Union of Mineworkers (NUM), Arthur Scargill, and the right-wing Prime Minister, Margaret Thatcher, together with the Chairman of the National Coal Board, Ian MacGregor. A confrontation of some sort between union power and elected government was almost inevitable, given the government's commitment to reduce subsidies to the nationalised industries in the face of any union opposition, outlined in the 1978 Ridley report. Ridley also recommended that various contingency plans such as the building up of coal stocks at power stations should be laid down, to prevent any power cuts in the event of a fierce challenge by power workers or coal miners. Meanwhile, the NUM hung on to the famous 1974 Plan for Coal which committed the government to a massive investment programme in new and existing pits. Thus, the NCB decision to close down 'uneconomic' pits, and thereby effectively destroying many traditional mining communities was enough to light the fuse of the conflict.

The announcement in March 1984 that Cortonwood pit in South Yorkshire was about to be closed was the final straw that caused the prevailing overtime ban to be escalated to a full strike, called by the NUM executive without balloting all its members. To secure solidarity, striking miners from the most militant areas such as Yorkshire and South Wales were sent in coachloads to other areas such as Nottingham to picket those miners who insisted on working until ballots were held. 'Flying pickets', as they were called, were almost matched in numbers by special units of police, drafted in from all over the country to protect the working miners. News of violent clashes – not only on the picket lines but also intimidation of miners and their families – hit the headlines throughout the summer of 1984. Thousands of miners were arrested and many were sacked by the Coal Board for their behaviour. There were numerous attempts by politicians, the TUC and the conciliatory services to get both the NCB and the NUM to agree on a formula to solve the issues behind the dispute, but each one ended in deadlock.

As the strike dragged on past Christmas, financial hardship among the striking miners led to a very gradual drift back to work. (The NUM could no longer support its members since its assets had been sequestrated after the failure to pay a £200,000 court fine for breach of the laws against secondary picketing). Finally, on 3rd March 1985, realising that there would be no satisfactory agreement while the strike continued, an NUM delegate conference agreed on an orderly return to work. The South Wales miners were among the first to go back – to the accompaniment of their traditional brass marching bands.

Although there was never any formal agreement, the collapse of the strike gave the go-ahead to the NCB to implement their original plan for pit closures.

1. Suggest some ways in which the dispute might have ended sooner.
2. Do you think that the miners had a good case? Give reasons for your answer.
3. How should the management of nationalised concerns be selected?
4. List the advantages and disadvantages of a union ballot of its members when making the decision to take industrial action.

Follow-up

1. Tabulate the following under *needs* and *wants*:
   (a) food (e) cigarettes
   (b) TV set (f) water
   (c) camera (g) clothing
   (d) house (h) radio
2. What are the full names of these large joint stock companies?
   (a) EMI
   (b) GEC
   (c) ICI
   (d) ICL
   (e) M & S
3. (a) Give some reasons why teachers might be considered a 'special case' for a pay rise.
   (b) Do you think that a teacher's pay should be based upon his or her efficiency?
4. Write a *defence* of one of the following:
   (a) arbitration
   (b) closed shop
   (c) collective bargaining
   (d) picketing
   (e) working to rule

Activities

1. Make a survey of your local town centre noting the work done by five different types of partnership. List five companies in your town which are *private* and five which are *public*.
2. Use reference books to find the particular advantages of
   (a) ordinary shares
   (b) preference shares
   (c) debentures
3. When BT (British Telecom) was privatised in 1984, within three weeks the 50p (partly paid-up) shares were trading at 102p. Why did the Conservative Government fix the initial price of the shares so low? What has happened to the price of BT shares since?
4. Consult the Stock Exchange prices in a newspaper and note the share prices of Britoil over a period of one month. Try to account for any ups and downs.
5. List trade unions with offices in your area. You will find their names in the Yellow Pages.

Examination questions

1. (a) What is meant by a monopoly?
   (b) What methods can a government use to control monopolies?
   (*University of Cambridge Local Examinations Syndicate, GCE, Economics, 1984*)
2. Explain, with examples, how public companies differ from public corporations.
   (*University of Cambridge Local Examinations Syndicate, GCE, Economics, 1984*)
3. Identify the major areas of conflict between employers and employees. Describe how trade unions might be involved in these disputes. Illustrate your answer by giving relevant examples.
   (*Joint Matriculation Board, GCE, Social Science, 1984*)
4. Examine the ways in which forms of production and different types of employment affect
   (a) people's attitude to work
   and (b) their social relationships in the community.
   (*Joint Matriculation Board, GCE, Social Science, 1984*)
5. (a) What are trade unions and state the reasons for their continued existence?
   (b) Explain the terms:
   (i) closed shop,
   (ii) collective bargaining.
   (c) What do you see as the role of a trade union in a time of high unemployment and increasing technological change?
   (*The West Midlands Examination Board, CSE, Social Science, 1983*)
6. What is a nationalised industry and how does it differ from other forms of business organisation?
   (*The West Midlands Examination Board, CSE, Social Science, 1983*)
7. *Either*, (a) Britain is said to have a mixed economy. Explain what this means with regard to ownership and control of industry.
   Or, (b) Explain what is meant by a *profession* and comment on any *two* of its characteristic features.
   (*Welsh Joint Education Committee, GCE, Sociology, 1981*)
8. What do we mean by the term 'public sector of employment'?
   (*North West Regional Examinations Board, CSE, Social and Community Studies, 1983*)

9. (a) What do Trade Unions do for their members?
   (b) Why do some people not like Trade Unions?
   (Associated Lancashire Schools Examining Board, CSE, Social Studies, 1983)

10. (a) What is 'Collective Bargaining'?
    (b) Explain the main stages through which wage negotiations normally pass.
    (c) What measures might be taken if agreement cannot be reached?
    (Associated Lancashire Schools Examining Board, CSE, Social Studies, 1984)

11. (a) Must a person join a trade union when starting work? Give a suitable explanation for your answer.
    (b) What do trade unions seek to do for their members?
    (c) Why do some people oppose trade unions?
    (d) How might trade unionism be affected by large numbers of unemployed people in society?
    (Associated Lancashire Schools Examining Board, CSE, Social Studies, 1984)

12. Describe how Trade Unions protect the interest of their members.
    (East Anglian Examinations Board, CSE, Social Studies, 1984)

UNIT 12
Money

GCSE Aims and Objectives

Knowledge — to promote comprehension of the functions of money, the money supply, the rate of inflation, the balance of payments and public spending upon the welfare state.

Application — to apply knowledge learnt to a study of the banking system, taxation, national income, Britain's trading position and their effect upon social welfare.

Evaluation — to make critical assessments of some modern economic problems.

TOPIC 12.1 WHY IS MONEY SO IMPORTANT?

To ask why money is important may seem a silly question. You would probably answer, 'We must have money to buy things.' Yet man could exist without money. Primitive peoples managed without money by using a system of **barter** whereby they exchanged surplus goods.

In a complicated industrial community there are thousands of goods and we could not waste our time on lengthy swapping expeditions. Besides, most people do not make anything they could swap. We have seen that the 25 million working people of Britain are engaged in *joint* productive efforts. Those who work with others to produce motor cars are not able to take a few at the end of the week and barter them for all the things they want.

A man can only live without money in two circumstances, and in both of these life would be very limited indeed:

1. If he were self-sufficient and was content to survive on goods he produced himself.
2. If he were part of a small primitive community where there were very few goods.

People who live in an advanced community demand sophisticated goods and services which are produced by complicated industrial techniques. Producers need money to purchase raw materials and pay wages. Workers require money to buy the goods produced. So workers sell their labour for money and use the money to buy the goods. Money is the oil which lubricates the economy and keeps it running smoothly. A complicated industrial economy could not exist without it.

Availability of durable goods in Britain, 1972 and 1982

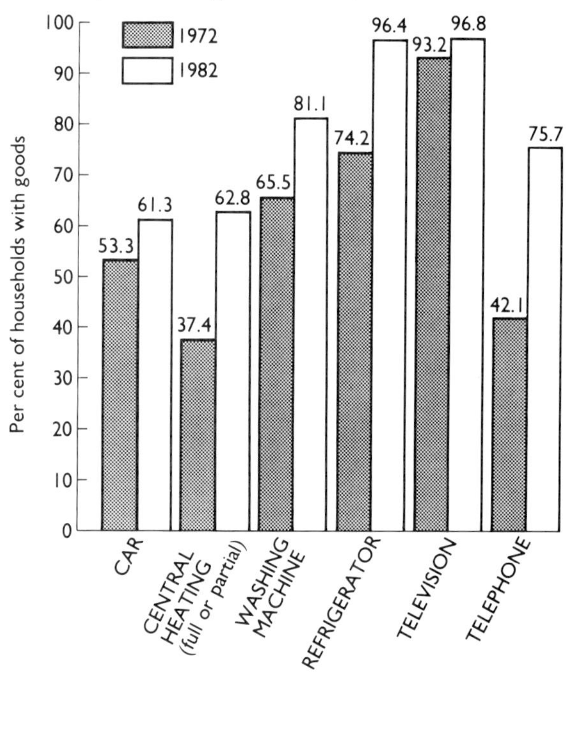

Consumers' expenditure in Britain, 1972 and 1982 (Source: *Britain*, 1984)

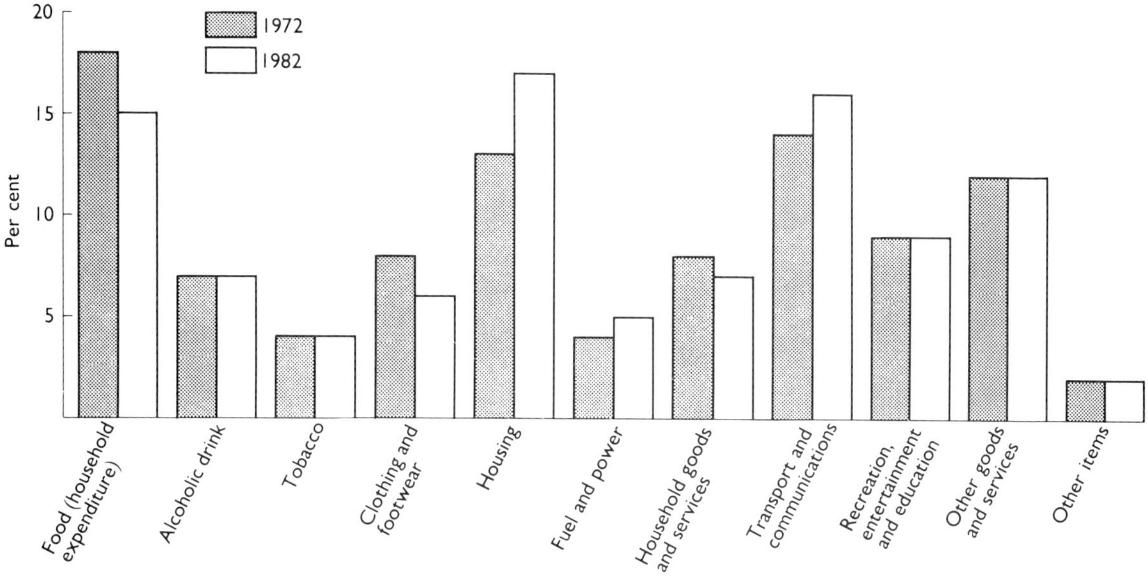

Ordinary citizens need money to make their purchases and pay their bills. Business organisations require money or capital so that production can be carried on. Workers want to be paid their wages in money and not in goods. The government collects its taxes in the form of money so that it can pay for services provided for the community. Welfare services cannot be provided without money. Finally, there must be sufficient money in the form of foreign exchange so that we can participate in international trade and balance our payments. In this unit we shall look at these different aspects of the use of money.

Prices are expressed in terms of money. We say that one thing is worth a certain amount of money, while another thing is valued at a different sum. In a barter economy values would be expressed in terms of goods. How many sheep for a horse? If the sheep were in good condition, perhaps four sheep would be a reasonable exchange. What could be bought for two sheep? Perhaps half a horse? Which half? And what if one of the sheep was worth less because it was thin and mangy? Life is much easier with money. The three important **money functions** are shown in the table below.

Functions of money

Measure of value – one pound note is worth as much as another
Medium of exchange – money can be changed for goods without people haggling
Store of wealth – it is possible to save money and use it later

TOPIC 12.2 THE MONEY SUPPLY

Politicians frequently talk about the importance of the **money supply**. Some people, known as *monetarists*, think that the sharply rising prices of recent years have been caused simply by a great increase in

The Bank of England in Threadneedle Street, London

the supply of money. So we must examine where money comes from. The money supply may be divided into three parts:

1. *Coins* worth about £500 million circulate throughout the economy. They are used mainly for small transactions. These coins are made by the *Royal Mint* at Llantrisant in Wales. Since the ½p coin went out of circulation on December 31st 1984, there have been seven coins in circulation. What are they?
2. *Notes* are printed mainly by the *Bank of England* although there are notes issued by banks in Scotland and Northern Ireland. Over £12,000 million worth of British currency is in the form of notes.
3. *Bank money* is the most important part of the money supply. By 'bank money' we mean money held by financial institutions. This money can be used to buy goods in the same way as you use coins or notes. A person who has an account at a bank can write a cheque which the bank will honour. **A cheque** is not money but serves as money because it is a claim upon money in a bank. Cheques are not legal tender and creditors may refuse them, but it is customary for cheques to be accepted, especially when backed by a *banker's cheque card*, which provides evidence of the reliability of the person presenting the cheque.

Economists use several different definitions of the money supply.

M0 = coins and notes in circulation
M1 = coins and notes in circulation *plus* sterling sight deposits held by the private sector only.
Sterling M3 = coins and notes in circulation *plus* all sterling deposits held by UK residents in both the public and private sectors.
M3 = Sterling M3 plus all deposits held by UK residents in other currencies.

Money: sterling M3 and M0 (percentage 6 months annualised growth rates)

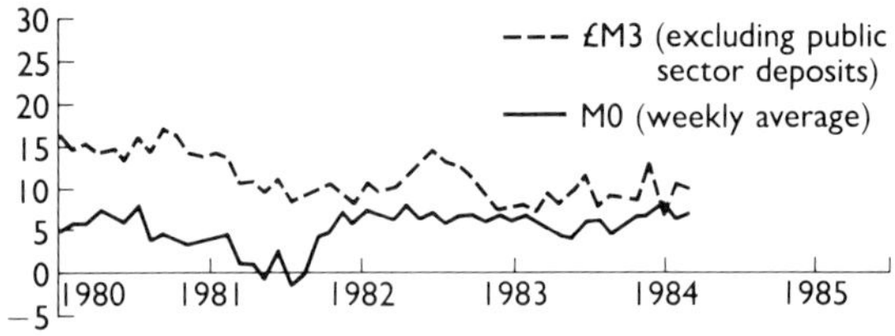

Although the value of the notes and coins in circulation amounts to over £12,000 million, the six **clearing banks** have deposits of over £100,000 million. This strange state of affairs is made possible because banks have learnt from experience that about 90 per cent of their deposits will not be drawn out in cash. There are two main ways in which banks create **credit** and subsequently increase the money supply:

1. By *loans*. Banks lend money for a certain period, e.g. they might lend a garage proprietor £50,000 to extend his business.
2. By *overdrafts*. This is an arrangement allowing a customer to overdraw (i.e. spend more than he has in his account).

Banks charge interest on loans and overdrafts; whether money is borrowed in the form of a loan or an overdraft depends on the circumstances of each case. If a person wanted to borrow £3,000 to buy a car the bank would probably suggest a loan to be paid back over two years, say at about £150 a month to cover the principal (£3,000) and interest charges. The bank would want the borrower to start paying something back each month as evidence of his intention to repay the sum within the specified period.

Cheques are an easy and safe way of making payments. A *crossed cheque* is crossed with two parallel lines and is safer than an *uncrossed cheque* because it must be passed through a bank account. A *credit transfer* allows an individual to settle a *number* of bills with *one* payment.

The increase in the quantity of bank money is one of the greatest problems facing the economy. The use of *credit cards* increases the danger of having too much money chasing too few goods. The six main clearing banks issue credit cards to those whom they consider to be creditworthy. An individual possessing an Access card or Barclaycard has a credit limit placed upon him, e.g. £600. You will have seen shops displaying signs indicating that they are willing to accept certain credit cards. Three sections of the community gain by credit cards:

1. Companies issuing credit cards charge shopkeepers a commission, e.g. if the article sold costs £20, the credit card company will gain £1 if it charges 5 per cent.
2. Shopkeepers attract customers who would not purchase articles at that time if they had to pay cash.
3. Shoppers benefit because they can purchase goods now and pay later.

As long as the credit card holder pays in the time allowed, he will not be charged any interest.

But there are dangers attached to credit cards:

1. Shoppers may overspend and find it difficult to pay later.
2. Competition is restricted; credit card holders may pay a high price for goods at a store offering credit facilities.
3. If interest is charged it is relatively high (varying from 18 to 30.6 per cent per annum).
4. If credit cards are lost or stolen they may be used dishonestly by somebody forging the holder's signature.
5. Credit cards increase the money supply and are therefore inflationary.

Finance companies also enable people to buy goods on credit by means of hire purchase, credit sales and personal loans.

TOPIC 12.3 INFLATION

Have you heard your parents grumble about **inflation**? Ever since human beings settled down to live in communities people have complained of rising prices. The trade cycle of booms and slumps was one of the cycles we looked at in the first unit. There are longer-term price cycles of inflation (rising prices) and deflation (falling prices), but the trend through the ages has been for prices to keep on increasing. A small annual percentage increase in prices is known as **creeping inflation**. This is not too serious, but galloping or **hyper-inflation** can destroy a society.

It happened in Germany after World War I. In 1914 a German mark was worth a shilling (5p) but by 1923 one English pound could be exchanged for about 20,000,000,000,000 (20 billion) marks.

Could it happen here? The rate of British inflation in 1974 was 20 per cent. It is frightening to think that even if inflation was at 15 per cent, during an average lifetime prices would increase 18,000 times! One of the main economic policies of the Conservative Governments, under Mrs Thatcher, has been to curb inflation and by 1985 the annual rate of inflation had been held to about 5 per cent. To understand how inflation can be curbed you must understand its causes.

Inflation (annual % increase in retail prices index)

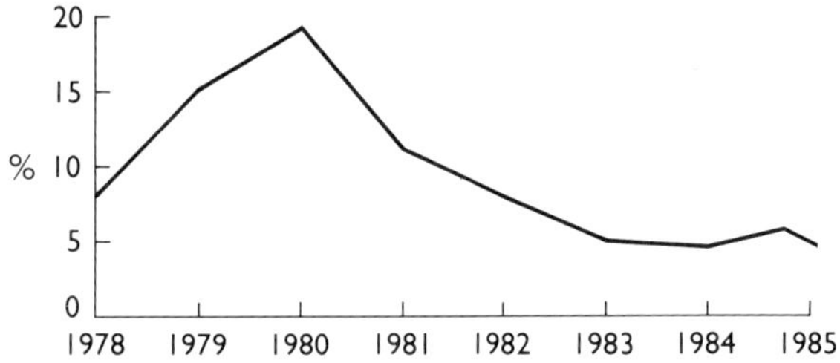

There are three main causes of inflation:

1. A large increase in the money supply, which is likely to cause a large rise in prices. Therefore the government has to control carefully the value of coins manufactured at the Royal Mint and the notes printed by the Bank of England. It also has to limit bank loans, overdrafts and other credit facilities.
2. An excessive increase in the demand for goods brought about by the extra purchasing power available if the money supply is increased. Economists call this **demand-pull inflation**.
3. An increase in the cost of production, e.g. in wages and raw materials. This is called **cost-push inflation**. The increased costs push up prices, so trade unions struggle to push up wages. It is a vicious **inflationary spiral**.

Ways of curbing inflation

- Limit the money supply by not issuing new money and by controlling the creation of credit by financial institutions.
- Maintain a prices and incomes policy upheld either by law or by voluntary persuasion (social contract)
- Raise income tax so that people have less spending money
- Increase productivity, especially of goods for export
- Reduce government expenditure on projects we can do without for the time being, e.g. a third London airport or the Channel Tunnel
- Impose severe conditions on hire purchase agreements, e.g. increase initial deposits and reduce the length of repayment periods
- Persuade trade unions and employers to work together in industrial peace
- Make people realise the tremendous dangers of inflation

What, then, are the main dangers of a high rate of inflation?

1. If money becomes worthless there will be an economic disaster because people will not accept money as a means of payment.
2. There is no incentive to save, because money put by declines in actual value.
3. If investment falls off so will production and jobs.
4. The prices of our export goods will be so high that other countries may not be able to afford them.
5. If other nations do not buy our exports we will not be able to buy their goods; therefore imports will be reduced and the standard of living will fall.
6. Pensioners, and others on relatively fixed incomes, will suffer greatly because incomes will lag behind rising prices.

TOPIC 12.4 NATIONAL FINANCE

The government needs money to provide services for the community, and most of this money is taken from the people by **taxation**. In the spring of each year the Chancellor of the Exchequer presents his annual Budget. This is an estimate of **government expenditure** (or spending) for the coming year and the ways in which it is proposed to raise the money.

The Budget is only an estimate and it never balances exactly. In fact the Chancellor does not intend it to balance in any one year. If he thinks it is necessary for the economy to be boosted, and for jobs to be created, the Chancellor may decide to plan for a **budget deficit**, i.e. the government will spend more money than it receives in taxation. The additional spending is made good by government borrowing.

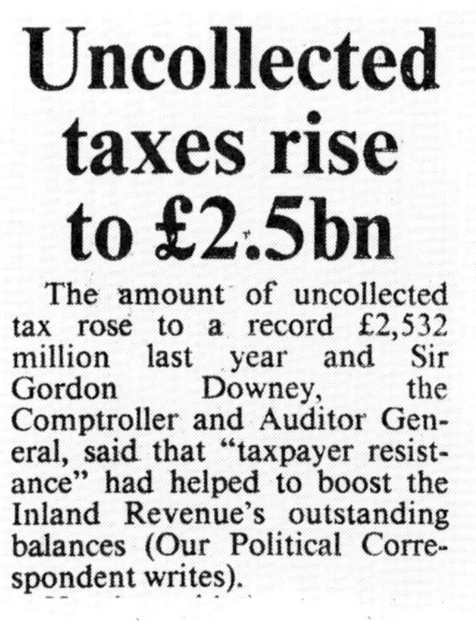

Uncollected taxes rise to £2.5bn

The amount of uncollected tax rose to a record £2,532 million last year and Sir Gordon Downey, the Comptroller and Auditor General, said that "taxpayer resistance" had helped to boost the Inland Revenue's outstanding balances (Our Political Correspondent writes).

The money may be spent on schools, hospitals, defence or less important projects. However it is spent, it is bound to provide employment opportunities. If the Chancellor wants to dampen down the economy he may plan for a **budget surplus**, i.e. the government will take more from taxpayers than it intends to spend on community services. We have already seen that increased income tax may curb inflation. The Chancellor has five main aims for the economy but he may find it very difficult to achieve all five at one time.

Main budgetary aims

- Control inflation
- Promote economic growth
- Correct an unfavourable balance of payments
- Secure social justice between different classes of the community
- Safeguard employment

Sometimes these aims may work in opposing directions. If the Chancellor tries to control inflation by increasing income tax, people will have less purchasing power and the decreased demand may result in more unemployment. Compare it with a football team that attacks to try to score a goal but consequently weakens its defence so that it concedes a goal. The Chancellor of the Exchequer's job is not an easy one. If his spring estimates prove inaccurate, or if circumstances change, he may introduce *interim budgets* or '*mini-budgets*', as they are called. The Chancellor will consult other cabinet ministers before he decides on tax changes or any new taxes he wishes to introduce.

There are two main types of tax:

1. **Direct taxes** on income, wealth and savings, collected by the Inland Revenue Commissioners.
2. **Indirect taxes** on consumer goods, collected by the Customs and Excise.

The main direct tax is income tax which provides a larger yield than any other tax in the United Kingdom. Indirect taxes which yield very large sums include taxes on tobacco, alcoholic drinks, betting, cars and petrol.

The ways in which the government gets money from taxpayers are frequently changing. Several new taxes have been introduced in recent years.

All the money that is earned by the citizens of a country is known as the **national income**. It is reckoned by adding together the incomes of every person and institution. The national income includes all wages, salaries, fees, dividends, rents, profits, etc. The largest amount is made up of the earnings of people in employment (about 66 per cent of the total). In 1984 the national income of the United Kingdom was over £275 billion and included

1. incomes from employment
2. incomes from self-employment
3. profits of companies
4. profits of public corporations
5. rents.

The best way of increasing the national income is by improved productivity and greater output. The national income will also be increased by inflation but this only reflects the increased amount of money in the economy and is of no real advantage. **Real national income** is a term economists use when the national income is adjusted to make allowances for price increases. The real national income provides a good indication of people's welfare. To secure a just

Advantages of indirect taxation

- Taxes may be imposed on goods the government wishes the public not to buy, e.g. the tax on petrol has increased greatly in recent years in an attempt to reduce oil imports
- People may decide to work harder to pay the increased cost of the commodity
- Indirect taxes are 'hidden' taxes (i.e. the taxpayer rarely knows exactly the amount of tax he pays), cannot easily be evaded, and allow the individual 'to tax himself' – e.g. if you want to pay tax then buy cigarettes and alcoholic drink!

Advantages of direct taxes

- Allowances can be made so that those with extra responsibilities may secure concessions, e.g. income tax allowances for wife, children, dependent relatives, etc.
- There is a graded scale so that tax rates get progressively higher the more income a person receives. Therefore the greatest burden falls on the well-off
- They help to check inflation because they take money from would-be-consumers and dampen down demand

society it is important that the national income should be distributed as fairly as possible. We have seen that some inequalities can be ironed out by progressive income tax. The government uses this method to act as a modern Robin Hood, taking from the rich and giving to the poor. It does this through the welfare state, which is one of the most important means of working towards social justice.

Some modern taxes

Capital gains tax	First used in 1962 as a tax on the profit made from the sale of valuable assets such as stocks and shares
Corporation tax	Introduced in 1965 to replace the old company profits tax, it is intended to encourage firms to plough back more of their profits into the business
Value added tax	Introduced in 1973–4, it is a sales tax imposed at percentage rates on certain goods and services
Capital transfer tax	Has replaced estate duty since 1974. It is a tax on gifts of over £15,000 and is a way of charging death duties by instalments

TOPIC 12.5 THE WELFARE STATE

A **welfare state** exists when the government spends money to provide a large range of social services and benefits for its citizens instead of leaving them to care for themselves. We are including the welfare state in this unit because it costs money to look after the welfare of the community. Have you ever heard anybody say something like 'Why don't *they* provide more hospitals?' Who are *they*? The speaker probably means 'the government', but he has not realised that the government cannot do anything without money.

The money for welfare services comes mainly from

1. taxes collected by the central government
2. rates collected by local authorities (see unit 4)
3. national insurance contributions.

Although communal help for the poor can be traced back to the Elizabethan Poor Law of 1601, not very much was done to improve their lot until the twentieth century. The Liberal governments of 1906 to 1922 introduced old age pensions and also gave monetary help to the unemployed, sick, disabled and mothers-to-be. In the inter-war years very slow progress was made towards extending a state welfare system, although there was the beginning of orphans' and widows' pensions, school meals and milk and rented houses for the working class.

It was not until after World War II that services were such as to warrant the term 'welfare state'. From the time of the *Beveridge Report* of 1942 the state has looked after its citizens from the cradle to the grave.

Let us trace some of these benefits by considering an average person's life-span. Look up the present figures and see what increases there have been in recent years. A *maternity benefit* is paid for eighteen weeks before the birth of a baby; then the mother receives a *maternity grant*. Ante-natal clinics are provided before the birth and post-natal clinics with check-ups and cheap foods are available after the birth. The local authority pays health visitors to watch the progress of mother and child. *Guardians' allowances* are payable to people who have in their family a child who has lost both parents.

If the family is on a very low income it may be eligible for a *family income supplement* (*FIS*) which is paid when the wage-earner is in full-time employment but the family income falls below a certain level. The FIS amounts to half the difference between the family income and the prescribed level.

The welfare state does not merely include payments of money to those in need. In the broadest sense the welfare state includes such things as free education, local authority housing, employment exchanges, the probation service and, most important of all, the National Health Service. The plank of the whole welfare state for over thirty years has been the NHS. It includes three main services:

1. hospitals administered, since 1974, through twenty regional hospital boards
2. local authority services provided by the council's social services department
3. doctors, dentists and pharmacists under the supervision of family practitioner committees.

When a young person starts work he is helped by the employment services. An unemployed person receives *unemployment benefit*, provided he or she qualifies by means of stamps on a national insurance card. A similar sum is paid in the case of sickness benefits. There are too many welfare benefits to

Despite welfare benefits many old people can barely afford to keep themselves warm

mention here, but the care of the welfare state is extended to widows and old people (*retirement pension*) and even includes a *death grant* to help with burial or cremation expenses.

However, despite the range of welfare benefits available, there are still a large number of people in Britain who live below the poverty line. Old people are especially vulnerable because they may be too proud to claim the allowances that are their due or they may be confused by all the form-filling. In severe weather they can barely afford to keep warm, and each winter takes its toll in the number of old people dying of hypothermia (low body temperature). Many of the fringe benefits such as the maternity grant and the death grant are now under threat as the government makes cutbacks in an effort to reduce public expenditure. With increases in the dependent population (see Unit 5) and unemployment, even greater strains are being placed on an already overstretched welfare service. Fortunately, politicians on all sides recognise that the welfare state is popular with the people of Britain and so it is not likely that it would be abandoned altogether.

'We'll have a bed for you in a day or two.'

One side-effect of the cutbacks in welfare services is the growth of private medicine. Many higher-paid workers now contribute to private health schemes which ensure that they receive immediate medical care when they need it. There are no waiting lists for operations in the private sector of medicine. It is bad luck for those who cannot afford it, and, unfortunately, they are usually the very people who need the most urgent care and attention.

Money is necessary for the welfare state to provide all its vital benefits and services. We have seen how life in Britain would be much more difficult without money. It just remains for us to consider how we are affected by money transactions between our country and other countries, i.e. the balance of payments.

Health groups claim £250m needed to stop NHS cuts

TOPIC 12.6 BALANCE OF PAYMENTS

The **balance of payments** is made up of all receipts from people abroad balanced against all payments made to foreigners. The balance of payments includes all visible and invisible items. The main visible items are foodstuffs, raw materials and manufactured goods; visible items make up what is called the **balance of trade**. The main invisible items are the earnings of banking, insurance and other financial services.

Balance of payments = visible *plus* invisible trade
Balance of trade = visible trade only

If the UK's income from abroad is greater than its total expenditure abroad, then there is a surplus on our balance of payments. If expenditure abroad exceeds our income, then there will be a balance of payments deficit. Consult the table of UK Balance of Payments 1970–1984 and decide which were the best and worst years regarding the current balance.

It is almost impossible for anybody to live in modern Britain and not hear about the country's balance of payments position. Radio, television and other instruments of the mass media report each month upon our foreign trade figures.

UK Balance of Payments 1970–1984 (£ million)

Year	*Visible Balance*	*Invisible Balance*	*Current Balance*
1970	−34	+857	+823
1975	−3,333	+1,810	−1,523
1980	+1,361	+2,116	+3,477
1981	+3,360	+3,569	+6,929
1982	+2,055	+2,868	+4,923
1983	−1,165	+4,411	+3,246
1984	−4,255	+4,879	+624

(*Source*: *An Economic Profile of Britain, 1985*, Lloyds Bank Group)

The more foreign trade, the more prosperous we are likely to be – as long as our imports, over the long term, do not greatly exceed our exports. Foreign trade allows people to enjoy a greater variety of goods.

What did you have for breakfast today? You may have enjoyed cornflakes produced from US maize, bread from Canadian wheat, butter from Ireland or bacon from Denmark. Make a list of all the foreign-produced goods which you have used today. Now tick those which could not be produced in Britain. Some of the items on your list could probably be produced either at home or abroad. If you are using an imported good, then it is probably produced more cheaply abroad. In accordance with what economists call the principle of comparative advantage (or costs), it is in the best interests of the peoples of the world if each country produces goods that it can produce most cheaply, and then trades with other countries.

Retail prices and average earnings

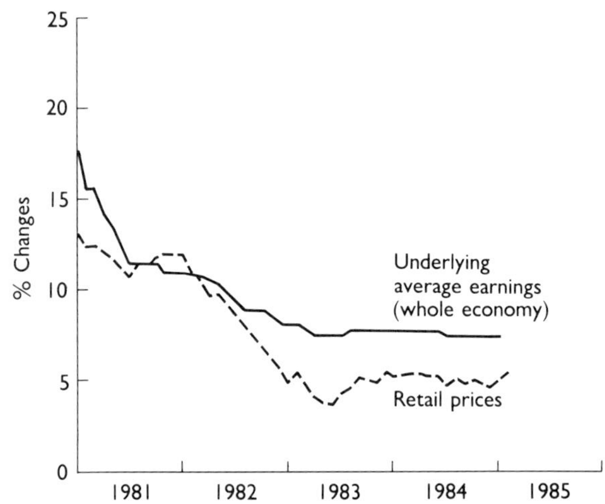

Geographical distribution of trade in Britain, 1982 (Source: *Britain*, 1984)

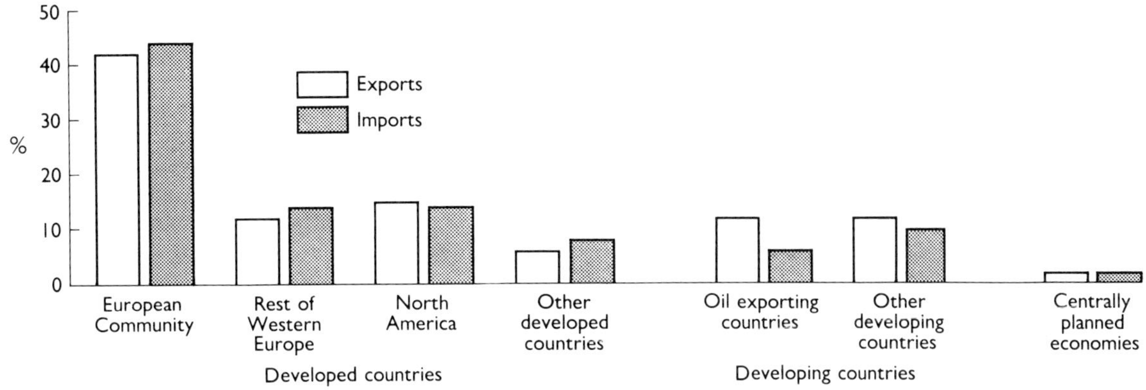

So the UK exists by importing and exporting. Our main imports can be divided into five sections. See table below.

Britain's main imports

1. Raw materials
2. Semi-manufactured goods
3. Fully manufactured goods
4. Food and beverages
5. Petroleum and petroleum products

We have to export to pay for our imports. Our four main categories of exports are shown in the table below.

Britain's main exports

1. Engineering products, e.g. machinery and vehicles
2. Other manufactures, e.g. metals, chemicals and textiles
3. Food and drink
4. Miscellaneous items

One of our most serious problems in recent years has been the great increase in the proportion of imported manufactured and semi-manufactured goods. Since 1980, however, we have been self-sufficient in oil.

Continual changes are taking place in the pattern of the UK's import and export distribution. Trade with the Commonwealth has decreased, but trade with the EEC has increased.

Explain what is meant by

barter
money functions
money supply
cheque
clearing banks
credit
inflation
creeping inflation
hyper-inflation
demand-pull inflation
cost-push inflation
inflationary spiral
taxation
government expenditure
budget deficit
budget surplus
direct taxes
indirect taxes
national income
real national income
welfare state
balance of payments
balance of trade

More state support for export business urged

The CBI was sending a clear message to the Government to allow industry to compete on equal terms in world markets until such time as international negotiations successfully ended all forms of government intervention, **Sir James Cleminson**, the president, said at the end of a debate on support for exports.

Case study

LIFE ON A MILLION POUNDS A WEEK

When Hans returned to Berlin after World War I he found a job in a printing factory. Life was more difficult than it had been before the war began in 1914, but he was able to make a reasonable living. Prices were rising and Germany owed her old enemies millions of pounds in reparations, but little of this affected Hans as he settled down to a normal life with his wife, Monika.

It was in 1921 that Hans and Monika began to realise that all was not well. Before the war the rate of exchange had been 20 marks to the pound sterling. A mark was worth 1 shilling (5p) and it had been that way as long as anybody could remember. When Hans came back from the war he earned £2 a week. By 1921 the value of the mark had depreciated so much that Hans' wage was £40 a week, but although Monika was a good manager in the home they did not seem any better off. In March 1922 Hans brought home his first £100 pay packet. A massive pay rise of 50 per cent put them in a better position for a few weeks. Hans was then on a wage of £150 a week.

Their gain was short-lived. In January 1923, life on a weekly wage of £5,000 was a bitter struggle. The rent, which had been 10 shillings (50p) in 1918, was now £800 a week. Workers were demanding pay increases of 60 per cent, and getting them. In February 1923 Hans said to Monika, 'We've got our pay rise.' He was on £10,000 a week.

Hans and Monika realised that their money would be worth much less in a few weeks' time so they spent it as quickly as possible. Only a fool tried to save money in the Germany of 1923.

Industry came almost to a halt when the French occupied the Ruhr. The government was forced to print more money as the mark was devalued again.

In April 1923 Hans' weekly wage was £50,000. Bread was £250 a loaf and cabbages £50 each, but a week later all prices doubled again. By July 1923 Hans was earning over £1,000,000 a week. In the mad rush of hyper-inflation the price of a loaf rose to £2,000 while a tin of beans cost £10,000.

Trying to hold on to a mark was like grasping a snowflake. The exchange value was changed twice a day at 11 a.m. and 5 p.m. People chased from one shop to another trying to find a shopkeeper who had not increased his prices.

Monika got up early in the morning to pack Hans' bus-fare in a suit-case. Hans' boss hired a lorry to carry the wages for his employees. People began to think in millions. A father, when asked how many children he had, might casually reply, 'Five million.'

But it was no joke. In August 1923 Hans received his first £50,000 note. It made his pay packet of two suitcases less of a burden, but he and Monika were as badly off as ever. The factory where Hans worked joined with fifty others in printing new notes.

In September 1923 life became ridiculous. Hans went to buy a suit for £8 million. He left his change of £2 million in the shop but by the time he returned it would only buy a pair of laces. When he went to a restaurant he asked for the bill after each course because prices changed while he ate. Monika told him of a friend who bought a pair of gloves for £4 million but picked up two left gloves in error. By the time she realised her mistake the gloves had shot up to £8 million; she had to pay another £4 million to make up a pair.

People papered their walls with bank notes and trundled pianos around the street to pay the grocer's bill. Even old scrap iron was more valuable than money. They wheeled their wages home in barrows.

In November 1923 the great inflation went into its last lunatic spasm and £1 exchanged for 20,000,000,000,000 (20 billion) marks. Dr Luther, the Minister of Finance, announced that the mark was to be replaced by the rentenmark which was supposedly backed by mortgaging all the land of the country. The price of inflation had been terrible: mass unemployment, suicides, bankruptcies and the rise to power of Adolf Hitler.

1. Who loses and who gains during a time of inflation?
2. Why are people discouraged from saving and investing during inflation?
3. How many years would it take for prices to double if inflation continued at an annual rate of 25 per cent? (This was about the rate for Britain in mid-1975.)

Follow-up

1. Imagine you are the Chancellor of the Exchequer. What taxes would you alter? (If you take off some taxes, make good the revenue in other ways.)
2. Give two arguments for and against:
   (a) capital transfer tax
   (b) petrol tax
   (c) tobacco tax
   (d) VAT
   (e) wealth taxes
3. How would you attempt to check inflation?
4. What are the difficulties of a prices and incomes policy?
5. The following actions will affect the balance of payments. Divide into two columns according to whether they are intended to
   (i) deal with a surplus
   (ii) correct a deficit
   (a) borrow from the International Monetary Fund
   (b) increase direct taxes
   (c) give awards to exporters
   (d) devalue the pound
   (e) decrease direct taxes
   (f) make HP more difficult
   (g) increase overseas aid
   (h) borrow from foreign banks
   (i) pay back foreign debts
   (j) increase customs duties

Activities

1. A native living in a barter economy goes down-river in a boat with a load of wood he wishes to exchange for a supply of skins. He makes four swaps before securing his skins. Draw a cartoon to illustrate his journey.
2. Use reference books to look up
   (a) fiduciary issue
   (b) finance companies
   (c) gold standard
   (d) Maundy money
3. Call at a post office for a pamphlet on either the PO Giro or National Savings. Summarise the services provided.
4. List the main advantages and disadvantages of
   (a) a current account
   (b) a deposit account

Examination questions

1. If every customer went to his bank and withdrew all his money the banking system would collapse.
   (a) Why should this be so?
   (b) Does it imply the banking system is unsound?

   (*University of Cambridge Local Examinations Syndicate, GCE, Economics, 1984*)
2. (a) Explain the main difference between direct and indirect taxation.
   (b) Briefly explain the following taxes:
   (i) Income Tax
   (ii) V.A.T.
   (iii) Capital Gains Tax
   (iv) Capital Transfer Tax
   (v) Corporation Tax.
   (c) Explain briefly what the Regulator is and how it works.

   (*Associated Lancashire Schools Examining Board, CSE, Social Studies, 1983*)
3. (a) What is the Welfare State?
   (b) What are the main arguments in favour of the Welfare State?
   (c) What are the main arguments against the Welfare State?
   (d) What future do you think there is for the Welfare State in the UK?
   Give suitable reasons for your answer.

   (*Associated Lancashire Schools Examining Board, CSE, Social Studies, 1983*)
4. Explain why, in a Welfare State, there are still squatters.

   (*East Anglian Examinations Board, CSE, Social Studies, 1984*)
5. What are the purposes of the Budget?

   (*East Anglian Examinations Board, CSE, Social Studies, 1984*)
6. Describe in detail the services offered by a clearing Bank.

   (*East Anglian Schools Examinations Board, CSE, Social Studies, 1983*)
7. (a) Give *five* reasons why the government is involved in managing economic affairs.
   (b) Describe the following methods of controlling inflation:
   (i) prices and incomes policy;
   (ii) monetary policy.

(c) Explain why state intervention in economic and social affairs has grown since 1945.

(The Associated Examining Board, GCE, Social Science, 1983)

8. Explain the place of voluntary organisations within the Welfare State and why they are needed.

(East Anglian Examinations Board, CSE, Social Studies, 1984)

9. (a) Briefly explain, with examples, the difference between indirect and direct taxation.
   (b) What are the advantages and disadvantages of increasing consumer spending by:
      (i) lowering taxation;
      (ii) increasing public expenditure?
   (c) Why is it difficult for governments to reduce public expenditure? Your answer should make reference to social, economic and political factors.

(The Associated Examining Board, GCE, Social Science, 1984)

10. Method of payment by households for purchase of goods and services in Great Britain.

Millions of transactions

	1971	*1976*	*1980*
NON CASH			
Cheque	450	650	1050
Standing Order	120	180	230
Direct Debit	40	100	170
Postal Order	270	150	125
Credit Card	10	60	130
Credit Transfer	40	60	100

Study the table above carefully. Use the information contained in the table to help you write an essay on the growth of the banking services.

(East Midland Regional Examinations Board, CSE, Social Studies, 1984)

11. (a) Name *three* groups of people which the Welfare State was designed to help.
    (b) Briefly describe what help is available from the Welfare State for any *two* of the groups you have mentioned in part (a).
    (c) What arguments can be used to justify cuts in expenditure in the Welfare State? Do you agree with them? Give reasons for your answer.

(The South-East Regional Examinations Board, CSE, Civics – Citizenship and Social Affairs, 1981)

12. 'The Chancellor announced that Budget Day would be on 10th March. Sir Geoffrey Howe replied to an onslaught on the Government's economic policies by Mr. Peter Shore, the Shadow Chancellor, and predicted that the overall figures for inflation this year would be in single figures.' (From a newspaper report early in 1981.)
    (a) What is the full title of 'the Chancellor'? Who is the present Chancellor?
    (b) What is the main task of the Shadow Chancellor?
    (c) What are the two main parts of a Budget speech?
    (d) What has this Government done to try to reduce inflation?
    (e) What objections have been made to the Government's policies?

(The South-East Regional Examinations Board, CSE, Civics – Citizenship and Social Affairs, 1982)

UNIT 13
Government

GCSE Aims and Objectives

Knowledge — to clarify understanding of the importance of authority and power, the role of the monarch, Lords and Commons and the workings of modern government.

Application — to apply knowledge gained about government to a study of types of government, the British Constitution, the passing of legislation and the importance of the Prime Minister, Cabinet and Opposition.

Evaluation — to interpret and assess the part played by government in the modern world.

TOPIC 13.1 AUTHORITY AND POWER

The chief priests and elders said to Jesus: 'By what authority are you acting like this? Who gave you this authority?' Throughout an individual's life-cycle many people and organisations exercise **authority** over him.

As a little child you obeyed your parents without thinking. They were older and wiser (and stronger!) than you. Parents naturally assume responsibility and they are also legally responsible for you. If they fail to exert authority over their children they may break the law. If parents fail to **control** young children then they themselves may feel the authority of the court.

Mankind has long exercised authority over animals. Authority is not always a matter of physical force. A shepherd's authority over his sheep involves protecting them and caring for them. If your family's dog bites the postman your father may be ordered to keep the animal under control. If he still fails to exercise his authority the court may order the dog to be destroyed. So it will die because there has been insufficient authority.

Animals living in the wild state exercise control over their young. What is it that gives a male lion authority over his cubs? It is mainly that he is stronger and can use the **power** of force. This state of affairs prevails throughout most of the animal world. In many cases the male has authority over the female because he is a more powerful animal. Is this why, for thousands of years, it has been customary for men to have authority over women? And does this mean that it will always be so? What argument could supporters of women's lib put forward against these ideas? How can one person have power over another except by using force?

'All those in favour of putting women in their places before it's too late say, "Aye".'

Church and Crown: The Royal Family leaving church after a Christmas service

In some cases a female animal exerts authority over the male of the species. You may remember that we referred to female rule over the family as matriarchal. Authority is linked with power. Any person or organisation recognised as having authority has power over others. The leader of a herd of elephants may force an old, weak elephant away from the herd and leave it to die. The leaders of certain Eskimo groups decide upon a similar fate for their old people and leave them on the ice to die. Do you think this is an example of power being used in the wrong way? Or may this be kinder than deciding that old people should be kept alive for long periods in geriatric hospitals?

It does not matter if the power is imaginary, so long as it is accepted. Atheists do not believe in God and therefore reject his authority. But this does not prevent God, or his representatives, having power over those who do believe. Many people accept the authority of the church because of deeply held **religious beliefs**. The ardent Roman Catholic usually accepts the authority of the Pope because he believes his power comes from God.

Throughout life there are authorities we have to recognise. At school, if we do not recognise the authority of teachers we may be punished. School governors have overall control of teachers – of what is taught and how it is taught. Workers may be dismissed if they do not obey the authority of the foreman. A boss may be called the 'Guvnor' because he governs what goes on at the place of work. Many organisations have governing bodies which exercise authority over their members.

The governing body of the Church of England is the General Synod. The Church of England is the state church or the established church. It still voices concern on social problems such as abortion and divorce, but it has lost much of its authority in recent years. Its archbishops, bishops and deans are appointed by the sovereign on the advice of the prime minister. Thus the prime minister, as leader of the government of the day, has some authority over the church. The Church of England must obtain Parliament's consent to any changes in its form of worship, and all clergy take an oath of allegiance to the sovereign.

The government has authority over the whole country. In times of national emergency the established church is expected to support the state. In this country the church may find a double pull on its loyalty. Ought it to accept the authority of Christ as the Prince of Peace and condemn war, or should it accept a government's war policy and pray for victory over the state's enemies? We have seen from the beginning of this book that humans are sometimes motivated by love and sometimes by hate.

The government expects Anglican officials to play their part in national affairs. For example, at the coronation of a new monarch or the funeral of his predecessor, church and government exercise a **joint authority**.

Advances towards democracy	*Date achieved*
• Abolition of the property qualification: you do not have to own property to vote in parliamentary elections	1858
• Secret ballot: nobody else knows how you vote	1872
• Adult male suffrage: votes for *all* men	1884
• Equal electoral districts: each MP should represent roughly the same number of people	1885
• Payment of MPs: a Member of Parliament need not have any private income	1911

TOPIC 13.2 WHO SHALL GOVERN?

At one time it was fairly easy to answer the question 'Who shall govern?' Most countries of the world were under the authority of a **monarch**. This king, queen or emperor had **absolute power**. Henry VIII had little difficulty in finding reasons (excuses) for having two of his wives beheaded. But peoples throughout the world became tired of being pushed around by one person whose right to govern rested on an accident of birth. During the French Revolution, King Louis XVI and his queen Marie Antoinette were guillotined and members of the king's government suffered the same fate. In 1917 the Russian Revolution deposed Tsar Nicholas II, who was probably shot in 1918 with the Tsarina and their children. In Britain changes came with less bloodshed, and more gradually.

We no longer believe in the divine right of the monarch to rule. What is the best form of government to replace absolute monarchy? Most modern Western governments are elected by the people. People's government is known as **democracy**, from the Greek word *demos* meaning the people and *krateo*, to rule.

The people of Britain struggled for a long period to set up a democratic form of government. Look up for yourself the details of Magna Carta (1215) and the 1689 Bill of Rights. The People's Charter of 1839 set out six points; the five points of the Chartists, which were eventually conceded, are all essential to the principle of democracy (see below).

The Chartists were not far enough ahead of their time to demand votes for women. After a desperate struggle by the **suffragettes** women over the age of thirty were granted the vote in 1918, and since 1928 all women have been allowed to vote.

Britain has been a democracy for about fifty years. When the legal age for adults was changed, eighteen-year-olds were able to vote for the first time in 1970. The only Chartist point never granted was their demand for annual parliaments. The Chartists thought it would be very democratic to be able to remove unpopular governments quickly, but annual parliaments would be unworkable. A government would not have time to carry out a worthwhile policy.

Main conditions of British democracy

- All people over the age of eighteen can vote, with a few exceptions such as peers, criminals and lunatics
- Free elections conducted by secret ballot with severe punishments for anyone found guilty of bribery or other corrupt practices
- Electors vote for candidates who represent a political party and the party which secures the most seats forms the government
- The leader of the majority party becomes prime minister
- A parliament cannot last for more than five years and then voters have an opportunity to elect a new government
- Everybody has to keep laws passed by Parliament, but people cannot be punished for opposing the government
- Her Majesty's Opposition has the official task of criticising the Government and the leader of HM Opposition is paid a salary for this purpose
- Minority groups are allowed to put their views freely and time is allotted to them on radio and television

Centre of government (i) Moscow's Red Square with the Kremlin on the right

What occasions can you think of when it is necessary even for a democratically elected government to be unpopular? What unpopular measures did the Chancellor of the Exchequer introduce in his last budget?

A country ruled by a government which is not elected is known as a **totalitarian state**. Franco's Spain, a right-wing totalitarian state, was ruled by a **fascist government**. The People's Republic of China, a left-wing totalitarian state, is ruled by a **communist government**. If you lived in such countries you might find that elections very rarely took place.

In a totalitarian state or dictatorship *the party* is supreme. If people in Hitler's Germany were not members of the Nazi Party, a visit from the Gestapo might 'persuade' them to join.

Newspapers in a totalitarian state will be owned by the state. *Isvestia* and *Pravda* are organs of the Russian government. The radio and press are also controlled by the government. One of the essentials of democracy is a 'free press' in the sense that anyone has the right to publish so long as the law of libel is not broken and no vital state secrets are revealed.

Totalitarian governments have used concentration camps and forced labour camps for political prisoners. Opponents of the government may disappear overnight and never be seen again. A feeling of **nationalism** is stirred up so that people feel intensely loyal to their country and its government. Nationalism often leads to a hatred of foreigners. Prejudice against Jews was fostered in pre-war Germany and millions of Jews were exterminated in gas chambers.

The evolution of a democratic state is similar to the growth of a tree; it does not take place overnight. There must be progressive cycles before democracy matures. It is often difficult to transplant the democratic concept to nations which have recently been granted self-government after years of rule by foreign governments. Where the occupying power had not prepared the people for self-government, as in some parts of Africa, they have experienced severe difficulties in establishing a democracy.

We can be proud that Britain, on the whole, has a good record of training 'colonial' people to rule themselves. The first step was to grant **dominion status** to Australia, New Zealand and Canada.

Centre of government (ii) The Houses of Parliament

Britain has been gradually relinquishing its rule over its 'colonies' and many independent nations have now been established. But it has not always been easy. There have been problems in Northern Ireland since the turn of the century. What do you think Britain should do to see that Northern Ireland is governed peacefully? Good government means that people can live their lives freely, fully and peacefully. Although a democratic government is sometimes criticised for progressing too slowly, it is better to move slowly, with the consent of the people, than to move rapidly against their wishes.

TOPIC 13.3 MONARCH OR PRESIDENT?

The United Kingdom (England, Wales, Scotland and Northern Ireland) is a **monarchical state**. The monarch is the head of the state and the government is carried on in the sovereign's name. We talk about **Her Majesty's Government**.

The Queen reigns but she does not rule. She is a **constitutional monarch** and her powers are limited by the **British constitution**. Our constitution is not a written one like that of the United States. The British constitution is a flexible one. It can be modified by Acts of Parliament.

Parliament can alter the law relating to the monarchy. At present the law states that the monarch must not be a Roman Catholic. Parliament also controls the monarch's financial allowance by voting a 'civil list' to provide a personal income for the monarch and the expenses of the royal household.

When newspapers refer to the **British Government** they mean the ruling party led by the prime minister and supported by his or her cabinet of senior ministers. The monarch still has some powers and they depend on the **royal prerogative**. These powers go back to a time before our modern Parliament and do not depend on Parliament. Today we have a constitutional monarchy and prerogative powers are exercised on the advice of a minister of the Crown. Ministers are responsible to Parliament which is responsible to the people who elected it. Democratic principles operate because the sovereign acts according to the wishes of the people.

The British Royal Family

The monarch's prerogatives

- The monarch appoints as prime minister the leader of the party which commands a majority in the House of Commons
- Cabinet ministers are chosen by the Prime Minister but accept their seals of office from the monarch
- The monarch dissolves (i.e. dismisses) Parliament on the advice of the prime minister
- The monarch creates peers and confers honours on the advice of the prime minister

A democracy can have a **president** or a monarch as head of state. The monarch's role in a democracy is restricted by elected representatives, as in Britain. A president is elected but usually by a different process from that used to elect the law-making body. Under the American constitution the president and the Houses of Congress are elected separately. It is possible to have a president who represents one party and one or both of the Houses of Congress representing a different party. The president might then be impeded in his duties. The president may pick his cabinet from non-elected people and he may not be answerable in the short run to the elected Houses; it was a long time before the Americans were able to dismiss President Nixon.

Why do we retain a monarchy in Britain while most other democracies have a president?

Monarch versus president

Advantages of a monarchy

- The monarch is trained from birth in the art of democratic government
- The monarch is above politics and does not support any party
- National unity is symbolised by the monarch
- Traditional pageantry and ceremony provide a popular spectacle
- The British monarchy finds support throughout most of the Commonwealth
- It provides a unifying symbol for the nation
- The monarch uses her private fortune to bolster up the Civil List money
- Other members of the Royal Family perform functions such as opening new schools
- The British monarch must be a Protestant and provides leadership in religion and family life
- A monarchy provides continuity of government and can warn, encourage and advise

Advantages of a presidency

- The president is elected with the support of the majority of the people
- A president owes his position to merit and personal ability
- It is possible to remove an inefficient or corrupt president
- It is more logical in a democracy that people should vote for the head of State
- A president does not rely upon an accident of birth
- He is respected for his own worth
- It is claimed that there have been fewer bad presidents than bad monarchs
- A president does not have a large number of relatives who have to be kept by the State
- It is possible to have a president of any religion
- A president is active in government and not just an ornamental figurehead

TOPIC 13.4 COMMONS AND LORDS

The **House of Commons** consists of 650 elected Members of Parliament. The two main parties claim about 600 seats, so it can be called a two-party system, despite the existence of a strong third party – the SDP/Liberal Alliance.

The political party which wins a majority of seats, or is able to command a majority of supporters in the House of Commons, usually forms the Government.

If you study the table you will see that the Alliance (of Liberals and SDP) did not gain the number of MPs they deserved in the 1983 General Election. Nearly 8 million people voted for the Alliance but they only secured 23 seats out of 650 in the House of Commons. This is largely because we use the 'first past the post' electoral system rather than **proportional representation**. Thus, the Conservative Party with the support of only 31·6 per cent of the electorate was able to gain nearly 400 seats. The present voting system punishes political parties whose votes are evenly distributed but rewards parties whose votes are concentrated geographically. For example, the Labour Party gained only 700,000 more votes than the Alliance, but won 186 more seats. The Alliance received 338,000 votes for every MP elected, whereas the Conservative Party gained one seat for every 33,000 votes cast and Labour secured an MP for every 40,000 votes cast. Conservative and Labour combined gained 93.2 per cent of seats in 1983, although only 52.2 per cent of the electorate voted for the two main parties. Alliance supporters are obviously in favour of proportional representation. What do you think about it?

The *speaker* is elected to preside over the House. He is impartial and does not involve himself in party politics. A speaker holds office until he retires or dies.

The General Election, 1983

Party	*Total Votes*	*Number of Seats*
Conservative	12,991,377	397
Labour	8,437,120	209
Alliance	7,775,048	23
SNP	331,975	2
Plaid Cymru	125,309	2
Others		17
Total MPs		650

The Government sits on the right of the speaker and the Opposition on the left. If you look at the diagram on page 170 you will see that the prime minister and his or her cabinet colleagues sit on the government front bench facing the leader of the Opposition and members of the shadow cabinet. Behind them sit the mass of *back-benchers* who make up the majority of MPs. You will see that there are two *voting lobbies*. Those who agree with a motion go into the '*aye*' *lobby* and those who disagree enter the '*no*' *lobby*. They are counted by MPs known as *tellers*. Can you think of any ways by which the voting procedure for MPs could be modernised?

One important task of Parliament is to pass Acts which alter the laws of the land in the light of changing conditions of life. This is done by introducing a Bill which goes through eleven stages. The action of the monarch in signing the Bill is known as the Royal Assent: at this moment the Bill becomes an Act. The passing of laws, or **legislation** as it is called, is a very thorough business. It is important that no ill-advised Act should be passed hastily.

The eleven stages that make a bill into an act

Stage	*Action*
1. First reading	The minister responsible presents a 'dummy Bill' to a Clerk who reads its title. This is a formality before the Bill is printed in full and distributed to MPs
2. Second reading	This is the most important stage. A debate takes place on the Bill's principles followed by a **division**, i.e. a vote
3. Committee stage	The Bill is discussed in detail by a committee of MPs who amend it where they think it necessary
4. Report stage 5. Third reading	These are usually taken together. The amended Bill is reported back to the House
6 to 10. House of Lords	The Bill passes through the same stages in the House of Lords
11. Royal Assent	The monarch signs the Bill after it has been passed by both Houses of Parliament

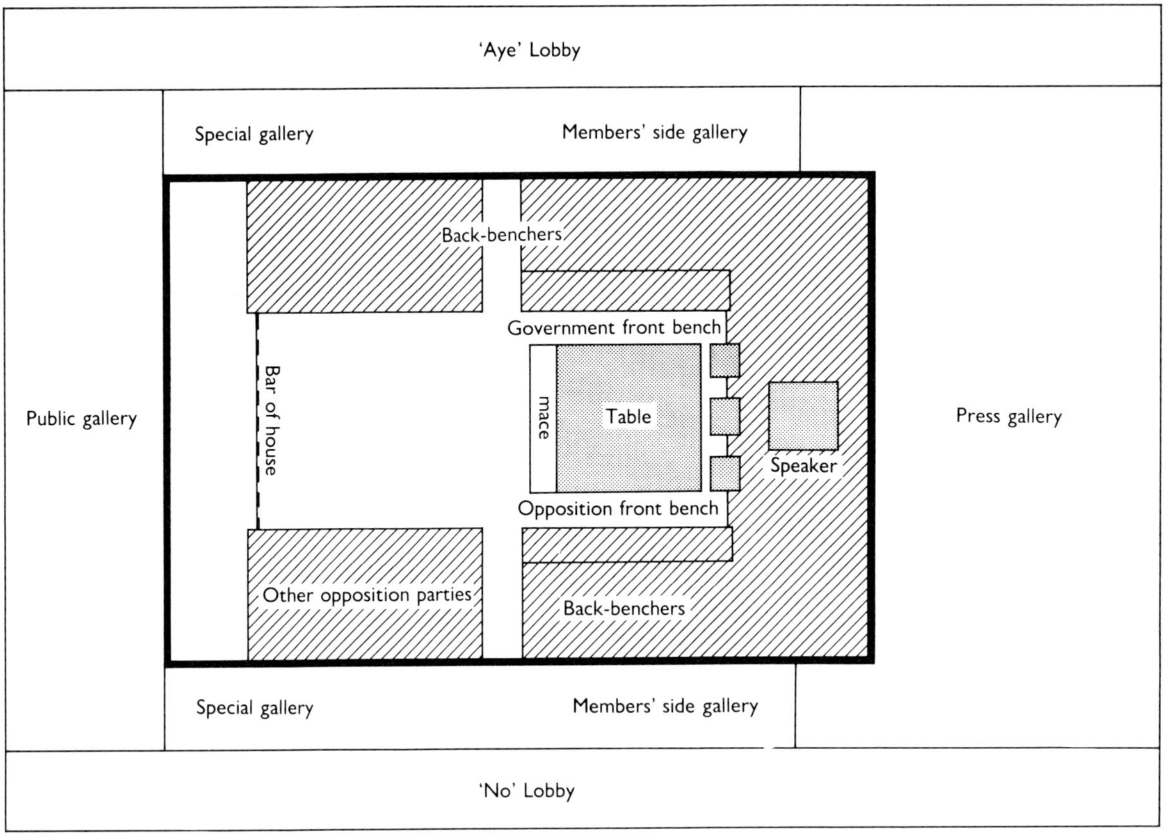

The layout of the House of Commons

Let us imagine a day in the life of a Member of Parliament. As the House of Commons does not sit until 2.30 p.m. an MP's morning is occupied in a variety of ways. He answers letters from constituents with problems and writes letters on their behalf to *government departments*. He may receive a *delegation* of people from a particular **pressure group** who hope to secure his support on a particular matter. MPs are members of *standing committees* which examine Bills in detail. If an MP is interested in education he may well be a member of the committee which deals with educational Bills. An MP who intends to take part in a parliamentary debate might work in the Commons library. Many MPs have jobs outside Parliament and could be engaged in work as journalists, lawyers, lecturers, trade union leaders, etc. After lunch an MP will probably attend the House. The daily timetable starts with prayers and preliminary business, before going on to question time. This is the main chance for an MP to raise grievances and secure answers from government ministers. It lasts for about an hour. Public business begins at about 4 p.m. and goes on until 10.30 p.m. except on Fridays, when the Commons sits from 11 a.m. to 4.30 p.m. Much of the work of the Commons is done in the evenings, so it is a tiring job. There have been all-night sessions.

An MP cannot be expected to listen to eight hours of talk every day; also there are evening functions and parliamentary delegations to other countries. An MP may arrange a *pairing* with a member of the opposite party so that each side will be one member short. Such pairings must be registered with the Whips' Office. The 'whips' enforce party discipline. An MP is expected to support party policy and vote in the way indicated by the whips. If an MP ignores a *three-line whip* (i.e. an urgent summons to be present for a vital vote) then he is likely to be disciplined by his party and his chances of advancing his political career will be damaged. MPs tend to specialise in certain subjects and to be present in the chamber when these subjects are discussed. At other times an MP may be ready to vote when reminded by the division bell. Many of his weekends will be spent in his constituency where he has responsibilities to those who elected him.

Membership of the Commons is more important than membership of the **House of Lords**, but the second chamber still plays an important part in the government of the country. We have seen that a Bill has to pass through the Lords, although the Lords cannot reject **money Bills**. Any Bill which the Speaker classes as a money Bill becomes law within one month of being sent to the House of Lords.

The part played by the Lords today

- Some non-controversial Bills are introduced in the Lords and pass through that House first
- Amendment of Bills which are sent up by the Commons, although if the amendments are not accepted the Lords normally give way
- The Lords, in exceptional circumstances, can delay a Bill for twelve months
- Provide a high standard of debate, with views expressed free from rigid party discipline
- It is the final court of appeal in legal matters

There are over 1,000 members of the House of Lords, although the average attendance is only about 250. There are four main groups of members:

1. 900 hereditary peers
2. 26 bishops (including two archbishops)
3. 150 life peers created under the 1958 Life Peerages Act
4. 20 law lords.

Members of the House of Lords receive no salary but can claim travelling expenses provided they attend at least one-third of the sittings. The granting of an expense allowance is just one of several recent changes. Others include the admission of life peers and women members and televising the proceedings. Reform of the House of Lords is believed by many to be overdue. In 1968 the Labour Government put forward proposals which were quite well received by both main political parties, but as governments have staggered from one political and economic crisis to another, time has not been available to put the reforms into effect. The changes put forward in the 1968 White Paper included:

1. A new House of Lords to be two-tiered.
2. The first tier to have about 230 voting members created for life by the monarch on the advice of the prime minister.
3. The second tier of non-voting peers at first to consist of hereditary peers, but their successors would not have the right to attend and their place would be taken by leading public figures from science, industry, trade unions, the professions, etc.
4. The delaying power would be cut to six months.

It is generally admitted that the membership and work of the House of Lords should be improved. One day a government will find time for this necessary reform.

The House of Lords

Lords defeat for Government on top pay awards

By Philip Webster, Political Reporter

The Government suffered another severe rebuff over the top people's pay award last night when it was defeated in the House of Lords by five votes.

At the end of a three-hour debate about the pay increases granted to senior military officers, top Civil Servants and judges, an opposition amendment regretting the Government's "insensitive" implementation of the Top Salary Review Body's report was carried by 140 votes to 135.

TOPIC 13.5 MODERN GOVERNMENT

Britain has a parliamentary government but the real power lies with the **prime minister**. There has been a growth in the importance of the prime minister in recent years.

He or she alone has the authority to choose the most important ministers who form the **cabinet**, and can dismiss any member who steps out of line. The prime minister's first job, after assuming office, is to pick a cabinet. Then he or she chooses less important ministers who will not have cabinet rank.

The leader of the party with the second largest number of seats in the Commons becomes the **leader of the Opposition** and selects a **shadow cabinet**. It is a fairly modern custom to select members of the Opposition to 'shadow' the important government ministers. The system has two important advantages:

1. An opposition MP specialises in a particular subject and is able to face his opposite member on the government front bench with expert knowledge.
2. If it wins the next election the Opposition is more ready to form the government, although it is by no means certain that the MP who has been acting as the spokesman on a certain subject (e.g. education) will become the secretary of state for education and science.

There are usually about twenty members of the cabinet, although on occasion an **inner cabinet** of about six ministers meets to discuss important issues. The cabinet meets once or twice a week, at 10 Downing Street, under the chairmanship of the prime minister. Cabinet policy is drawn up. The principle of **collective responsibility** means that all members of the Government must support cabinet policy.

To return to the concept of power with which we began this unit, consider for yourself who really holds the power in modern Britain. Is it the people or Parliament, or the cabinet, or the prime minister, or do you think they all have some say?

Britain is really ruled by a cabinet government with the prime minister having far more say than anybody else because he or she selects and sacks other ministers. MPs will nearly always obey the party whips and go into the division lobby which the whips indicate. If MPs hope for promotion they will toe the line, and the prime minister always has the power to advise the monarch to dissolve Parliament and make members face an election. Perhaps it is rather harsh to refer to MPs as 'lobby fodder', but nevertheless modern government is mainly in the hands of the prime minister and a few members of the cabinet.

Explain what is meant by

authority
control
power
religious beliefs
joint authority
monarch
absolute power
democracy
suffragettes
totalitarian state
fascist government
communist government
nationalism
dominion status
monarchical state
Her Majesty's Government
constitutional monarch
British constitution
Parliament
British Government
Royal prerogative
president
House of Commons
proportional representation
legislation
division
pressure group
House of Lords
money bills
Prime Minister
Cabinet
Leader of the Opposition
shadow cabinet
inner cabinet
collective responsibility

Case study

CUTTING THROUGH THE RED TAPE

Brenda Jenkins and her three children lived in a damp, old terraced house in Blackdene. Her husband had left her two years ago, and since then she had been living – or rather existing – on Social Security. Brenda muddled from day to day as best she could but she always had trouble paying the bills. It wasn't just because it was hard to budget her weekly payments to pay quarterly bills such as electricity and gas, but also because Brenda was illiterate. Since she couldn't read she didn't understand the bills and final demands that kept arriving. She got so worried that she went to the local DHSS office. She was a bit worried about going as she always felt like a criminal when confronted with someone in authority. However, it was the right thing to do as she was given some very good advice. Apparently there was a benefit direct-payment scheme available, which meant that all Brenda's bills would be paid automatically. It all seemed very complicated, but the clerk assured her that everything would be all right.

Sure enough, for six months everything worked out fine. Various bills and letters kept arriving but Brenda just ignored them. However, one day, as Brenda went to switch on a light, nothing happened. The electric heaters didn't work either. Slowly she realised that the electricity supply had been cut off. The next day Brenda went back to the DHSS office to ask them what had happened. The clerk at the desk assured her that he would sort it out as soon as possible. For the next week Brenda had to light the house with candles and she and the children huddled round the gas cooker to keep warm. After a week she still had no electricity so Brenda went into the local electricity board shop. A rather grumpy cashier took down all the details and said she would contact the DHSS to sort it all out. But days passed, weeks passed, months passed and still Brenda had no electricity. She had got so used to living without it that she gave up going to the DHSS to see what they were going to do about it.

But one winter's day, one of Brenda's children became quite ill; a neighbour thought it looked like pneumonia so she called a doctor in. When the doctor arrived and saw the conditions the family was living in, he immediately said that he would contact the DHSS and sort it all out. Within 24 hours the electricity supply was reconnected. The next day, the doctor explained to Brenda that her electricity had been cut off because she was in arrears with paying her bills; his investigations revealed that some bureaucratic bungling at the local DHSS had caused an administrative error in the operation of the benefit direct-payment scheme, and he had now referred her case to the local ombudsman in order to get just compensation. Brenda didn't understand what all that 'gobbledygook' meant – she was just grateful that she had light and heat again in her home.

1. What is bureaucracy?
2. How do you think the doctor managed to cut through the 'red tape'?
3. Collect your own examples of official 'gobbledygook'.
4. What does the local ombudsman do?

Follow-up

1. Would you rather live in a democracy or under a dictatorship? Give reasons.
2. Give reasons for and against a woman prime minister.
3. Prepare a dialogue to be taped between a person who believes Britain is a real democracy and one who says 'Britain is a democracy once every five years. After a General Election the people are puppets.'
4. Find out all you can about different systems of proportional representation. What are the advantages and disadvantages of PR?
5. What limits are there upon the powers of a modern prime minister?

Activities

1. Collect press cuttings about the activities of pressure groups. In each case, say whom they are trying to influence, and why.
2. Find out the names of the present
   (a) Chancellor of the Exchequer
   (b) Foreign Secretary
   (c) Home Secretary
   (d) Secretary of State for Education and Science.
3. Draw a cartoon strip to illustrate a day in the life of an MP.
4. Collect cuttings about your local MP. Do you think he or she does a good job?
5. Draw a map of your region and shade in red the areas represented by Labour MPs, blue the areas represented by Conservative MPs and in orange the areas represented by the Alliance.

Examination questions

1. 'The Monarch reigns but does not govern'.
   (a) Explain the above statement.
   (b) Name FIVE important functions performed by the monarch and the person who advises in performing each function.
   (c) The monarch does not govern; nevertheless government is carried out in the name of the monarch.
   For example we refer to His/Her Majesty's Ship (HMS).
   Give FIVE further examples of this system.
   (*Associated Lancashire Schools Examining Board, CSE, Social Studies, 1983*)
2. (a) List *five* ways in which you might seek to influence parliamentary legislation.
   (b) Describe the relationships between central government and pressure groups.
   (c) Assume that you live in an area which is to be demolished to make way for a shopping precinct and that you and your neighbours would like to stop this development. The Conservative controlled local authority which is responsible for initially taking the decision is broadly in favour of the scheme. The opposition parties, including the Labour Party, are against the scheme. The national government is Labour controlled. Explain how you would attempt to stop this scheme being implemented and assess the likely effectiveness of your actions.
   (*The Associated Examining Board, GCE, Social Science, 1983*)
3. (a) Why do you think that there are so many political parties and why is their existence a threat to constant government?
   (b) Explain the purpose of the vote and the differences between and the advantages and disadvantages of 'majority' and 'proportional' representation.
   (*The West Midlands Examinations Board, CEE, Social Science, 1983*)
4. EITHER
   (a) Describe any political system that you have studied and compare it with the British system.
   OR
   (b) Describe the role of Parliament in our society and explain how this role is carried out.
   (*The West Midlands Examinations Board, CEE, Social Science, 1983*)
5. (a) Explain the procedure which occurs once a voter has registered his/her vote by dropping the voting slip into the ballot box.
   (b) What qualifications are needed for a person to stand as a candidate in Local Government Elections?
   (c) What is the Electoral Register and how is one's name placed upon it?
   (*Associated Lancashire Schools Examining Board, CSE, Social Studies, 1984*)
6. (a) What is the Cabinet?
   (b) How is it formed?
   (c) How is it organised?
   (d) How useful is it in British politics?
   (*Associated Lancashire Schools Examining Board, CSE, Social Studies, 1984*)
7. Compare the system of communism in the USSR with that in Yugoslavia.
   (*East Anglian Examinations Board, CSE, Social Studies, 1983*)
8. Explain how and why the political rights and status of women have changed since 1900.
   (*East Anglian Examinations Board, CSE, Social Studies, 1983*)
9. (a) Outline the stages through which an individual must pass between deciding that he would like to be an MP and actually being elected.

(b) A person complains to his MP that in spite of sending letters that he has declared all his income, the Local Inspector of Taxes continues to threaten him with prosecution. Describe *three* ways in which the MP can deal with the person's complaint.

(c) Briefly describe the other duties which are common to all MPs. Which do you think are the most important? Give reasons for your answer.

(*The South-East Regional Examinations Board, CSE, Civics – Citizenship and Social Affairs, 1981*)

10. (a) List the stages which a Bill must pass through in the House of Commons. Describe carefully what happens at each stage. Briefly describe what is necessary before the Bill becomes law.

(b) How might a pressure group, such as a Trade Union or an Employers' Association, attempt to change or prevent a Bill becoming law?

(*The South-East Regional Examinations Board, CSE, Civics – Citizenship and Social Affairs, 1981*)

11. The United Kingdom is a democracy. Democracy means that the government is elected by the people, and rules with their agreement. In Britain this type of democratic government works through Parliament.

(a) 'The government is elected by the people'. Who are 'the people'? i.e. who is entitled to have a vote in parliamentary elections?

(b) Briefly describe how parliamentary elections are organised.

(c) Describe how parliament makes decisions. Your answer should include the making of policy, Bills and debates.

(d) Give *two* criticisms that are made of our system of parliamentary democracy. How would you answer these questions?

(*The South-East Regional Examinations Board, CSE, Civics – Citizenship and Social Affairs, 1983*)

12. 'Most of my work isn't done at the House of Commons, but in my constituency. I am a kind of local ombudsman to whom everybody comes when they are in difficulties. The work of an MP is hard and never ending. . . . It would be very easy to lose sight of the reasons which led me into the Labour Party.'

Adapted from chapter by S. Newens in *Work Two* edited by R. Fraser.

(a) Why does 'everybody' go to an MP when they are in difficulties? What does he mean when he says he is a kind of local ombudsman?

(b) What might be the 'reasons' which would cause a person to join *either* the Labour Party *or* the Conservative Party?

(c) Explain the various ways in which a back bench MP can influence a Bill presented to Parliament. Describe how outside organisations can attempt to persuade the MP to seek changes.

(*The South-East Regional Examinations Board, CSE, Civics – Citizenship and Social Affairs, 1982*)

UNIT 14

World Problems

GCSE Aims and Objectives

Knowledge — to develop a fuller awareness of world problems such as the population explosion and the Third World.

Application — to analyse methods of dealing with world problems and in particular the work of the United Nations Organisation and the European Economic Community.

Evaluation — to evaluate the usefulness of different methods of dealing with world problems.

Ethiopian famine victims with part of an aid consignment from the United States of 6,000 pairs of jeans which were unwearable in the desert heat

TOPIC 14.1 THE POPULATION EXPLOSION

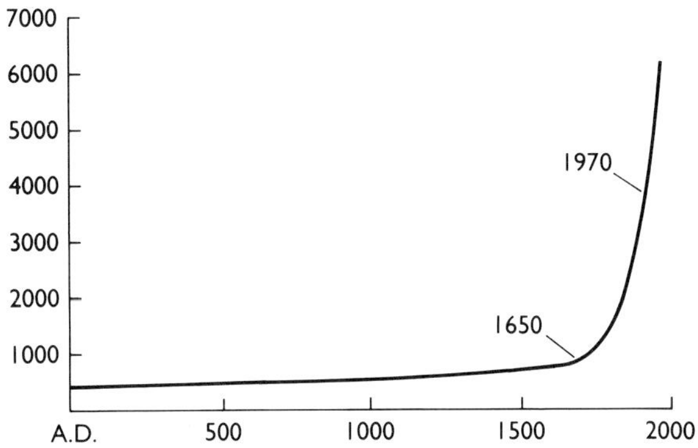

A

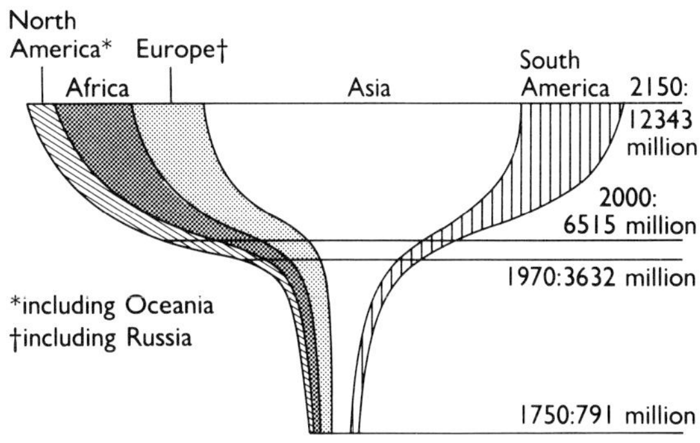

B

It is doubtful if in 1685 there were 800 million people living on earth. We have only a rough idea of the number because in those days there were no large-scale censuses or registrations of births and deaths. We do know that by 1985 the world population had increased to over 4,800 million. If the present rate of population increase continues, there will be over 6,000 million people by the end of this century. Population estimates may be very inaccurate because of changes in birth and death rates. If power gets into the wrong hands and a nuclear war takes place, there might be nobody left at all. It is up to all of us to see that this does not happen.

One of the main aims of Social Studies is to look for solutions to problems of poverty, power and population pressures. It is more important to value the subject for its own sake, and the useful application that we can make of our knowledge, than to be concerned merely with passing an examination.

Look at the two illustrations above. From the first (A) you will see that world population began to increase greatly in the seventeenth century. Now look at the second illustration (B) and trace the steep increase in the twentieth century, especially after 1960. What are the reasons for these rapid changes?

It is difficult to account for the **population explosion** that has taken place in the last 300 years. Why didn't the population begin to increase rapidly before? Many factors had to combine before the take-off in the population explosion could begin.

The elimination of the *bubonic plague* was one vital factor. In 1665 the last dreadful plague epidemic came to Britain from the east. The disease was brought to this country by black rats that carried the plague fleas. They probably arrived in ships from Holland. Nobody is certain how many people died of the plague, but it may well have been about half of the population. Records of the time are found in the diaries of Samuel Pepys and John Evelyn. In September 1665 Evelyn recorded that in London 'there [were] perishing 10,000 poor creatures weekly'.

This was probably an exaggeration but we shall never know because of inadequate records. In the latter part of the seventeenth century the brown rat got the better of the black rat and human beings were saved from visitations of the plague.

Factors assisting the population explosion in Europe

- No serious outbreaks of bubonic plague after the seventeenth century
- Improved food of greater variety because of advances in agriculture and transport
- Progress in preventive medicine, e.g. Jenner discovered a vaccination against smallpox
- Industry provided the means to keep clean, e.g. increased supplies of soap and washable cotton clothes. Crockery and cutlery became fashionable
- Curative medicine was advanced, e.g. use of the antiseptic carbolic acid spray (Lister)
- New medical knowledge spread from abroad, e.g. the ideas of Pasteur and Koch about bacteria
- People were able to afford larger families
- More hygienic midwifery and improved health care brought a drop in the infant mortality rate

The first economist to draw attention to the population problem was the Reverend T.R. Malthus, whose book *An Essay on the Principle of Population* appeared in 1798.

Malthus realised that large population increases could bring social and economic evils because the forces of nature would control the number of people. Certain 'checks' would cause people to suffer great hardship. These **Malthusian checks** included

1. starvation
2. disease
3. war.

We call them Malthusian checks because he was the first to appreciate an inevitable cycle of events and the consequent misfortunes. He did not like the idea of these misfortunes any more than anybody else.

In a later edition of his book Malthus wrote about **moral restraint**, meaning that people might refrain from sexual intercourse and so limit population numbers.

In the late nineteenth century it appeared that the ideas of Malthus had been proved wrong. The population of the world increased rapidly but people did not seem worse off. In spite of occasional famines, people were better fed and better clothed than when Malthus wrote his famous book. The general improvement in living conditions was particularly obvious in Britain. It was easy to blame Malthus, but there are many things he could not have foreseen.

Things that Malthus could not have foreseen

- New scientific farming methods would bring about a great increase in food production
- Food-producing regions would be opened up in the USA, Canada, Argentina, Australia and New Zealand
- Railways and ships would be used to transport huge quantities of wheat from the prairies
- The industrial revolution would enable industrial countries to increase exports of manufactured goods and buy imported foodstuffs
- Refrigerated foodstuffs could be transported thousands of miles
- The adoption of **contraception** or birth control methods would allow family planning to be practised

The four stages of the population cycle in Britain

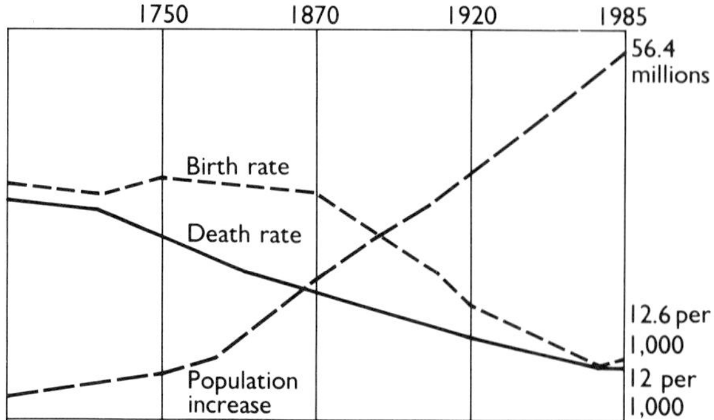

STAGE ONE	STAGE TWO	STAGE THREE	STAGE FOUR
Until 1750	1750–1870	1870–1920	1920 to today
High birth rate and high death rate	High birth rate and falling death rate	Falling birth rate and falling death rate	Low birth rate and low death rate

The diagram above shows how the population explosion in Britain was triggered by a declining death rate, and that we have now passed through all four stages of the **population cycle** (see also Unit 5). We are now experiencing a low birth rate combined with a low death rate and the population is just about replacing itself. However, there are some countries in Europe such as West Germany and Denmark where there is actually a decreasing population, and this is obviously a matter of concern to their governments.

Yet, despite the low fertility rates in Europe, the world population is still growing at a phenomenal rate. In fact, the main world population explosion is taking place right now in the twentieth century. In 1900 the world population was 1.85 billion and the estimated population in the year 2000 is 6.10 billion. The annual increase in population in 1900 was 13 million compared to the estimated annual increase of 90 million in the year 2000. The reason for this rapid growth is that the newly developing countries in Africa, Asia and South America have experienced a very sudden drop in the death rate, mainly due to disease control, but the birth rates have often remained unchanged. Today 34 developing countries have population growth rates of over 3 per cent. An example of the population growth momentum can be drawn from China. If all Chinese couples had only two children from now on, the present population of 1 billion would grow to 1.2 billion by the year 2000, and would not stabilise until the year 2075 at around 1.6 billion. That is a prospect which

Chinese planners view with horror, hence China's campaign for the 'one-child family'.

Globally, there is evidence that the population growth rate has peaked and is now beginning to slow down. With the gradual introduction of contraceptive methods, the birth rates in developing countries is beginning to fall, but the annual population increase is still projected to grow until the turn of the century. Africa is the fastest growing continent; the 1980 population of 476 million is projected to grow to over 875 million in 2000 and to reach 1,643 million by 2025. The question we have to ask is whether these countries will have the resources to feed, house and employ their projected populations at any acceptable standard of living. With the growing dependence of the developing nations on foreign aid, it is a question which concerns us all.

TOPIC 14.2 THE THIRD WORLD

The **Third World** is the name given to the poorest parts of the world. These are countries which have not benefited from the wealth brought by either capitalist or communist economies.

The world is divided into three 'camps'. Countries such as the Soviet Union and the People's Republic of China have made great progress with planned economies. Others, such as the United States and the United Kingdom, have gained great material wealth from capitalist or mixed economies. The countries in these two groups are known as the **developed countries**. Finally there is the Third World of **less-developed countries**; these are the poor nations that make up three-quarters of the world's population.

These countries are usually politically non-aligned, i.e. not committed to either the United States bloc or the Soviet bloc. The less-developed countries are often prepared to accept assistance from both capitalist and communist groups. The so-called super-powers of today compete with others to supply **foreign aid** in an attempt to influence the **emergent nations**.

China is experiencing a population explosion. Couples are now only allowed to have one child per family

However, foreign aid is crucial to the survival of many millions of African people who would otherwise face starvation. Ethiopia has suffered from famine conditions for years, but as a result of a particularly severe drought and a subsequent 80% crop failure in 1983, over 6 million people were placed at risk. In October 1984, it was estimated that over 1,000 people were dying of starvation each week in the northern provinces. In such a situation thousands of people leave their villages in the search for food and usually end up in relief centres, which are totally unable to cope with the huge influxes of sick, starving and exhausted people. Television pictures of the harrowing scenes at these relief centres brought the crisis of Ethiopia to the world's attention. It immediately mobilised massive emergency aid programmes in most of the developed countries of the world. Foodstuffs, medical supplies and transport, funded by governments, charities and public fund-raising schemes such as Live Aid, are all now being sent to the famine-stricken areas, but the crisis is likely to continue until long-term improvements can be made to food production techniques and the people can be taught how to cope should another drought occur in future.

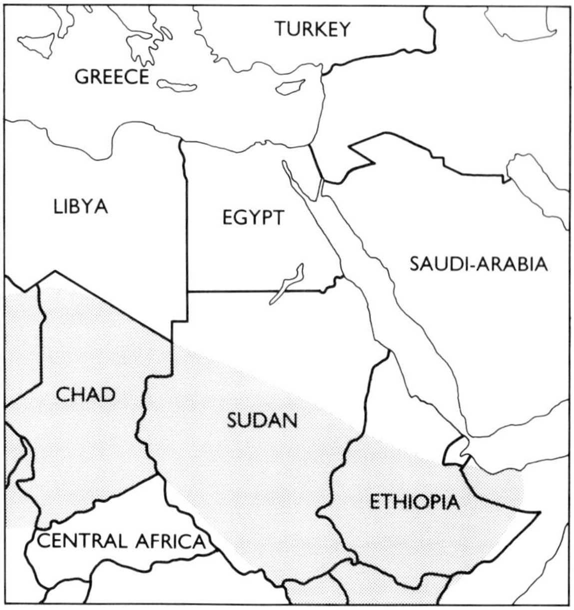

The famine-stricken areas of Africa

A famine victim and her child in an Ethiopian relief centre ▷

Third World aid programmes may 'prolong food problems'

By Michael Simmons

President Mitterrand opens the 11th ministerial session of the UN World Food Council in Paris today with more people facing starvation and malnutrition in the world than probably ever before.

Heated exchanges are expected when delegates examine the effectiveness of emergency efforts during the past year to relieve famine in sub-Saharan Africa. Officials will emphasise that, if allowed to continue, these programmes could postpone real solutions to the continent's underlying food problem and make matters worse.

"The bottom line," one council official said at the weekend, "is that private capital inflows to the needy countries are drying up. Sub-Saharan Africa is now losing about $1 billion a year in net investment. This is in addition to demand for higher debt repayments and worsening terms of trade."

Ministers from donor countries will be specifically warned against complacency about recent relief efforts. In many cases aid which has arrived has not always been properly coordinated, "going off in about 25 different directions."

The council, set up by the UN in 1974 to coordinate the necessary political action to carry out General Assembly food directives, has no executive authority but it can pillory certain governments for inappropriate or non-existent aid and development policies.

European and North American ministers, for instance, are likely to be castigated for the adverse long-term effects of their large agricultural subsidy programmes.

Third World governments, who are also participating, will be criticised for failing to give enough weight or financial back-up to food production of their own. The "extremely low level" of spending per head in this area, one conference document points out, is inconsistent with public policy declarations.

A second document dwells on the "poor progress" made so far in eliminating hunger and outlines a four-point programme for ending starvation by the year 2000.

How can the richer countries best help the Third World? There are two controversial schools of thought which you ought to consider:

1. Should most of the aid be provided in *food parcels* to feed the starving millions?
2. Should aid be concentrated upon sending *technical assistance* in the form of experts and equipment such as tractors so that they stand on their own feet?

If aid from the developed countries is centred on helping the less-developed countries to help themselves, millions may die in the short run, but in the long run many more millions may be saved.

Perhaps the best answer lies in reconsidering our basic concept of life-cycles. It is necessary to speed the advance of Third World countries so that we telescope the process of development. How long does it take a less-developed country to become developed? In Sri Lanka (formerly Ceylon) the expectation of life at birth was increased in ten years (1946 to 1956) from 43 to 61 years. The advantages of imported medicine and knowledge allowed Sri Lanka to achieve something that had taken 200 years in Britain.

Many forget that Britain was once a less-developed country. It will be helpful to see how and why we developed. Britain had no super-powers willing to send her large supplies of food so that they could exert political influence upon her. We lifted ourselves by our own boot-straps. Britain quickly adopted ideas from abroad to help in the development of industry and technology.

The countries of the Third World must develop their own resources, industries and power supplies. Many of them have the advantage of being primary producers. Their raw materials could provide the basis for manufacturing industries. The concept of continual change means that the present pattern of less-developed countries and developed countries will not always remain.

It would be better for the peace of the world if the less-developed countries were encouraged to help themselves to become developed. The rival super-powers must be wary of indulging in a bitter competition to supply aid in return for political influence. The hopes of mankind may rest eventually in a world government where people work in harmony. In the immediate future the best hope lies in the **United Nations Organisation**.

'Poor devils - another year under the threat of nuclear war.'

TOPIC 14.3 THE UNITED NATIONS ORGANISATION

By 1984, the United Nations had 158 members. Only 12 countries are not members of the United Nations today. How many of these have you heard of? Where are they?

Sovereign countries not members of the UN (1984)

Andora	Nauna
China (Taiwan)	San Marino
Kiribati	Switzerland
Dem. People's Republic of Korea	Tonga
Liechtenstein	Tuvalu
Monaco	Vatican City

More than 3,500 million people are nationals of countries which are members of the United Nations, co-operating in the search for peace, justice and social and economic progress.

Points of the UN Charter

- To maintain peace and security
- To develop friendly relations among nations
- To achieve co-operation in solving world problems
- To be a centre for harmonising the actions of nations in achieving these ends

From 1945 until 1971, the UN Chinese seat was occupied by the Republic of China which was confined to the island of Taiwan from 1949. Since 1971, the People's Republic of China has taken its rightful place in the United Nations – with nearly 1,000 million people, China has the largest population in the world. Membership of the UN is open to all peace-loving states whose admission is approved by the General Assembly upon the recommendation of the Security Council. The official languages of the UN are Arabic, Chinese, English, French, Russian and Spanish. The UN uses an emblem (See Figure) in white centred on a light blue background.

At present the United Nations is not a **world government**. It is an organisation of sovereign nations. Many people think it should always remain so. They do not want individual nations to lose their sovereignty, i.e. the power to rule themselves. There is much to be said for nations keeping their own identity and culture. We return to the concepts of freedom and relationships. How much are the peoples of the various nations willing to sacrifice in the interests of the people of the whole world?

Headquarters of the UN in Manhattan, New York

Principles of the UN

- All members are sovereign and equal
- All members are pledged to solve international disputes by peaceful means
- No member shall use force or threaten to use force against the independence of any country
- All members pledge themselves to assist actions taken in accordance with the UN Charter
- The United Nations shall not interfere in the domestic affairs of any country except to enforce world peace

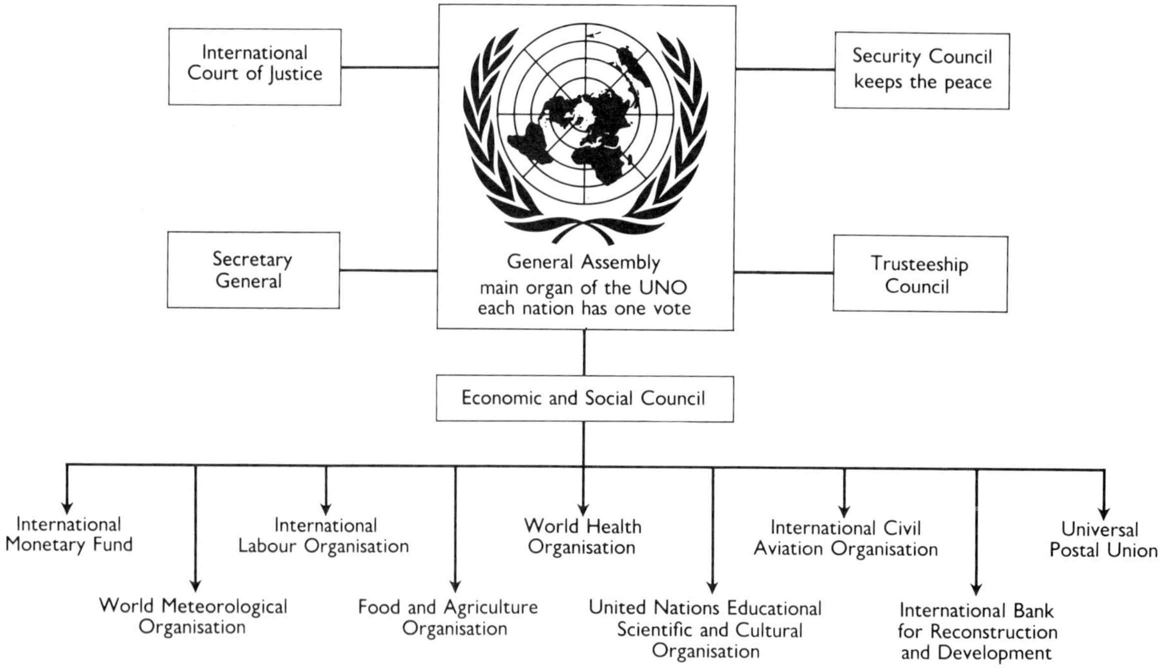

The United Nations Organisation

The six main organs of the United Nations are shown in the diagram. All are based in New York except the International Court of Justice which is located at The Hague.

The **General Assembly** is the nearest thing to a world parliament. All member countries are represented and each has one vote. On important issues a two-thirds majority is necessary, but most matters are settled by a simple majority. The Assembly can discuss and make recommendations about any matters within the scope of the UN Charter. Its recommendations carry the moral weight of world opinion, but it has no power to compel governments to take action.

In recent years, special sessions of the General Assembly have been held on Afghanistan and Palestine (1980) and Namibia (1981). The fact that the problems of these countries are still with us indicate the limited power of the UN Assembly.

The **Security Council** is the main UN body dealing with the peace of the world. It meets any time peace is threatened. The Council has fifteen members, five of whom are permanent members (China, France, the Soviet Union, the United Kingdom and the United States). The other members are elected by the General Assembly and serve for two years. Decisions require nine votes, except on matters of important procedure when any permanent member can use the **veto** and stop the council acting. On occasions, the veto has prevented speedy action and has been the cause of criticism. If fighting breaks out anywhere in the world the Security Council has the power to order military action by a UN force drawn from member countries.

In recent years the Security Council has been responsible for sending observers or peace forces to Beirut in the Lebanon and other troubled areas of the Middle East. Although the Security Council has the power to order military action by a force drawn from U.N. member countries, there are people who think that a permanent UN task force could be rushed to any troubled spot and the problem could be nipped in the bud. However, such hopes are dampened by the permanent members' use of the veto and the fact that other countries can do little in face of action by either of the two super powers, U.S.A. and U.S.S.R. Consider, for example, the unilateral actions of the U.S.S.R. in Afghanistan and the U.S.A. in Grenada. What was the attitude of the U.N.O. to the British military task force to the Falklands?

The Secretary General is head of the *Secretariat* which works for the other organs of the UN and carries out the day-to-day tasks of the organisation.

Previous Secretary Generals have had considerable influence upon world affairs. They include Trygve Lie (Norway) 1946–53, Dag Hammarskjold (Sweden) 1953–61, U Thant (Burma) 1961–71 and Kurt Waldheim (Austria) 1972–81. In 1982 Javier Perez de Cuellar was appointed for a five year term.

The Economic and Social Council aims to build a just and prosperous world; its activities include efforts to improve trade, transport, technology, housing, crime prevention and family planning.

The **Trusteeship Council** oversees governments which are responsible for supervising trust territories. The UN Charter provided for an international trusteeship council to safeguard the interest of inhabitants of territories which were not yet self-governing. All of the original territories except one, Mictonesia (Pacific Islands) which is administered by the USA have become independent. The **International Court of Justice** is composed of independent judges elected regardless of their nationality. There are fifteen such judges, no two of whom may be nationals of the same state. They are elected by the Security Council and the General Assembly sitting independently. These judges give decisions in case of international disputes. The International Court has its seat at The Hague but may sit elsewhere if it is considered desirable.

French back Argentina on Falklands at UN

From Zoriana Pysariwsky, New York

France has decided to throw its support behind an Argentinian-inspired draft resolution at the United Nations seeking a far-reaching dialogue between Britain and Argentina, including negotiations on the Falkland Islands.

The move could provide momentum for further support from within the European Community.

According to diplomatic sources, France informed the British Foreign Office of its decision to depart from its traditional abstention after Argentine dropped any mention of the word "sovereignty" in the draft resolution. Instead, the measure simply refers to "all aspects of the future" of the Falkland Islands.

The French move comes after intense lobbying by Argentina and is certain to strain Anglo-French relations. Since 1982, when the first post-Falklands war resolution came before the General Assembly, France has wanted to vote with Argentina but has been held back by British pressure.

The debate is due to begin on Tuesday.

The names of most of the **UN agencies** suggest the work that they do; e.g. the *International Civil Aviation Authority* deals with regulations and problems relating to aircraft and airports, and the *International Labour Organisation* is concerned with work, wages, trade unions, etc. We only have space to look at two of the agencies in detail, so we will concentrate upon the *Food and Agriculture Organisation* and the *World Health Organisation*. These are two of the most vital agencies and their work is very important in respect of the population explosion and the less-developed nations.

The Food and Agriculture Organisation has its headquarters in Rome. Its purpose is to increase the production of the world's farmlands, forests and fisheries. The FAO was the first permanent agency and began its work in 1945.

There are two main ways of coping with the population explosion:

1. In the long term the number of people must be limited, probably by methods of birth control.
2. In the short term the increasing population must be fed. The Food and Agriculture Organisation is the most important body concerned with feeding the peoples of the world.

Some ways of feeding more people

- Improved strains of rice and wheat with increased yields. (How successful is the 'Green Revolution'?)
- Aid in the form of agricultural machinery, seeds, fertilisers and advice
- Irrigation schemes to reclaim desert-land
- Soil conservation methods to combat soil erosion
- Effective pesticides to destroy insects which eat so much of the food produced
- Propaganda to persuade people to eat different foodstuffs, e.g. soya beans, red kangaroo meat, seaweed, etc.
- Industrialised animal production similar to battery-hen rearing. (What moral objections could be made to this practice?)
- Synthetic foodstuffs, e.g. from petroleum
- Education to teach people to reject anti-social customs and beliefs (e.g. India's sacred cows which are disease-ridden)
- Fish-farming combined with control of catches so that fish breeding grounds are not destroyed

The World Health Organisation was founded in 1946 and has its headquarters at Geneva. For over thirty years it has worked to cut mortality and improve medical standards. One of the reasons why the world population continues to explode at an increasing rate is that the World Health Organisation is better at keeping people alive than the Food and Agriculture Organisation is at feeding them. Mass immunisation and vaccination plus the use of penicillin, DDT, antibiotics and other new drugs have given people a longer and healthier life.

The WHO has been particularly successful in attempts to eliminate 'killer diseases'. Until the WHO started a serious campaign, malaria was the world's worst killer disease, but from 1953 to 1959 malaria deaths fell from 800,000 to 10,000 a year. The WHO then turned its attention to smallpox. When its anti-smallpox campaign started, the disease was endemic in thirty countries and many others were reporting importation of the virus. Vaccination against smallpox had been beneficial for a very long time, but the difficulty was to get people to use it. WHO vaccination teams fanned out across vast areas of Ethiopia, Bangladesh and India, drawing the net ever closer around remaining pockets of smallpox infection. At the end of the campaign the population still at risk is but a minute fraction of that over which smallpox held sway twenty years ago.

Since immunisation is one of the most cost-effective of all health services, the WHO, in conjunction with the children's charity UNICEF, has adopted an Expanded Programme of Immunisation (EPI) with the goal of providing immunisations for all children of the world by 1990. Over 3 million children still die annually from measles, tetanus and whooping cough, while over a quarter of a million children are crippled by polio, mainly in developing countries. It may be easier to grasp these numbers if one realises that for every breath one takes a child dies from a vaccine-preventable disease. In 1982 this ambitious programme was accelerated to ensure that the goal date of 1990 is met. The UN is particularly committed to the success of EPI to show the aspirations and effectiveness of the UN system.

We have seen that the UN organs and agencies are far from being just talking-shops. The UN has often been subject to press criticism and not enough credit has been given for all the work it has done in so many spheres. Local branches of the UN do their best to make people realise that the UN needs sustained support more than carping criticism. It is useful to have a reminder of a few of the achievements of the organisation.

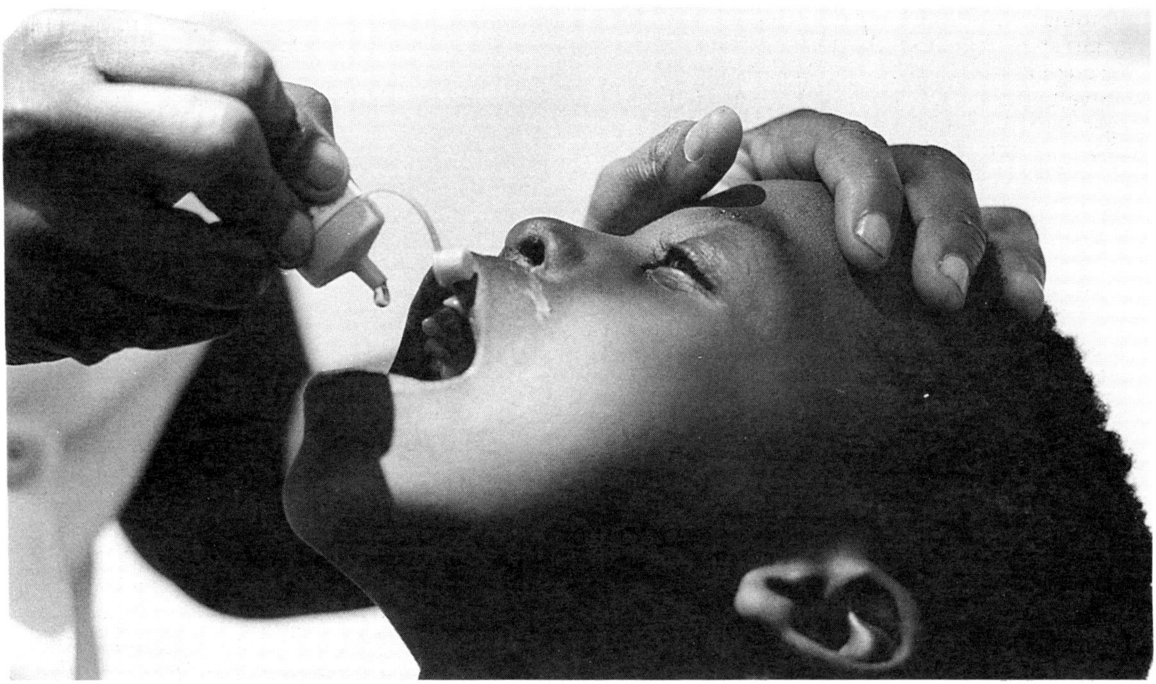

It takes just two drops of vaccine dropped into a child's mouth three times to immunise him or her against polio for life

UN achievements

For peace

- A UN Emergency Force has been used in the Middle East on two occasions – 1956 and 1973
- From 1960 to 1964 UN troops helped to maintain law and order in Zaïre (previously the Belgian Congo) after independence had been granted
- UN observers supervised the ceasefire after India–Pakistan hostilities in 1948–9 and 1965
- A UN peace-keeping force took up positions in Lebanon in 1978 and has played a role in attempting to bring stability to that troubled country in more recent times

For justice

- South Africa's **apartheid** (racial segregation) policy has been denounced by the General Assembly as a 'crime against humanity' and actions against it have included an arms embargo and a boycott of sports events which separate black and white people
- The Security Council demanded that South Africa should vacate the vast trust territory of Namibia. (So far the South African government has not responded. What further action do you think should be carried out?)
- The Security Council ordered all nations to stop trading with the minority white regime in Zimbabwe (formerly Rhodesia). These **economic sanctions** helped to bring about a government representing the majority of the people of the new Zimbabwe

TOPIC 14.4 THE EUROPEAN ECONOMIC COMMUNITY

The **European Economic Community** (EEC) or Common Market is a community of twelve member countries. In 1957 six European countries agreed to co-operate in a single economic and social unit. By the Treaty of Rome they agreed to remove trade restrictions between them. A common policy is being developed on agriculture, postal rates, social security, transport and the free movement of money and people. It will be a long time before complete integration is achieved but the eventual aim is to have a common monetary policy and a European parliament elected by the peoples of Europe. Because of the success of the EEC, other countries applied to join and three new members, including the United Kingdom, were accepted in 1973, and Greece was accepted in 1981. Two more, Spain and Portugal, were admitted in 1986.

Membership of the European Community

EEC countries

Since 1958	*Since 1973*
Belgium France Italy Luxembourg Netherlands West Germany	Denmark Ireland United Kingdom
	Since 1981 Greece
	Since 1986 Portugal Spain

What have been the successes of the Common Market?

1. A higher standard of living: prices have increased but wages have increased more, so people living in the EEC can buy extra goods with their wages.
2. Increased production: the gross national product (i.e. the total value of a country's production) increased nearly twice as fast in the original six countries of Europe from 1958 to 1973 as it did in Britain.

The European Parliament in Strasbourg

3. Increased trade between the countries of Western Europe: as about 40 per cent of Britain's imports and exports are with the Common Market it is in our interest to be able to trade freely in this nearby area of 250 million people.
4. Organisations such as *Eurorail*, *Euratom*, *European Coal and Steel Community* (ECSC) have encouraged large-scale enterprises.

The establishment of the ECSC in 1952 marked the beginning of modern efforts at large-scale European co-operation. A common market in coal and steel products was set up. The ECSC High Authority now supervises all EEC industrial co-operation.

There are four other main institutions:

1. **The Council of Ministers** is the most important political institution of the EEC. It consists of the foreign ministers from the national governments of the member states. The Council meets in Brussels and makes decisions under the Treaties which led to the formation of the present EEC group. Although legally its decision should be made by a majority vote, the Council of Ministers has, since 1966, sought to make unanimous decisions wherever possible.
2. **The European Commission** is composed of 16 members in 1986 (two each from France, Italy, West Germany and the UK and one each from the other eight members). The Commission is a very important EEC body and initiates all the policy matters from its office in Brussels; however, no vital changes can be made without the authority of the Council of Ministers.
3. **The European Parliament** meets at Strasbourg and consists of 434 members. It could one day be the most significant institution in Europe, but at present it has far less power than the Council of Ministers, although it does have the right to be consulted and it can initiate discussion on vital matters. France, West Germany, Italy and the UK each return 81 MEPs while other EEC members have small representations. Who is your present MEP?
4. **The European Court of Justice** meets in Luxembourg and is composed of 11 judges and 5 advocates-general. It decides whether acts of member countries and the EEC institutions are in accordance with the Treaty of Rome.

Although Britain has enjoyed considerable increases in trade with the EEC since we became full members, nevertheless there are still problems to be ironed out, particularly in respect to agriculture, the EEC Budget, fisheries and the full admission of Spain and Portugal. Nearly two thirds of the EEC Budget is spent upon agriculture, but in Britain agricultural workers make up less than 3 per cent of the working population. Other European countries, such as France and West Germany, gain far more from the Common Agricultural Policy (CAP) than we do, and there are some who believe that the EEC subsidises inefficient farmers. The Common Agricultural Policy of guaranteed prices for farmers has certainly made for high food prices, over production and difficulties in the redistribution of so-called 'butter mountains' and 'wine lakes'.

In 1984, the EEC faced its greatest financial crisis and the almost complete exhaustion of its funds. Britain, as a major contributor to the EEC Budget, wanted to decrease the annual rate of expansion of the Budget and in particular the amount earmarked for CAP. After many discussions, in October 1984, Britain agreed to a supplementary budget of £600 million for the remainder of 1984 and to a draft 1985 Budget of £15 billion. In return for Britain's agreement, the other EEC members accepted the fact that expenditure on agriculture would become a progressively smaller amount of the total EEC Budget.

Since 1982, the EEC Common fishery policy has meant that Britain has shared a 200 mile fishing zone with other EEC countries. The entry of Spain into the EEC is likely to increase the problems faced by Britain's declining fishing industry. Spain's modern fishing fleet is the fourth largest in the world and is larger than the entire fishing fleet of the other EEC countries. The EEC Commission is attempting to produce a package of measures relating to fishing which would allow Spain's fishing industry to enter gradually, thus reducing the immediate effect of Spain's entry and giving some support to the fishing industries of Britain and France in particular. Spain and Portugal joined the EEC on January 1st 1986. They had been applying for membership for seven years. There were difficulties in the negotiations (for example they are large wine producers and the EEC Budget already pays out £800 million to wine growers). However, most other EEC countries welcomed the membership of Spain and Portugal because the EEC will be larger, stronger and more influential. Besides, they are comparatively new democracies whose governments and peoples will benefit from membership of a wider community.

A barley 'mountain' surplus to EEC requirements

Explain what is meant by

population explosion
Malthusian checks
population cycle
moral restraint
contraception
Third World
developed countries
less-developed countries
foreign aid
emergent nations
United Nations Organisation
world government
General Assembly
Security Council
veto
Economic and Social Council
Trusteeship Council
International Court of Justice
UN agencies
apartheid
economic sanctions
European Economic Community
Council of Ministers
European Commission
European Parliament
European Court of Justice

Case study

RARE FEAST

The Rev. A. Goodbody was delighted when he was chosen by the Missionary Society to go to a very isolated part of South America. For years he had been begging his church to send him to the overseas mission field. It would be a welcome change from his quiet rural living in Suffolk.

Alan Goodbody knew that he was being sent to a dangerous part of the American continent. The climate was equatorial, the Amazon basin unhealthy for a white man, while the natives were reputed to be far from friendly. Years ago they had been cannibals. These people lived in a part of the Third World which was very under-developed. Famine was a common occurrence, and it was by no means certain that they would not resort to cannibalism again, if the need arose.

After months of preparations and travelling, Alan Goodbody arrived in his new 'parish'. He was well received by the missionary who had completed a year's service and knew that Goodbody's arrival meant his departure. The Missionary Society had given Alan Goodbody a crash course in the rudiments of the native language and he was soon trying it out. The tribal chiefs were interested in everything connected with the white man's way of life.

The Rev. Goodbody had never been asked so many questions about his religion. His parishioners back in Suffolk had not cared about religious beliefs. They had been far more interested in the weather and its effect on their crops.

Some of the questions which the natives asked were far from easy. The new missionary found it very difficult to explain why so-called Christian countries were making huge bombs which could kill millions of people. Did not God tell them to love one another?

'If your God say you forgive enemy, is there war in your land?' asked one of the tribal chiefs.

'Oh, yes,' said the Rev. Goodbody, 'I am afraid we have had two long wars this century.'

'Are many people killed?'

'Unfortunately, yes,' answered the new missionary. 'Millions and millions.'

'Do you eat them?' questioned the chiefs.

The Rev. Goodbody broke into a hearty laugh. These people seemed so simple. How could they think such wicked thoughts?

'Of course we don't eat them,' he answered.

The chiefs paused to reflect.

'What a waste,' murmured the oldest chief, 'what a waste.'

1. What reasons can be put forward for and against intelligent beings using their resources to manufacture H-bombs capable of killing millions of people?
2. Try to find out what the following main religions of the world teach about peace and war:
   (a) Christianity
   (b) Judaism
   (c) Hinduism
   (d) Buddhism
   (e) Islam
3. Suggest ways in which a world government would help to abolish war.
4. List the most important peaceful productive aims that could be achieved if the United Kingdom did not spend so much on defence.

Follow-up

1. Suggest 5 reasons why people in the developed areas have a responsibility for those in less-developed areas.
2. What do you think is a reasonable size for a family? Give reasons for your answer. Why might you have answered differently if you lived in Ethiopia, Australia or Libya?
3. Write a story about the folly of war beginning 'It was the day some idiot pressed the button . . .'.
4. Man has been described as 'a naked ape'. List some of man's violent characteristics and say how you think they could be adapted for peaceful ends.
5. How can the EEC help
   (a) the peace of the world?
   (b) less-developed countries?

Activities

1. Do some research on birth rate and death rate movements to find which of the four stages in the population cycle have been reached by
   (a) Brazil
   (b) China
   (c) Zaïre
   (d) The United States
   (Use the *UN Statistical Year Book* where possible)
2. Compare the past, present and likely future population statistics for
   (a) Africa
   (b) America
   (c) Asia
   (d) Europe
   (e) Oceania
3. Design a poster to illustrate *one* of the following slogans:
   (a) 'Each one. Teach one.' (To read and write)
   (b) 'Give a man a fish and you feed him for a day. 'Teach a man to fish and you feed him for life.'
   (c) 'Two will do.' (Children)
4. On a blank map of the world indicate the main 'flashpoints' since 1945 (e.g. Falklands, 1982).
5. Collect pictures and make a display showing distinctive features of the EEC countries.

Examination questions

1. (a) What is meant by 'The Third World'? What are the other two worlds?
   (b) Briefly describe a problem which is shared by many Third World countries.
   (c) Describe ways in which the UN and its Agencies are trying to help the Third World.
   (d) Why is a solution to the problems of the Third World a matter of vital importance to other nations which do not seem to be directly affected?
   (*The South-East Regional Examinations Board, CSE, Civics – Citizenship and Social Affairs, 1981*)
2. (a) Name the *three* main institutions of the European Economic Community (the Common Market).
   (b) What were *two* of the aims of the founders of the European Economic Community?
   (c) In what ways is the European Economic Community considered to be successful?
   (d) What are considered to be the main weaknesses of the European Economic Community? Your answer should contain examples.
   (*The South-East Regional Examinations Board, CSE, Civics – Citizenship and Social Affairs, 1981*)
3. 'A large number of people in some countries suffer from starvation and sickness. In Bangladesh, 40% of the children die before they reach the age of five. One of the causes of this poverty and misery is the large growth in the number of people. It has all happened so quickly that we call it "the population explosion".'
   From: J. Nobbs, *Introducing Social Studies*, Macmillan 1979.
   (a) Name *two* countries (apart from Bangladesh) that have rapidly growing populations.
   (b) Give *two* reasons why the population has grown so rapidly in some countries.
   (c) Briefly describe *two* problems (apart from starvation and sickness) that a rapidly growing population can cause.
   (d) Describe the ways in which other countries or organisations are trying to help poor countries. Explain the difficulties involved in giving successful aid.
   (*The South-East Regional Examinations Board, CSE, Civics – Citizenship and Social Affairs, 1983*)

4. 'Once, in the early post war years, the world was presented as a two-man band, with America on one side and Russia on the other. Later, the international system was seen as more of a three ring circus. The Third World competed with the NATO countries and the communist bloc for the crowd's attention.

Everything has now got more complicated. With the rise of OPEC and the newly industrialised countries, the Third World begins to seem a less useful idea.'

Adapted from '*New Society*', 15.1.81.

(a) What do you think is meant by the first paragraph?
(b) Why is 'the idea of the Third World' now regarded as less useful?
(c) Briefly describe the main differences between the type of government and political system of a country which is a member of NATO and a member of the communist bloc.
(d) Why do some powerful nations consider it important to give aid to Third World Countries?

(*The South-East Regional Examinations Board, CSE, Civics – Citizenship and Social Affairs, 1982*)

5. 'Fighting and warfare between nations have taken place at all periods in man's history, but now we have reached a turning point. The world no longer dares have a full-scale war.'

Adapted from: K. Lambert, *Life in our Society*, Nelson 1974.

(a) Why is it that the world 'no longer dares have a full-scale war'?
(b) Why do some people argue that even if we do not have another world war we should still have a large defence force?
(c) What are the objections to a high level of expenditure on armaments?
(d) Describe how the UN tries to keep peace in the world. How successful has it been? Give examples to illustrate your answer.

(*South East Regional Examinations Board, CSE, Civics and Social Affairs, 1983*)

6. Explain fully why the work of UNICEF is so important.

(*East Anglian Examinations Board, CSE, Social Studies, 1983*)

7. From the British point of view, what would seem to be the advantages and disadvantages of belonging to the EEC?

(*East Anglian Examinations Board, CSE, Social Studies, 1983*)

8. What factors have caused a food shortage in *three* different parts of the world? Describe the work being done to improve the situation in each case.

(*East Anglian Examinations Board, CSE, Social Studies, 1983*)

9. Describe the arguments for and against the idea of World Government.

(*East Anglian Examinations Board, CSE, Social Studies, 1984*)

10. World population is increasing by two per cent every year. If it continues to increase at this rate world population will double in the next thirty-five years.

(a) Give possible reasons for the world population explosion.
(b) Discuss the consequences for a country which has a high rate of population growth.

(*East Anglian Examinations Board, CSE, Sociology, 1984*)

11. The military dispute over the Falkland Islands in 1982 between the Argentine and the United Kingdom illustrated and highlighted a series of major world issues.

Explain the following:

(a) Democracy.
(b) Sovereignty.
(c) The role and power of the United Nations. Refer especially to the Security Council and the General Secretary.
(d) The role and power in a dispute of this nature of organisations such as the European Economic Community, the Organisation of American States and the Commonwealth.

(*Associated Lancashire Schools Examining Board, CSE, Social Studies, 1984*)

12. The Divided World

THE COST OF ARMED FORCES

Country	*Total forces*	*Defence budget (billion US dollars)**	*Percentage of gross national product*
USSR	3,425,090	90	20.5
China	2,900,000	12	10
United States	2,252,900	82.5	7.4
India	948,000	2.4	4
France	503,600	8.5	4.2
Britain	361,500	8.7	5.8
Egypt	298,000	1.7	22.6
Israel	115,000	1.5	22

* 1974 figures (1 US billion = one thousand million)

(a) Examine the table above and answer the following questions.
  (i) Giving the figures, say which **two** countries spend most on defence.
  (ii) Giving the figures, say which **two** countries spend the biggest percentages of their wealth on defence.
  (iii) Why do these two countries spend the biggest percentages on defence?
  (iv) How much does Britain spend on defence and what percentage is it of the country's wealth?
  (v) Giving the reasons for your answer, say what effect a large defence budget has on a country's standard of living.

(b) In a paragraph, explain what are NATO and the Warsaw Pact, their membership and when and why they were formed.

(c) Describe 'the nuclear arms race' and the balance of power that has been created during the post-war period.

(*West Midlands Examinations Board, CSE, Social Studies, 1984*)

◁ Reproduced from *'The Divided World'* by Lionel Grigson, Macdonald Educational Limited, 1975.

Index